Ram Charitra Maurya
Bioinorganic Chemistry

Also of Interest

Inorganic Chemistry.
Some New Facets
Ram Charitra Maurya, 2021
ISBN 978-3-11-072725-8, e-ISBN 978-3-11-072728-9

Molecular Symmetry and Group Theory.
Approaches in Spectroscopy and Chemical Reactions
R. C. Maurya, J. M. Mir, 2019
ISBN 978-3-11-063496-9, e-ISBN 978-3-11-063503-4

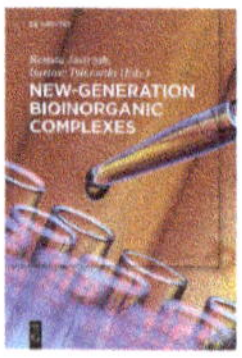

New-Generation Bioinorganic Complexes
Renata Jastrzab, Bartosz Tylkowski (Eds.), 2016
ISBN 978-3-11-034880-4, e-ISBN 978-3-11-034890-3

Organoselenium Chemistry
Brindaban C. Ranu, Bubun Banerjee (Eds.), 2020
ISBN 978-3-11-062224-9, e-ISBN 978-3-11-062511-0

Ram Charitra Maurya

Bioinorganic Chemistry

Some New Facets

DE GRUYTER

Author
Prof. Dr. Ram Charitra Maurya
Rani Durgavati University
B-95 Priyadarshini Colony,
Dumna Road
Jabalpur 482001
Madhya Pradesh
India
rcmaurya1@gmail.com

ISBN 978-3-11-072729-6
e-ISBN (PDF) 978-3-11-072730-2
e-ISBN (EPUB) 978-3-11-072741-8

Library of Congress Control Number: 2021937867

Bibliographic information published by the Deutsche Nationalbibliothek
The Deutsche Nationalbibliothek lists this publication in the Deutsche Nationalbibliografie; detailed bibliographic data are available on the Internet at http://dnb.dnb.de.

Cover image: Christoph Burgstedt/Gettyimages
Typesetting: Integra Software Services Pvt. Ltd.
Printing and binding: CPI books GmbH, Leck

www.degruyter.com

Dedicated to
My wife
Mrs. Usha Rani Maurya,
who has always been a source of inspiration for me
throughout my growth
&
My sons, Ashutosh and Animesh, and daughter Abhilasha and son-in-law Adarsh
for
encouragement

Preface

Bioinorganic chemistry is an interdisciplinary subject developed through the cross-fertilization of knowledge emerged from studies in chemistry, biochemistry, physics, biology, medicine, environmental science and so on. Although the subject is of vital interest to researchers, many areas of the subject are matured enough for textbook presentation. Looking over this, University Grants Commission in India has recommended the topics of bioinorganic chemistry as part of the curriculum for undergraduate and postgraduate courses in chemistry. Moreover, in abroad too, topics of bioinorganic chemistry are included in syllabi of most of the universities and institutes.

During my long journey of teaching span of more than 40 years in three different universities, I had always realized the lack of textbooks in bioinorganic chemistry that covers most of the topics for in-depth teaching of the subject matters. Moreover, I was also in quest of reading materials in the form of research papers, reviews and books, so that curriculum at postgraduate, M.Phil. and Ph.D. levels in any University of India and abroad too can be revised for updating man powers in chemistry for manifold applications. This book entitled *Bioinorganic Chemistry: Some New Facets* comprising four chapters is designed to meet out the objectives mentioned above. Although many more topics are still not covered in this book as it may only increase the bulkiness of the book, these will be given attention in my next book.

The book also aims at assisting students in preparing competitive examinations, namely, NET, GATE, SLET, Doctoral Entrance Test (DET) and others, particularly in India. The end of each chapter has multiple-choice, short-answer and long-answer questions covering the topics discussed in the chapter for self-evaluation of students.

In completing a book of this nature, one accumulates gratefulness to the previous authors and editors of books, research papers, reviews and monographs on the relevant topics. I have consulted these sources freely and borrowed their ideas and views with no hesitation in preparing this manuscript. These sources are mentioned at several places in the text and also listed in bibliography, for which I am highly thankful to these authors.

Moreover, this book is the outcome of my teaching of the subject for more than two decades for several batches of M.Sc. and M.Phil. students at the Department of Chemistry, Rani Durgavati University, Jabalpur (M.P.), India. I have benefitted enormously from the response, questions and criticisms of my students.

Looking over the problems of students in learning the subject, I have tried my best to present the subject with student-friendly approach that is expressing it in an interactive manner and in simple language with many illustrative examples. Therefore, I hope that the book will serve as a textbook for M.Sc., M.Phil. and Ph.D. students of chemistry.

My endeavour will be amply rewarded if the book is found helpful to the students and teachers. In spite of serious attempts to keep the text free of errors, it would be presumptuous to hope that no error has crept in. I shall be grateful to all

https://doi.org/10.1515/9783110727302-202

those who may care to send their criticism and suggestions for the improvement of the book to my e-mail ID (rcmaurya1@gmail.com).

The writing of this book was initiated in September 2018 at 4515 Wavertree Drive, Missouri City-77459, Texas, during our stay with my daughter Dr. Abhilasha and son-in-law Dr. Adarsh, for which they deserve thanks.

Last but not least, I thank my wife Mrs. Usha Rani Maurya for her patient understanding of the ordeal which she had to undergo due to my almost one-sided attention during the completion of this challenging task. I am also indebted to my students Dr. J. M. Mir, Dr. P. K. Vishwakarma and Dr. D. K. Rajak for their encouragement and cooperation.

October 2, 2020 R. C. Maurya
B-95, Priyadarshini Colony, Dumna Airport Road, (Author)
Jabalpur (M.P.), India

Contents

Chapter I
Coordination chemistry of chlorophylls/ bacteriochlorophylls and its functional aspects in photosynthesis

1.1 Introduction

Organisms that have persisted over millions of years could do so by making use of one or both of the two choices for energy requirements of living cells.

(a) Heterotrophs/chemotrophs

Heterotrophs, including animals, fungi and most types of bacteria, also called chemotrophs, use the chemical option to meet cellular energy needs. They ingest other plants and animals and oxidize organic compounds (carbohydrates and fats). The final stage of energy metabolism, the oxidative phosphorylation, generates ATP (adenosine triphosphate) using energy from electron transport. In other words, heterotrophic organisms survive by degrading complex organic molecules provided by other organisms.

(b) Phototrophs

The other way to generate cellular energy is the photosynthetic option, used by phototrophs. They absorb energy from solar radiation and divert the energy through electron transport chains to synthesize ATP and generate reducing power in the form of NADPH. These energetic products of photosynthesis (ATP and NADPH) are used to make carbohydrates (glucose) from two molecules of CO_2 and H_2O and release O_2 molecules as by-products into the atmosphere. This released O_2 of the atmosphere is now used by aerobic heterotrophs for degradation of energy-rich organic molecules to CO_2 and H_2O.

Phototrophs comprised both prokaryotic and eukaryotic organisms. Prokaryotic organisms contain mostly bacteria and blue-green algae having no clearly different cell nucleus or internal cellular compartmentation, while eukaryotic organisms consist of plants and animals with cells that have a clearly different membrane-enclosed nucleus and well-distinct internal compartments. Higher plants, algae and photosynthetic bacteria come under eukaryotic organisms.

A basic question now arises that where from CO_2 comes in atmosphere for photosynthetic option of energy generation by phototrophs. In fact, heterotrophs generate CO_2 through breathing that comes in the atmosphere and this is taken by phototrophs/photosynthetic organisms in photosynthesis to make glucose. Thus, it is the solar energy providing a driving force for non-stop cycling of atmospheric CO_2 and O_2 passing across our biosphere (Figure 1.1).

https://doi.org/10.1515/9783110727302-001

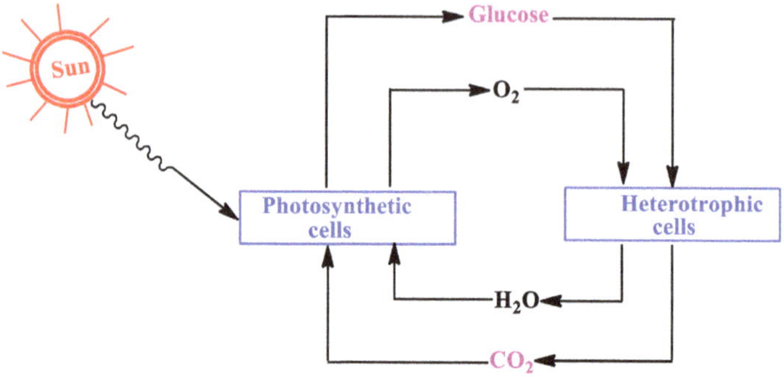

Figure 1.1: Photosynthetic cells using the energy of sunlight to produce glucose from CO_2 and H_2O along with release of O_2. Released O_2 is taken by heterotrophic cells for degrading energy-rich glucose to CO_2 and H_2O which are again converted to glucose by photosynthetic cells. This cycle continues further.

Thousands of glucose molecules are further combined to form one or the other two giant carbohydrate molecules: (i) cellulose which makes up the framework of the plants and (ii) starch which is stored in the seeds. In fact, the life on earth obviously could have continued in this manner by synthesizing such complex molecules from the simple ones.

Plants and other photosynthetic organisms fix about 10^{11} tons of carbon from CO_2 into organic compounds annually. This represents storage of over 10^{18} kJ (kilojoules) of energy. Indeed, except for nuclear fuel and geothermal energy, our energy sources ultimately are solar. Wood, peat (deposit of organic debris), natural gas and oils represent solar energy stored through photosynthesis. In spite of enormous magnitude of the above-said conversion (fixation) of carbon into organic compounds (stored solar energy through photosynthesis), the total amount of fixed carbon on the Earth is decreasing as a result of consumption. As our reserves of energy diminish, it becomes increasingly important to understand how photosynthesis works and how our activities affect it.

1.2 Diverse photosynthetic organisms

Photosynthesis occurs not only in familiar green plants we see around us but also in lower eukaryotic organisms not visible to the naked eyes. Moreover, photosynthesis also occurs in prokaryotic organisms. The photosynthetic prokaryotes include the cyanobacteria (formerly called the blue-green algae), the green sulphur bacteria mostly present in lakes around the mountain and the purple sulphur bacteria generally present in sulphur springs. Probably, the most versatile photosynthetic organisms are the cyanobacteria normally present in both freshwater and saltwater. Cyanobacteria are amongst the most self-sufficient organisms in our biosphere because these can fix atmospheric nitrogen also. While 50% of the photosynthetic activity, that is, carbon fixation on the Earth takes place in ocean (large sea), rivers and lakes brought about by the many different organisms that constitute the phytoplankton (tiny free floating aquatic plants).

1.2.1 Photosynthetic organisms: dependence on different hydrogen donors

Photosynthetic organisms can be divided into two classes:
(i) those that produce O_2 and
(ii) those which do not produce O_2.

Green leaf cells of higher plants are oxygen producers. They use water as the hydrogen donor for the reduction of CO_2 and in this process yield molecular oxygen. An overall equation for CO_2 fixation as it occurs in plants is

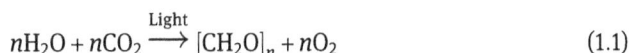

$$nH_2O + nCO_2 \xrightarrow{\text{Light}} [CH_2O]_n + nO_2 \tag{1.1}$$

In the above reaction, n is often assigned the value 6 to correspond with the formation of glucose $\{[CH_2O]_6\} = C_6H_{12}O_6$ as the end product of CO_2 reduction.
 Or more generally,

$$H_2O + CO_2 \xrightarrow{\text{Light}} [CH_2O]_n + O_2$$

Here [CH_2O] represents part of a glucose molecule.
 As an exception, the cyanobacteria are O_2-generating photosynthetic system and thus bear a resemblance to that of green plants. The other photosynthetic bacteria do not generate O_2. Most of them are, in fact not requiring oxygen, that is, anaerobic and hence cannot tolerate oxygen. As an example, hydrogen sulphide is used as a hydrogen donor by the green sulphur bacteria as shown in the following equation:

$$2H_2S + CO_2 \xrightarrow{\text{Light}} [CH_2O]_n + H_2O + 2S$$

Using H_2S as hydrogen donor, these bacteria produce elemental sulphur as the oxidation product of H_2S rather than giving off O_2.
 Instead, organic compounds such as lactate and isopropanol are used as organic donors by some non-sulphur photosynthetic bacteria. The respective oxidation products are pyruvate and acetone as shown in the following equations:

$$2\left(\begin{array}{c} CH_3 \\ | \\ H-C-OH \\ | \\ COO^- \end{array}\right) + CO_2 \xrightarrow{h\nu} [CH_2O] + H_2O + 2\left(\begin{array}{c} CH_3 \\ | \\ C=O \\ | \\ COO^- \end{array}\right)$$

$$\text{Lactate} \qquad\qquad\qquad\qquad \text{Puruvate}$$

$$2\left(\begin{array}{c} CH_3 \\ | \\ H-C-OH \\ | \\ CH_3 \end{array}\right) + CO_2 \xrightarrow{h\nu} [CH_2O] + H_2O + 2\left(\begin{array}{c} CH_3 \\ | \\ C=O \\ | \\ CH_3 \end{array}\right)$$

$$\text{Isopropanol} \qquad\qquad\qquad\qquad \text{Acetone}$$

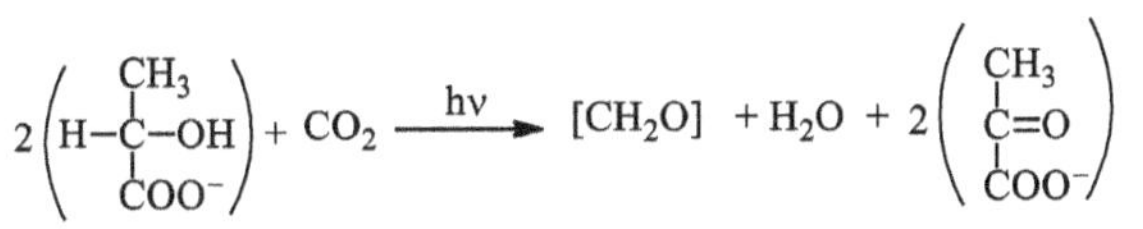

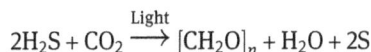

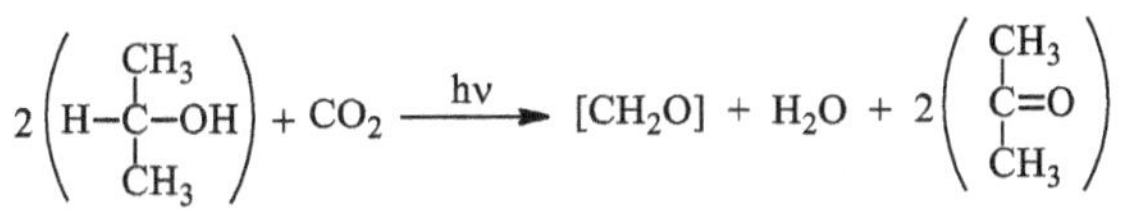

Though CO_2 is the major electron acceptor in all the photosynthetic autotrophs (organisms that can form their organic constituents from CO_2), few higher plants can also use nitrate as an electron acceptor, which is reduced to NH_3. Nitrogen-fixing photosynthetic organisms use both molecular nitrogen (N_2) and CO_2, of which N_2 is reduced to NH_3 and CO_2 is reduced to carbohydrates. Certain photosynthetic bacteria can even use proton (H^+) as the ultimate electron acceptor, which is reduced to H_2. Some bacteria use sulphate as the electron acceptor:

$$H_2O + 2H^+ \xrightarrow{hv} 2H_2 + [O]$$

$$3\,H_2O + N_2 \xrightarrow{hv} 2NH_3 + [3O]$$

$$9\,H_2O + 2NO_3^- \xrightarrow{hv} 2NH_3 + 6H_2O + 9[O]$$

Cornelis Van Niel, a pioneer worker in the study of comparative metabolism of plants and bacteria, postulated that plants and bacterial photosynthesis are basically similar processes regardless of the dissimilarities in the hydrogen donors they use. This resemblance is apparent if one writes photosynthesis equation in a more common way as follows:

$$2H_2D + CO_2 \xrightarrow{\text{Light}} [CH_2O]_n + H_2O + 2D$$

where H_2D stands for hydrogen donor and D is its oxidized form. Depending upon the photosynthetic organisms, H_2D may be H_2O, H_2S, lactate or other organic compounds. Thus, photosynthesis in plants could be formulated as a reaction in which CO_2 is reduced to carbohydrate (organic material) by hydrogen derived from water. The essence of this view of photosynthesis is that water is split by light. It is estimated that photosynthesis splits 1.5×10^9 km^3 of water on the Earth, and reconstituted by respiration and combustion once every 2 millions of years or so.

Van Neil also predicted that the molecular oxygen formed during plant photosynthesis is derived exclusively from water and not from CO_2, which was later on established by isotope tracer experiments with ^{18}O labelled water:

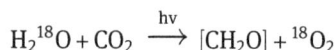

$$H_2{}^{18}O + CO_2 \xrightarrow{hv} [CH_2O] + {}^{18}O_2$$

1.3 Light and dark reactions in photosynthesis

In green plants, photosynthesis completes via two stages. In the first stage (a) (given below), light is used to oxidize H_2O:

$$2H_2O \xrightarrow{hv} [4\dot{H}] + O_2$$

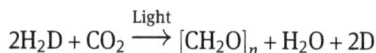

and in the second stage (b) (shown below), the resulting reducing agent, here symbolized as [H˙], subsequently reduces CO_2 to carbohydrate:

$$[4H^{\cdot}] + CO_2 \longrightarrow \underset{\text{Carbohydrate}}{[CH_2O]} + H_2O$$

The two stages (a) and (b) of photosynthesis are called as **photo-phase** or **light reactions** and **synthesis-phase or dark reactions**, respectively. The synthesis phase (b), where CO_2 is converted to carbohydrate, is called dark reactions because light is not directly needed in this.

1.3.1 Photo-phase or light reaction

Chlorophyll and other specialized pigment molecules, in the photo-phase reaction, capture light energy and are thus oxidized. A series of electron transfer reactions occur, which end with the reduction of $NADP^+$ (nicotinamide adenine dinucleotide phosphate) to NADPH. This generates a trans-membrane proton gradient whose energy is used to synthesize ATP (adenosine triphosphate) from ADP + Pi (inorganic phosphate). The oxidized pigment molecules are reduced by H_2O, thus generating O_2. So, in the light-phase reaction, the photosynthetic cells absorb light energy and save it in the chemical form as the two energy-rich products ATP and NADPH. The structures of ADP (adenosine diphosphate), ATP, $NADP^+$ and NADPH are shown in Figure 1.2.

1.3.2 Synthesis-phase or dark reactions

The ATP and NADPH so formed in the light reaction are used to reduce CO_2 to glucose and other organic products like pyruvate and acetone in the dark-phase reaction.

The generation of O_2 taking place in the photo-phase reactions only, and the reduction of CO_2 to glucose and carbohydrates that need no light, are thus two distinct and separate routes. These are shown in Figure 1.3.

The two reaction phases can be combined into a single redox reaction:

$$2H_2A + CO_2 \xrightarrow{hv} [CH_2O] + 2A + H_2O$$

The identity of the electron donor H_2A depends on the photosynthetic organisms. Photosynthetic bacteria use inorganic molecules, namely H_2S, H_2 or NH_3 or organic compounds such as lactate or isopropanol. For higher plants and algae, the electron donor H_2A is water, which is oxidized to O_2 (product 2A in the above reaction).

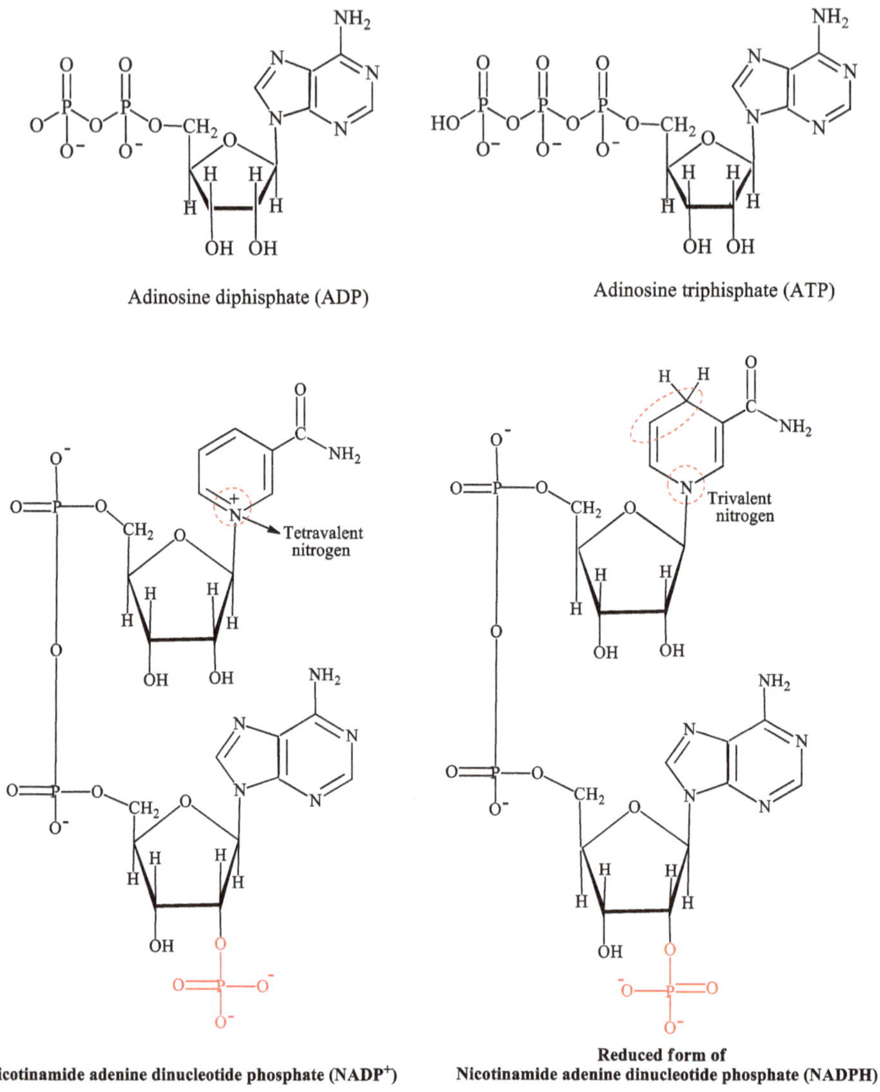

Figure 1.2: Structures of ADP, ATP, $NADP^+$ and NADPH.

Dark reactions do not imply that these take place in the dark, that is, at night. In fact, these reactions do not require light energy. In living plants and bacteria, dark reactions take place with light reactions in the daytime. The green leaf cells respire when they utilize oxygen to consume glucose and organic nutrients generated by photosynthesis during the daytime. Hence, light and dark reactions can also be termed as light-dependent and light-independent reactions, respectively.

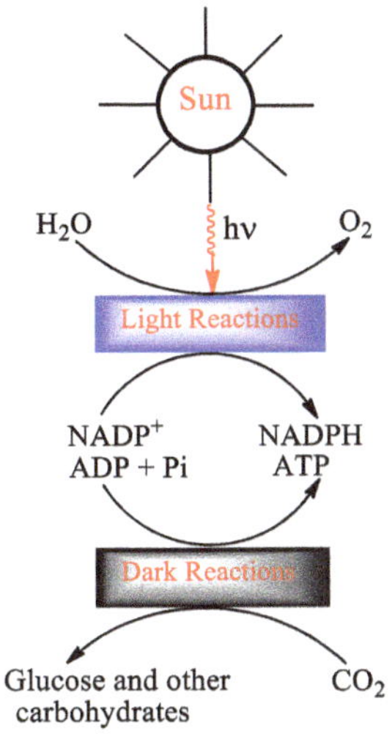

Figure 1.3: Two phases of photosynthesis linked together in a cycle. In the course of photo-phase reaction, NADPH and ATP are formed using light energy. These energetic products are used to make glucose and other dark reactions.

1.4 Chloroplasts: the photosynthesis location

Chloroplast is the place of photosynthesis in eukaryotes (algae and higher plants). Most of the plant cells and some algae contain an organelle called the chloroplast. The chloroplast permits plants to gather energy from sunlight to complete a process known as photosynthesis. In eukaryotic photosynthetic cells, both light and dark reactions happen in chloroplasts, which may be regarded as the major solar power plants in such cells. The green leaf cells also contain mitochondria. At night, when solar energy is not accessible, mitochondria generate ATP as per requirement of the cells using oxygen to oxidize carbohydrate generated in chloroplasts during the daytime hours. Thus, mitochondria are chemical power plants in dark.

Photosynthetic cells have 1–1,000 chloroplasts in number, which significantly differ in size and shapes in different species, but are usually ~5 μm long ellipsoids. They are generally much larger in volume than mitochondria. Similar to mitochondria, which they resemble in many ways, chloroplasts have a highly permeable outer membrane and a nearly impermeable inner membrane separated by a narrow intermembrane space. The inner membrane encircles the stroma, a concentrated solution of enzymes in addition to those required for carbohydrate synthesis in dark reactions. The DNA, RNA and ribosomes involved in the synthesis of several chloroplast proteins are also present in stroma.

The stroma sequentially surrounds a third membranous compartment called **thylakoid** (Greek: *thylakos*, a sac or pouch). The thylakoid is possibly a single highly folded vesicle (small bag), although in most organisms it appears to consist of stacks (heaps of things) of disc-like sacs (small bags) called **grana**. These (grana) are interconnected by un-stacked **stromal lamella** (a thin flat structure of bone or tissue).

A chloroplast generally holds 10–100 grana. The thylakoid membranes contain all the photosynthetic pigments (harvester of light energy) of the chloroplasts, and all the enzymes essential for the foremost light-dependent reactions. The structure of chloroplast found in corn is shown in Figure 1.4.

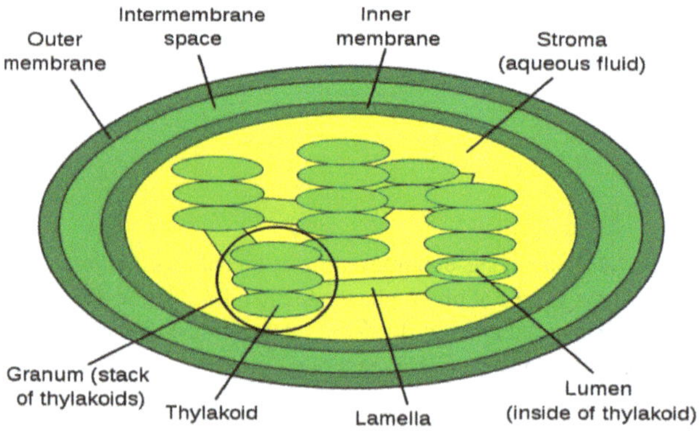

Figure 1.4: Structure of chloroplast from corn.

1.5 Light-harvesting pigments: chlorophylls

Photosynthetic cells contain light absorbing compounds called pigments that are physically located in thylakoid membranes. With the exception of certain halophilic (salt loving) bacteria, all photosynthetic organisms possess one or more types of the class of green pigments, chlorophylls. Figure 1.8 shows the structures of several of the different types of chlorophyll that occur in nature. In chloroplast of most of the green plants, the primary photoreceptor is chlorophyll *a*.

Chlorophylls resemble haemes, the prosthetic groups of haemoglobin, myoglobin and the cytochromes, and they are derived biosynthetically from protoporphyrin IX. Haeme belongs to the class of metal porphyrins. The basic framework of all porphyrins is built of four pyrrole rings, linked together in a large ring-like structure called porphin (Figure 1.5).

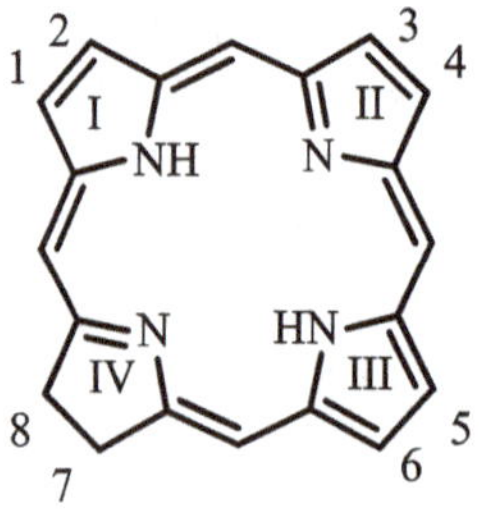

Figure 1.5: Structure of porphin.

By substitution of eight hydrogen atoms about the edge of porphin structure with various side chains, namely, acetyl, methyl, vinyl, propionyl, and others, various porphyrins found in nature are produced. The particular compound found in haeme is known as protoporphyrin IX. It has methyl group ($-CH_3$) at 1, 3, 5 and 8, propionyl group ($-CH_2CH_2COOH$) at 6 and 7 and vinyl group ($-CH{=}CH_2$) at 2 and 4 positions of the parent compound porphin. For each unit of cytochromes or haemoglobin, coordinated positions of Fe(II) are occupied by four nitrogens of protoporphyrin IX (Figure 1.6).

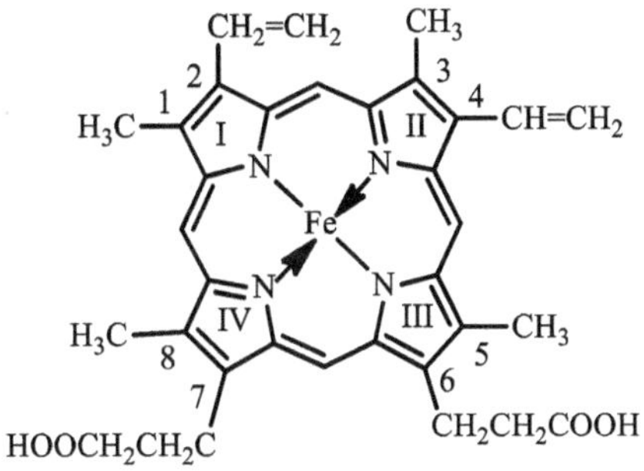

Figure 1.6: Structure displaying coordination of four nitrogens of protoporphyrin IX with Fe(II).

However, chlorophyll molecules vary from haeme molecule in several respects:

(i) Chlorophylls have an additional ring (ring V) with carbonyl and carboxylic ester substituents, that is, the methylidine carbon between pyrrole rings III and IV is connected to $CH(COOCH_3)(C{=}O)$ moiety to form cyclopentanone ring. This moiety is susceptible to keto-enol tautomerism. This characteristic five-ring porphyrin derivative is called **pheoporphyrin** (pheo = a plant; Figure 1.7).

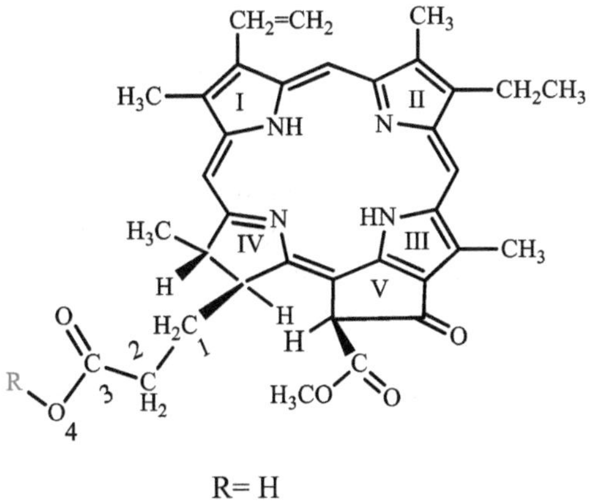

Figure 1.7: Structure of pheoporphyrin.

(ii) Instead of Fe(II), the central metal is Mg(II) in chlorophylls.
(iii) The two key chlorophylls, chlorophyll *a* and chlorophyll *b*, present in plants and cyanobacteria have **one of the pyrrole rings (ring IV) reduced** with addition of two hydrogens. Moreover, chlorophyll *a* and chlorophyll *b* differ in terms of substituent at position 3 (CH_3 group in chlorophyll *a* and CHO group in chlorophyll *b*) as shown in Figure 1.8. The bacteriochlorophyll *a* and bacteriochlorophyll *b* occurring in purple and green bacteria contain **two reduced pyrrole rings, namely rings II and IV**.
(iv) Finally, ring IV having propionyl side group is esterified with a long-chain isoprenoid alcohol. Chlorophylls *a* and *b* have the alcohol phytol while the bacteriochlorophylls *a* and *b* have either phytol/phytyl or geranylgeraniol/geranylgeranyl, subject to the species of bacteria. Actually, the presence of this alcoholic chain makes chlorophyll highly hydrophobic and soluble in non-polar media. Photosynthetic organisms also contain small amount of **pheophytins** or **bacteriopheophytins**, which are the same as the corresponding chlorophylls or bacteriochlorophylls except that two 2H's replace Mg. Pheophytins or bacteriopheophytins play a special role as electron carriers in photosynthesis.
(v) The macrocyclic ring in chlorophylls is referred to as **chlorin** ring.

The structures of various types of chlorophylls are shown in Figure 1.8.

1.5.1 Accessory pigments in thylakoids (photosynthetic cells)

Besides chlorophylls, secondary or accessory pigments are also present in photosynthetic cells that absorb light in wavelength ranges where chlorophyll is not as effective. Thus, they supplement the photochemical action spectrum of chlorophylls. Accessory pigments are of two types: (i) carotenoids and (ii) phycobilins (Figure 1.9). The carotenoids are represented by β-carotene (a precursor of vitamin A) and xanthophyll. All carotenoids contain 40 carbon atoms and display extensive conjugation.

Most of the carotenoids absorb in the 400–500 nm range [vide electronic spectrum of carotenoids given in (Figure 1.11b)] and thus are coloured yellow, red or orange. Xanthophyll's colour is a bright yellow. Generally, the bright colours of flower petals, fruits and vegetables are owing to the presence of carotenoids therein. The yellow and red colours of fall leaves result from the damage of green chlorophylls preferentially, exposing the colour of the carotenoids.

Phycobilins, which are linear tetrapyrroles, absorb in the 550–630 nm range (chlorophylls are cyclic tetrapyrroles). A common red phycobilin is phycoerythrobilin. Two other pigments are the red phycoerythrin and the blue phycocyanin, which contain a set of linear tetrapyrroles (vide electronic spectral patterns of these pigments given in Figure 1.11b). The variation in colours among plant species is due to differences in the proportional abundance of chlorophylls and other accessory

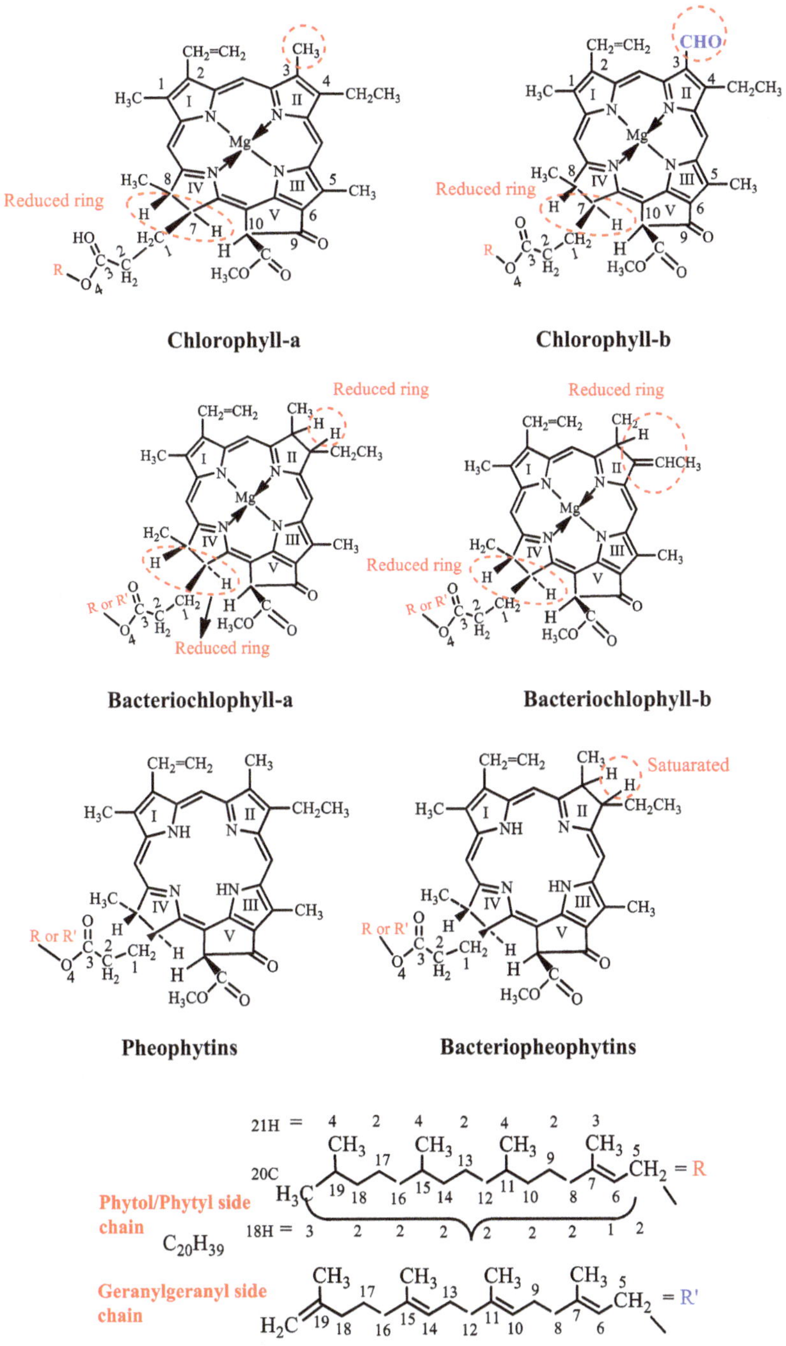

Figure 1.8: Structures of various types of chlorophylls found in various photosynthetic organisms.

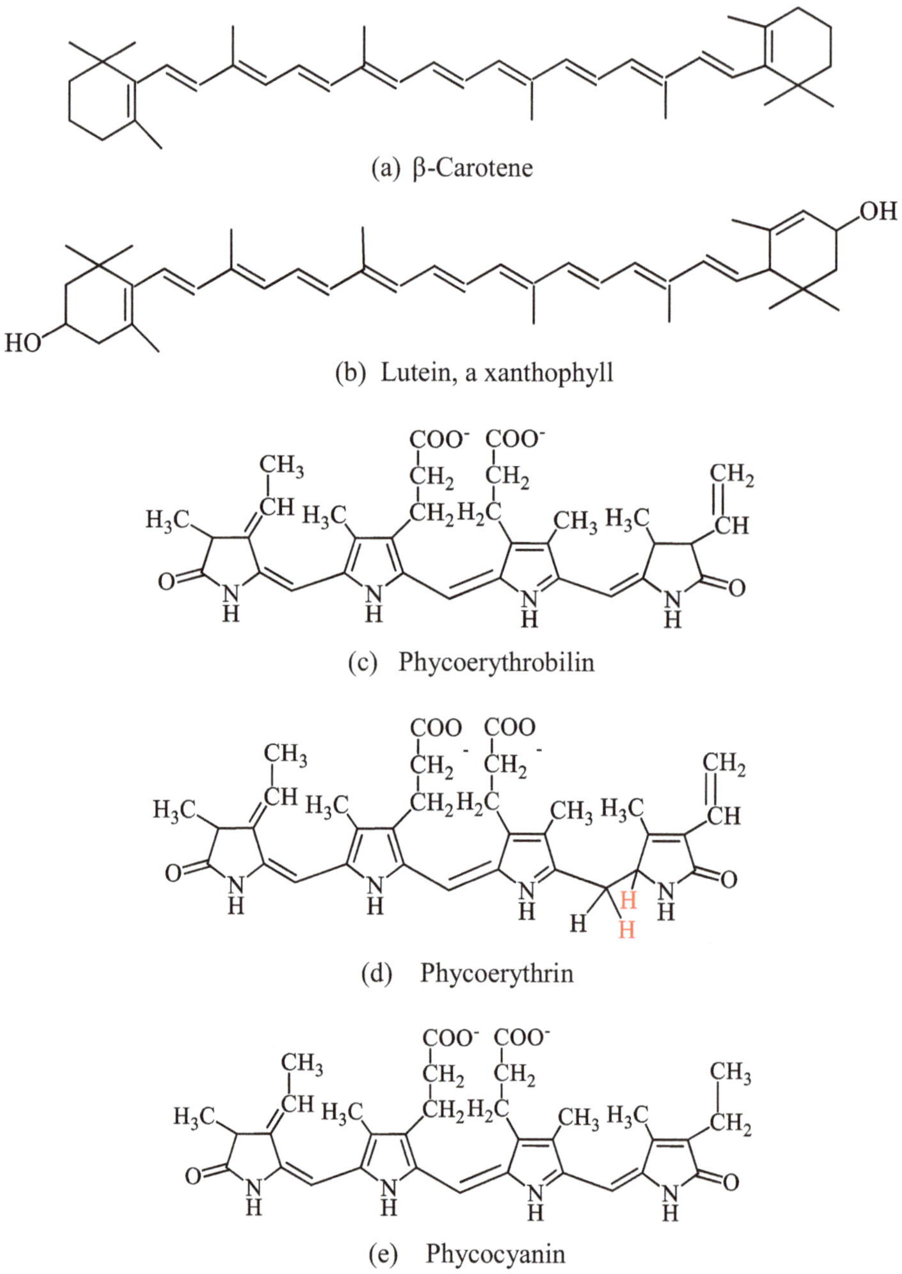

Figure 1.9: Molecular structures of accessory pigments: (a) β-carotene; (b) lutein, a xanthophyll; (c) the red phycobilin, phycoerythrobilin; (d) the red phycoerythrin; and (e) the blue phycocyanin. These biomolecules assist chlorophylls *a* and *b* in absorbing light for photosynthesis.

pigments. Green colours in plants resulted mainly by a greater proportion of chlorophylls, whereas other colours, namely red and purple are due to an abundance of accessory pigments.

1.6 Chlorophylls in photosynthetic light absorption: absorption spectra of various photosynthetic pigments and their functions

The experimental support/evidence that chlorophylls play the major role in photosynthetic light absorption comes from the construction of a photochemical action spectrum for a typical green plant as shown in Figure 1.10. The solid line represents the biological action spectrum, which is a plot of the efficiency of each wavelength in sustaining plant growth as measured by CO_2 uptake and O_2 evolution, that is, the relative rate of photosynthesis. The dotted line is the combined absorption spectrum of chlorophylls *a* and *b*. The two lines (solid and dotted) are almost similar in shape, indicating the importance of chlorophyll *a* and *b* molecules in supporting photosynthesis.

It is worthwhile to mention here that the photosynthetic efficiency of chlorophylls in the lower wavelength region (400–500 nm range) is supplemented by the accessory pigments present in the chlorophylls.

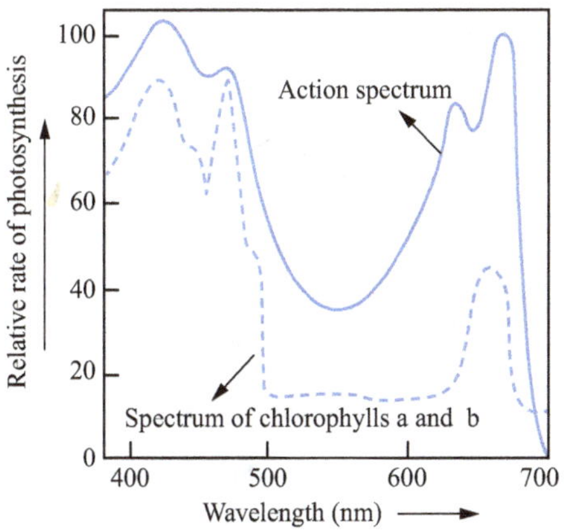

Figure 1.10: (a) The solid line represents biological action spectrum of a typical green plants. (b) The dotted line is the combined spectrum of chlorophylls *a* and *b*.

The chlorophyll and other photosynthetic pigment molecules strongly absorb radiation and undergo the electronic transitions, π(HOMO) $\rightarrow \pi^*$(LUMO), in the visible region, that is, the most intense form of solar radiation reaching the Earth's surface. The comparatively small chemical differences among various chlorophylls greatly affect their absorption spectral patterns. The effective photoreceptor property of chlorophylls and other photosynthetic pigments is because they have a number of single and double bonds, that is, they are highly conjugated molecules.

The reduction of ring IV in chlorophyll *a* or *b* makes the conjugated aromatic ring system unambiguously asymmetrical. This changes the electronic absorption spectrum of the chlorophyll molecule significantly. While the long-wavelength α-absorption band at ~ 550 nm of the reduced form of symmetrically conjugated cytochrome *c* (Figure 1.11a) in 10 μM solution is relatively weak, asymmetrically conjugated chlorophyll *a* (Figure 1.11b) has an intense absorption band [molar extinction coefficient (ε) ~ 10^5 cm^{-1} M^{-1}, among the highest observed for organic compounds] at 676 nm. Chlorophyll *b* has a similar band at 642 nm (Figure 1.12a and b). Bacteriochlorophylls *a*, in which the asymmetry of the conjugated system is even more pronounced, have extremely strong absorption bands in the region 770 nm (Figure 1.13). All chlorophylls thus absorb very well, particularly at long wavelengths.

While in the lower wavelength region, light is not appreciably absorbed by chlorophyll *a* at 460 nm, it is captured by chlorophyll *b*, which has an intense absorption band at that wavelength (Figure 1.11a and b). The spectral region from 500 to 600 nm is only weakly absorbed, but it does not pose a problem for most green plants.

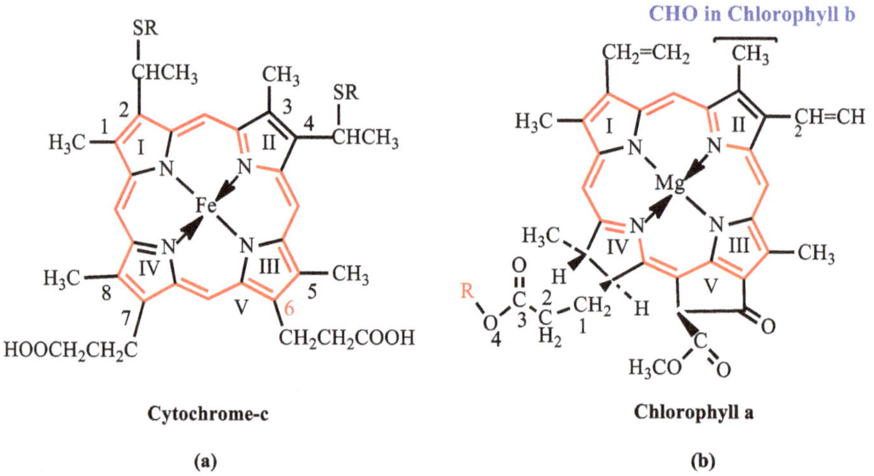

Figure 1.11: (a) Red colour showing symmetric conjugation in the haeme group of cytochrome *c*. (b) Red colour showing asymmetric conjugation in chlorophylls.

Contrary to this, light is often a limiting factor for blue-green algae (cyanobacteria) and red algae. They possess accessory light collecting pigments (such as phycocyanobilin and phycoerythrobilin) that enable them to trap light which is not absorbed strongly by the chlorophylls of photosynthetic plants.

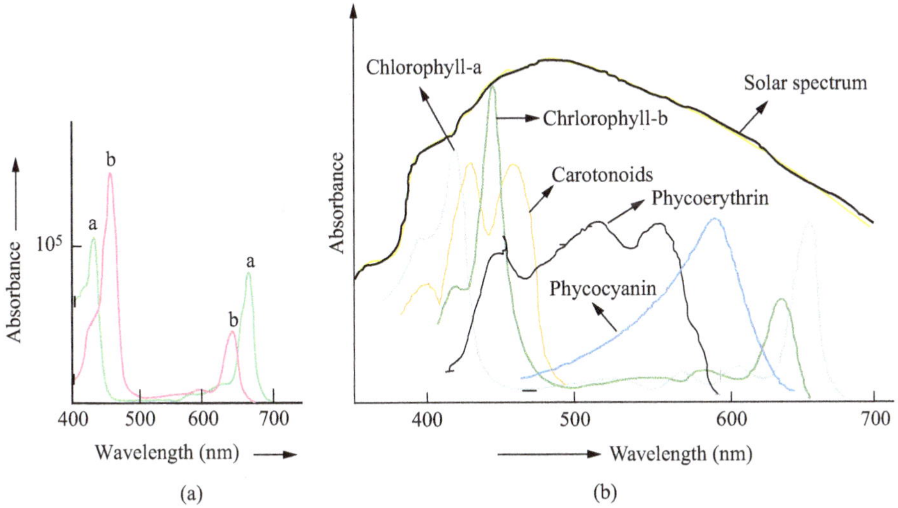

Figure 1.12: (a) Absorption spectra of chlorophyll *a* and *b*. (b) Absorption spectra of various photosynthetic pigments.

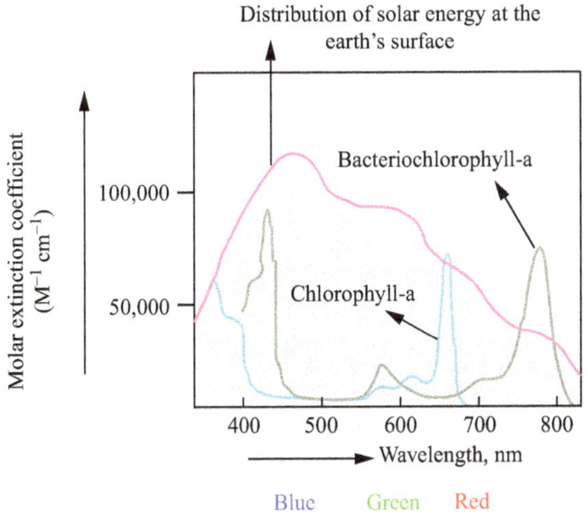

Figure 1.13: Absorption spectra of chlorophyll *a* and bacteriochlorophyll *a*.

1.7 The role of Mg(II) in chlorophylls

(i) The choice of Mg(II) in chlorophyll is really unique. In fact, without Mg, chlorin ring is fluorescent (i.e. the absorbed light energy is emitted back immediately). But after incorporation of Mg(II), chlorophyll becomes phosphorescent. This change in the nature of chlorophyll from fluorescent to phosphorescent due to Mg incorporation is biologically very important.

(ii) If the fluorescence occurs exclusively, the absorbed light energy is lost immediately and it will not be available for chemical transformation in a chemical reaction. Hence, the absorbed light energy must be stored for some while so that it can be utilized in a chemical reaction for the transformation of light energy to chemical energy.

(iii) For this phosphorescence behaviour to take place, there must be an excited state of finite lifetime. Possibly, the mixing of the excited singlet and triplet states through spin–orbit coupling in Mg gives a relatively stable triplet excited state. In fact, triplet to singlet transition is not allowed and it makes the triplet excited state stable.

(iv) Mg(II) (d^0 system) does not have any crystal field stabilization energy to prefer the square planar geometry, rather it prefers the tetrahedral geometry where the steric hindrance is less. However, the rigid chlorin ring enforces Mg(II) to have the planar geometry. Consequently, the Mg(II)–N bonds remain in strained condition, and the electron constituting the bond can be readily excited by absorption of light energy. This absorbed light can be utilized in the desired chemical reaction.

(v) Rigidity of the macrocyclic centre is further reinforced through coordination by chlorophyll to the Mg(II) centre. It is notable that the macrocyclic ring experiencing conjugation or delocalization of π-electron cloud (Figure 1.11b) is itself sufficiently rigid. The rigidity of the system minimizes the energy loss due to molecular vibration (i.e. thermal vibration).

(vi) Stacking of chlorophylls (i.e. polymerization) in antenna chlorophyll (Chl), that is, $(Chl)_n$ is attained through the bridging action of Mg(II) between the adjacent chlorophyll moieties. This polymerization of chlorophyll takes place through the coordination of the keto group at C-9 of cyclopentanone ring to another chlorophyll molecule giving an approximate square pyramidal geometry around Mg (II) (Figure 1.14).

(vii) The H_2O coordinated to Mg(II) at the axial position in the chlorophyll of active reaction centre, that is, Chl · H_2O · Chl experiences the photoinduced H-atom transfer to the keto group. This radical formation provides the electron required in the photosynthetic process. Thus, coordination of water molecule to the Mg(II) centre plays a crucial role (Figure 1.15).

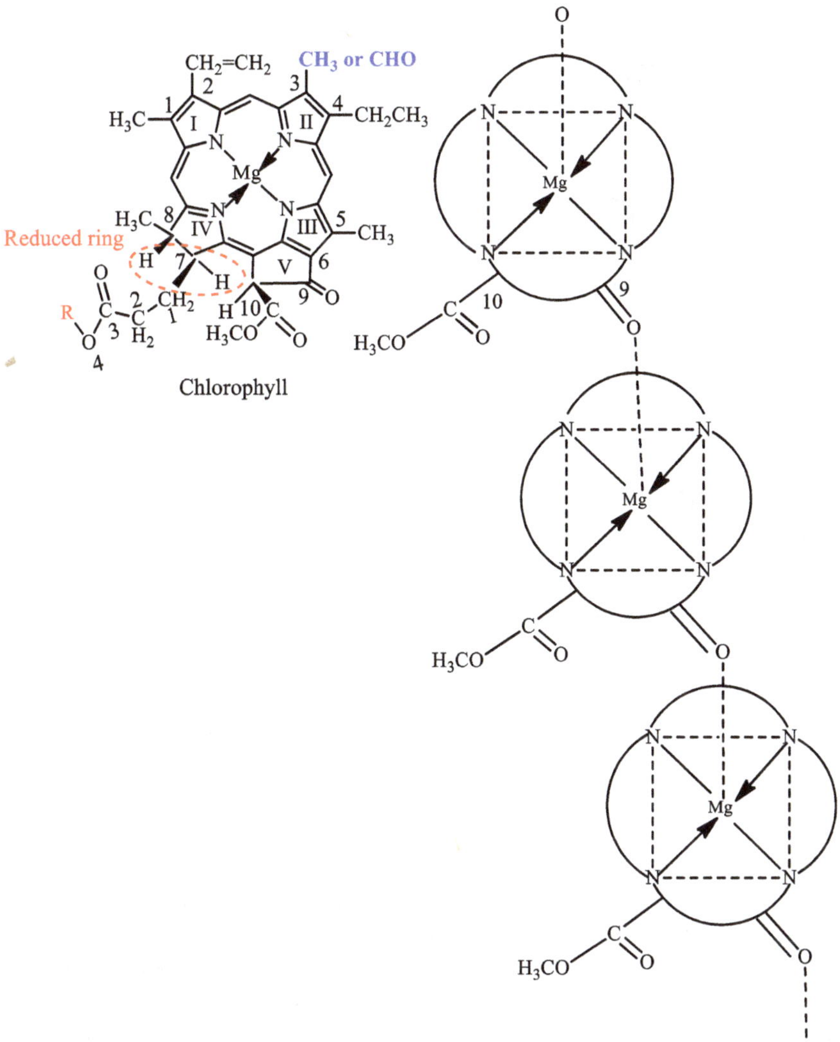

Figure 1.14: Pictorial representation of polymerized chlorophyll $(Chl)_n$ acting as antenna chlorophyll.

1.8 Hill reactions: illuminated chloroplasts evolve O_2 and reduce electron acceptors

How does light drive the synthesis of ATP and NADPH (chemically energetic compounds) that are used in synthesis phase? In thermodynamic term, this requires the transduction of electromagnetic energy (light) into chemical bond energy. It was ultimately explained by the experimental evidence carried out by Robert Hill in

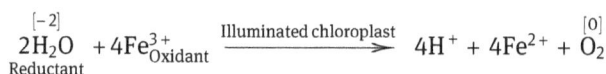

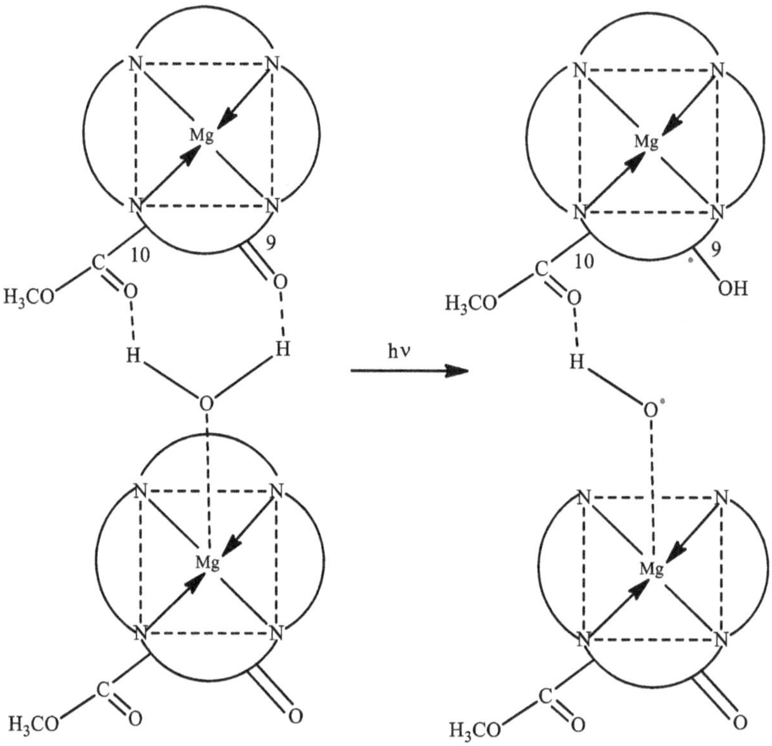

Figure 1.15: Pictorial representation ofChl · H_2O · Chl present at the active reaction centre.

1937 using leaf extracts containing chloroplasts, for electron transport processes (oxidation–reduction) in photosynthesis. When chloroplasts of leaf extracts were illuminated in the presence of an appropriate artificial electron accepting compound, namely, potassium ferricyanide, Hill observed two simultaneous processes, neither of which occurred in dark:

(i) Concomitant reduction of the electron acceptor, ferricyanide to ferrocyanide.
(ii) Oxidation of H_2O with evolution of O_2.

$$\underset{\text{Reductant}}{\overset{[-2]}{2H_2O}} + 4Fe^{3+}_{\text{Oxidant}} \xrightarrow{\text{Illuminated chloroplast}} 4H^+ + 4Fe^{2+} + \overset{[0]}{O_2}$$

The Hill reaction is a landmark to explain/unveil the mechanism of photosynthesis for several reasons:

(i) It dissected photosynthesis by showing that O_2 evolution can occur without the reduction of CO_2. Synthetic electron acceptors such as $[Fe(CN)_6]^{3-}$ can substitute for CO_2. It confirmed that the evolved O_2 comes from H_2O rather than from CO_2, because no CO_2 was present.

(ii) It showed that the isolated chloroplasts can perform a significant partial reaction of photosynthesis.
(iii) It revealed that a key happening in photosynthesis is the light-driven transfer of an electron from one substance to another in the thermodynamically uphill (ascending/climbing) way (because light energy is required). The reduction of ferricyanide to ferrocyanide by light is a transformation of light energy into chemical bond energy.

The overall schematic diagram of Hill reaction is shown in Figure 1.16.

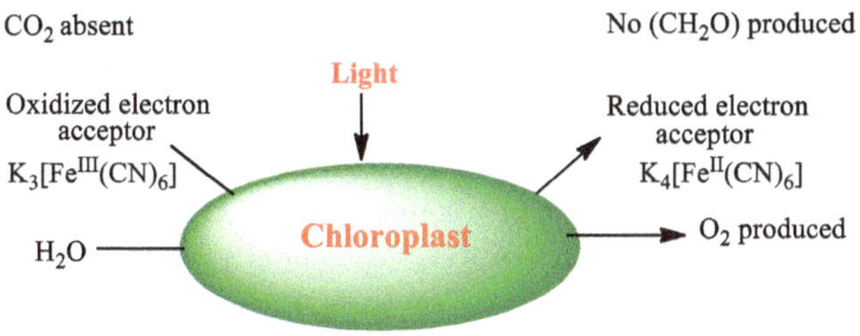

Figure 1.16: Schematic diagram of Hill reaction.

Thus, Hill was able to show that electron flowed (apparently through an electron transport chain) from H_2O (a reducing agent) to the electron acceptor. It was later shown that the physiological electron acceptor in green plant chloroplast was $NADP^+$ and the reaction occurring was

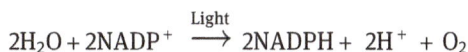

$$2H_2O + 2NADP^+ \xrightarrow{\text{Light}} 2NADPH + 2H^+ + O_2$$

The distinction between electron flow in chloroplasts and that in mitochondria now becomes evident. The above reaction is reverse of the process in mitochondria: the electron flows from the electron donating molecule H_2O to the electron accepting $NADP^+$ species. As the direction of electron flow is thermodynamically uphill, light energy is required.

1.9 Photosystems I and II in photosynthesis by green plants

(i) Chlorophyll catalyses the reduction of $NADP^+$ to NADPH and oxidation of H_2O to O_2 in the presence of light. In this process, the electron flows from H_2O to $NADP^+$ through an electron transport chain (P-680 to P-700*), which looks like Z when electron carriers are placed in the order of their reduction potentials. Thus, the chain is very often described as **Z-scheme**.

(ii) The above process is carried out in photosynthetic organisms by two kinds of functional units called photosystems present in thylakoid membranes of chloroplasts. The existence of two such specific photosystems, so-called **photosystem I** and **photosystem II**, was established by the phenomenon of **red drop** and the dependence photosynthetic rate or quantum yield/quantum efficiency of photosynthesis (i.e. the rate of O_2 evolved per photon) on the wavelength of light.

1.9.1 Red drop

Investigation based on the reliance of photosynthesis rate on the wavelength of incident light led to the discovery that chloroplast contains two photosystems. Here, the photosynthetic rate means the rate of O_2 evolution in the photosynthesis process. When it is divided by the number of quanta absorbed gives the relative quantum efficiency/quantum yield of this process. For a single kind of photoreceptor, the quantum efficiency is expected to be independent of wavelength over its entire absorption band. This is not the case in photosynthesis. In fact, the quantum efficiency of photosynthesis drops sharply at wavelength >680 nm, even though chlorophyll absorbs light in the range from 680 to 700 nm (Figure 1.17). The photosynthesis rate can, however, be improved by taking long wavelength light making addition of light of a shorter wavelength, such as 600 nm. Using both 600 and 700 nm of light together, the photosynthetic rate is greater than the sum of the rates when the two wavelengths are used independently.

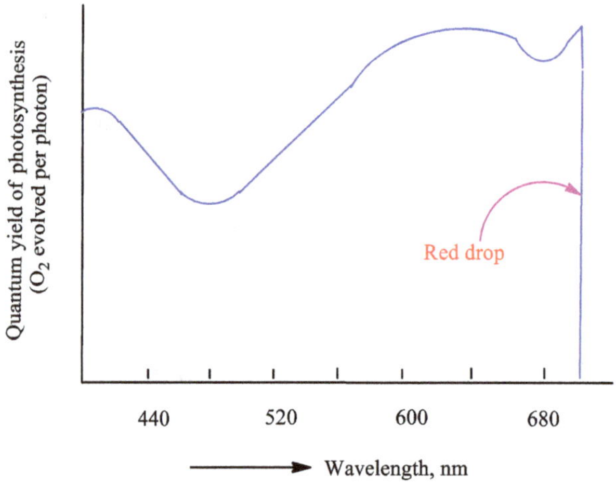

Figure 1.17: The sharp drop in quantum yield at wavelength >680 nm.

The above observations are called the red drop. Based on the enhancement phenomenon, **Robert Emerson (University of Illinois)** proposed that photosynthesis entails the interactions of two light reactions. Both of them can be driven by light of wavelength <680 nm, but only one of them by light of longer wavelength.

1.9.2 Photosystem I (also called PS-I or P-700, P stands for pigment)

It comprises the key acceptor **chlorophyll *a*,** and other accessory pigments that absorb light in the region 700 nm (or lower; 600–700 nm) which generates a strong reductant to bring about the reduction of $NADP^+$ to NADPH.

1.9.3 Photosystem II (abbreviated as PS-II or P-680)

It contains **chlorophylls *a* and *b*** plus accessory pigments and absorbs light primarily at 680 nm (or lower) to produce a very strong oxidant to oxidize H_2O to O_2. Each of the two photosystems contain ~250 light-harvesting pigments (~200 chlorophylls and ~50 carotenoids) in addition to different electron carriers.

All photosynthetic cells have **PS-I.** Photosynthetic organisms, namely, higher plants, algae and cyanobacteria which evolve O_2 contain both **PS-I and PS-II.**On the other hand, **PS-I** is found in photosynthetic bacteria, which do not evolve O_2.

In terms of standard reduction potential, under biological conditions, pH ~ 7, of the involved redox couples, O_2/H_2O ($E_0{'}$ = 0.82 V) and $NADP^+$/NADPH ($E_0{'}$ = −0.34 V), the electron flow from H_2O to $NADP^+$ to produce NADPH is a thermodynamically uphill process. But the photo excitation of PS-I and PS-II can make the flow of electron downhill as shown in the Z-scheme.

When the chlorophyll present in PS-I and PS-II is excited, distribution of its electron pattern changes. Chlorophyll on excitation can serve as both a better reducing agent (because excited electrons can be easily removed) and also a better oxidizing agent (as the positive hole generated due to excitation of electron can accept electron favourably). Thus, excited chlorophylls can initiate a series of redox reactions.

The two photosystems in green plants (Figure 1.18) are linked together by electron transport chain. The coordinated action of the two photosystems begin with absorption of light by photosystem-I.

1.9.4 Photosystem I

Light absorbed by chlorophylls and accessory pigments is transmitted to reaction centre designated as P-700 (**PS-I**). Upon excitation by light, an electron in a P-700 molecule is moved to a higher energy level, generating an excited molecule, P-700*.

When P-700 (ground state) is excited to P-700* (excited state), its reduction potential changes from +0.4 V to about −1.3 V. In fact, the uphill reaction is favoured by the absorption of 700 nm photon (of ~171 kJ mol^{-1} energy) and the photon energy is utilized to elevate the electron. The activated electron, now at a higher energy level, is passed via a chain of electron carriers starting with A_O, (a form of chlorophyll), and ending with $NADP^+$. In the reaction form, the first electron transfer step is

$$\text{P-700}^* + A_0 \rightarrow \text{P-700}^+ + A_0^-$$

In other words, electron transfer from P-700* results in a deactivated electron-deficient electron centre P-700^+ and a reduced electron acceptor A_O^-. The resulting A_O^- transfers its electron to phylloquinone (A_1), which forward an electron to Fe–S complex. Next in line to accept an electron from reduced Fe–S complex is another Fe–S protein, ferridoxin (Fd). The best known ferredoxin is that of spinach chloroplast, which has a molecular weight of 10,700 dalton and a Fe_2S_2 redox centre. Finally, electrons are transferred from reduced ferredoxin to $NADP^+$ by flavoprotein catalyst, ferridoxin–$NADP^+$ oxidoreductase resulting in NADPH:

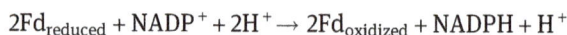

$$2Fd_{reduced} + NADP^+ + 2H^+ \rightarrow 2Fd_{oxidized} + NADPH + H^+$$

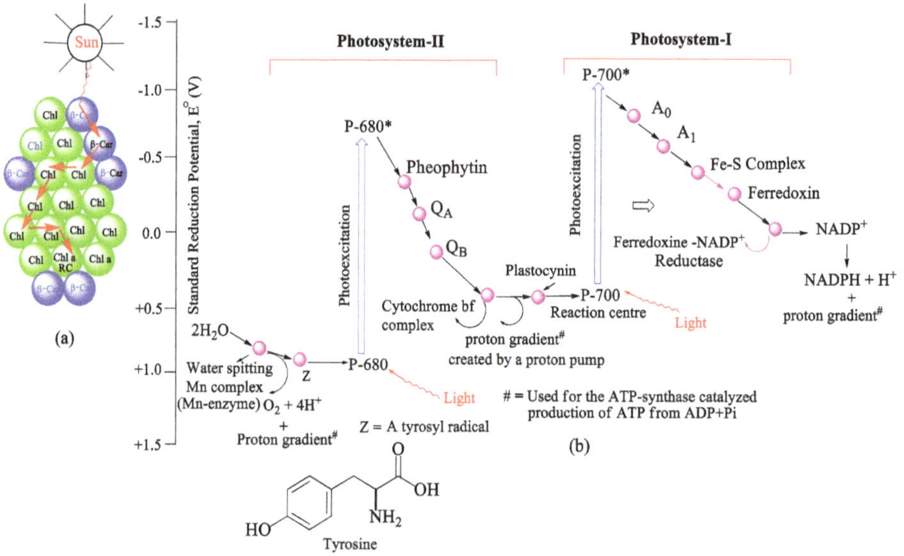

Figure 1.18: (a) Light absorption by accessory pigments in photosystem and (b) pictorial representation of PS-I and PS-II into the Z-scheme.

Excitation and transfer of electron from P-700 leaves an unstable, electron-deficient species, P-700*. The electron hole in P-700* is refilled by another transport chain which is driven by **PS-II**. Illumination of **PS-II** with a chlorophyll reaction centre,

P-680, produces an activated form, P-680*. The energetic electron in P-680* is quickly passed on to pheophytin (Ph), a chlorophyll-like electron acceptor:

$$P680^{*} + Ph \xrightarrow{\text{Light}} P680^{+} + Ph^{-}$$

Reduced pheophytin, Ph^{-}, transfers an electron to plastoquinone Q_A, a protein-bound species and, Q_B, a free form protein. Q_A and Q_B have structures similar to coenzyme Q in mitochondria. The next part of the electron transport chain connecting PS-I and PS-II is comprised of a cytochrome *bf* complex, which is an association of several important membrane proteins using haeme groups and iron–sulphur centres to transfer electrons from reduced Q_B (i.e. plastoquinol Q_BH_2) to a blue copper protein, plastocyanin (PC). The reduced copper centre in PC transfers an electron to P-700^{+}, the electron-deficient (strong oxidizing) form of P-700. Figure 1.19 shows the role of plastocyanin in the context of the thylakoid membrane in which this process operates.

1.9.5 Water splitting reactions assisted by a water splitting complex (Mn_4 cluster)

The oxidized form of **PS-II** oxidizes water and the released electrons enter into the photochemically active centre. A water splitting complex, which is possibly a manganese containing enzyme (Mn_4 cluster)catalyses this process. The oxidation of H_2O to O_2 involves the net transfer of four electrons as follows:

$$2H_2O \rightarrow 4H^{+} + 4e^{-} + O_2$$

The mechanistic model of the water splitting complex/enzyme for water oxidation is given as follows. The P-680^{+} formed in PS-II is a strong oxidant and it extracts electron from H_2O through the Z-factor (i.e. $H_2O \rightarrow$Mn complex$\rightarrow$PS-II). It is suggested that the Mn-based enzyme is first oxidized in **four one electron transfer steps** and then this oxidized form of enzyme leads to O_2 evolution. If Mn(IV)/Mn(II) cycle operates, then the enzyme must contain at least two Mn centres to accommodate the four-electron transfer process.

In order to explain the observation of O_2 evolution, several models are proposed time to time. One such popular model (Figure 1.20; known as **Kok cycle)** is presented here. The enzyme contains four Mn centres. This Mn cluster along with four to five Cl^{-} ions and two to three Ca^{2+} ions constitute a catalytically active complex known as **oxygen evolving complex** (OEC) which can bind two molecules of water. During extracting of electrons from H_2O, protons and O_2 are released. OEC cycles through a series of states designated by S_O, S_1, S_2, S_3 and S_4. These states are actually the different combinations of manganese (III) and manganese (IV) centres. The OEC is oxidized by P-680^{+} sequentially to different states. The lower states, that is, S_O, S_1, and S_2 are **cubane** (C_8H_8)-like Mn_4O_4 complexes while the higher states, S_3 and S_4 are **adamantane**-like Mn_4O_6 complexes (Figure 1.21). The sequential electron withdrawal

$$2H^+$$

Plastoquinone, Q_B

[2,3-dimethyl-5-(3-methylbut-2-enyl)cyclohexa-2,5-diene-1,4-dione]

Plastoquinol, Q_BH_2

[2,3-dimethyl-5-(3-methylbut-2-enyl)-benzene-1,4-diol]

$$QBH_2 + 2PC\ (Cu^{2+}) \longrightarrow QB + 2PC\ (Cu^+) + 2H^+$$

His 57
Meth 92
II
Cu
CH$_3$
Cys 84
His 57

Structure of Coordinated copper ion in Plastocynin (PC)

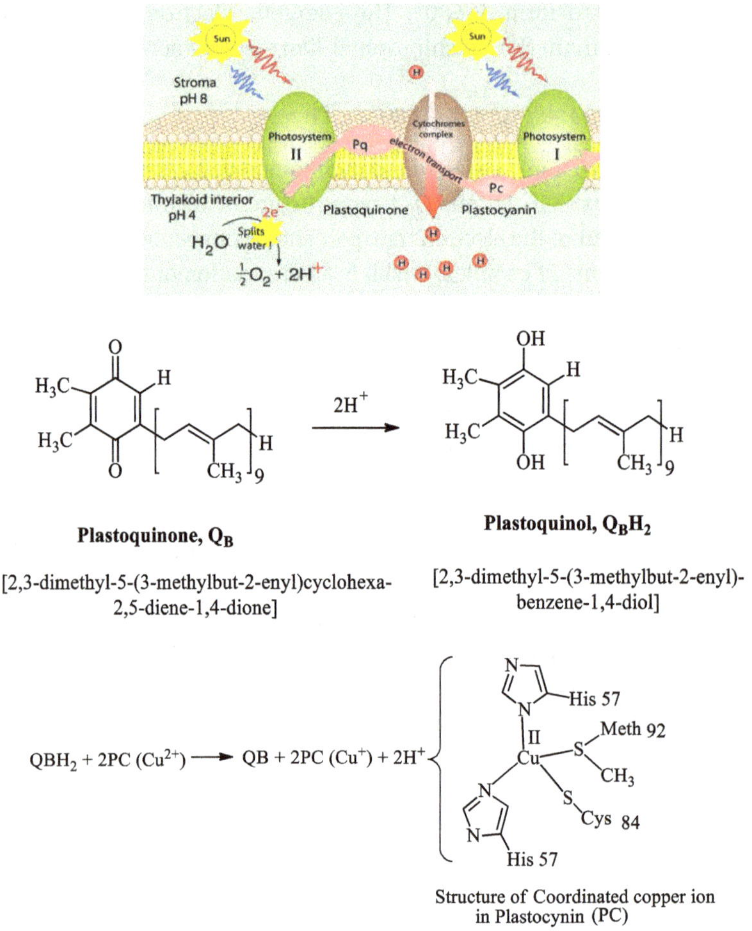

Figure 1.19: Role of plastocyanin in the context of the thylakoid membrane.

and release of protons and O_2 are shown in Figure 1.20. In the conversion of S_4 to S_0, the adamantine-like structure adopts the cubane-like structure with the release of O_2. This step (S_4 to S_0) is light independent.

The possible oxidation states of Mn in S_0, S_1, S_2, S_3 and S_4 states are given in Figure 1.22.

It is suggested that for the final step, $S_3 \rightarrow [S_4] \rightarrow S_0$ (where S_4 is probably a transient only), the reduction of a tyrosyl radical occurs with the evolution of O_2. Here, it is notable that Ca^{2+} (which may be replaced by Sr^{2+} to restore the activity) and Cl^- are essential cofactors of the Mn-OEC. Their actual role is not well established. However, it is suggested that some bridging ligands (like carboxylate) bridge the Mn and Ca sites and Cl^- acts as a **gate keeper** to control the substrate (H_2O) accessibility to tetra-manganese core of Kok cycle. The Cl^- may also serve as a bridging ligand

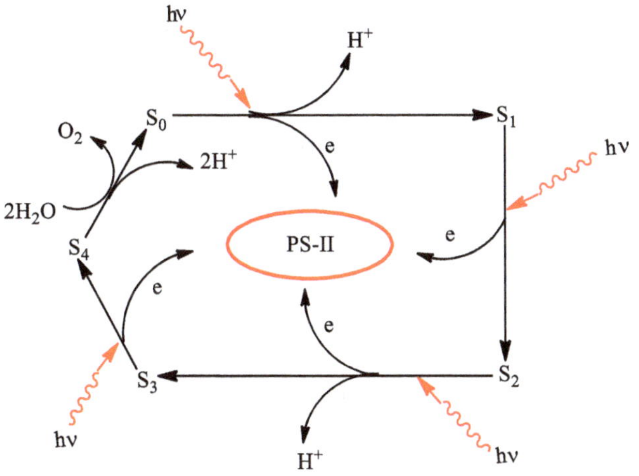

Figure 1.20: The schematic mechanism of oxidation of H_2O in chloroplast catalysed by the **oxygen evolving complex,** a Mn_4 cluster.

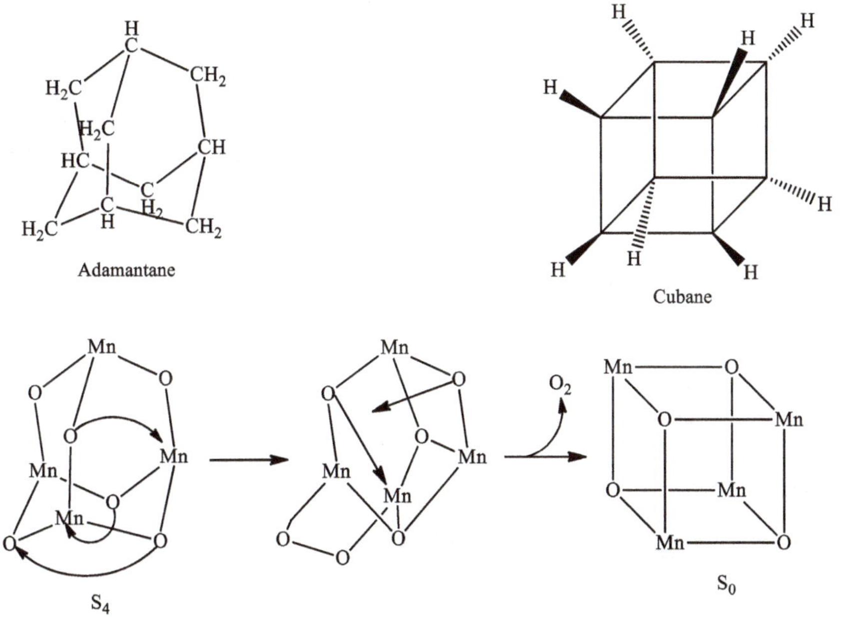

Figure 1.21: Adamantane-like structure of S_4 and cubane-likestructure of S_0 and conversion of S_4 to S_0.

between Ca and Mn centres. Ca-bound H_2O molecule remains with the oxo-sites of Mn_4 cluster as H-bonded and these H-bonded H_2O molecules (bound to Ca-site) participate in the activity of the Kok cycle.

S_0: $Mn^{III}(\mu\!-\!OH)_2Mn^{II}$

O

$Mn^{IV}(\mu\!-\!O)_2Mn^{IV}$

S_1: $Mn^{III}(\mu\!-\!O)_2Mn^{III}$

O

$Mn^{IV}(\mu\!-\!O)_2Mn^{IV}$

S_2: $Mn^{III}(\mu\!-\!O)_2Mn^{IV}$

O

$Mn^{IV}(\mu\!-\!O)_2Mn^{IV}$

S_3: $Mn^{III}(\mu\!-\!\overset{\bullet}{O})(\mu\!-\!O)Mn^{IV}$

O

$Mn^{IV}(\mu\!-\!O)_2Mn^{IV}$

S_4: $Mn^{III}(\mu\!-\!\overset{\bullet}{O})_2Mn^{IV}$

O

$Mn^{IV}(\mu\!-\!O)_2Mn^{IV}$

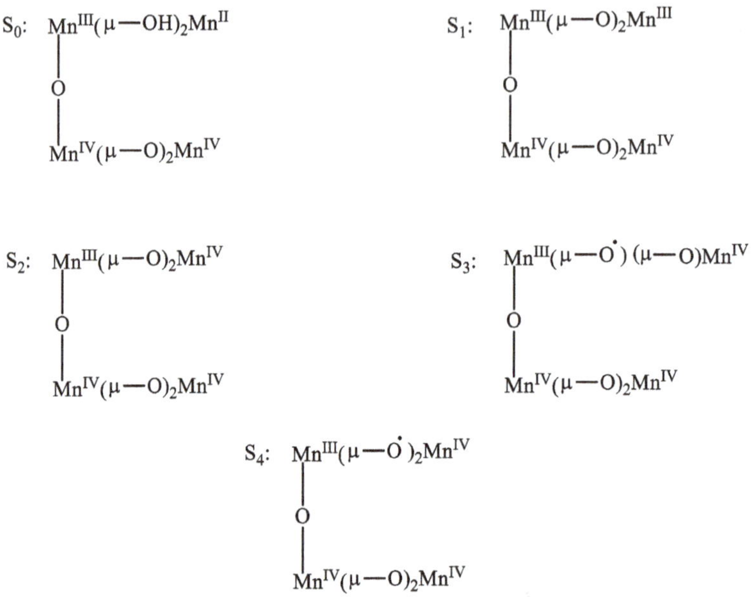

Figure 1.22: Possible oxidation states of Mn in S_0, S_1, S_2, S_3 and S_4 states.

Based on X-ray absorption fine structure and ESR spectroscopy, Yachandra et al. [1] proposed a structural model for the O_2-evolving manganese complex. This model is pictorially presented in Figure 1.23.

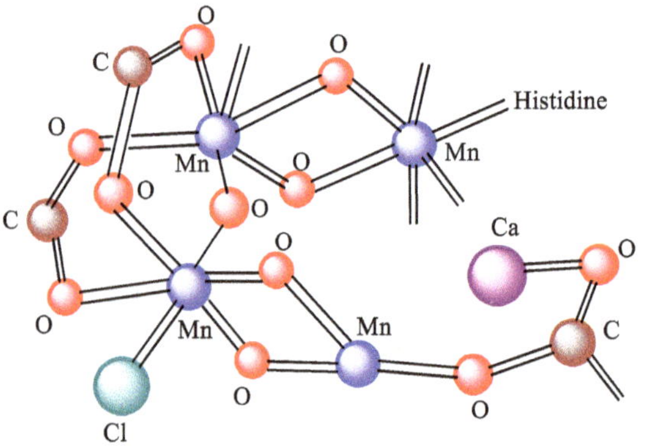

Figure 1.23: Structural model for the O_2-evolving Mn_4 cluster of PS-II. The ligands of the Mn atoms probably include oxygen (μ-oxo) bridges, the oxygen of the carboxylic amino acid side chain, a histidine side chain N and a Cl⁻ ion.

Reactions occurring in Z-scheme are summarized as follows:

$$\textbf{Photosystem-II}: 2H_2O \rightarrow O_2 + 4H^+ + 4e \quad (Mn_4 \text{ cluster})$$

$$\textbf{Photosystem-I}: 2NADP^+ + 2H^+ + 4e \rightarrow 2NADPH$$

The electrons released in **PS-II** are transmitted to **PS-I** as discussed in Z-scheme. Then, the dark reaction produces $CH_2O/C_6H_{12}O_6$ by reducing with the help of NADPH.

1.9.6 Dark reaction

$$6CO_2 + 12NADPH + 12H^+ \rightarrow 12NADP^+ + 6H_2O + C_6H_{12}O_6$$

Thus, the overall reaction is

$$6CO_2 + 6H_2O \rightarrow C_6H_{12}O_6 + 6O_2$$

1.9.7 Cyclic photophosphorylation in photosynthesis

In the electron transport chain, P-700* to $NADP^+$, if there is insufficient $NADP^+$ to accept the electrons from the reduced ferredoxin (Fd) because of very high ratio of NADPH to $NADP^+$, then the high potential electrons flow back to the oxidized form of P-700 through cytochrome bf complex and plastocyanin. It creates a cyclic flow of electron and during the cyclic flow, ATP from ADP + Pi is also generated. This is called cyclic phosphorylation. In this cyclic flow, PS-II is not involved and consequently no O_2 is produced from H_2O. However, ATP is generated, but no NADPH is formed (Figure 1.24).

The flow of electrons from the excited **PS-II** to **PS-I** (as downhill)is combined with phosphorylation of ADP to ATP as shown in the following equation:

$$2H_2O + 2NADP^+ + 2ADP + 2Pi \xrightarrow{\text{Light}} O_2 + 2NADPH + 2H^+ + 2ATP$$

NADPH results from the transfer of 2e and one H^+ form a water molecule to $NADPH^+$ in the light phase of photosynthesis. From NADPH, the electrons flow to CO_2 and other acceptors. The required energy for the reduction of CO_2 to glucose during the dark-phase photosynthesis of green plants is supplied by ATP. The complete reaction of the conversion of CO_2 to glucose in the presence of NADPH assisted by ATP is

$$CO_2 + 3ATP + 2NADPH + 2H^+ + 2H_2O \rightarrow 1/x(CH_2O)_x + 3ADP + 3Pi + 2NADP^+$$

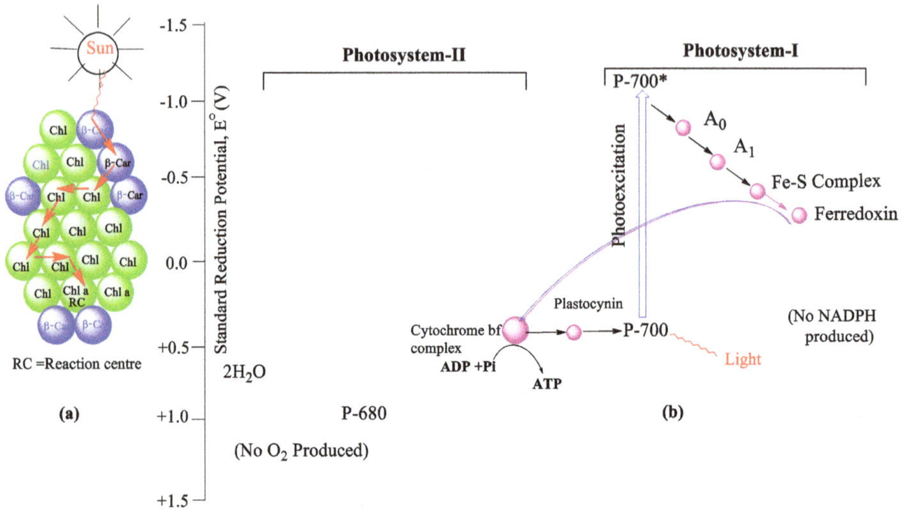

Figure 1.24: (a) Photosystem showing absorption of light by accessory pigments. (b) Schematic diagram of photophosphorylation due to cyclic electron flow in photosystem-I producing ATP.

1.10 Antenna chlorophylls and reaction centres in chloroplasts

The key reactions of photosynthesis occur at photosynthetic reaction centres. Photosynthetic congregations comprised of chlorophyll molecules and a set of accessory pigments, such as other chlorophylls and carotenoids. These are contained in reaction centres. This is due to the fact that most chlorophyll molecules and accessory pigments do not participate directly in photochemical reactions but function to collect light, that is, they act as light-harvesting antennas. These antenna chlorophylls/pigments carry on the energy of absorbed photons (units of light) from molecule to molecule until it reaches a photosynthetic reaction centre (Figure 1.25).

The pass on of energy from antenna chlorophylls to the reaction centre takes place in a very small time interval of $<10^{-8}$ s having an efficiency of >90%. This much high efficiency depends on the chlorophyll molecules with proper spacing and orientations. It has been observed that even in bright sunlight, a reaction centre directly captures only ~1 photon per second. This is a metabolically insignificant rate. A complex of **antenna pigments or light-harvesting complex** is, therefore, essential.

The chlorophyll molecule bears a highly non-polar phytyl chain and a polar moiety containing Mg(II)–chlorin complex. Thus, the chlorophylls can act as surfactants. In non-polar solvent it may undergo polymerization leading to $(Chl)_n$ through the coordination by the keto group at C-9 of cyclopentanone ring to another

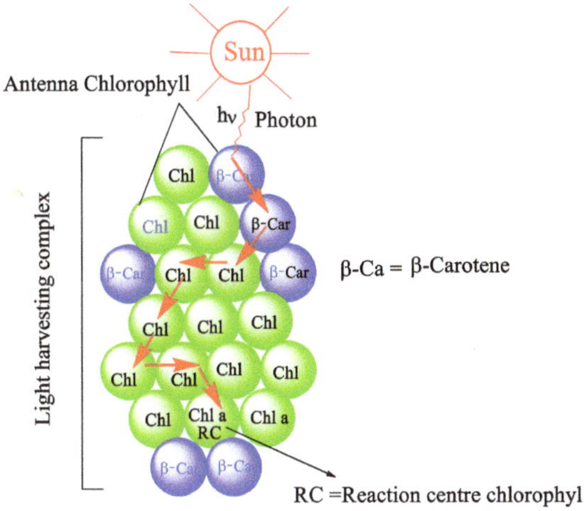

Figure 1.25: Pass on of energy of absorbed photons from antenna chlorophylls to the reaction centre.

chlorophyll molecule giving an approximate square pyramidal geometry (Figure 1.14). In the presence of water it may form $(Chl \cdot H_2O)_n$ oligomers, where H_2O occupies the fifth coordination site of Mg(II) and this H_2O molecule remains hydrogen bonded with the ester group oxygen at position C-10 and keto group oxygen at position at C-9 of cyclopentanone ring of another chlorophyll molecule. A dimer unit is shown in Figure 1.15.

It is currently believed that **antenna chlorophyll** remains as $(Chl)_n$, which terminates with a $(Chl \cdot H_2O \cdot Chl)$,and is an **active reaction centre** in the photosynthetic system. The antenna chlorophyll molecules harvest the light energy or sensitized by energy transfer from carotenes, which are previously photo-excited. This energy of excitation is ultimately shifted to the reaction centre $(Chl \cdot H_2O \cdot Chl)$, which is also associated with different electron carriers. It is suggested that on photo-excitation, the hydrogen bonded H_2O molecule in the active unit experiences a photo-induced H atom transfer to the keto group (Figure 1.14). This radical formation provides the electron required in the photosynthetic process. In fact, oxidation of $(Chl \cdot H_2O \cdot Chl)$ at PS-I is initiated by quinone which diffuses away with electron to cause reduction of CO_2 to glucose. This photo-induced H atom transfer is not possible in non-hydrogen bonded chlorophyll monomers, or dimers or polymers (i.e. antenna chlorophyll).

The complete electron transport system leading to reduction of CO_2 to glucose in photosynthesis is schematically shown in Figure 1.26.

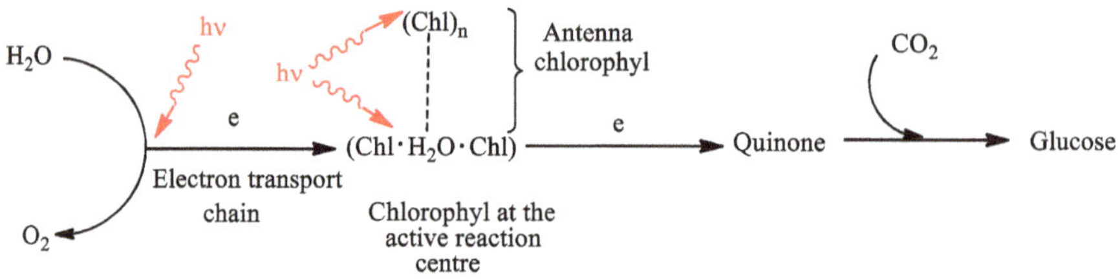

Figure 1.26: Schematic representation of photoelectron transfer in photosynthesis for the reduction of CO_2 to glucose.

Reference

[1] V. K Yachandra, V. J. DeRose, M. J. Latimer, I. Mukerji, K. Sauer, and M. P. Klein. Where plants make oxygen: A structural model for the photosynthetic oxygen-evolving manganese cluster, Science, 260 (1993) 675–679.

Exercises

Multiple-choice questions/fill in the blanks

1. Which of the following come in purview of heterotrophs?
   (a) Fungi
   (b) Most types of bacteria
   (c) Animals
   (d) All of these

2. CO_2 comes in atmosphere for photosynthesis by phototrophs from:
   (a) CO_2 formed by respiration in heterotrophs that returns to the atmosphere
   (b) Bacteria
   (c) Fungi
   (d) All of these

3. The phototrophs include:
   (a) Prokaryotic organism
   (b) Eukaryotic organism
   (c) Both prokaryotic and eukaryotic organisms
   (d) None of these

4. Thousands of glucose molecules formed by photosynthesis in plants further combine to form giant carbohydrate molecules:
   (a) Cellulose which makes up the framework of the plants
   (b) Starch which is stored in the seeds
   (c) Both cellulose and starch
   (d) None of these

5. Which of the following represent solar energy stored through photosynthesis?
   (a) Wood (b) Natural gas (c) Oils (d) All of these

6. The photosynthetic prokaryotes include:
   (a) Cyanobacteria (formerly called the blue-green algae)
   (b) Green sulphur bacteria found in mountain lakes
   (c) Purple sulphur bacteria common in sulphur springs
   (d) All of these

7. The cyanobacteria or blue-green algae are the most versatile photosynthetic organisms because they can fix:
   (a) Atmospheric N_2
   (b) Atmospheric CO_2
   (c) Both atmospheric N_2 and CO_2
   (d) None of these

8. Sort out the photosynthetic system that instead of giving off O_2, extrude elemental sulphur as the oxidation product:
   (a) Cyanobacteria
   (b) Green sulphur bacteria
   (c) Non-sulphur photosynthetic bacteria lactate
   (d) All of these

9. In the light-phase reaction, the photosynthetic cells absorb light energy and conserve it in chemical form as the energy-rich product called:
   (a) ATP (b) NADPH (c) Both ATP and NADPH (d) None of these

10. Sort out the correct statement with regard to the difference in structures of chlorophyll *a* and chlorophyll *b*:
   (a) At position 3 in chlorin ring, CH_3 group in chlorophyll *a* and CHO group in chlorophyll *b*
   (b) At position 3 in chlorin ring, CHO group in chlorophyll *a* and CH_3 group in chlorophyll *b*
   (c) At position 3 in chlorin ring, CHO group in chlorophyll *a* and C_2H_5 group in chlorophyll *b*
   (d) None of these

11. Conjugated aromatic ring systems in chlorophyll *a* and chlorophyll *b* are:
    (a) Symmetrical
    (b) Asymmetrical
    (c) Both (a) and (b)
    (d) None of these

12. Geometry of Mg(II) in chlorophyll is
    (a) Tetrahedral
    (b) Square planar
    (c) Strained square planar geometry
    (d) Geometry cannot be predicted

13. After incorporation of Mg(II) in chlorin ring, chlorophyll becomes
    (a) Fluorescent
    (b) Phosphorescent
    (c) both (a) and (b)
    (d) None of these

14. Red and purple colours in plants are due to an abundance of
    (a) Chlorophyll
    (b) Accessory pigments
    (c) Both (a) and (b)
    (d) All of these

15. Which one of the following contains cyclic tetrapyrroles?
    (a) Phycoerythrobilin
    (b) Phycoerythrin
    (c) Phycocyanin
    (d) Chlorophyll

16. Heterotrophs are also called chemotrophs because they use the chemical option to meet cellular needs.

17. Phototrophs generate cellular energy by ____.

18. The chloroplast allows plants to harvest energy from sunlight to carry on a process known as ____.

19. When solar energy is not available at night, mitochondria present in green leaf cells generate ATP using oxygen to oxidize carbohydrate generated in chloroplasts during the daytime hours. Thus, mitochondria are chemical power plants in ____.

20. Photosynthetic cells contain secondary or accessory pigments, which absorb light in wavelength ranges where it is not as efficient.
21. Chlorophyll catalyses the reduction of $NADP^+$ to and oxidation of O_2.

Short-answer-type questions

1. Differentiate between heterotrophs and phototrophs.
2. Where from CO_2 comes in atmosphere for photosynthetic option of energy generation by phototrophs? Explain through a diagram.
3. Briefly highlight about the photosynthetic bacteria that do not produce oxygen but give other oxidized product.
4. Briefly explain light and dark reactions in photosynthesis.
5. Show the structural difference between chlorophyll *a* and chlorophyll *b* diagrammatically.
6. Draw the structures of accessory pigments that contain linear tetrapyrroles.
7. Present the pictorial representation of polymerized chlorophyll $(Chl)_n$ acting as antenna chlorophyll.
8. Highlight the salient features of Hill reactions to unveil the mechanism of photosynthesis.
9. What is red drop in photosynthesis? Explain with a plot of quantum yield versus wavelengths of light. What information one gets from this model?
10. Present the schematic diagram of photosystems I and II into the Z-scheme.

Long-answer-type questions

1. Present a detailed view of photosynthetic organisms that produce O_2 and do not produce O_2.
2. What is photosynthesis? Describe in detail light and dark reactions in photosynthesis.
3. Present the structural details of several of the different types of chlorophyll that occur in nature.
4. Discuss the role of accessory pigments in photosynthetic cells. Also draw their structures.
5. Present a detailed account of absorption spectra of various photosynthetic pigments and their functions.
6. Highlight the role of Mg in chlorophyll.
7. Describe the role of photosystem I and photosystem II during photosynthesis in green plants.

8. Present a detailed view of mechanistic model of the water splitting manganese containing enzyme for water oxidation to evolve O_2 during photosynthesis.
9. Explain the role of cyclic photophosphorylation in photosynthesis.
10. Describe the role of antenna chlorophylls and reaction centres in chloroplasts in photosynthesis for the reduction of CO_2 to glucose.

Chapter II
Complexes containing nitric oxide: synthesis, reactivity, structure, bonding and therapeutic aspects of nitric oxide–releasing molecules (NORMs) in human beings and plants

2.1 Introduction

2.1.1 Discovery of nitric oxide (NO)

Nitric oxide (NO), which is a colourless gas, was known during the thirteenth century when it was first time prepared from nitric acid. Its first synthetic recognition was credited to Johann Glauber who identified its formation when potassium nitrate was added with sulphuric acid. Its immediate transformation to brown fumes of NO_2 in the presence of O_2 was observed from its beginning. It was first time characterized as a distinct chemical species by Joseph Priestley (1733–1804). He also described its disproportionation to N_2O and NO_2 when NO_2 when heated over iron powder. The formation of iron–NO complex by the reaction of NO with $FeSO_4$ as a black solution was first time reported by Joseph Priestley. This black colouration formed the basis of the brown ring test used by so many chemists for the qualitative test of nitrate ion. It was later on characterized as $[Fe(NO)(H_2O)_5]^{2+}$. The accepted chemical formula of nitric oxide as NO, wherein nitrogen and oxygen are present in equal proportions, was established by Henry Cavendish (1731–1810) and Sir Humphrey Davy (1778–1829).

2.1.2 Importance of nitric oxide complexes

For more than a century, metal nitrosyl complexes or complexes containing NO grouping are known to chemists, and since then it has been the source of tremendous interest to scientific community working in this field. The nitrosyl complexes became more important in recent past on account of the discovery that NO is useful in our body for smooth muscle relaxation, tumour regulation, long-term memory formation and so on. However, nitrosyl complexes have been very little investigated compared to metal carbonyls. The possible reasons behind this are:

(i) Carbon monoxide as such can be taken in the synthesis of majority of metal carbonyls. In case, excess carbon monoxide if somehow taken in the synthesis, it is rarely harmful. Moreover, for kinetically slow transformations, high-pressure, high-temperature conditions are also accessible. On the other hand, these later

https://doi.org/10.1515/9783110727302-002

conditions are rarely acceptable with NO. This is because of its thermodynamic instability and its tendency to serve as an oxidizing agent:

$$\begin{gathered}3NO \rightarrow N_2O + NO_2, \qquad \Delta H = -37.2\ \text{kcal/mol} \\ (\text{disproportionation reaction})\end{gathered} \tag{2.1}$$

$$\begin{gathered}NO \rightarrow 1/2N_2 + 1/2O_2, \qquad \Delta H = -21.6\ \text{kcal/mol} \\ (\text{decomposition reaction})\end{gathered} \tag{2.2}$$

Because of the above restrictions, coordination chemists have inspired to develop new methodology for the synthesis of metal nitrosyl complexes (vide infra).

(ii) The lack of reactivity of metal nitrosyl complexes seems to be another reason for inattentiveness towards the synthesis of nitrosyl complexes. Because of the stability of the MNO bond in metal nitrosyls, ligand displacement was not found in metal nitrosyls, while such reaction is very common and also important in the chemistry of metal carbonyls.

In recent past, there has been considerable upsurge in the study of transition metal nitrosyls and the reactions of nitrosyl ligands. This is partly due to the better understanding of the way in which NO binds to a metal, which is subtly different to the situation involving CO, principally because NO has an additional electron.

There has been significant enhancement in the study of metal nitrosyl complexes and reactions of the nitrosyl ligand in the past decades or so. This is somewhat owing to the better awareness of the way of binding of NO to a metal. In fact, the binding of NO to metal is a bit different compared to CO in metal carbonyls. This is primarily due to the presence of an additional electron in NO as shown in the Lewis structures of CO and NO as follows:

:C:::O:	$:\dot{N}::\ddot{O}:$
Carbon monoxide	Nitric oxide

Accordingly, NO can bind to a metal by donating two electrons in two ways: (i) as an electron donor, forming NO^+ and then coordinating with metal and (ii) as an electron acceptor forming NO^- and then coordinating with metal, and even forming $N_2O_2^{2-}$. Another incentive for exploring NO reactivity has arisen due to the advancement in pollution control for removing or reducing the concentration of NO in exhaust gasses coming out from the internal combustion engines. Additional interest in nitrosyl chemistry has come from the possible synthesis of organo-nitrogen compounds in reaction assisted or moderated by transition metal catalysts. Transition metal nitrosyls have recently been used in several industrially important organic syntheses.

The biggest stimulus to investigating NO chemistry is the declaration of role of NO in 1980 as one of the extremely significant molecules to regulate the functioning of living organisms. This put the whole scientific community in wonder. The crucial function that this molecule plays in signal transduction and cytotoxicity is considered as one of the prime wonders nowadays in biochemistry. The biological importance of NO molecule is well pronounced in neuroscience, physiology and immunology. The compliments discovered in biochemical concern of NO bagged the vote in 1992 as "Molecule of the Year" by the journal *Science*, published by the American Association for the Advancement of Science (AAAS). Nowadays NO is believed to be linked with several physiological pathways, including platelet aggregation and adhesion, neurotransmission, synaptic plasticity, vascular permeability, hepatic metabolism, senescence and renal function. The vital role of NO at higher (μM) concentration in host immunity and tumour suppression is also recognized. Robert F. Furchgott, Louis J. Ignarro and Ferid Murad shared the 1998 Nobel Prize in Medicine and Physiology for discovering the role of NO as an important signalling molecule in cardiovascular system. In cardiovascular system, smooth muscles are often the target of NO action, leading to vasodilatation in blood vessels and thereby controlling the blood pressure. In microbial world, NO plays a mediator role in denitrification (Figure 2.1). In addition, the molecule is tailored in various ways due to its important therapeutic potential:

$$NO_2^- \longrightarrow NO \longrightarrow N_2O \longrightarrow N_2$$

Figure 2.1: Denitrification involving NO as an intermediate.

Eye is one of the primary sense organs that is very sensitive. Any defect in eye function needs ultra care and meticulous way of treatment. Among eye health problems, intraocular pressure (IOP), cataract and retinal hypertension continue to remain as the primary risk factors amenable to treatment. There are numerous evidences supporting eye adaption with NO connected with the NO–guanylate cyclase (GC) pathway.

2.2 Metal nitrosyl complexes?

Compounds containing NO groups(s) are usually referred to as nitrosyl compounds when addendum is inorganic in nature and as nitroso compounds when addendum is organic in nature.

Nitrosyl compounds	**Addendum**
NaNO	Inorganic
NOCl	Inorganic
$Na_2[Fe(CN)_5(NO)] \cdot 2H_2O$	Inorganic
$K_3[Cr(CN)_5(NO)] \cdot H_2O$	Inorganic

$[PPh_4]_2[Mn(NO)_2(CN)_4] \cdot 2H_2O$	Inorganic
$[Mo(NO)_2(tsc\text{-}dha)(H_2O)]$	Inorganic

Nitroso compounds	**Addendum**
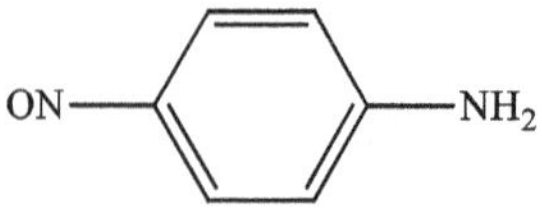 4-Nitrosoaniline	Organic
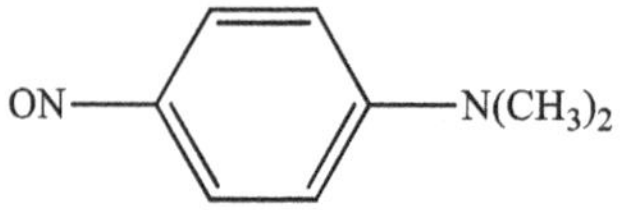 *N,N*-Dimethyl-4-Nitrosoaniline	Organic
 4-Nitrosophenol	Organic

Nitrosyl complexes belong to the class of coordination compounds containing nitrogen monoxide (NO) as one of the ligands.

2.3 Synthetic methods of metal nitrosyls

Metal nitrosyl can be synthesized using different methods and employing various nitrosylating agents as illustrated further.

2.3.1 Nitric oxide gas as the nitrosylating agent

Because of the thermodynamic instability of the stored NO gas commercially available in a cylinder, it is recommended to use freshly prepared NO gas in the preparation of metal nitrosyls. It can be generated by the following two reactions along with respective reactants:

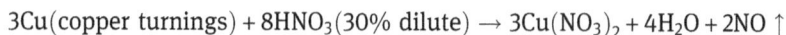

$$2KNO_3 + 6FeSO_4 + 4H_2SO_4 \rightarrow 3Fe_2(SO_4)_3 + K_2SO_4 + 4H_2O + 2NO\uparrow$$

$$3Cu(\text{copper turnings}) + 8HNO_3(30\%\ \text{dilute}) \rightarrow 3Cu(NO_3)_2 + 4H_2O + 2NO\uparrow$$

The NO gas so generated in both the reactions can be dried by passing through a column of anhydrous calcium chloride.

When one considers a reaction of some metal complex/salt with NO gas, the most useful rule will be those appropriate to the neutral diatomic molecule. Thus, we shall assume that the reagent NO functions as (i) either a one-electron donor or (ii) a three-electron donor (Figure 2.2) when it adds to or substitutes on a metal complex.

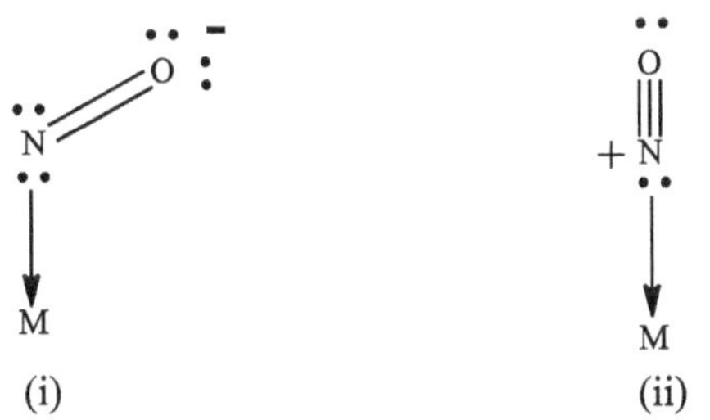

Figure 2.2: (i) NO functions as a one-electron donor and (ii) NO functions as a three-electron donor.

(a) Synthesis of [Ru^{II}(NO)(dha-tsc)(Cl)(H_2O)]complex

Maurya et al. [1] synthesized this complex by dissolving $RuCl_3 \cdot 3H_2O$ (1 mmol, 0.207 g) in 20 mL of 1:1 v/v HCl:H_2O mixture. The resulting solution was bubbled with NO gas (30 min) formed by adding 30% of HNO_3 to copper turnings. The solution changes to red from brown colour. The ethanolic solution of Schiff base (dha-tsc) (1 mmol, 0.241 g) was then added and the resulting solution was refluxed for 3 h. This resulted in a whitish brown suspension, and after cooling down to room temperature, a solid mass was obtained (Figure 2.3). The product so formed was filtered and washed 2–3 times with methanol and dried under vacuum.

Simple adduct formation, $[M(L)_n] + xNO \rightarrow [M(L)_n(NO)_x]$

The reaction of adduct formation with NO is likely when the substrate complex [ML_n] is having 15 or 17 electrons, that is, deficient with 3 or 1 electron(s), respectively, from inert gas configuration. For example:

$$\left[Co^{II}(Cl)_2(L)_2\right](17e) + NO \rightarrow \left[Co^{III}(NO)(Cl)_2(L)_2\right]$$

$$\left[Co^{II}(chel)_2\right]\ (17e) + NO \rightarrow \left[Co^{III}(NO)(chel)_2\right]$$

For the last reaction, so many examples are known, wherein “chel” being a dithiocarbamate, dithiolate, dimethylglyoximate, a Schiff base or porphyrin. Some of them have been characterized crystallographically and found to contain bent CoNO groups:

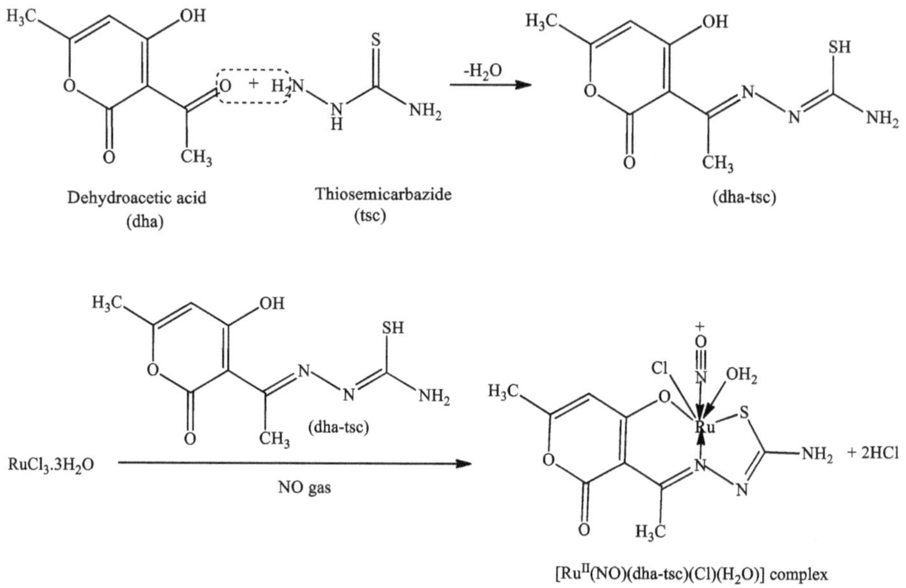

Figure 2.3: Synthesis of Schiff base (dha-tsc) and its ruthenium nitrosyl complex $[Ru^{II}(NO)\text{-dha-tsc})(Cl)(H_2O)]$ using nitric oxide gas as the nitrosylating agent.

$$[Cr^{II}(EPPh_3)_2X_2] + 2NO \rightarrow [Cr^{IV}(NO)_2(EPPh_3)_2X_2]$$
$$E = OorNH$$

(b) $[M(L)_n] + xNO \rightarrow M(NO)_x(L)_m + (n-m)L$-type substitution

The guiding principle behind these types of substitution reactions is that in an 18-electron complex, a 1 electron donor, namely, halide, alkyl, metal–metal bond will be replaced by a bent nitrosyl and a 3-electron donor by a linear nitrosyl. In other words, *n* two-electron donors are replaced by (2*n*)/3 nitrosyls.

(i) $[Co(NO)(CO)_3](18e) + 2NO \xrightarrow[hv]{} [Co(NO)_3](3\times 2 \text{ donor CO} \approx 2\times 3/3 = 2NO)$

$[Cr(CO)_6](18e) + 4NO \xrightarrow[hv]{} [Cr(NO)_4](6\times 2 \text{ donor CO} = 2\times 6/3 \approx 4NO)$

$[Fe(CO)_5](18e) + 2NO \rightarrow [Fe(NO)_2(CO)_2](3\times 2 \text{ donor CO} = 2\times 3/3 \approx 2NO)$

$[Mn(CO)_5I](18e) + 3NO \rightarrow [Fe(NO)_2(CO)_2](4CO + I = 9e \approx 3NO = 9e)$

$[Mn(NO)(CO)_4](18e) + 2NO \rightarrow [Mn(NO)_3(CO)](3\times 2 \text{ donor CO} = 2\times 3/3 \approx 2NO)$

(ii) Replacement of ligand(s) equivalent to three electrons

When the precursor metal complex has saturated coordination, the replacement reactions are basically straightforward:

$$[Mn(CO)_5I]\,(18e) + 3NO \longrightarrow [Mn(NO)_3(CO)]$$

$$[Ir(H)(CO)(L)_3]\,(18e) + NO \longrightarrow [Ir(NO)(CO)(L)_2]$$

$$[Ru(H)_4(L)_3]\,(18e) + 2NO \longrightarrow [Ru(NO)_2(L)_2]$$

$$[Ir(H)\{P(OPh_3)_3\}_4] + NO \longrightarrow [Ir(NO)\{P(OPh_3)_3\}_3]$$

(iii) Substitution reactions with NO in metal clusters are rather more complex due to the possibility of cleavage of metal–metal bonds

$$[Fe_3(CO)_{12}]\,(18e) + \xrightarrow{NO} [Fe(NO)_2(CO)_2]\,(18e)$$

$$[Co_2(CO)_8]\,(18e) + \xrightarrow{NO} [Co(NO)(CO)_3]\,(18e)$$

$$[Mn_2(CO)_{10}]\,(18e) + \xrightarrow[hv]{NO} [Mn(NO)(CO)_4]\,(18e)$$

$$[Mn_2(CO)_8(L)_2]\,(18e) \xrightarrow{NO} [Mn(NO)(CO)_4] + [Mn(NO)(CO)_3(L)] + L$$

(iv) In case a precursor complex differs from the inert gas configuration by 1-electron, then the 18-electron shell can be attained by substitution of a 2-electron donor

$$[Os(SbPh_3)_3Cl_3]\,(17e) + \xrightarrow{NO} [Os(NO)(SbPh_3)_2Cl_3]\,(18e)$$
$$[RuCl_3(H_2O)_3]\,(17e) + \xrightarrow{NO} [RuCl_3(NO)(H_2O)_2]\,(18e)$$
$$[Co(L)_4]\,(17e) + \xrightarrow{NO} [Co(NO)(L)_3]\,(18e)$$
$$[V(CO)_6](17e) + \xrightarrow{NO} [V(NO)(CO)_5]\,(18e)$$
$$[V(CO)_4(L)_2]\,(17e) + \xrightarrow{NO} [V(NO)(CO)_4(L)]\,(18e)$$

(c) Reductive nitrosylation

Before coming to the reductive nitrosylation, we have to first look over the molecular orbital (M.O.) diagram (Figure 2.3) of an NO molecule.

From the M.O. diagram shown in Figure 2.4 of NO molecule, it is well clear that the unpaired electron on NO accommodates in the π^*2p_y orbital. As the NO molecule has a low ionization potential of 9.26 eV, it is expected that it can function as a reducing agent by losing/donating its odd electron. In fact, this has been observed:

$$MF_6 + NO \longrightarrow NO^+\,MF_6^-$$

(here, M = Mo, Tc, Re, Ru, Os, Ir or Pt).

In aprotic (without proton) solvents, reductive nitrosylation of higher halides of molybdenum and tungsten under mild conditions has been observed as shown in the following equations:

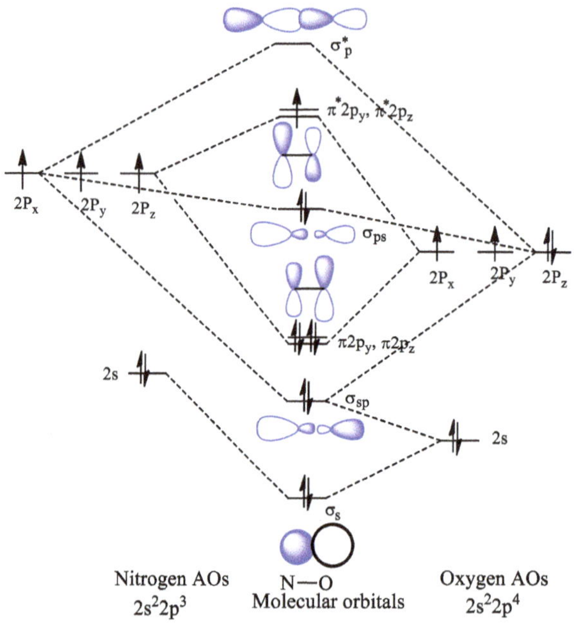

Figure 2.4: M.O. diagram of nitric oxide molecule.

$$Mo^{V}Cl_5 + 5NO \rightarrow [Mo^{0}(NO)_2Cl_2] + 3NOCl$$

$$Mo^{IV}Cl_4(CH_3CN)_2 + 4NO \rightarrow [Mo^{0}(NO)_2(CH_3CN)_2Cl_2] + 2NOCl$$

$$W^{VI}Cl_6 + 6NO \rightarrow [W^{0}(NO)_2Cl_2] + 4NOCl$$

These reactions are similar to the carbonylation of $RhCl_3$

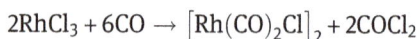

$$2RhCl_3 + 6CO \rightarrow [Rh(CO)_2Cl]_2 + 2COCl_2$$

2.3.2 NO^+ as the nitrosylating agent

The NO^+X^- salts, where X represents non-coordinating anions such as BF_4^-, PF_6^- or HSO_4^-, offer a good source of NO^+ (7 + 8 – 1 = 14e), which is isoelectronic with carbon monoxide (6 + 8 = 14e). Similar to CO, NO^+ may react with metal complexes in two distinct ways: (i) addition reactions or (ii) substitution reactions.

(i) Addition reactions

Some of the addition reactions are as follows:

$$[Rh(CNR)_4]^+ + NO^+ \rightarrow [Rh(CNR)_4(NO)]^{2+}$$

$$[Rh(PPh_3)_3Cl] + NO^+ \rightarrow [Rh(PPh_3)_3(NO)Cl]^+$$

$$[Ir(CO)(L)_2Cl] + NO^+ \rightarrow [Ir(CO)(NO)(L)_2Cl]^+$$

$$[Ru(NO)(L)_2Cl] + NO^+ \rightarrow [Ir(NO)_2(L)_2Cl]^+$$

$$[Rh(PPP)_3Cl] + NO^+ \rightarrow [Rh(PPP)_3(NO)Cl]^+$$

$$PPP = PhP[(CH_2)_3PPh_2]_2$$

(ii) Substitution reactions

These types of reactions commonly occur in metal carbonyls where NO^+ replaces CO ligand most often. Examples are

$$[Fe(CO)_3(PPh_3)_2] + NO^+ \xrightarrow{MeOH/C_6H_6} [Fe(NO)(CO)_2(PPh_3)_2]^+$$

$$[M(CO)_4(dppe)] + NO^+ \xrightarrow{MeOH} [M(NO)(CO)_3(dppe)]^+ (M = Mo, W)$$

$$[Cr(CO)_4(dppe)] + NO^+ \xrightarrow{CH_3CN} [Cr(NO)_2(CH_3CN)_4]^{2+}$$

$$dppe = 1,2-bis(diphenylphosphino)ethane$$

$$[Mo(CH_3CN)_3(CO)_3] + NO^+ \xrightarrow{CH_3CN} cis-[Mo(CH_3CN)_4(NO)_2]^{2+}$$

$$[W(CH_3CN)_3(CO)_3] + NO^+ \xrightarrow{CH_3CN} cis-[W(CH_3CN)_3(CO)(NO)_2]^{2+}$$

$$[Ir(NO)(CO)(L)_2] + NO^+ \xrightarrow{CH_3CN} [Ir(NO)_2(L)_2]^+$$

$$[Ni(CO)_2(L)_2] + NO^+ \xrightarrow{MeOH} [Ni(NO)(CO)(L)_2]^+$$

$$[Co(NO)(CO)(L)_2] + NO^+ \xrightarrow{MeOH} [Co(NO)_2(L)_2]^+$$

$$[W(CO)_5X]^- + NO^+ \xrightarrow{MeOH} [W(NO)(CO)_4X]^-$$

$$X = halideion$$

2.3.3 Nitrosyl halide (NOX) as the nitrosylating agent

Although the covalent-bonded nitrosyl halides (NOX) undergo as part of the following equilibrium, yet they normally react with metal complexes by simple oxidative addition:

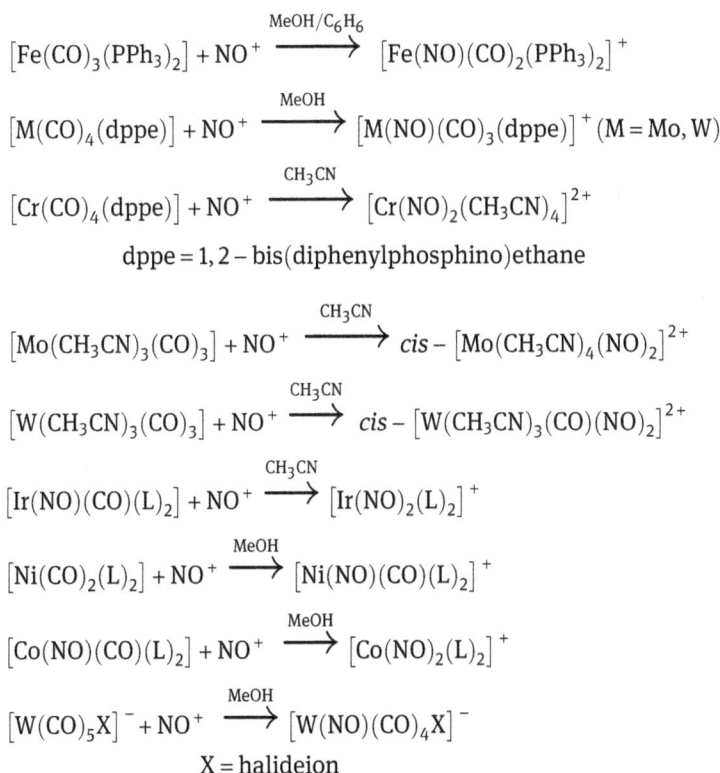

$$2NOX \rightleftharpoons 2NO + X_2$$

Examples of oxidative addition by elemental halogen have been reported in reactions of metal carbonyls with NOX. NOX is supposed to be the reagent in this section:

$$[M(CO)_4(PPh_3)_2] \rightarrow [M(NO)Cl_3(OPPh_3)_2] \quad (M = Mo, W)$$

$$[M(CO)_3(PPh_3)_3] \rightarrow [M(NO)Cl_2(OPPh_3)_3] \quad (M = Mo, W)$$

$$[Mo(CO)_3(PPh_3)_2Cl_2] \rightarrow [Mo(NO)Cl_2(OPhP_3)_3]$$

Several cases are described in which oxidative addition occur in coordinatively unsaturated substrates. For example,

$$[RuCl_2(NH_3)_4] \rightarrow [Ru(NO)Cl_3(NH_3)_2]$$

$$[PtX_4]^{2-} \rightarrow [Pt(NO)X_4Cl_2]^{2-} \quad (X = Cl^-, NO_2^-, CN^-)$$

$$[Rh(NO_2)(CO)(NH_3)_2] \rightarrow [Rh(NO)(NO_2)Cl(NH_3)_2]$$

$$[Pt(NH_3)_4]^{2+} \rightarrow [Pt(NO)(NH_3)_4Cl]^{2+}$$

When reacting substrates are coordinatively saturated, addition of 1 mol of NOX displaces ligands from the substrate equivalent to four electrons with addition:

$$[Ni(CO)_4] \xrightarrow{HCl} [Ni(NO)_2Cl_2]$$

$$[Ni(PPh_3)_4] \rightarrow [Ni(NO)Cl((PPh_3)_2]$$

$$[W(CO)_2(NO)(bipy)Cl] \rightarrow [W(NO)_2Cl_2(bipy)]$$

$$[Mo(CO)_4L_2] \rightarrow [Mo(NO)_2Br_2L_2]$$

$$[Re(CO)_4LCl] \rightarrow [Re(NO)(CO)_2LCl_2]$$

$$[Re(CO)_3L_2Cl] \rightarrow [Re(NO)(CO)_2LCl_2]$$

In the presence of nitrosyl chloride (NOCl), cyano complexes undergo simple oxidation. Here NOCl functions as one electron oxidant:

$$[Fe^{II}(CN)_6]^{4-} \xrightarrow{DMF} [Fe^{III}(CN)_6]^{3-}$$

$$K_3[Mn^{III}(CN)_6] \rightarrow K_2[Mn^{IV}(CN)_6]$$

$$K_2Mn^{II}[Mn^{II}(CN)_6] \rightarrow KMn^{II}[Mn^{III}(CN)_6]$$

Using NOCl as oxidant, complexes of cobalt and nickel of composition, $MX_2(PR_3)_2$, oxidize to $MX_3(PR_3)_2$. This is one of the few routes to the synthesis of Ni(III) complexes.

2.3.4 *N*-Nitrosoamides as the nitrosylating agents

N-Methyl-*N*-nitrosourea (I) and *N*-methyl-*N*-nitroso-*p*-toluene (II) belong to *N*-nitrosoamides. They react with a range of metal hydrides to form metal nitrosyls, and presumably the parent amide. In this reaction, hydrogen atom, which is formally a one-electron donor, is replaced by the nitrosyl group. This reaction might be expected to form a bent nitrosyl. But it is found that extrusion of a two-electron donor ligand accompanies this reaction and linear nitrosyl complexes are obtained:

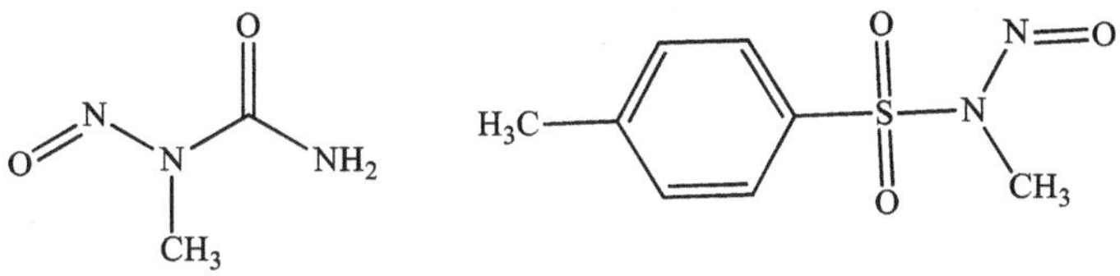

(I) (II)

For example, when $[HMn(CO)_5]$ reacts with *N*-methyl-*N*-nitrosourea, formation of $[Mn(CO)_4NO]$ results with displacement of CO along with H from $[HMn(CO)_5]$:

$$[HMn(CO)_5] \rightarrow [Mn(CO)_4NO] + CO$$

Other examples of linear metal nitrosyl formation assuming *N*-methyl-*N*-nitroso-*p*-toluene sulphonamide as a reagent are

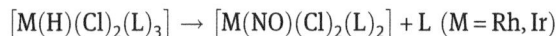

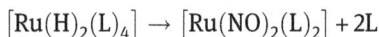

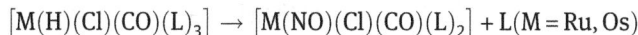

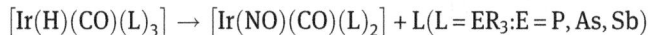

$$[Mo(H)(Cp)(CO)_3] \rightarrow [Mo(NO)(Cp)(CO)_2] + CO$$

$$[M(H)(Cl)_2(L)_3] \rightarrow [M(NO)(Cl)_2(L)_2] + L \ (M = Rh, Ir)$$

$$[Rh(H)(L)_4] \rightarrow [Rh(NO)(L)_3] + L$$

$$[Ru(H)_2(L)_4] \rightarrow [Ru(NO)_2(L)_2] + 2L$$

$$[M(H)(Cl)(CO)(L)_3] \rightarrow [M(NO)(Cl)(CO)(L)_2] + L(M = Ru, Os)$$

$$[Ir(H)(CO)(L)_3] \rightarrow [Ir(NO)(CO)(L)_2] + L(L = ER_3{:}E = P, As, Sb)$$

$$[Co(H)_3(PPh_3)_3] \rightarrow [Co(NO)(PPh_3)_3] + H_2$$

In the last reaction, the elimination of molecular hydrogen takes place firstly followed by NO/H interchange.

The only example in which ligand extrusion is not observed involves the reaction of tricyclohexylphosphine (PCy_3) complex with *N*-methyl-*N*-nitroso-*p*-toluene sulphonamide:

$$[Ru(H)(Cl)(CO)(PCy_3)_2] \rightarrow [Ru(NO)(Cl)(CO)(PCy_3)_2]$$

2.3.5 Coordinated NO as the nitrosylating agents

The nitrosyl complex $[Co(NO)(DMG)_2]$ (DMG = mono anion of dimethylglyoxime) is soluble in several solvents. This has been reported to transfer attached NO group to a number of metal complexes producing respective metal nitrosyls. The following two types of common reaction routes that can occur are as follows:

(i) In the first type of reaction, transfer of coordinated NO from $[Co(NO)(DMG)_2]$ to other complexes forming metal nitrosyl complexes is shown below:

$$[Co(NO)(DMG)_2] + [M(L)_n] \rightarrow [Co(DMG)_2] + [M(NO)(L)_n]$$

(ii) Sometimes, halogen transfer is a possible secondary reaction route. Co^{II} is a potential halogen acceptor. So, in this case, the net reaction is formally "NO/halogen interchange", or a type of redistribution reaction. The reaction of $[Co(NO)(DMG)_2]$ with $[M(Cl)(L)_n]$ observing this reaction route is given below:

$$[Co(NO)(DMG)_2] + [M(Cl)(L)_n] \rightarrow [Co(Cl)(DMG)_2] + [M(NO)(L)_n]$$

For identifying only simple NO transfer from $[Co(NO)(DMG)_2]$, $[Co(Cl)_2(L)_2]$ was taken as a substrate and the observed reaction is as follows:

$$2[Co(NO)(DMG)_2] + [Co(Cl)_2(L)_2] \rightarrow [Co(NO)_2(L)_2]^+Cl^- + [Co(DMG)_2] + [Co(Cl)(DMG)_2]$$

Some of the reactions showing simple NO/halogen interchange occur as follows:

$$[Co(NO)(DMG)_2] + [Ni(Cl)_2(L)_2] \rightarrow [Ni(NO)(Cl)(L)_2] + [Co(Cl)(DMG)_2]$$

$$[Co(NO)(DMG)_2] + [Rh(Cl)(L)_3] \rightarrow [Rh(NO)(L)_3] + [Co(Cl)(DMG)_2]$$

$$[Co(NO)(DMG)_2] + [Ru(Cl)_2(L)_3] \rightarrow [Ru(NO)(Cl)(L)_2] + [Co(Cl)(DMG)_2(L)]$$

2.3.6 Hydroxylamine (NH_2OH) as the nitrosylating agents

Hydroxylamine undergoes a wide range of complex redox chemistry, and depending upon the pH, several species such as NH_3, NH_4^+, N_2 or N_2O may be formed (D. M. Yost and H. Russell, Jr., *Systematic Inorganic Chemistry*, Prentice Hall, New York, 1946). The synthesis of metal nitrosyls using hydroxylamine was neglected previously. It is now used as a nitrosylating agent in basic medium. Hieber and co-workers [2] employed hydroxylammonium chloride ($NH_2OH{\cdot}HCl$) first time as a nitrosylating agent.

Hydroxylammonium chloride on reaction with molybdate, cyanide and KOH in aqueous medium produces a violet compound of the formula, $K_4[Mo(NO)(CN)_5]\cdot 2H_2O$. Because of some controversy in formulation, Wilkinson et al. [3] proposed another octa-coordinated structure having composition, $K_4[Mo(NO)(OH)_2(CN)_5]$ for this violet-coloured compound. Later on, X-ray single crystal studies finally confirmed the hexa-coordinated structure of composition $K_4[Mo(NO)(CN)_5]\cdot 2H_2O$ for this compound. This compound can also be synthesized directly from $K_4[Mo(CN)_6]$.

Wilkinson et al. [3] did extensive work using hydroxylamine in strongly alkaline/cyanide medium with other oxometallate anions. Thus, with CrO_4^{2-} (K_2CrO_4), a green complex, namely, potassium pentacyanonitrosylchromate(I) monohydrate, $K_3[Cr^I(NO)(CN)_5]\cdot H_2O$, was isolated. Likewise using the same procedure with VO_3^- [ammonium vanadate (NH_4VO_3)], an orange complex, $K_5[V(NO)(CN)_5]\cdot H_2O$ had been isolated which is structurally characterized as $K_3[V(NO)(CN)_5]\cdot 2H_2O$ [4]. Interestingly, use of a slightly different procedure, that is, with VO_3^- and $NH_2OH/KOH/KCN$, passing of H_2S leads to the isolation of a yellow complex of composition, $K_4[V(NO)(CN)_6]\cdot H_2O$ [5]. While ν(CN) and ν(NO) in $K_4[V(NO)(CN)_6]\cdot H_2O$ are closely similar to that of $K_3[V(NO)(CN)_5]\cdot 2H_2O$, the cell constants differ. Moreover, the compound is diamagnetic. Nitroprusside, $[Fe(NO)(CN)_5]^{2-}$, the best known cyanonitrosyl, has not been synthesized in this way. The reason behind this is most probably the lack of suitable precursor cyano complexes.

In 1984, Bhattacharya et al. [6] reported the synthesis of $K_3[Re(NO)(CN)_5]\cdot 2H_2O$ by taking potassium perrhenate(VII), $KReO_4$, excess of KCN, KOH and $NH_2OH\cdot HCl$. The same authors in 1985 reported [7] the synthesis of some new molybdenum(II) cyanonitrosyl complexes of composition, $R_2[Mo(NO)(CN)_5]\cdot 2H_2O$ (R = Ph_4P or Bu_4N), using $Na_2MoO4\cdot 2H_2O$, KCN and $H_2OH\cdot HCl$ in alkaline medium maintaining pH 8.

The hydroxylamine method can also be applied to some cyano complexes as precursor to be nitrosylated. Thus, $K_3[Mn(CN)_6]$ can be converted into purple $K_3[Mn(NO)(CN)_5]\cdot 2H_2O$ by this method. Similarly, purple $K_2[Ni(NO)(CN)_3]$ can be prepared by the reaction of hydroxylamine and $K_2[Ni(CN)_4]$.

Besides cyanide as a co-ligand, nitrosyl complexes can be generated by alkaline hydroxylamine in another way as follows:

$$[Co(CNR)_5X] + 2NH_2OH \xrightarrow{OH^-} [Co(NO)(CNR)_3] + 2CNR + H_2O + NH_4X$$

$$[Fe(CNR)_4X] + 4NH_2OH \xrightarrow{OH^-} [Fe(NO)_2(CNR)_2] + 2CNR + 2H_2O + NH_4X$$

From the above synthetic examples of nitrosyl complexes using hydroxylamine method, it is well clear that this method requires a strong alkaline medium. Hieber et al. [8] proposed the demand of alkali in the following way:

$$2NH_2OH \rightarrow NOH + NH_3 + H_2O$$

$$\left.\begin{array}{l} NOH \rightleftharpoons NO^- + H^+ \\ H^+ + OH^- \rightarrow H_2O \end{array}\right\} \text{Le Chatelier's principle}$$

In this method, the requirement of the alkaline medium is just for the displacement of the equilibrium of the second reaction towards right direction to produce more and more "NO^-."

In fact, in strong alkaline medium (KOH), the OH^- so produced will abstract proton from the right-hand side of the above reversible reaction to give feebly ionized H_2O molecule. Hence, according to Le Chatelier's principle, more and more NO^- will be produced, which serve as a reductive nitrosylating agent.

Maurya et al. have recently reported [9] the preparation of some dinitrosylmolybdenum(0) complexes of composition, $[Mo(NO)_2(L)(OH)]$ (L = a monobasic tridentate Schiff base obtained from 4-aminoantipyrine and 4-acyl-3-methyl-1-phenyl-2-pyrazolin-5-one derivatives), taking ammonium molybdate and excess of $NH_2OH \cdot HCl$ in mild alkaline medium of *N,N*-dimethylformamide:

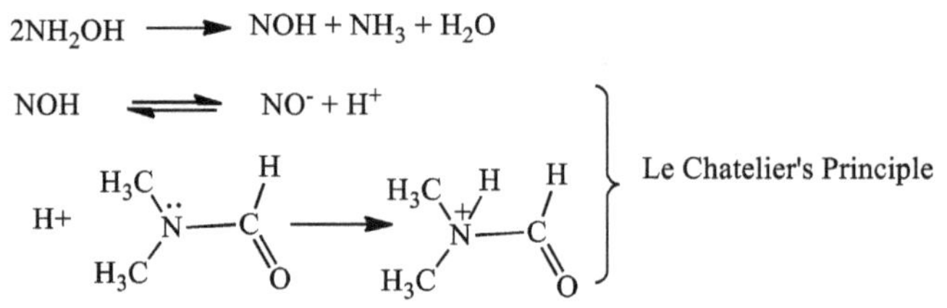

N,N-Dimethyformamide

The kinetic study conducted by Malatesta and Sacco [10] for the reaction of $[Ni(CN)_4]^{2-}$ with NH_2OH in alkaline medium recommends that in the presence of oxygen, the reaction is

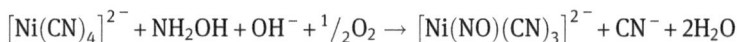

$$[Ni(CN)_4]^{2-} + NH_2OH + OH^- + {}^1/_2O_2 \rightarrow [Ni(NO)(CN)_3]^{2-} + CN^- + 2H_2O$$

The resulting equation is explained with the opinion that no NH_3 is produced and that encompasses a pre-equilibrium in which NH_2OH replaces CN^-. Further, deprotonation of coordinated NH_2OH takes place followed by oxidation of the resultant $[Ni(NO)(CN)_3]^{4-}$ by oxygen completes the reaction. In the absence of oxygen, the reaction is very slow and kinetically less manageable where the mechanism is supposed to be as follows:

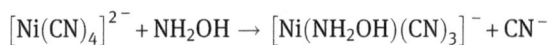

$$[Ni(CN)_4]^{2-} + NH_2OH \rightarrow [Ni(NH_2OH)(CN)_3]^- + CN^-$$

$$[Ni(NH_2OH)(CN)_3]^- + NH_2OH + OH^- \rightarrow [Ni(NO)(CN)_3]^{2-} + NH_3 + 2H_2O$$

The stoichiometry regarding the ammonia produced per mole of $[Ni(NO)(CN)_3]^{2-}$ formed and NH_2OH used cannot be set properly because of the side reactions like

$$[Ni(NH_2OH)(CN)_3]^- + CN^- + NH_2OH \rightarrow [Ni(CN)_4]^{2-} + N_2 + H_2O + OH^- \text{ and}$$

$$3NH_2OH \rightarrow NH_3 + N_2 + 3H_2O$$

Thus, this work contradicts the suggestion made earlier for the involvement of NO^- in these reactions. If NO^- is not the species formed by the reaction of NH_2OH with strong alkali for the reductive nitrosylation, then hydroxylamine can nitrosylate in neutral and even in acidic medium. To put it to test, Sarkar and Müller [11] attempted the reductive nitrosylation of several oxometallate anions using NH_2OH. Taking SCN^- as a co-ligand, they were able to isolate species like $[Cr(NO)(SCN)_5]^{3-}$, $[Mo(NO)(SCN)_4]^{2-}$ and $[Os(NO)(SCN)_5]^{2-}$. It is interesting to note that $NH_2OH \cdot HCl$ is used as such in these cases which suggest that the reaction medium is slightly acidic in nature.

Weighardt et al. [12] performed the reaction of pentavalent and trivalent vanadium to get $[V(NO)(dipic)(NH_2O)(H_2O)]$. On the other hand, they observed that MoO_4^{2-} in the presence of terpyridine reacts with NH_2OH in acidic medium to form $[Mo(NO)(terpy)(NH_2O)(H_2O)]$ [13]. Both the above reactions suggest that NH_2OH functions as a nitrosylating reagent in alkaline as well as acidic medium. The interesting part of this work lies in the fact that NH_2OH can be deprotonated to give a mononegative bidentate hydroxylamido (NH_2O^-) ligand. The same authors have shown that this hydroxylamido group can be further deprotonated to function as bidentate hydroxyamido

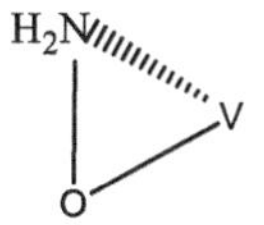

Bidentate coordination of NH_2O^-

(−2) group. The kinetic studies made on the formation of $[Ni(NO)(CN)_3]^{2-}$ have suggested that NH_2OH first get coordinated and then deprotonated to generate the nitrosyl group [10]. The complexes synthesized by hydroxylamine method in any pH condition invariably contain a linear M–N–O group whenever structural data are obtained.

2.3.7 Acidic solution of nitrite salts (NO_2^-/H^+) as the nitrosylating agents

The following equilibrium is the basis of the nitrosylation reaction occurring through this method:

$$NO_2^- + 2H^+ \rightleftharpoons NO^+ + H_2O$$

In an attempt to classify reported nitrosylation reactions, there is some difficulty in differentiating reactions of the above type (attack of NO^+, formed by acidic solutions of nitrite salts, on metal complexes) from coordination of NO_2^-, and thereafter by oxide abstraction. Godwin and Meyer [14] have successfully distinguished between these in reaction given below:

$$[Ru(L)_2(X)(H_2O)]^+ \xrightarrow{NO_2^-/H^+} [Ru(NO)(L)_2X]^{2+} \quad (L = o-\text{phenorbipy})$$

In the absence of added H^+, no anation by NO_2^- is detected in 1 h in this reaction. Instead, the nitrosylation reaction took place instantaneously.

Reactions of this type with other substrates are:

$$K[Fe(NO)(CO)_3] + KNO_2 + CO_2 + H_2O \rightarrow [Fe(NO)_2(CO)_2] + 2KHCO_3$$

$$Na[Fe(H)(CO)_4] + 2NaNO_2 + 3CH_3COOH \rightarrow [Fe(NO)_2(CO)_2] + 2CO + 3CH_3COONa$$

$$K[Co(CO)_4] + NaNO_2 + 2CH_3COOH \rightarrow [Co(NO)(CO)_3] + 2CH_3COONa(K) + H_2O + CO$$

$$K[Co(CO)_4] + KNO_2 + 2CO_2 + 2H_2O \rightarrow [Co(NO)(CO)_3] + 2KHCO_3$$

2.3.8 Alkyl nitrites as the nitrosylating agents

The use of alkyl nitrites as nitrosylating agents is based on the following equilibrium in protic solvents:

$$RONO + H^+ \rightarrow NO^+ + ROH$$

The following reaction was performed by the addition of metal complex to a solution of isopentyl nitrite and HPF_6 in benzene/methanol:

$$[Fe(CO)_3(PPh_3)_2] \rightarrow [Fe(NO)(CO)_2(PPh_3)_2]^+$$

Alcoholic solutions of metal halides on reaction with *n*-pentyl nitrite in the presence of phosphines form metal nitrosyls, probably through hydride intermediates:

$$MCl_3 \rightarrow [M(NO)Cl_3(PPh_3)_2] \quad (M = Ru, Os)$$

$$MCl_3 \rightarrow [M(NO)Cl_2(PPh_3)_2] \quad (M = Rh, Ir)$$

2.3.9 Nitric acid as the nitrosylating agent

The synthesis of red crystalline salts of petacyanonitrosylferrate(II), of which sodium nitroprusside (SNP) is very popular, was reported by Lionel Playfair (1818–1898) in

1848. This synthesis followed the routes given below taking either concentrated nitric acid or nitrite salts:

$$K_4[Fe(CN)_6]\cdot 3H_2O + 6HNO_3 \rightarrow H_2[Fe(NO)(CN)_5] + 4KNO_3 + NH_4NO_3 + CO_2$$

$$H_2[Fe(NO)(CN)_5] + Na_2CO_3 \rightarrow Na_2[Fe(NO)(CN)_5] + H_2O + CO_2$$

$$[Fe(CN)_6]^{4-} + NO_2^- + H_2O \rightarrow [Fe(NO)(CN)_5]^{2-} + CN^- + 2OH^-$$

Afterwards, the corresponding V, Cr, Mn and Co salt anions $[M(NO)(CN)_5]^{2-}$ (where M = V, Cr, Mn or Co) were reported.

The analogous ruthenium and osmium complexes have been prepared using the identical method. When $Na_3[Re(CN)_5(H_2O)]$ is treated with moderately strong nitric acid, a cyano-nitrosyl complex of rhenium of composition, $Na_2[Re(CN)_5(NO)]$, is said to be formed. $K_3[Re(CN)_8]$ on warming with 2 M nitric acid yields a diamagnetic cyanonitrosyl complex of rhenium of composition, $Ag_3[Re(CN)_7(NO)]$. It is found to be contaminated with AgCN.

2.3.10 Synthesis of nitrosyl complexes using redox reaction

We have seen above that using nitric acid as a nitrosylating agent, the synthesis of $Na_2[Fe^{II}(NO)(CN)_5]$ and its Ru and Os analogous has been successfully carried out in the same formal oxidation state of +2. Though the oxidation state of the central metal linked with NO has got no sense in the real sense of the term, yet a notation $(MNO)^{n+}$ would avoid an extreme formalism of NO either as NO^+ or NO^- (vide infra). In this sense, using the same synthetic approach, Fe, Ru and Os give $(MNO)^{3+}$ in their complexes. On the other hand, using alkaline hydroxylamine method of preparation, chromium gives $(CrNO)^{2+}$ group, whereas molybdenum gives $(MoNO)^+$ moiety. This motivated Griffith to reduce the compound $K_3[Cr(NO)(CN)_5]$ (having $(CrNO)^{2+}$ group) following polarographic method of reduction to isolate a blue reduced species $K_4[Cr^0(NO)(CN)_5]\cdot H_2O$ analogous to $K_4[Mo(NO)(CN)_5]$. Several attempts have been made to isolate one electron oxidation product of $K_4[Mo(NO)(CN)_5]$ in order to obtain $[Mo^I(NO)(CN)_5]^{3-}$. But the pure green-coloured compound, $[(PPh_4)]_3[Mo(NO)(CN)_5]\cdot H_2O$, was synthesized by Sarkar and Müller [15] in 1978 by aerial oxidation. Likewise, the cyanonitrosyl complex, $K_3[Mn(NO)(CN)_5]\cdot 2H_2O$ (purple coloured), on oxidation using bromine or HNO_3 forms yellow-coloured complex compound, $K_2[Mn(NO)(CN)_5]\cdot 2H_2O$. SNP on reduction with sodium in liquid ammonia gives ochre yellow-coloured compound of composition, $Na_3[Fe(NO)(CN)_5]\cdot 2NH_3$. This compound is found to be an unstable solid. The corresponding tetraethylammonium salt, when treated with acetic acid in acetonitrile changes it into a blue-coloured complex, $(Et_4N)_2[Fe(NO)(CN)_4]$.

2.3.11 Synthesis of metal nitrosyl by substitution of cyano groups in parent cyanonitrosyl complex

Some cationic dinitrosyl complexes of vanadium in (–I) oxidation state were synthesized by Sarkar and Maurya [16]. For this, the parent compound $K_3[V(CN)_5(NO)] \cdot 2H_2O$ with *o*-phenanthroline/dipyridyl in 1:4 mole ration was taken in a reaction vessel, and shaking was continued for 24 h. The resulting compounds of red-violet colour were characterized as $[V(NO)_2(o\text{-phen})_2] \cdot CN$ and $[V(NO)_2(dipy)_2] \cdot CN$, respectively. The same group also reported the synthesis of some hepta-coordinated mixed-ligand cyanonitrosyl $\{VNO\}_4$ complexes of vanadium(I) having composition $K_2[V(NO)(CN)_4(L)_2]$ (L = 2,3-Lut, 2,4-Lut, 2,6-Lut, 3,5-Lut, 2,3,6-Coll, 2,4,6-Coll or nicotine) taking the parent compound, $K_4[V(NO)(CN)_6]$, in aqueous medium and the organic base in methanol. The desired compounds were obtained as yellow mass after constant stirring of the reaction mixture at 50–60 °C under nitrogen atmosphere.

Maurya and co-workers [17] have successfully carried out the synthesis and characterization of some novel cyanonitrosyl $\{CrNO\}^5$ complexes of chromium(I) having the general formulae, $[Cr(NO)(CN)_2(L\text{-}L)]$ (L-L = *o*-phen or dipy), $[Cr(NO)(CN)_2(L)_2]$ (L = py or qu), $[Cr(NO)(acac)_2]$, $[Cr(NO)(DTC)_2(H_2O)]$ and $[Cr(NO)(CN)_2(L)_2(H_2O)]$ (L = pyridine, aniline or benzimidazole derivatives by the interaction of potassium pentacyanonitrosylchromate(I) monohydrate, $K_3[Cr(NO)(CN)_5] \cdot H_2O$ with the said ligands in aqueous acetic acid medium.

Bhattacharya et al. [18] reported the synthesis of molybdenum(0) cyanonitrosyl complex of composition $[Mo(NO)_2(CN)_2(o\text{-phen})] \cdot 2H_2O$. This compound was synthesized by the addition of the hot aqueous solution of *o*-phenanthroline into a freshly prepared solution of $[Mo(NO)_2(CN)_4]^{2-}$ with stirring at 50 °C.

2.4 The $\{M(NO)_m\}^n$ formalism for metal nitrosyl complexes

For the transition metal complexes with ligands, such as NH_3, SCN^-, Cl^- or CN^-, the assignment of formal oxidation state is normally unquestionable and leads to a single label. However, Kaim [19] suggested that CN^- in some low oxidation state compounds may behave in a non-innocent manner. Complexes having other ligands, namely, dipyridyl, dithiolene and π-bonded organic ligands, may be represented by resonance forms. This results in the assignment of alternative formal metal oxidation states. For instance, the dithiolene ligand in complexes may be denoted either as a neutral or dianionic ligand (Figure 2.5), and thereby the formal metal oxidations states will obviously be M^{n+} and $M^{(n+2)+}$.

The idea of *innocent* and *non-innocent* ligands was given first time by C. K. Jørgensen [20] in the 1960s, which sought to resolve these ambiguities. NO is also a non-innocent ligand, and as discussed earlier, this may lead to two different oxidation states, M^{n+} and $M^{(n+2)+}$, based on the assignment of the charge on the ligand.

(i) (ii)

Figure 2.5: (i) Neutral and (ii) dianionic coordination of dithiolene.

The ambiguities in oxidation state for NO as a non-innocent ligand directed Enemark and Feltham [21] to propose a widely accepted and cited notation $\{M(NO)_m\}^n$. This made no suppositions concerning the formal charge on the ligand or on the metal. In this notation, the electrons occupying the $\pi^*(NO)$ orbitals are added to the n electron count for the remainder of the complex containing the metal and the "innocent" spectator ligands. For example, $\{MNO\}^n$ for $[Cr(NO)(CN)_5]^{3-}$ assumes that the cyanide ligands are "innocent" and are associated with a charge of −1, each giving −5 total. To balance the fragment's overall charge, the charge on {MNO} is +2 (i.e. −3 = −5 + 2). Since Cr^{2+} has four d electrons and neutral $NO^{\cdot}$ has one electron in $\pi^*(NO)$, the total electron count n is 5, that is, it is a $\{MNO\}^5$ complex. Since CN^- is a strong field ligand, the resultant complex would be expected to have a low-spin configuration. The advantage of this notation is its simplicity and that the d-electron count is the same no matter whether the nitrosyl ligand is formulated as NO^+, NO or NO^-. Table 2.1 provides some examples of this notation.

Table 2.1: $\{M(NO)_x\}^n$ notation for some nitrosyl complexes.

Compound	Formal oxidation state of the metal	No. of d electrons	No. of electrons in π *NO orbital	$\{M(NO)_x\}^n$
$K_3[Cr(NO)(CN)_5]$	Cr(II)	d^4	1	$\{CrNO\}^5$
$Na_2[Fe(NO)(CN)_5]$	Fe(III)	d^5	1	$\{FeNO\}^6$
$K_3[Mn(NO)(CN)_5]$	Mn(II)	d^5	1	$\{MnNO\}^6$
$K_2[Co(NO)(CN)_4(2\text{-}CP)]$	Co(II)	d^7	1	$\{CoNO\}^8$
$[Mn(NO)_2(CN)_2(AMPPHP)$	Mn(II)	d^5	2	$\{Mn(NO)_2\}^7$
$[Mo(NO)_2(CN)_2(MPHP)_2$	Mo(II)	d^4	2	$\{Mo(NO)_2\}^6$
$[Ru(NO)(Cl)_3(H_2O)_2]$	Ru(III)	d^5	1	$\{RuNO\}^6$
$[Mn(NO)(CO)_4]$	Mn(0)	d^7	1	$\{MnNO\}^8$
$[Ru(NO)(H)(PPh_3)_3]$	Ru(I)	d^7	1	$\{RuNO\}^8$
$[Ir(NO)(H))(PPh_3)_3]^+$	Ir(II)	d^7	1	$\{IrNO\}^8$
$[Os(NO)_2(PPh_3)_2(OH)]^+$	Os(II)	d^6	2	$\{Os(NO)_2\}^8$
$[RuCl(NO)_2(PPh_3)_2]^+$	Ru(II)	d^6	2	$\{RuNO\}^8$

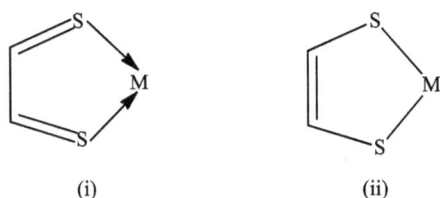

2.4.1 Limitations of the $\{M(NO)_m\}^n$ formalism

The followings are limitations of Enemark and Feltham formalism/notation:

(i) This formalism does not provide any suggestion for the M–N–O geometry in metal nitrosyls and also does not link the notation to the 8 and 18 effective atomic number (EAN) rules.

(ii) On EAN viewpoint, the ruthenium nitrosyl complex, $[RuCl(NO)_2(PPh_3)_2]^+$ is consistent with two linear three-electron nitrosyl ligands, that is, two Ru–N–O bonds are linear but in the solid state, one Ru–N–O bond is linear and the other bent (Figure 2.6). In solution, the two ligands exchange on the ^{15}N NMR (nuclear magnetic resonance) timescale. Enemark and Feltham's formalism designates the complex as $\{Ru(NO)_2\}^8$ and does not provide an indicator to distinguish the bonding modes of the nitrosyls, that is, whether Ru–N–O bond is linear or bent. It is notable that the $\{RuNO\}^8$ (see Table 2.1, last complex) is the same as d^8 [Ru(0); d^6s^2)] derived from the NO^+ formalism regardless of whether a nitrosyl ligand is linear, intermediate or bent.

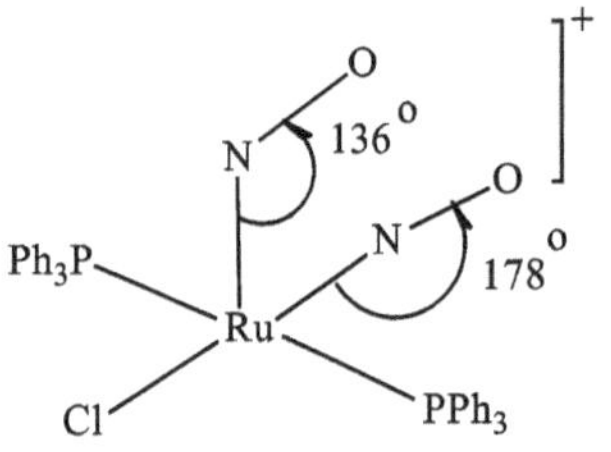

Figure 2.6: One linear and one bent Ru–N–O bond in solid state $[RuCl(NO)_2(PPh_3)_2]^+$.

(iii) Enemark and Feltham formalism basically does not state about the oxidation state of the metal in the complex as well as the coordination number of the complex. Recent studies based on independent spectroscopic and structural data may provide convincing evidence for a definite metal oxidation state, which gives indirect suggestion for the metal–nitrosyl bonding. Hence, we see a need to communicate this information from a notation or formalism.

(iv) The idea of antiferromagnetic coupling between unpaired electrons of the NO group and the metal unpaired electrons weakens the fundamental assumptions of the Enemark–Feltham model. Since this model implies weaker interactions within the MNO moiety. Moreover, the antiferromagnetic coupling models necessitate a description of the oxidation states and spin states of the metal and the NO ligand.

(v) For polynitrosyl complexes, this model has found no wide applicability.

2.5 Alternative formalism/notation for metal nitrosyl complexes

We are aware that most of the metal nitrosyl complexes follow 16- and 18-electron rules. Therefore, it appears rational to focus the notation for metal nitrosyls on these parameters instead of the modified *d*-electron count as suggested by Enemark and Feltham. Moreover, the availability of single-crystal X-ray measurements these days and the probability of accurately estimating the M–N–O bond angle from spectroscopic data means that this parameter can be incorporated in the notation using the shorthand description given in Table 2.2, that is, 180–160° (linear, *l*), 140–160° (intermediate, *i*), 110–140° (bent, *b*).

Table 2.2: Analytical data of metal nitrosyls*.

Bond angle (°)	No. of structures	Description
180–170	2,058	Linear (*l*)
170–160	461	Linear (*l*)
160–150	76	Intermediate (*i*)
150–140	45	Intermediate (*i*)
140–130	44	Bent (*b*)
130–120	54	Bent (*b*)
120–110	15	Bent (*b*)
110–100	2	Bent (*b*)
110–90	1	π-Bonded

*Based on Crystallographic Database, Cambridge; www.ccdc.cam.ac.uk/products/csd/

Incorporating the facts given above, the alternative notation in the general form is

$$[L_mM(NO)_n l/i/b]^y$$

Here $n + m$ = coordination number of complex comprising the NO grouping and y = EAN count considering NO donating three electrons for linear (*l*) and one electron for bent (*b*) and either three, two or one for intermediate (*i*) bond angles. The electron-donating capabilities for **linear, intermediate** and **bent nitrosyl** complexes follow directly from the **valence bond** and M.O. depictions of bonding as given in Figures 2.7 and 2.8. For metal nitrosyls having M–N–O angles between 140° and 160° (intermediate *i*), it is essential to take other factors depending on additional spectroscopic, magnetic and structural data into consideration before making a final assignment.

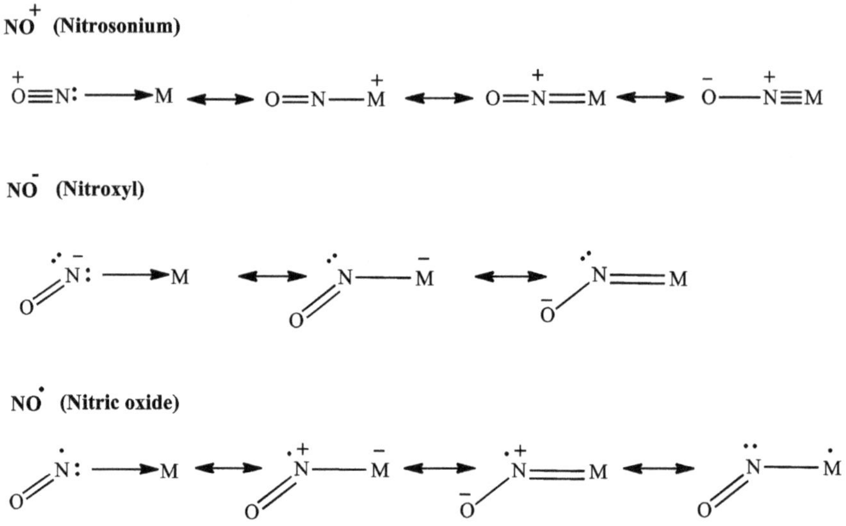

Figure 2.7: M–NO bonding: valence bond approach.

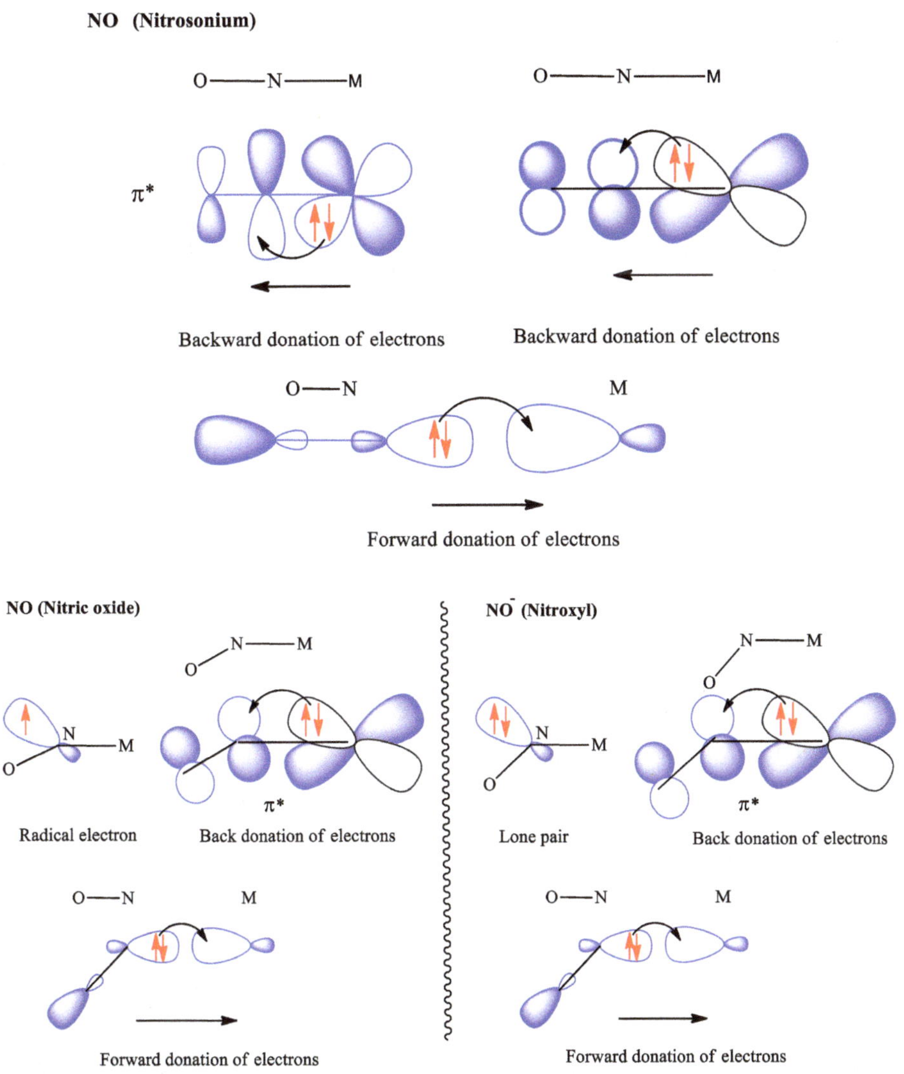

Figure 2.8: M.O. diagrams for M–NO bonding: linear, intermediate and bent metal nitrosyls.

Examples The EAN notation is well demonstrated with regard to some particular examples given in Figures 2.9 and 2.10.

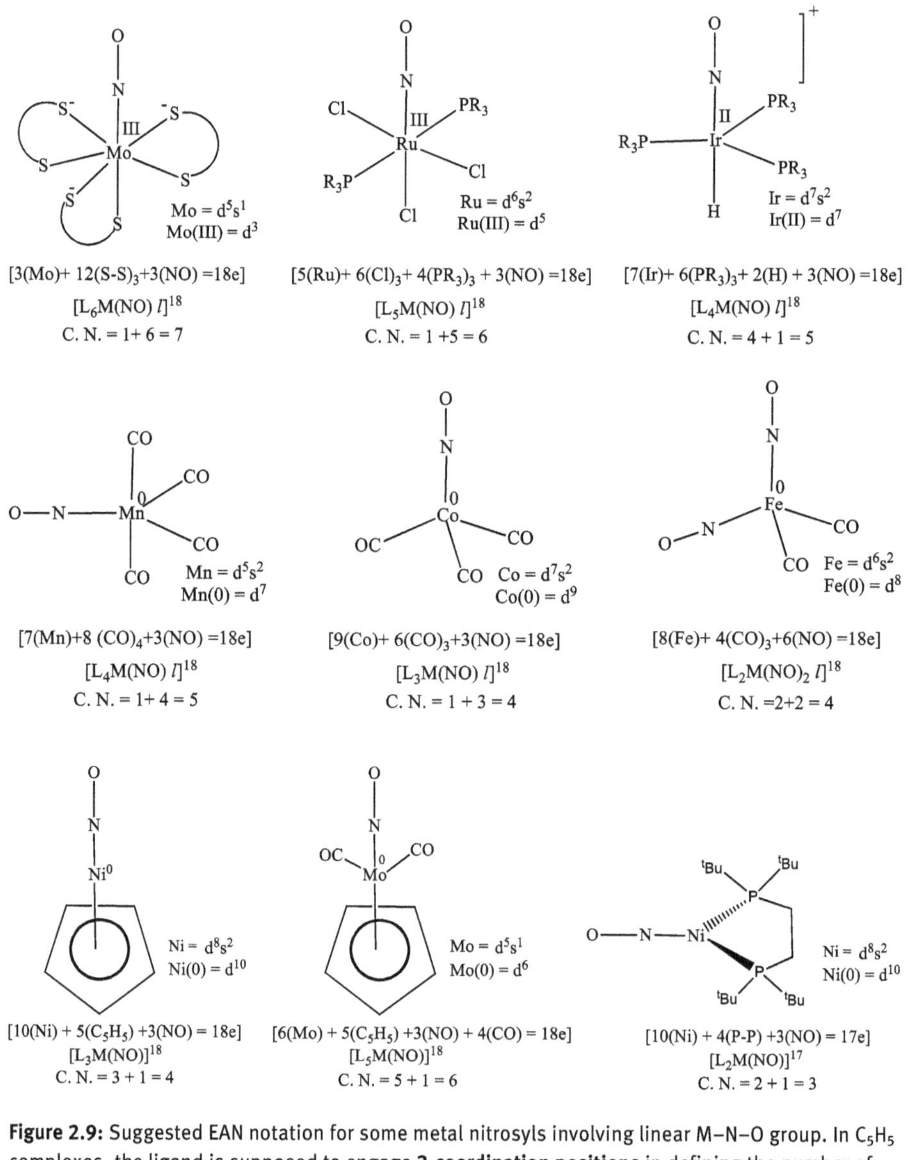

Figure 2.9: Suggested EAN notation for some metal nitrosyls involving linear M–N–O group. In C_5H_5 complexes, the ligand is supposed to engage **3-coordination positions** in defining the number of spectator ligands.

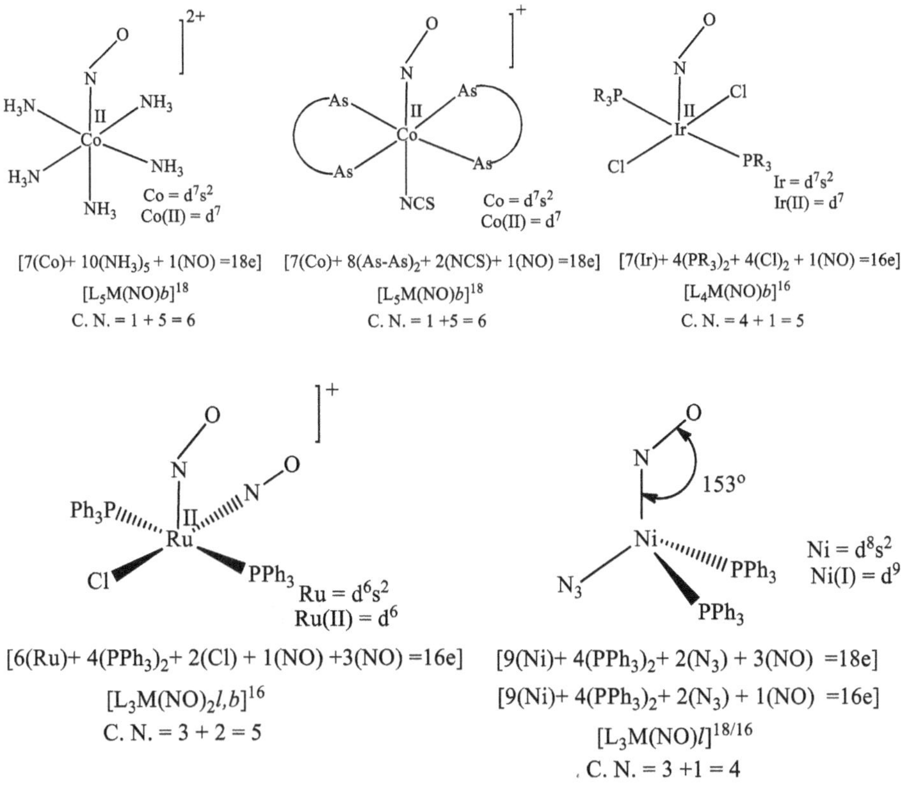

Figure 2.10: Suggested EAN notation in some metal nitrosyl complexes having bent M–N–O grouping.

Figure 2.10 gives examples of metal complexes with bent M–N–O grouping falling into either 18- or 16-electron EAN categories. Also they have either octahedral (18-electron) or square-pyramidal (16-electron) geometries. Following the descriptions given in Figures 2.7 and 2.8, the bent M–N–O grouping is demarcated having NO as a one-electron donor in these examples. The last example in Figure 2.10 is a tetrahedral nickel complex having a Ni–N–O bond angle 153°. Therefore, this falls into the intermediate *i* category. As mentioned above, this ambiguous category necessitates additional considerations in order to decide whether the NO is acting as a one-, two- or three-electron donor. This is reflected in the two alternative electron counts (18 and 16) shown. The 17-electron count with 2-electron donor is omitted because of the diamagnetic nature of the complex. When the complex holds both linear and bent M–N–O groupings, then this may be shown using both *l* and *b* as given for the ruthenium complex shown in Figure 2.6.

Figure 2.11a,b demonstrates the structures of two closely associated osmium nitrosyl complexes. In complex (a), the innocent spectator ligand is OH while in complex (b) it is Cl. As per the Enemark–Feltham notation, both are $\{Os(NO)_2\}^8$ complexes (vide Table 2.1), but following the new notation two dissimilar Os–N–O bond angles/geometries are shown. The resulting electron count clearly distinguishes both the complexes.

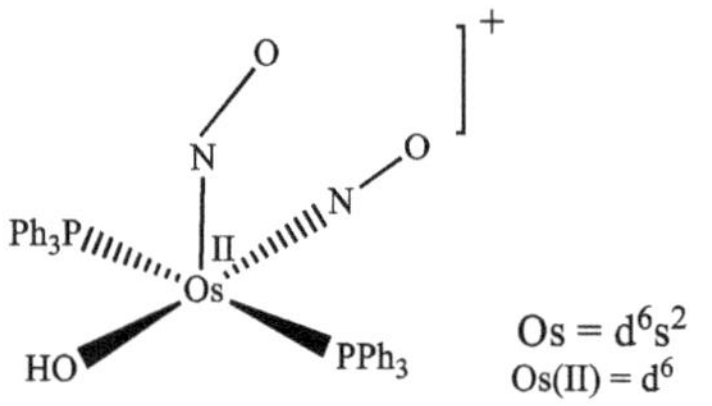

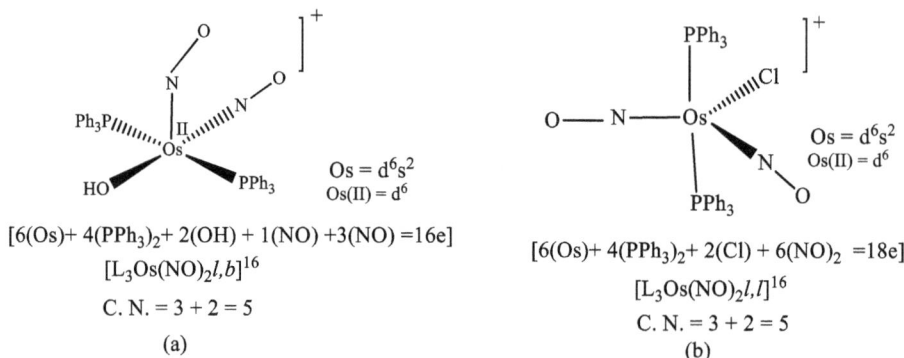

$[6(Os)+ 4(PPh_3)_2+ 2(OH) + 1(NO) +3(NO) =16e]$

$[L_3Os(NO)_2 l,b]^{16}$

C. N. = 3 + 2 = 5

(a)

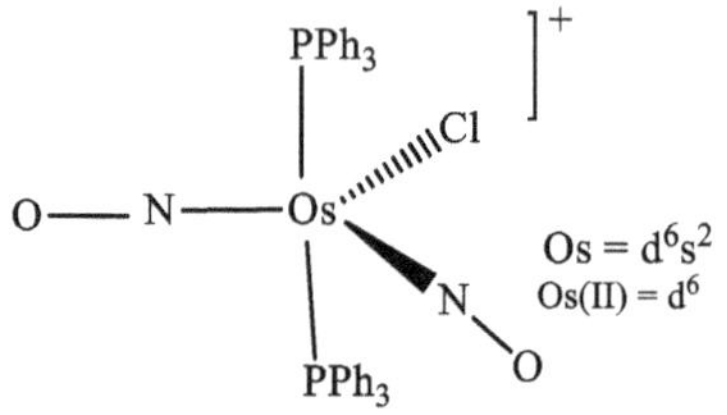

$[6(Os)+ 4(PPh_3)_2+ 2(Cl) + 6(NO)_2 =18e]$

$[L_3Os(NO)_2 l,l]^{16}$

C. N. = 3 + 2 = 5

(b)

Figure 2.11: Different M–N–O geometries of the two closely related Os nitrosyl complexes.

2.6 Simplified procedure for calculation of EAN of metal nitrosyl complexes

In calculating EAN of metal nitrosyl complexes, a similar method is applied as that one uses in case of metal carbonyls. Since in metal nitrosyl complexes, metal displays variable oxidation states, nuclear charge is taken as $(Z + n)$ for the calculation of EAN, instead of Z as in case of metal carbonyls. In fact, here "n" stands for the number of electrons gained or lost by the metal to accomplish the oxidation state observed in the metal nitrosyl.

The EAN of metal nitrosyls can be calculated in a manner similar to that of carbonyls. In metal nitrosyls, the metals exhibit a variable oxidation state. Hence, instead of considering the nuclear charge Z for the calculation, $(Z + n)$ should be used. Here, "n" indicates the number of electrons gained or lost by the metal to attain the oxidation state observed in the nitrosyl complex. The way of calculation of EAN in pentacyanonitrosyl-chromate(I) $[Cr(NO)(CN)_5]^{3-}$ is as follows:

$$\text{EAN} = Z + n + a + b + c$$

In the present complex anion, Cr (Z = 24) is in (+1) oxidation state, and that is attained by losing one electron. Hence, $n = -1$. Furthermore, all the six ligand groups (five cyanide and one nitrosyl) being terminal, each of these ligands donates a pair of electrons, and thus, $a = 6 \times 2 = 12$. As there are no bridging bonds (b) or M–M bonds (c) in the complex anion, both b and c are equal to zero.

Thus,

$$\text{EAN} = 24 - 1 + 12 + 0 + 0 = 35$$

Hence, this complex anion $[Cr(NO)(CN)_5]^{3-}$ does not follow the EAN rule. Calculation of EAN for some selected nitrosyl complexes are given in Table 2.3.

Table 2.3: Effective atomic number of some nitrosyl complexes.

Nitrosyl complexes	(*Z* + *n*)	(*a*)	(*b*)	(*c*)	EAN = (*Z* + *n* + *a* + *b* + *c*)
$[Fe^{2-}(CO)_2(NO^+)_2]^0$	26+2 =28	4×2 = 8	0	0	36 [Kr]
$[Co^-(CO)_3(NO^+)]^0$	27+1 =28	4×2 = 8	0	0	36 [Kr]
$[Mn^{3-}(CO)(NO^+)_3]^0$	25+3 =28	4×2 = 8	0	0	36 [Kr]
$[Fe^+(NO^+)(H_2O)_5]^{2+}$	26−1 =25	6×2 = 12	0	0	37
$[Mn^+(NO^+)(CN)_5]^{3-}$	25−1 =24	6×2 = 12	0	0	36 [Kr]
$[Fe^{2+}(NO^+)(CN)_5]^{2-}$	26−2 =24	6×2 = 12	0	0	36 [Kr]
$[Mo^0(NO^+)(CN)_5]^{4-}$	24	6×2 = 12	0	0	36 [Kr]
$[Co^{3+}(NO^-)(CN)_5]^{3-}$	27−3 =24	6×2 = 12	0	0	36 [Kr]

2.7 New notation with the formal charges on the nitrosyl ligand and the formal metal oxidation state

Analogous to Enemark–Feltham formalism, the new notation makes no endeavour to state the formal charges on the nitrosyl ligand and the formal oxidation state of the metal. In fact, this notation gives emphasis on the geometry of the M–N–O bond, coordination number of the metal and the total number of electron count.

Based on the extensive analysis of the vibrational data for metal nitrosyl complexes, De La Cruz and Sheppard [22] have emphasized that the most of the nitrosyl complexes obey the 18- and 16-electron rules. Hence, this parameter is establishing whether the molecule has a closed shell, that is, coherent with the EAN or "18-electron" rule. The total electron count has important chemical implications since it points out whether the compound is likely to undergo electrochemical conversion or nucleophilic addition in order to achieve an 18-electron configuration.

Taking into consideration the many differences of opinion, which have arisen in the literature concerning the assignment of a formal charge to NO and the formal oxidation state of the metal, it is better to have a notation which is not rigid and which specifies only the total number of valence electrons.

So, in case of complexes having reliable spectroscopic or theoretical evidence that the metal has a clearly distinct oxidation state, then this information may be added in the notation. For example, notations, namely, $\{L_5Ru^{II}(NO^+)l\}^{18}$ and $\{L_5Co^{III}(NO^-)b\}^{18}$ taken from Figures 2.9 and 2.10, are shown:

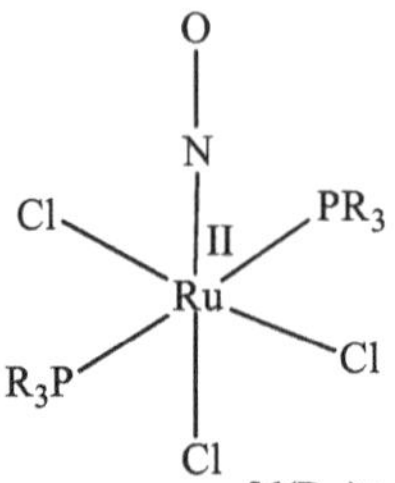

$Ru = d^6s^2$

$Ru(II) = d^6$

$[L_5Ru^{II}(NO^+)\ l]^{18}$

$[6(Ru)+ 6(Cl)_3+ 4(PR_3)_3 + 2(NO^+) =18e]$

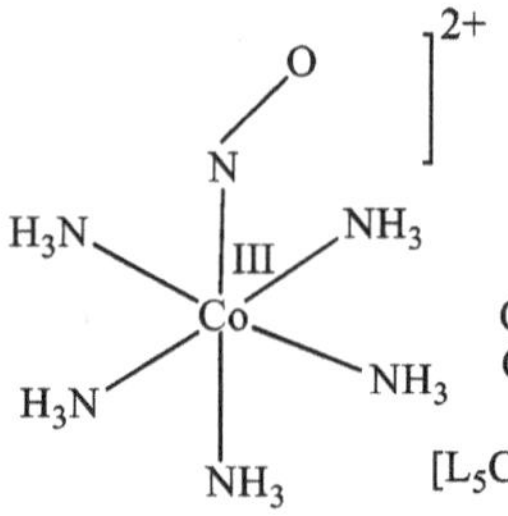

$Co = d^7s^2$

$Co(III) = d^6$

$[L_5Co^{III}(NO^-)b]^{18}$

$[6(Co)+ 10(NH_3)_5+ 2(NO^-) =18e]$

This addition validates that these two complexes have the low-spin d^6 configurations common for octahedral complexes.

When metal oxidation states and spin states in metal nitrosyl complexes are designated based on reliable theoretical or spectroscopic data and it is felt necessary to indicate that the bonding is best represented by antiferromagnetic coupling, then this may be indicted in the following manner:

(a) The nitrosyl complex, *trans*-$[Fe(NO)(cyclam)Cl]^+$, is designated as $\{L_5Fe^{III}(NO^-)\ l\ (S = 1/2\}^{19}$, and this suggests an intermediate spin Fe(III) centre ($S = 3/2$) antiferromagnetically coupled to NO^- ($S = 1$).

(b) The iron nitrosyl complex, *trans*-$[Fe(NO)(cyclam)Cl]^{2+}$, is labelled as $\{L_5Fe^{IV}(NO^-)\ l\ (S = 1\}^{18}$, and this endorses Fe(IV) ($S = 4/2$) coupled antiferromagnetically to NO^- (S =1).

(c) The nitrosyl complex, trans-[Fe(NO)(cyclam)Cl], has new notation as $\{L_5Fe^{III}(NO^{2-})\ b\ (S = 0\}^{18}$, which suggests a **low-spin** Fe(III) ($S = 1/2$) antiferromagnetically coupled to what is formally described as NO^{2-} ($S = 1/2$).

The number of unpaired electrons in the complex, determined by magnetic measurements, are indicated by $S = 1/2, 1, 3/2, \ldots$ and a multiplicity of $2S + 1$.

2.8 Transition metal nitrosyl complexes: bonding

2.8.1 Structural studies: X-ray study

X-ray structural studies have shown that while majority of transition metal nitrosyls principally contain linear M–N–O bond angles, a substantial number of metal nitrosyls exist with bond angles close to 120°. Before taking into consideration the M–N–O bonding either as linear or bent, it may be useful to discuss the older bonding concept.

In addition to acting as a doubly or triply bridging ligand, NO binds to a metal in one of the following ways:

(i) By donation of one electron from an antibonding NO orbital to the metal followed by additional electron pair donation from NO^+ to that metal.

(ii) By donation of two electrons to the metal from neutral NO.

(iii) By acceptance of one electron from a metal followed by electron pair donation from NO^- to that metal.

In these three cases, NO functions as (i) a three-electron donor, (ii) a two-electron donor and (iii) a one-electron donor.

The free NO bond length is found to be 1.154 Å. This lies between that of a double (1.18 Å) and a triple (1.06 Å) bond and is coherent with its bond order of 2.5. On oxidation, NO changes to NO^+ and leads to an increase in bond order to 3 and the bond distance reduces to 1.06 Å. On reduction, NO changes to NO^-, and this introduces an extra electron into π^*. Consequently, a reduction in the bond order to 2 and an increase in bond length (1.26 Å) occur. The changes in bond length narrated above are reflected in the stretching frequency of NO which decreases from 2,377 (NO^+) to 1,875 (NO) to 1,470 cm^{-1} (NO^-).

2.8.1.1 Linear and bent M–N–O in terms of NO^+ and NO^-

The above ideas regarding the bonding modes of NO have been valid for many years. The recent X-ray structural studies have, however, suggested that they must be modified. Thus, although nitrosyl complexes may still be regarded in terms of NO^+ and NO^-, the former label is now given to those species, which contain linear M–N–O bond angles, while the latter is assigned to nitrosyl compounds with an M–N–O bond angle close to 120°.

The manner in which NO^+ and NO^- bind to metals to give "linear" and "bent" nitrosyl, respectively, may be visualized in terms of the following simple **hybridization schemes** (Figures 2.12 and 2.13). The donation of one electron from NO to a metal affects the formation of NO^+. The nitrogen atom of NO^+ is sp-hybridized so that subsequent donation of an electron pair to the metal results in the formation of an M–N–O bond angle of 180°. On the other hand, donation of one electron from the metal functioning as a Lewis base to the NO gives in the formation of NO^-. In this species, the nitrogen atom is sp^2-hybridized, and so the donation of an electron pair by NO^- outcomes in an M–N–O bond angle of 120°. Structurally, NO^+ and NO^- may be represented in the following hybridization schemes.

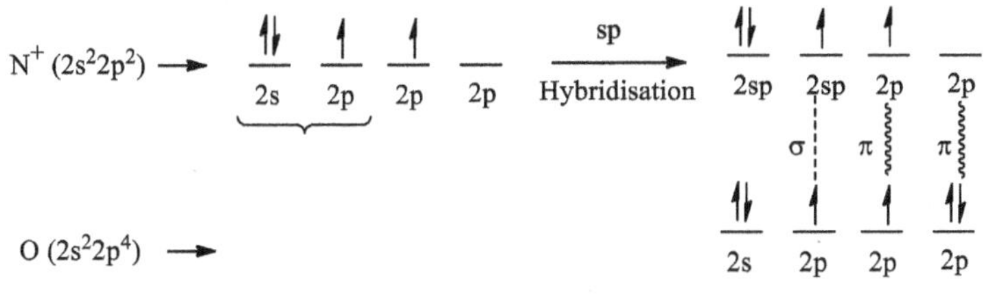

Figure 2.12: Sp hybridization in NO^+.

Let us compare the mode of bonding of NO in metal nitrosyls to the mode of bonding of CO in metal carbonyls. The mode of bonding of coordinated NO and CO has long

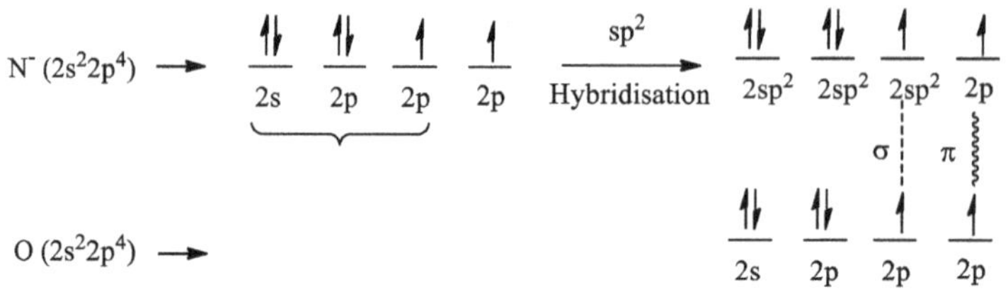

Figure 2.13: Sp^2 hybridization in NO.

been assumed as strictly similar and in that rationality CO, NO^+, CN^- and N_2 (isoelectronic ligands; 14 electrons in each) should coordinate in a similar manner. It has now become apparent that NO can bind to metals in ways a bit different to that of CO. This is well clear from the following bonding schemes of NO^+ and NO^-.

(i) In the very simple way, the bonding in metal nitrosyls is presumed as an electron transfer from the π*2py M.O. of NO (because of the low ionization potential, 9.5 ev of the NO) and proceeds with the coordination of NO^+ through the nitrogen lone pair. Backdonation of electron density from the filled metal d-orbital to the π*2p M.O.s of the NO^+ reinforces the lone pair electron of the nitrogen. Like CO, NO thus serves as a σ-donor and a π-acceptor. In this way, NO, thus functions as a three-electron donor while other isoelectronic ligands, namely, CO, N_2 and CN^- are formally two-electron donors. Considering NO as the nitrosylating agent, the overall modes of bonding of NO^+ are pictorially shown in Figure 2.14.

    In the conceptual sense, backdonation is an important parameter in explaining organometallic compounds. This leads to the change in NO bond and metal–NO bond, showing decreasing and increasing trend, respectively. Albeit, so many similar properties are shared by CO- and NO-tagged complexes, but the differences among such organometallic compounds is generally explained on the basis of chemical nature or reactive behaviour for NO being accepted as from more stable π-acceptor complex than CO. This, in other words, means justifies well remarkable or high order of electron delocalization in metal–NO core. The similar effect could be observed in the notation developed by Enemark and Feltham treating MNO as a covalent functionality assignable as $\{M(NO)_m\}^n$, wherein n refers to the total number of d-electrons in addition to π* (NO) orbitals.

(ii) Instead, transfer of an electron from the metal serving as a Lewis base to the NO results in the formation of NO^-. Thereafter, coordination of NO^- proceeds through the donation of a lone pair of electrons on nitrogen atom to the metal. Again, the backdonation is the main reason behind the strengthening of lone pair of nitrogen (Figure 2.15(i)). However, poor backdonation will occur when NO^- is bonded to the metal because of the half-filled π*2p M.O.s of the NO^- compared to NO^+ having fully vacant π*2p MOs (Figure 2.15(ii)). Considering NO as the nitrosylating agent, the overall modes of bonding of NO^- are pictorially shown in Figure 2.16.

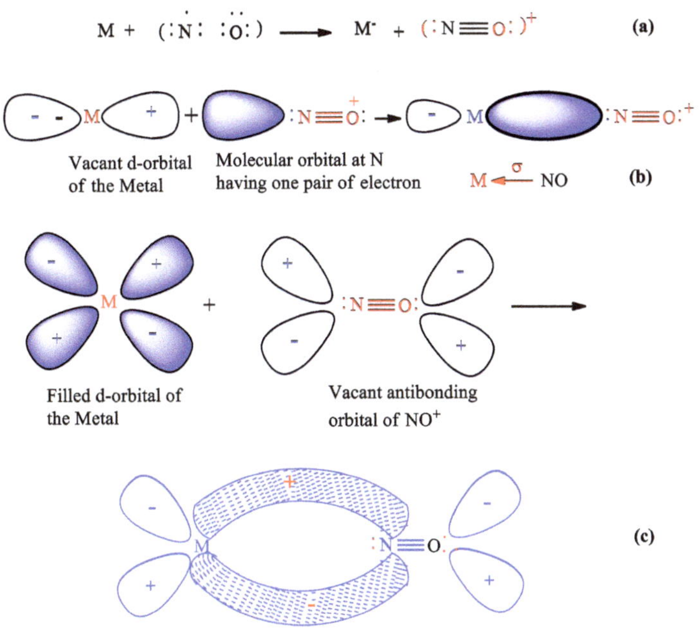

Figure 2.14: (a) Shift of one electron from the π*2py M.O. of NO to the metal d-orbital, (b) the making of metal nitrogen σ-bond and (c) the creation of metal nitrogen π-bond.

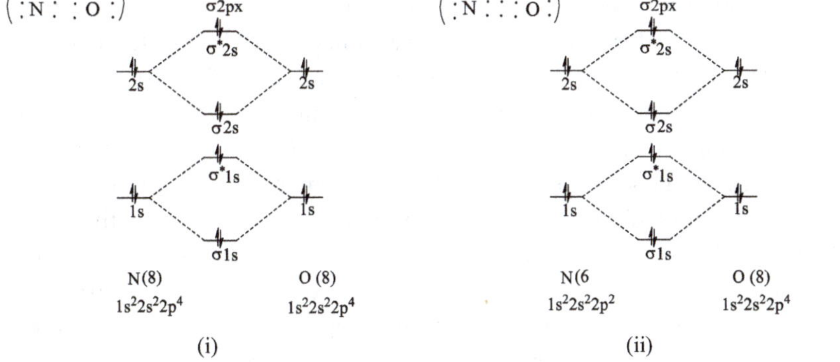

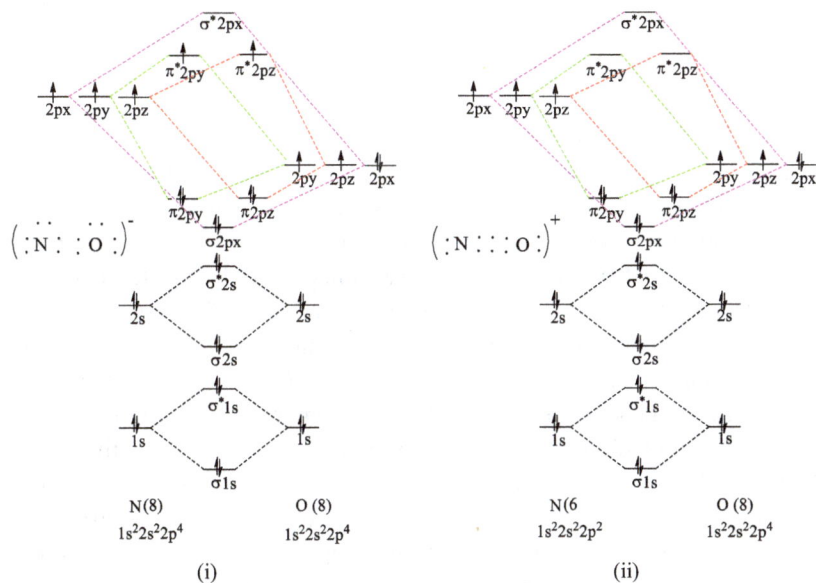

Figure 2.15: Molecular orbital diagram of (i) NO^- and (ii) NO^+.

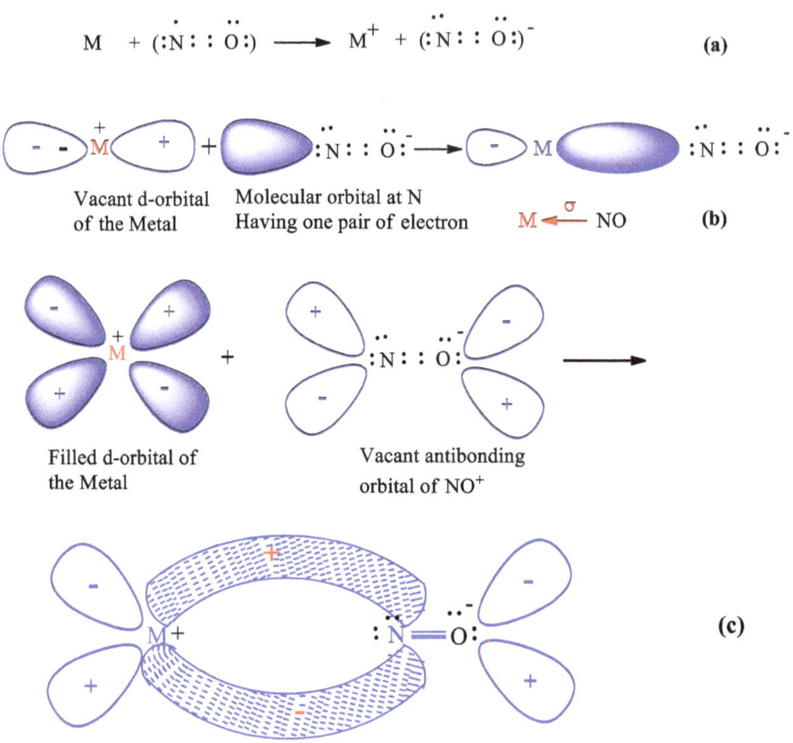

Figure 2.16: (a) Shift of an electron from the metal M to π*2py M.O. of NO, (b) the creation of metal nitrogen σ-bond and (c) the making of metal nitrogen π-bond. Other orbitals of NO are omitted for clarity.

2.8.1.2 M–N–O bond variation from 180° to 120°

In combination with various spectroscopic studies, X-ray structural evaluation of noted metal nitrosyl complexes has indicated that M–N–O angle in metal nitrosyls can be in and around 180° or 120° depicting a linear or bent M–N–O complex, respectively. This may be attributed to several reasoning. According to Kettle's view [22a], the two π*2p orbitals of each of the carbonyl ligand in $M(CO)_3$ systems are not of equal energy (non-degenerate). Hence, the central metal M will undergo back-donation in dissimilar way. Consequently, the ∠M–C–O in $M(CO)_3$ systems will deviate from linearity. Based on the same reasoning, Enemark [22b] reported such deviation (from linear fashion) criterion for mononitrosyl species also with the understanding that the overall symmetry is less than C_3. Remarkably, this process will not affect the bond angles, ∠M–N–O, in complexes having NO^- grouping to depart from 120°. Nevertheless, owing to the packing effect in the crystal lattice, this deviation is expected in both the types, namely, NO^+ and NO^- bonded centres.

2.8.1.3 Correlation of M–NO vibrational mode with ∠M–N–O

In metal nitrosyls, the existence of NO^+, NO and NO^- was firstly concluded from the location of the nitrosyl stretching frequencies [ν(NO)] in the infrared (IR) spectrum. The metal nitrosyls having NO^+ were said to absorb in the range 1,900 and 1,700 cm^{-1}, while NO^- species were said to appear in between 1,700 and 1,500 cm^{-1}. On the other hand, complexes containing neutral NO grouping were supposed to occur with IR bands in either of these regions. The X-ray structural studies have, however, concluded that a re-evaluation of the designation of NO^+ and NO^- must be made, and that the validity of assigning these assignments solely on the basis of IR spectroscopy is not always correct. A comparison of the ν(NO) of the nitrosyl complexes with the corresponding ∠M–N–O concludes that no useful correlation can be find out at least in the range 1,750–1,600 cm^{-1}.

Thus, it can be reasonably said that a lower value of ν(NO) in the IR spectrum of a nitrosyl complex may be used to ascertain the presence of a NO^- ligand only under limited conditions. For instance, if two complexes differ only due to the presence of the NO^+ ligand in one and the NO^- ligand in the other, then the complex displaying the lower ν(NO) will contain the NO^- ligand.

2.8.1.4 ∠M–N–O and isomerism in metal nitrosyls

It is well known that NO can bind to a metal in a "linear" or "bent" mode, as a three- and one-electron donor, respectively. This means that isomerism can occur in metal nitrosyl complexes exclusively owing to the presence of different ∠M–N–O in the isomers. The IR spectrum of nitrosyl complex, $[Cr(NO)(L)_2(Cl)_2]$ (L = PCH_3Ph_2), displays two ν(NO) bands at 1,750 and 1,650 cm^{-1}. The appearance of these two stretching bands is not because of dimerization, dissociation or ionization of the complex in question. It was explained with the view that the nitrosyl compound, $[Cr(NO)(L)_2(Cl)_2]$, exists in two isomeric forms. In one isomer, nitrosyl ligand is linked as NO^+, while in other isomer as NO^-. The chemical structures suggested for the two isomers are given in Figure 2.17.

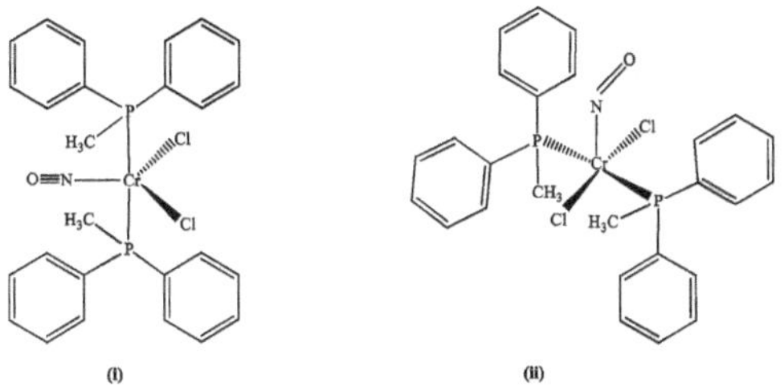

Figure 2.17: Proposed structures of the isomers of $[Cr(NO)(L)_2(Cl)_2]$ (L = PCH_3Ph_2).

As shown in Figure 2.17, the isomer (i) is trigonal bipyramidal containing a linear equatorial NO^+ ligand and the isomer (ii) is square pyramidal having a bent axial NO^- ligand. The lower ν(NO) at 1,650 cm^{-1} may reasonably be attributed to the isomer (ii) having bent Cr–N–O grouping.

The isomerism in complex (i) containing NO^+ to complex (ii) having NO^- may be depicted as follows:

$$Mn^n(NO)^+ \rightarrow M^{(n+2)}(NO)^-$$

Here n stands for the formal oxidation state of metal in NO^+ complex (i). Owing to the effective localization of two electrons (2e) into the N atom of nitrosyl grouping in the NO^- species, the formal oxidation state of the metal in the isomer (ii) is increased by two in numbers. Remarkably, here the M–N–O system serves as "electron sinks" as the isomerization is reversible. Therefore, it is expected that such nitrosyl complexes that can undergo such isomerization route may serve as valuable catalysts in favourable circumstances.

2.8.1.5 ∠M–N–O correlation with M–N bond lengths

It might be expected that the two bonding modes of NO as NO^+ and NO^- in metal nitrosyls not only give rise to different M–N–O bond angles but in dissimilar M–N bond lengths too. These two aspects are well explained as follows: (i) bonding of NO as NO^+ permits the backdonation of electrons from a metal d-orbital to the π*2p orbitals of the NO. This results in the M–N bond to achieve a substantial multiple bond character. Consequently, the M–N bond length will be much shorter compared to M–N single bond. Instead, backdonation occurs in a restricted way from the metal when NO^- is the targeted binding with the metal centre (as already explained in the bonding scheme of NO^-). In such a situation, the M–N bond lengths in "bent" nitrosyl complexes will also be short but not as short as those in "linear" nitrosyls. For the linear nitrosyls, the M–N bond lengths fall in the range 1.78–1.57 Å, whereas in bent nitrosyl species these bond lengths come in between 1.98 and 1.86 Å.

2.8.1.6 Some other parameters observed in bonding of metal nitrosyls

Based on the X-ray structural data of quite a good number of metal nitrosyl complexes of different coordination numbers, some other structural parameters observed in the bonding of these complexes are presented in Table 2.4.

Table 2.4: Structural components of some metal nitrosyls.

Complex	Geometry	Bond distance (Å) M–N	N–O	M–N–O bond angle	ν(NO) (in cm^{-1})
4-Coordinate					
1. $[Ir(NO)(PPh_3)_3]$	T_d	1.67	1.24	Linear	1,600
2. $[Ni(NO)(MeCN)(CH_3PEt_2)_2]^+$	T_d	–	–	Linear	–
3. $[Ir(NO)_2(PPh_3)_2]^+$	DT_d	1.771	1.213	163.5°	1,760, 1,715

Table 2.4 (continued)

Complex	Geometry	Bond distance (Å) M–N	N–O	M–N–O bond angle	ν(NO) (in cm^{-1})
4. $[Ni(NO)(PPh_3)_2(N_3)]$	DT_d	1.686	1.164	152.7°	1,710
5-Coordinate					
5. $[Mn(NO)(CO)_4]$	TBP	1.797	1.152	Linear	1,759
6. $[Mn(NO)(CO)_3(PPh_3)]$	TBP	1.78	1.15	Linear	1,713
7. $[Mn(NO)(CO)_2(PPh_3)_2]$	TBP	1.73	1.18	178°	1,661
8. $[Os(NO)(CO)_2(PPh_3)_2]^+$	TBP	1.89	1.12	177°	1,750
9. $[Ru(NO)(diphos)_2]^+$	DTBP	1.735	1.197	174°	1,673
10. $[Ru(NO)(H)(PPh_3)_3]$	TBP	1.795	1.180	176	1,640
11. $[Ir(NO)(H)(PPh_3)_3]^+$	TBP	1.68	1.21	175	1,780
12. $[Ru(NO)_2(PPh_3)_2Cl]^+$	SP basal	1.738	1.162	179.5	1,845
	apical	1.859	1.170	136.0	1,687
13. $[Os(NO)_2(PPh_3)_2(OH)]^+$	SP basal	1.71	1.25	Linear	1,842
	apical	1.98	1.12	127.5	1,632
14. $[Ir(NO)(CO)(PPh_3)_2Cl]^+$	SP	1.97	1.16	124.1	1,680
15. $[Ir(NO)(CO)(PPh_3)_2I]^+$	SP	1.89	1.17	125	1,720
16. $[Ir(NO)(PPh_3)_2Cl_2]$	SP	1.94	1.03	123	1,560
17. $[Ir(NO)(PPh_3)_2(Me)I]$	SP	1.91	1.23	120	1,525
6-Coordinate					
18. $[V(NO)(CN)_5]^{3-}$	O_h	1.662	1.294	171.4	1,530
19. $[V(NO)(CN)_6]^{4-}$	PBP	1.680	1.165	164.2	1,508
20. $[Cr(NO)(CN)_5]^{3-}$	O_h	1.71	1.21	176	1,660
21. $[Mo(NO)(CN)_5]^{4-}$	O_h	1.95	1.23	175.1	1,455
22. $[Mn(NO)(CN)_5]^{3-}$	O_h	1.66	1.21	174.3	1,725
23. $[Ru(NO)(NH_3)_5]^{3-}$	O_h	1.80	1.11	167	–
24. $[Ru(NO)(OH)(NO_2)_2(NH_3)_2]$	O_h	1.76	1.12	176.6	–
25. $[Os(NO)(PPh_3)_2(HgCl)Cl_2]$	O_h	1.79	1.03	178	1,820
26. *trans*-$[Co(NO)(en)_2Cl]^+$	DO_h	1.820	1.043	124.4	1,611
27. $[Fe(NO)(CN)_5]^{2-}$	O_h	1.653	1.124	175.7	1,940

DT_d, distorted tetrahedral; SP, square pyramidal; PBO, pentagonal bipyramidal; TBP, trigonal bipyramidal; DTBP, distorted trigonal bipyramidal; DO_h, distorted octahedral.

(i) Four coordinated complexes

Among the four coordinated metal nitrosyl complexes investigated, $[Ir(NO)(PPh_3)_3]$ and $[Ni(NO)(MeCN)(CH_3PEt_2)_2]^+$ are found to have tetrahedral geometry involving M–N–O bond angle linear. The two nitrosyl complexes, $[Ir(NO)_2(PPh_3)_2]^+$and $[Ni(NO)(PPh_3)_2(N_3)]$, also seem to be tetrahedral but the M–N–O bond angles in both are departed from linearity by approximately 27°. In $[Ni(NO)(PPh_3)_2(N_3)]$ complex, the non-linearity of the Ni–N–O bond angle was explained by applying Kettle's theory. The short Ni–N bond length of 1.686 Å in this compound suggests that the nitrosyl group is bonded as NO^+. This consecutively favours a linear Ni–N–O bond angle.

(ii) Five coordinated complexes

The three nitrosyl complexes of trigonal bipyramidal geometry, namely, $[Mn(NO)(CO)_4]$, $[Mn(NO)(CO)_3\text{-}(PPh_3)]$ and $[Mn(NO)(CO)_2(PPh_3)_2]$ may be taken as suitable examples of five coordinated complexes. They will demonstrate the results of substitution on the bonding between the metal and the NO^+ ligand. It has been observed that with increase in phosphine substitution in $[Mn(NO)(CO)_4]$, the M–N bond length decreases. This suggests that the weak π-accepting ability of PPh_3 ligand in comparison to CO permits increase in the backdonation of electrons from the metal d-orbital towards the π*2p orbitals of the NO^+.

The nitrosyl complex $[Os(NO)(CO)_2(PPh_3)_2]^+$ is electronically equivalent to trigonal bipyramidal nitrosyl complex, $[Mn(NO)(CO)_2(PPh_3)_2]$. Even though the osmium nitrosyl contains a linear Os–N–O grouping, its Os–N bond length of 1.89 Å is not in the range generally found in complexes containing NO^+ grouping. However, the exact reason for the long Os–N bond distance is not known.

Interestingly, the trigonal bipyramidal complexes, $[Ru(NO)(H)(PPh_3)_3]$ and $[Ru(NO)\text{-}(diphos)_2]^+$, having $\nu(NO)$ at 1,640 and 1,673 cm^{-1}, respectively, are designated as NO^- complexes. This notation is based on the old IR spectral criteria. However, both the complexes have principally linear Ru–N–O bond angles with small Ru–N bond lengths. This suggests the presence of NO^+ grouping in these complexes. The mixed ligand nitrosyl complex $[Ir(NO)(H)(PPh_3)_3]^+$ is electronically equivalent to $[Ru(NO)(H)(PPh_3)_3]$, which also contains a linear Ir–N–O bond angle.

(iii) Six coordinated complexes

Quite a good number of six coordinated metal nitrosyl complexes have been examined by X-ray techniques and found that they contain M–N–O bond angles linear. The ruthenium nitrosyl complex, $[Ru(NO)(NH_3)_5]Cl_3 \cdot H_2O$, was initially believed to contain NO^- grouping. However, X-ray crystal structural study of this compound concludes that it has Ru–N–O bond angle of 167° and the comparatively small Ru–N bond length of 1.80 Å. This indicates that the complex contains NO^+ grouping.

The cobalt (III) octahedral nitrosyl complex, $[Co(NO)(en)_2Cl]^+$, contains a bent nitrosyl ligand at axial position *trans* to the chloro ligand (Figure 2.18). It has a long Co–Cl bond length of 2.58 Å. This reasonably long Co–Cl bond distance is presenting the *trans* directing effect of NO^- ligand.

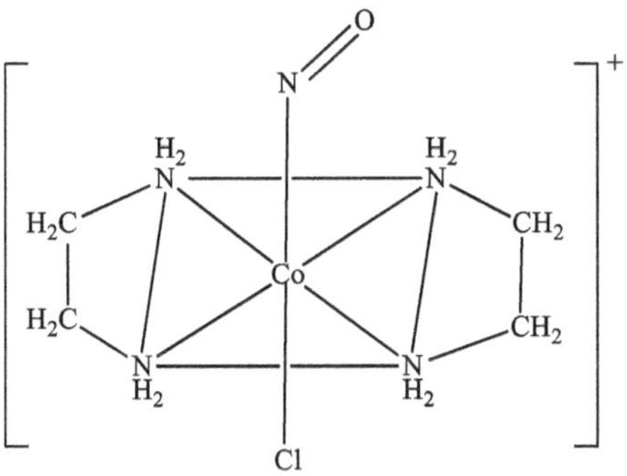

Figure 2.18: Structure of $[Co(NO)(en)_2Cl]^+$ having bent nitrosyl ligand.

Numerous pentacyanonitrosyl complexes, $[M(NO)(CN)_5]^{n-}$, have been examined for their structures. While no distinction could be made between cyano and nitrosyl ligands in $K_3[Cr(NO)(CN)_5]$, no such problem was faced with $[Cr(en)_3][Cr(NO)(CN)_5]$ in which the ∠Cr–N–O was found to be linear. The M–N–O grouping in cyanonitrosyl compounds, namely, $K_3[V(NO)(CN)_5]$, $K_4[Mo(NO)(CN)_5]$, $K_3[Mn(NO)(CN)_5]\cdot 2H_2O$, $Na_2[Fe(NO)(CN)_5]$ and $K_4[V(NO)(CN)_6]$ has also been established as linear. From the reported data of the involved bond length of C–N grouping in all these cyanonitrosyls have been shown to have a constant C–N length, but differences in N–O bond length do exist indicating that stabilization of a metal is achieved principally by the nitrosyl group. The studies based on electron spectroscopy for chemical analysis (ESCA) of a series of cyanonitrosyl complexes (vide infra) substantiate these findings.

The above discussions conclude that the use of the IR spectral criteria for differentiating between NO^+ and NO^- ligands in pentacyanonitrosyl complexes is limited and not perfectly correct. Even though the complexes examined contain NO^+ groupings, their IR spectra show very low $\nu(NO)$ values. These would had been taken as characteristic of the presence of NO^- ligands. In fact, the presence of the negative charge causes $\nu(NO)$ to be very low in these complexes. However, coordination of nitrosyl as NO^+ happens in these complexes.

2.8.1.7 General structural features of the mononitrosyl six coordinated metal complexes

The most common structural aspects in mononitrosyl octahedral metal complexes are the fairly short M–N bond lengths. This suggests the multiple bonding between metal and nitrosyl ligand in such complexes. The N–O bond lengths fall in the range 1.03–1.29 Å with average value of 1.159 Å from the whole group. The ∠M–N–O fall in the range 180–109° with assemblage close to the values of 120° and 180° (Table 2.5).

Table 2.5: ∠M–N–O in $\{MNO\}^n$ six coordinated complexes.

n	∠M–N–O (degree)	Average* ∠M–N–O
4	171–178	
5	169–178	175(7)
6	170–180	
7	138–159	146(11)
8	119–141	126(9)
9	Unknown	
10	Unknown	

*The average is the estimate of the mean. The number shown in parentheses is the estimated standard deviation of the mean.

2.8.1.8 Six-coordinated $\{MNO\}^n$ complexes and M–N–O angle projections

In structurally characterized monometallic nitrosyls, the most common coordination number encountered is 6. For complexes involving $\{MNO\}^n$ electron configuration, six coordination happens with n = 4, 5, 6, 7 and 8. The ∠M–N–O range came across for each of these six coordinated complexes of $\{MNO\}^n$ electron configuration is shown in Table 2.5. It can be seen from this table that when n is less than or equal to 6, ∠M–N–O is almost linear (±10°). In complexes of $\{MNO\}^7$ and $\{MNO\}^8$ electron configurations, the range of ∠M–N–O is rather larger and averaged to 146(11)° and 126(9)°, respectively. Thus, the tabular data show that the M–N–O bond angle of six coordinated complexes decreases when n rises to 7 and 8.

The deviation in ∠M–N–O with n enlisted in Table 2.5 resembles with the deviation in ∠O–N–O observed in NO_2^+, NO_2 and NO_2^- (Figure 2.19).

2.8.2 M.O. calculations of bonding in metal nitrosyls: linear to bent MNO bond angle transformation in hexa- and penta-coordinated nitric oxide complexes

The X-ray structural verification of metal nitrosyls in various coordination numbers as described above furnishes a qualitative aptitude to present the bonding picture of this class of complexes. In addition to these diffraction-based geometry parameters applicable for the determination of nature of bonding, M.O. approach can serve as more qualitative front in discussing these binding features supported by various published works from several pioneer workers in this field. These efforts have been presented and discussed further.

During 1960s, research groups of Harry Gray and Dick Fenske reported the first theoretical calculations that were used to explain the spectral properties of pentacyanonitrosyl complexes of 3d-transition metals. As per the semiempirical

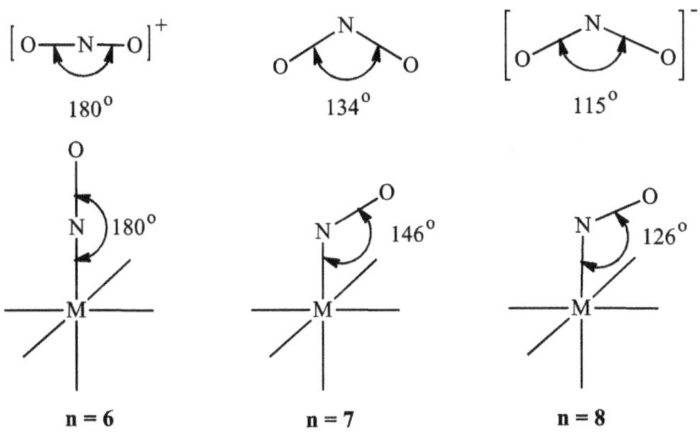

Figure 2.19: Deviation in ∠M–N–O with n = 6, 7 and 8 bears a resemblance with thedeviationin ∠DO – N – OinNO$_2$$^+$, NO$_2$ andNO$_2$$^-$.

approach laid in this study, NO has been suggested as a better π-acceptor as compared to the remaining two co-ligands, namely, CO and CN$^-$. Also, it was found that the N–O antibonding orbital possesses the energy of the order of involved d-orbitals. Later in 1971, the famous Walsh analysis (D. M. P. Mingos in 1971) proved more qualitative front for the justification of linear to bent fashioned NO complexes. More distinctive feature of carbonyl and nitride complexes as compared to nitrosyl complexes thus became more elaborative.

These advancements were further followed by another semiempirical approach in 1973, by Mingos elaborating linear versus bent M–NO. Later, 1974 saw bonding models of nitrosyl complexes consisting of five coordination number by Hoffmann and Mingos represented the first report forming both the linear and bent nitrosyls.

For the octahedral cobalt nitrosyl complexes, $[(NH_3)_5Co(NO)]^{2+}$, the Walsh diagram analysis is shown in Figure 2.20.

In Figure 2.20, e_g set of the related d-orbitals is able to be recognized easily, and is important for being referred to as antibonding orbitals of strong metal–ligand bond, whereas t_{2g} set becomes non-degenerate because of high pronunciation of π interactions existing (π*(NO) and d_{xz}, d_{yz} orbitals). The figure clearly shows this π interaction lying between the two sets of d-orbitals (parental structure).

In the case of low-spin complexes of the notation $\{L_5M(NO)\}^{12-18}$ having octahedral conformation, the metal–N–O attains a linearity because of the combined effect of e_g and π*(NO) orbitals, and hence causes the xz-plane to bend. From the reported calculations it is obvious that this bending tendency could be linked with one of the components of antibonding orbitals, that is, [(π*(NO)-d_{xz}] of the [e(π*(NO)-$d_{xz,yz}$)] [23] as shown in Figure 2.20. This occupancy is mainly found among low-spin forms of $\{L_5M(NO)\}^{18-22}$ complexes. Therefore, the fall in energy of this orbital is directly

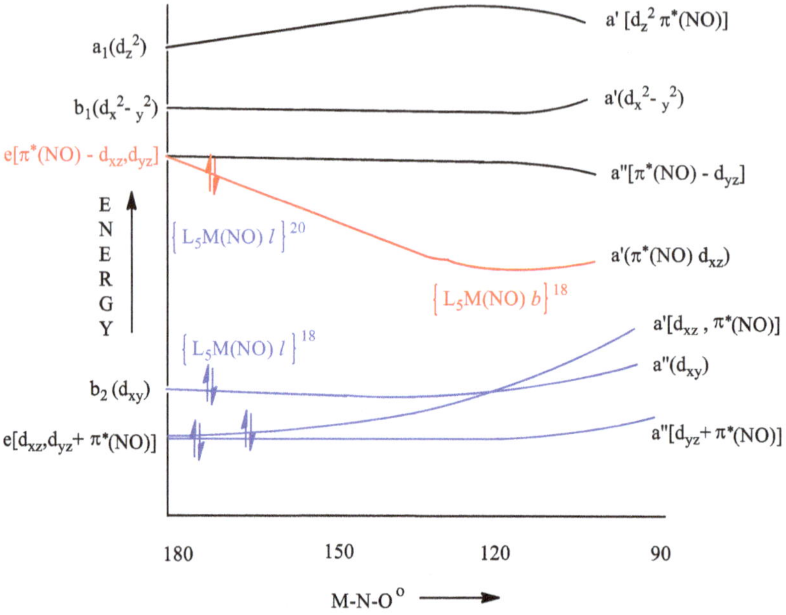

Figure 2.20: Calculated Walsh diagram for $[(NH_3)_5Co(NO)]^{2+}$. The main orbital (shown in **red** colour) is liable for bending distortion. (Linear fashion remarked by **20 electrons in number** (shown **blue**), while bent geometry by **18 electrons in number** (shown **red**)).

linked with the decrease in ∠M–N–O. For d_z^2 the bent fashion imposed in linear M–N–O indicates the intense mixing with d_{xz}-π*NO, becoming same symmetric towards this alignment. This can be clearly understood by the drawings shown in Figure 2.21(a) and (b) indicating the respective linearity deviation.

For more stable component as indicated by Figure 2.21(a), mixings of the nitrogen atom orbital result in an outdirecting hybrid resembling the effect of nitrogen lone pair, and this leads to a stability direction at the central metal lying spatially between π_{xz}^* of NO and d_z^2. Thus, the bent distortion rehybridizes to increase the expression of antibonding adhering with d_z^2 (Figure 2.21b). Here, it may also be noted that the out-of-plane distortion (bending) is especially favourable for NO (and O_2). This is due to the fact that the metal and π*(NO) orbitals have similar energies. On the other hand, the bending distortion in CO/N_2 is less favourable because of the respective π* energies have at higher values in metal orbital comparison.

Calculated Walsh diagram discussed above accounts for the following:

$[L_5M(NO)]^{12-18}$ linear ∠M–N–O (~180°)
$[L_5M(NO)]^{19}$ intermediate ∠M–N–O (140–160°) and longer M–N bond
$[L_5M(NO)]^{20}$ bent ∠M–N–O (~120°) and longer M–N bond
$[L_5M(NO)]^{20-22}$ bent ∠M–N–O (~120°) and longer M–N bond

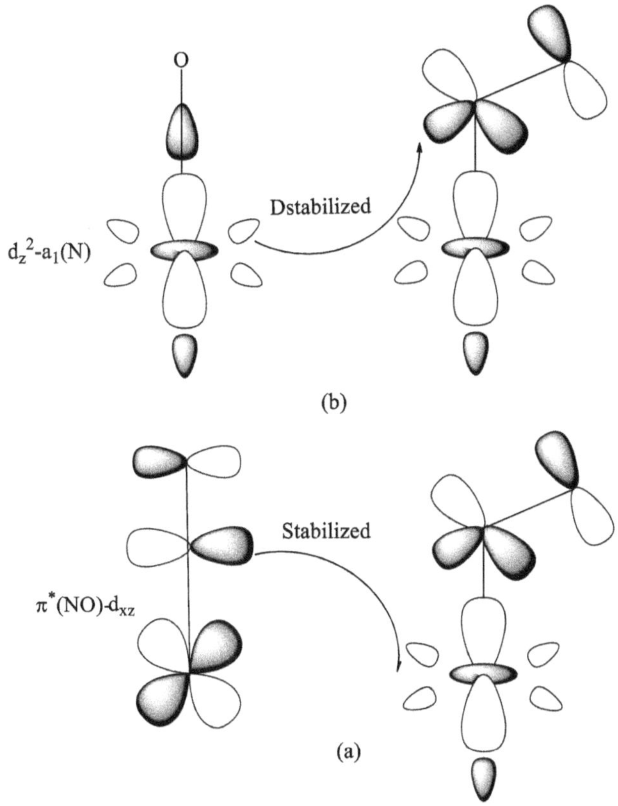

Figure 2.21: Mixing of $d_z{}^2$ orbital with d_{xz}-π*NO when the linear nitrosyl bends.

Examples of complexes following such a correlation have been given in Figure 2.22.

In view of the new formalism described above (vide Figure 2.20), the bent fashion adopted may be designated as a transformation from $\{L_5M(NO)l\}^{20}$ formalism to $\{L_5M(NO)b\}^{18}$. In fact, this relates to the electron pair shifting from the metallic centre resembling the nitrogen lone pair (Figure 2.21a). In Figure 2.22, the two examples of complexes belonging to the group $\{L_5M(NO)\}^{19}$ bear differences in the magnitude of their ∠M–N–O keeping them to follow a bent and intermediate conformation, thereby stresses the effect of changes that happened in the metal–ligand character. This is further supported by detecting increasing trend in the values of M–N bond lengths upon bending in connection with the weakening of π overlap existing between π_x*(NO) and d_{xz}, and this underlines for 20-electron species. The nitrosyl complex $[Cu(NO_2Me)_5(NO)]^{2+}$ is an example of a 22-electron complex showing ∠Cu–N–O of 121° despite lacking ground state spin-paired condition. The formulation mainly invokes the Jahn–Teller effect in the distorted structure along with the long Cu–NO signatures.

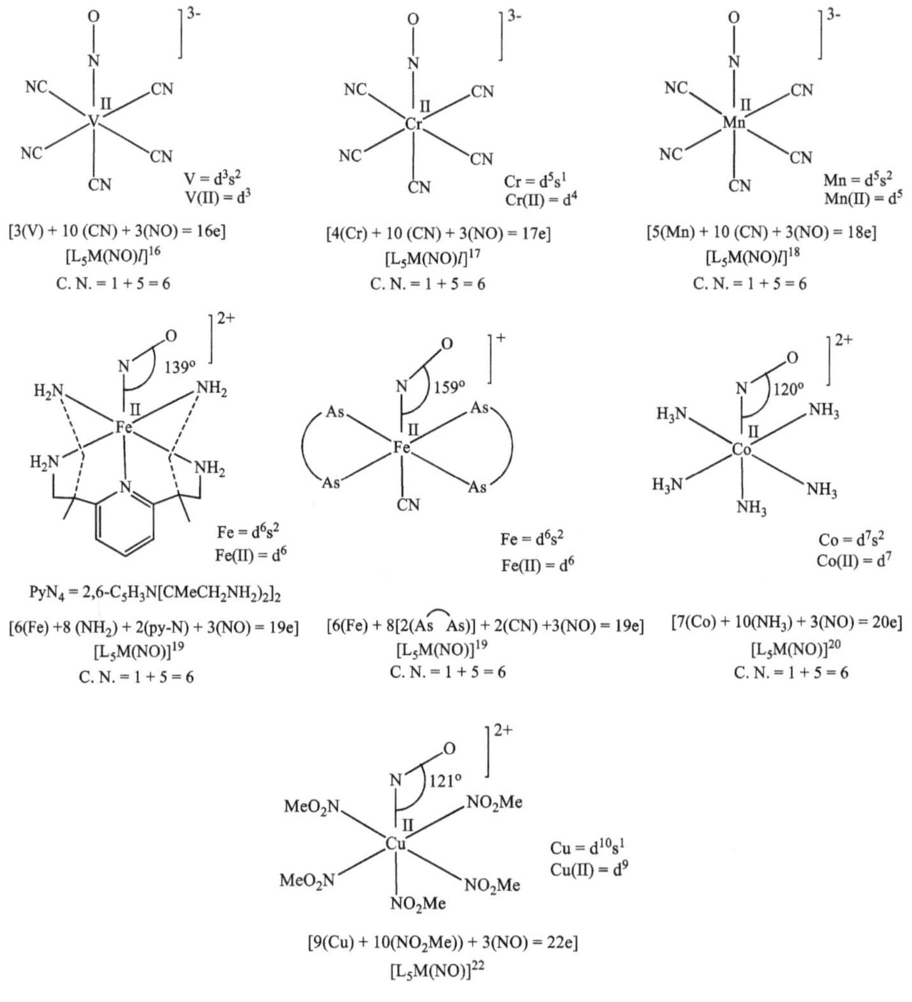

Figure 2.22: Structures of some O_h metal nitrosyls. In the entire notation given above, NO is considered as three-electron donor.

After successful application to hexa-coordinate complexes, the use of Walsh method was then applied to five-coordinate nitrosyl complexes provided with a M.O. interpretation under the following criteria:

(i) The case of bent fashion adopted by metal–NO indicates better basal character of ligands in terms of σ- or π-donoring capability.
(ii) In [$ML_2L'_2(NO)$]-type metal nitrosyls, the NO–M fashion would be bent for *trans*-effect of L *trans* to L, possessing weaker basic nature.
(iii) The nitrosyl grouping in the equatorial position of a trigonal bipyramidal nitrosyl complex is less likely to bend than the apical side of a square pyramidal nitrosyl complex.

(iv) Nitrosyl ligands have a preference to be linearly coordinated in axial positions in a trigonal bipyramidal nitrosyl compound and in equatorial positions of a square pyramidal nitrosyl compound.

(v) In case of trigonal bipyramidal conformation confined for $[ML_4(NO)]$ class of type nitrosyls, nitrosyl functionality at equatorial position when the colligation favours strong π-acceptability. Similarly, if L shows strong π-donoring behaviour, several forms of geometries are possible displaying variation in M–NO bent projections within the trigonal bipyramidal geometry.

2.8.3 Molecular orbital calculations: density functional theory approach

The above discussion concludes that the Walsh methodology provides an excellent tool for molecular energy level diagram with respect to bond angles. However, there are so many cases wherein a more refined approach other than the Walsh method is preferred. In the past several decades, density functionalized studies have opened more adequate/accurate elaborations for in sighting electronic/spin states of metallic compounds [24]. Density functional theory (DFT) calculations may also be used for the analysis of spectroscopic data accurately. The conjoint computational DFT calculations and spectroscopic studies on a number of nitrosyl complexes of biological and medicinal relevance have provided a stimulus for a bonding model.

The DFT calculations making use of suitable functionals are capable to imitate the geometries of closed shell molecules satisfactorily. For example, in case of $[RuNOCl_5]^{2-}$, different bond lengths based on DFT calculation using functionals B3LYP and SVWN5 are comparable with the results obtained from other methods (Figure 2.23).

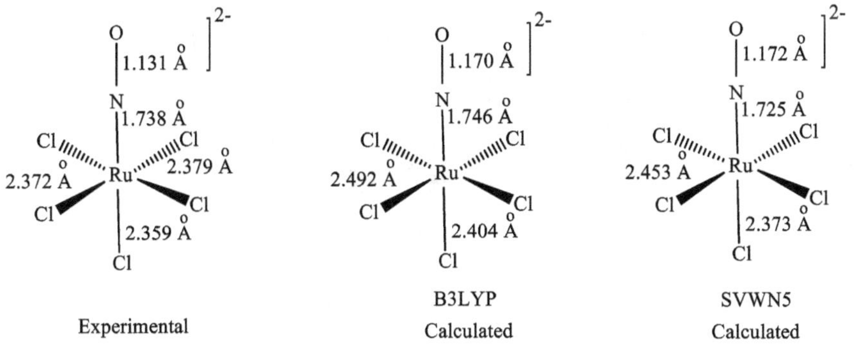

Figure 2.23: Experimental and DFT calculated bond lengths in $[RuNOCl_5]^2$.

Based on both experimental and calculated outcomes, it is concluded that the nitrosyl complex $[Ru(NO)Cl_5]^{2-}$ does not display *trans* influence on the chloride *trans* to the nitrosyl group. This has been clarified by considering the electronic structure of this complex.

Some novel NO containing molybdenum(0) pyrazolone Schiff base complexes have been prepared and thereafter characterized using different physicochemical techniques by Maurya and Mir [25]. The important bond lengths and bond angles produced from the optimized structure (Figure 2.24b) of one of the representative complexes, $[Mo(NO)_2(amphp\text{-}aap)(OH)]$ (amphp-aapH = *N*-(4-acetylidene-3-methyl-

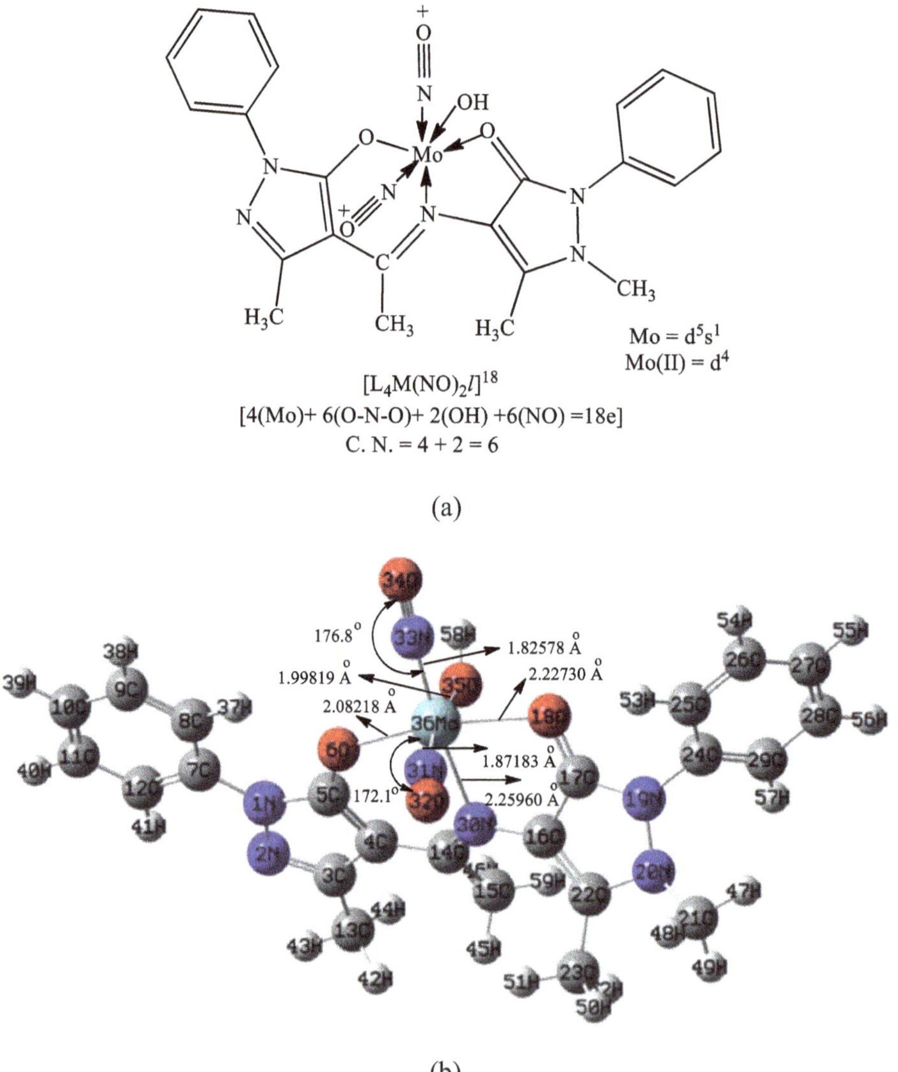

Figure 2.24: (a) Two-dimensional structure of the $[Mo(NO)_2(amphp\text{-}aap)(OH)]$complex and (b) 3D optimized structure of the $[Mo(NO)_2(amphp\text{-}aap)(OH)]$complex.

1-phenyl-2-pyrazolin-5-one)-4-aminoantipyrine) (Figure 2.24a) using Gaussian 09 software under 6-311G/LANL2DZ/RB3LYP molecular specification. The calculated bond lengths such as (36)Mo–(6)O, (36)Mo–(18)O, (36)Mo–(35)O, (36)Mo–(30)N, (36)Mo–(31)N and (36)Mo–(33)N for the target complex were found to be 2.08248, 2.22730, 1.99819, 2.25960, 1.87183 and 1.82578 Å, respectively.

The very interesting part of the calculated bond angles is that it predicts the linear M–N–O group for the exemplary complex based on the computed bond angles like (36)Mo–(33)N–(34)O, 176.807° [$\nu(NO)^+$:1,775 cm^{-1}] and (36)Mo–(31)N–(32)O, 172.092°[$\nu(NO)^+$:1,650 cm^{-1}]. Moreover, computed outcomes indicate that the dinitrosyl complex in question does exhibit a *trans* influence on the [(36)Mo–(30) N]*trans* to one of the nitrosyl groups [(N(33)–O(34)]. Further, the other nitrosyl group [(31)N–(32)O] does not display a *trans* influence on the [Mo–O(35)] because of the unknown reason.

Recently, Maurya et al. [26] carried out the DFT study of a dinitrosylmolybdenum(0) Schiff base complex, [$Mo(NO)_2$(tsc-dha)(H_2O)] (tsc-dha = Schiff base derived from dehydroacetic acid and thiosemicarbazide) (Figure 2.25).

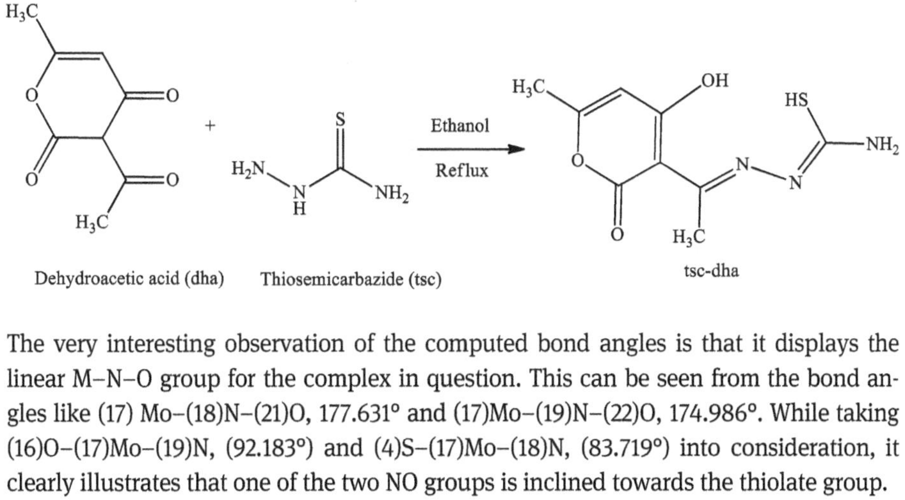

The very interesting observation of the computed bond angles is that it displays the linear M–N–O group for the complex in question. This can be seen from the bond angles like (17) Mo–(18)N–(21)O, 177.631° and (17)Mo–(19)N–(22)O, 174.986°. While taking (16)O–(17)Mo–(19)N, (92.183°) and (4)S–(17)Mo–(18)N, (83.719°) into consideration, it clearly illustrates that one of the two NO groups is inclined towards the thiolate group.

2.8.4 Linear versus bent nitrosyl ligands: Enemark–Feltham approach

In linear metal nitrosyl complexes, the internuclear bond distances for O–N and N–M bonds lie in the range 1.14–1.20 Å and 1.60–1.90 Å, respectively. However, the range for ∠M–N–O for such complexes is found as 180–160°. Moreover, the observed regions for N–O and M–N bond distances and ∠M–N–O are 1.16–1.22 Å, 1.80–2.00 Å and 140–110°, respectively, in most of the bent metal nitrosyls.

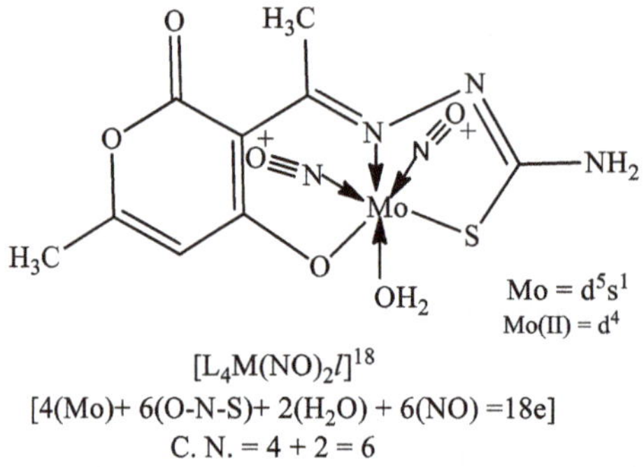

(i)

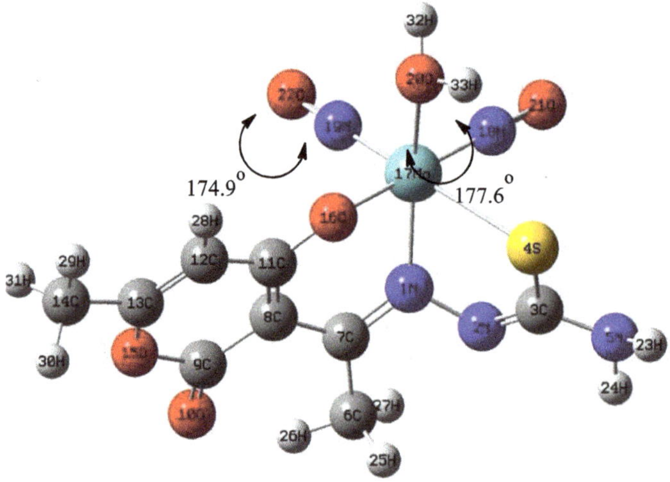

(ii)

Figure 2.25: (i) Two-dimensional structure of the [$Mo^0(NO)_2$(tsc-dha)(H_2O)] and (ii) 3D optimized structure of the [$Mo^0(NO)_2$(tsc-dha)(H_2O)].

It is notable that as per the earlier view on mononitrosyl complexes involving linear M–N–O groupings, such complexes were supposed to have coordinated NO^+. Additionally, mononitrosyl complexes with bent M–N–O groupings were supposed to have coordinated NO^-. But later on, such assumptions for linear and bent M–N–O groupings in mononitrosyl complexes have been found to be misleading. As per valence bond applicability over this observation considers sp hybridization for NO^+, whereas it is sp^2 fashioned for bent arrangement.

2.8.4.1 Six-coordinate octahedral complexes

As per the Enemark–Feltham methodology, for octahedral complexes, the aspect that decides the bent versus linear NO ligands is connected with the total number of π-symmetry electrons. In this sense, if π-electrons are found more than 6 in a nitrosyl complex, it is likely that the system will have bent-fashioned NO binding with the centre. For example, $[Co(ethylenediamine)_2(NO)Cl]^+$ have seven electrons belonging to π-symmetry (from t_{2g} and NO) is bent form of M–NO grouping, whereas in $[Fe(CN)_5(NO)]^{2-}$ bearing six π-symmetric electrons is supposed to have linear nitrosyl linkage.

Notably, in six-coordinated mononitrosyl octahedral complexes of $\{M(NO)\}^6$ electron configuration, ∠M–N–O is principally linear in most of the cases. This can be very well understood from a simplified M.O. treatment. This M.O. treatment requires that the M–N–O bond should lie on the *z*-axis. The other ligands may acquire positions either *trans* to the NO or on the other two Cartesian axes. In the present M.O. scheme, two orbitals, namely, $d_{x^2-y^2}$ and d_{z^2} of the metal interact with the ligand orbitals as well as that of NO. The other two orbitals, namely, d_{xz} and d_{yz} interact with the π^*(NO) orbitals only. The remaining metal d_{xy} orbital is rather unperturbed in this arrangement. Figure 2.26 displays the organizational sequence of M.O.s with rising energies.

According to the M.O. diagram referred to in Figure 2.26, $(e_1)^4(b_2)^2(e_2)^0$ is the expected electronic arrangement (configuration) usually found in case of nitrosyls having $\{M(NO)\}^6$ as core representation. Under such cases or in the case of less availability of d electrons in O_h complexes, the electronic filling starts with the occupancy of strongly bonded (e_1) and metal centred (b_2) orbitals (with the empty antibonding e_2 orbitals). This describes the multiple bond order along the linearly oriented MNO moiety of the complex indicative of the fact to eradicate the possibility of bent M–N–O bond.

For nitrosyl complexes involving $\{M(NO)\}^7$ formalism, presence of one more electron compared to $\{M(NO)\}^6$ notation outcomes in the new electronic configuration $(e_1)^4(b_2)^2(e_2)^1$. Consequently, occupation of a totally antibonding π-type orbital takes place, and this enforces bending according to Walsh's rules. This inevitably leads to distortions of the M–N–O bond angle, a change in symmetry and mixing of the a_1 and the *x* component of the previously designated e_2 orbital. This is demonstrated in Figure 2.26(ii). This mixing gives a more bonding a′ level, mainly π^*(NO) incorporated with d_{z^2}, and an equivalent antibonding level, which is mainly d_{z^2} in character. There are other smaller changes in energies and symmetries of the e_1 and b_2 levels, but these are insignificant with respect to the frontier orbitals. On account of the bending of the M–N–O bond, the electronic configuration of $\{M(NO)\}^7$ (ignoring the six electrons in the former e_1 and b_2 levels) will be $\{d_{yz}\pi^*(NO)\}^2\{d_{xz}\pi^*(NO)\}^2$ $\{d_{xy}\pi^*(NO)\}^2$–$(a')^1(a'')^0$. This is equivalent to describing the coordinated NO as $NO^\bullet$. Interestingly, in nitrosyl complexes of $\{M(NO)\}^8$ notation, the electronic configuration in the M.O. system will be $\{d_{yz}\pi^*(NO)\}^2\{d_{xz}\pi^*(NO)\}^2\{d_{xy}\pi^*(NO)\}^2$ $(a')^2(a'')^0$. This

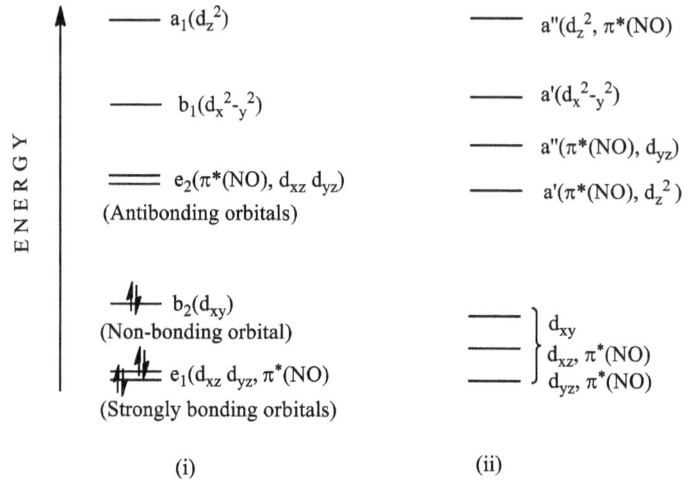

Figure 2.26: Organization of molecular orbitals in six-coordinate $\{M(NO)\}^n$ complexes: (i) where ∠M–N–O is 180° and (ii) where ∠M–N–O is 120°. The M–N–O bond is considered as the *z*-axis.

is equivalent to the coordination of singlet NO^- (Figure 2.27). The M.O. scheme shown in Figure 2.26 acquires stronger field ligation defined by energy separation between "t_{2g}" and "e_g" levels. Consequently, the M.O. arrangements for $\{M(NO)\}^8$ notation (Figure 2.27) reflect it.

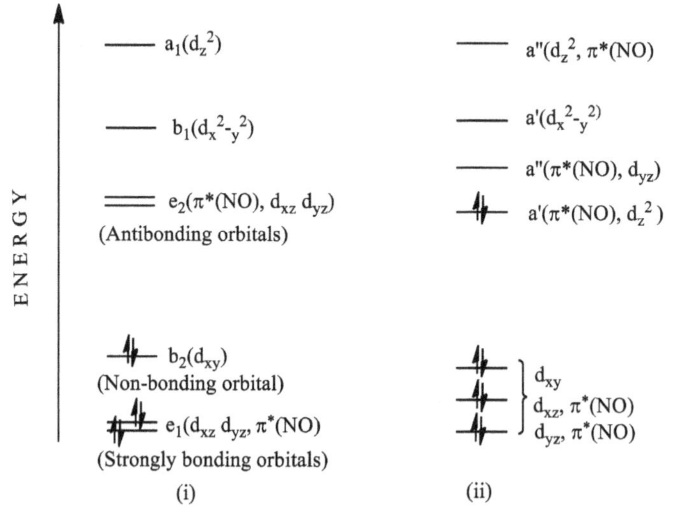

Figure 2.27: Electronic configuration of an octahedral complex involving $\{M(NO)\}^8$ electron configuration representing the coordination of singlet NO.

Considering the perspective of biologically relevant systems, six-coordinate complexes of $\{Fe(NO)\}^6$ notation would normally be anticipated to have principally a linear Fe–N–O bond. On the other hand, those having $\{Fe(NO)\}^7$ formalism would be likely to have a bent arrangement. Moreover, in cobalt nitrosyl complexes involving $\{Co(NO)\}^8$ notation, the Co–N–O bond would be considerably bent.

2.8.4.2 Five-coordinate complexes

Tetragonal pyramidal and trigonal bipyramidal are two possible geometries for five-coordinate complexes. Notably, five-coordinate macrocyclic nitrosyl complexes of Mn, Fe and Co almost have the tetragonal pyramidal geometry. It is notable that for such a structural organization, the simple M.O. picture outlined in Figure 2.26 is reasonably applicable for ascertaining the species having linear or bent M–N–O bond angles.

2.9 Characterization of metal nitrosyl complexes using spectroscopic and other physical methods

Similar physical methods referred for the general structural elucidation are employed for structural analysis of nitrosyl complexes. Thus, IR/vibrational spectroscopy, electronic, electron spin resonance (ESR), X-ray photoelectron, mass spectral, magnetic and kinetic studies are the most commonly used physical methods. Besides these, resonance Raman, Mössbauer (MB), X-ray absorption spectroscopic measurements, magnetic circular dichroism and magnetic resonance imaging are some other physical methods used in characterization of metal nitrosyl complexes.

2.9.1 Vibrational spectral studies

Application of Hook's law as is known provides instant speculation about expected vibrational modes of functional groups generally found in the composition of compounds. In the present case whether it is NO, NO^+ or NO^- each can be easily detected by knowing the vibrational mode of these bonds. Fourier transform (FT)-IR analysis provides main functional group identity of NO showing a single vibrational band at 1,860 cm^{-1}, 2,200 cm^{-1} for NO^+ and $\nu(NO^-)$ is assigned at 1,470 cm^{-1}. Whereas in metal coordination the respective vibrational modes fall in the range of 1,900–1,300 cm^{-1}. Figure 2.28 highlights the characteristic frequency ranges for metal nitrosyl complexes [27].

In usual practice regarding the FT-IR-based differentiation between linear and bent M–NO functionalization in nitrosyl complexes, 1,650–1,900 cm^{-1} refers to linear fashion, and 1,525–1,690 cm^{-1} is attributed to bent projection. This is here mentionable that these two distinctive regions may get overlap and hence may cause ambiguity in the affirmation of the two types of modes. The variational mode found for linear reflects triple bond existence and double bond for a bent NO grouping of NO–core complexes supported by Hook's law. In addition to bent and linear M–NO, bridging character of NO usually vibrates in the range 1,650–1,300 cm^{-1}, though a fine conclusive of referring to a specific vibrational mode in this connection is not possible because of the ambiguity generated by overlapping tendencies by these vibrational modes commonly having wavenumber values in compliment to one another (vide Figure 2.28).

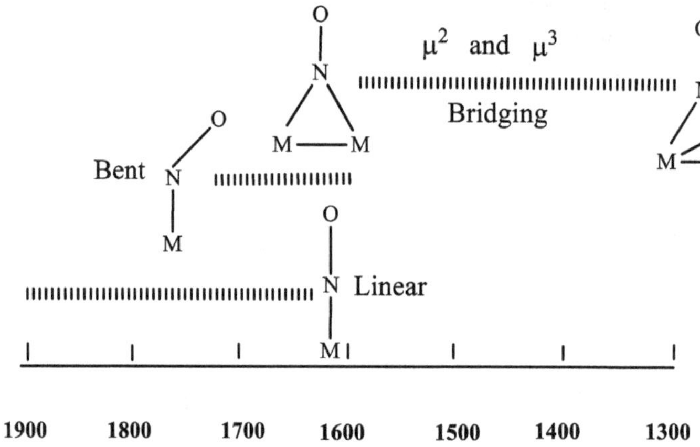

Figure 2.28: Array of frequencies observed in different categories of metal nitrosyls.

The *v*(NO) has been found to be dependent upon several factors including the nature of central metal atom/ion, nature of coordination sphere, the oxidation state and charge of inner and outer sphere, and the nature and type of frameworks selected as co-ligands other than NO. This led to an exhaustive work reported by Haymore and Ibers [28] formulating the estimation for adjustments of *v*(NO) in so many nitrosyl model complexes. The study reveals two main classes of wavenumbers, referring to a lesser or a greater than 1,606–1,611 cm^{-1} range allowing to divide the respective complexes as linear and bent. Also, the corrected *v*(NO) low wavenumbers could be made to fetch distinguishing mark for bridging and non-bridging M–NO groupings. Because of the involvement of several variables affecting on *v*(NO), it has not been widely accepted.

Based on the conspicuous fascination drawn towards vibrational analysis of several types of M–NO binding, De La Cruz and Sheppard [29] accumulated significant vibrational results for a number of nitrosyl complexes and established a correlation for EAN rule. Owing to inadequacy of binary nitrosyl compounds, it was found quite difficult to address appropriate model compounds of such a class. However, the common potentiality of possible existence of neutral NO and CO made it easy to use mixed carbonyl–nitrosyl form of complexes as reference compounds.

From their overall study, the following points may be pondered in view of the problems addressed so far:

(i) In general, a linear M–NO complex bearing a unit positive charge increases 100 cm^{-1} to expected ν(NO) value having linkage with one positive charge of the complex.

(ii) Similarly, a value of 145 cm^{-1} gets subtracted for negative unit charge in place of positive unit referred in (i).

(iii) Approximately 180 cm^{-1} lowering in ν(NO)/unit negative charge was observed for metal nitrosyl complexes having bent M–N–O groupings.

(iv) In case of halogens at terminal position, an increase in ν(NO) wavenumber by 30 unit/halogen is approximated, whereas bridging halogens result roughly with an increase of 15 cm^{-1}/halogen in the vibrational mode of NO. Similarly, a positive shift of 5 cm^{-1} in ν(NO) is theoretically correlated for enhanced electronegativity (for a halogen).

(v) When a terminal NO is substituted by a CN group, an increase of roughly 50 cm^{-1} is expected in the magnitude of ν(NO) value.

(vi) The trivalent phosphorus derivatives are found to affect the ν(NO) accordingly to their electron-donating or electron-withdrawing properties. A lowering of roughly 70 cm^{-1} per trialkylphosphine ligand, a reduction of about 55 cm^{-1} per triphenylphosphine ligand, a decrease in ν(NO) by 40 cm^{-1} per trialkoxy and 30 cm^{-1} in triphenoxy ligands and a rise of about 10 cm^{-1} per PF_3 ligand have been seen. It has been observed that the presence of more and more electron-withdrawing substituents on phosphorus reduces or eliminates the effective basicity of the phosphine ligands.

(vii) A let down in ν(NO) by about 60, 70 and 80 cm^{-1} is seen for the pentahapto-cyclopentadienyl groups, η^5-C_5H_4Me and η^5-C_5Me_5, respectively.

Based on consideration of the above points for corrections, the following corrected ranges are given by De La Cruz and Sheppard for linear M–N–O complexes:

First-row transition metals: 1,750 (Cr)–1,840 (Ni) cm^{-1}
Second-row transition metals: 1,730 (Mo)–1,845 (Pd) cm^{-1}
Third-row transition metals: 1720 (W)–1760 (Ir) cm^{-1}

A wide-ranging metal nitrosyl complexes containing linear and bent M–N–O groupings absorb IR radiation over the large array of 1,862–1,690 cm^{-1} and 1,720–1,525 cm^{-1}, respectively. However, an overlap in these regions (as mentioned above) ignores the classification of these two types of structures. Under this situation, help of another spectroscopic measure is taken for differentiating linear from bent M–NO grouping. Notably, it is the $\nu(^{14}NO) \rightarrow \nu(^{15}NO)$ isotopic shifts. For this, the isotopic band shift (IBS) [IBS = $\nu(^{14}NO)-\nu(^{15}NO)$] are plotted against M–N–O bond angle for penta-, hexa- and hepta-coordinated mononitrosyl complexes. This plot highlights that the IBS values are grouped between 45 and 30 cm^{-1} or between 37 and 25 cm^{-1} for linear or bent M–N–O groupings, respectively.

Since solid-state vibrational analysis (FT-IR spectroscopy) is one of the predominant analytical techniques in explaining the structural isomerization, in case of nitrosyl complexes targeting linear and bent fashioned M–NO shows 60–100 stretching frequency (cm^{-1}) difference of 60–100 cm^{-1} as signature of both the conformers. Some cobalt nitrosyl complexes exhibit structural isomerism wherein linear and bent nitrosyl isomers coexist in equilibrium. Several recent results supported by DFT calculations also suggest that among such functionalities the above assumptions are stable enhancing parameters. Also, various factors in addition to co-ligation, value of reduction zones and nature of central ion/atom justify the same. Table 2.5 demonstrates the examples of such complexes.

Table 2.5: Stretching frequency data of NO in some nitrosyl complexes of cobalt coexisting as M–NO angle isomers in equilibrium.

Compound	ν(NO) (linear) (cm^{-1})	ν(NO) (bent) (cm^{-1})	$\Delta\nu$ (cm^{-1})
$[Co(NO)(PPhMe_2)_2Cl_2]$	1,760	1,650	110
$[Co(NO)(PMe_3)_2Br_2]$	1,750	1,670	80
$[Co(NO)(PMe_3)_2Cl_2]$	1,750	1,655	95

It is observed that in transition metal dinitrosyl complexes that have both linear and bent M–N–O groupings, the stretching frequencies of the two ν(NO) are strongly separated. This can be firmly assigned to the bond stretching of the linear that have higher frequency, and bent which have lower frequency. Moreover, a significant difference of more than 150 cm^{-1} in the two ν(NO)s may also suggest the presence of linear and bent M–N–O groups. Notably, dinitrosyl complexes containing only linear M–N–O grouping exhibit two ν(NO) stretching frequencies having small difference of 30–60 cm^{-1}. These two cases are demonstrated with pertinent data shown in Table 2.6.

Table 2.6: Structural parameters in some of the dinitrosyl complexes.

Compound	ν(NO) (1) (cm^{-1})	ν(NO) (2) (cm^{-1})	$\Delta\nu$ (cm^{-1})	∠MNO (1)	∠MNO (2)	Δ∠MNO
M–N–O **groups: linear and bent**						
$[Ru(NO)_2(PPh_3)_3(Cl)]^+$	1,850	1,687	163	179.5	136	43.5
$[Os(NO)_2(PPh_3)_2(OH)]^+$	1,842	1,632	210	177.6	133.6	44.0

Table 2.6 (continued)

Compound	v(NO) (1) (cm^{-1})	v(NO) (2) (cm^{-1})	Δv (cm^{-1})	∠MNO (1)	∠MNO (2)	Δ∠MNO
M–N–O **groups: linear**						
$[Fe(NO)_2)Cl]_2(\mu\text{-dppe})$	1,786	1,724	62	169.6	165.8	3.8
$[Fe(NO)_2(dppe)]$	1,707	1,657	50	178	176	2.0
$[PPN][S_5Fe(NO)_2]$	1,739	1,695	44	172.8	165.9	6.9
$[PPN][(SPh)_2Fe(NO)_2]$	1,727	1,692	35	169.5	168.6	0.9

PPN, bis-triphenylphosphineiminium cation; dppe = 1,2-bis(diphenylphosphino) ethane.

Biologically important metal nitrosyls, especially haeme nitrosyls, have motivated the development of additional modern spectroscopic techniques for precise characterization of metal nitrosyls. For the five-coordinate ferrous haeme nitrosyls, vibrational spectra display very typical vibrational signatures of Fe–NO functionality. The ν(NO) stretching frequency is found detectable at 1,670–1,700 cm^{-1}. The metal bonded NO, that is, Fe–NO mode generally appears at 520–540 cm^{-1} to ensure the metal coordination. Moreover, it can be identified in a more clear way by resonance Raman spectroscopy using Soret excitation and nuclear resonance vibrational spectroscopy (NRVS). From the available data, the in-plane vibrational mode of M–N–O bent fashion is difficult to read out. However, employing NRVS confirms it with the indication of 370–390 cm^{-1} vibration.

Octahedral iron nitrosyl compounds having axial N-donor coordination display interesting spectral changes. The NO stretching frequency [ν(NO)], easily detected in IR spectroscopy, moves towards lower wavenumbers and appears in the range 1,610–1,640 cm^{-1}. On the other hand, the assignment of Fe–NO stretching and Fe–N–O bending modes has caused in a long-lasting dispute. Employing resonance Raman spectroscopy, only one low-energy isotope-sensitive band is detected typically around 550–570 cm^{-1} from the spectra. For the first time, Chottard and Mansuy reported this feature at 549 cm^{-1} for the NO adduct of deoxy-Hb and assigned it as ν(Fe–NO) [30]. This assignment was subsequently supported by other groups.

Recent spectral studies based on resonance Raman and, especially, NRVS finally detected a second vibrational mode connected with the Fe–N–O unit at approximately 440 cm^{-1} in both the six coordinate exemplary complex [Fe(TPP)(MI)(NO)] (TPP^{2-} = tetraphenylporphyrin dianion; MI = 1-methylimidazole) and related derivatives, and in M_b(II)–NO. A careful examination of earlier resonance Raman data on M_b(II)–NO and corresponding changes shows that this feature is, in fact, present in these data as a very weak band. These results express further doubt on the initial assignment of the 550–570 cm^{-1} Raman feature to the ν(Fe–NO). This issue was ultimately resolved using single-crystal NRVS experimental results on the model complex [^{57}Fe(TPP)(MI)(NO)].

Remarkably, observed NRVS results on powder samples of [^{57}Fe(TPP)(MI)(NO)] and the corresponding$^{15}N^{18}O$-labeled complex detected two isotope-sensitive features at 437 and 563 cm^{-1}. These move to 429 and 551 cm^{-1} in the $^{15}N^{18}O$-labeled complex. Comparable vibrational features are observed in the NRVS spectra of Mb (II)–NO at 443 and 547 cm^{-1}.

In ruthenium mononitrosyl complexes of the general composition $[Ru(NO)X_5]^{2-}$, where X = Cl, Br, I or CN, the trans effect of the NO was first time identified by Durig and his co-workers [30a] in 1966. In association with other structural details, the ν (Ru–X) stretching frequency for *trans* positioned NO grouping with respect to X functionality shows 30–40 cm^{-1} lower wavenumber than the *cis*-form of the same analogue. Similar findings have been reported for $[Ru(NO)(NH_3)_4X]^{n+}$ form of complexes. These results are presented in Table 2.7.

Table 2.7: Spectral data of *trans* and *cis* effect in ruthenium nitrosyl complexes.

S. no.	Complex	ν(Fe–NO) (cm^{-1})	ν(Fe–OH) (cm^{-1})
Trans effect			
1.	$K_2[Ru(NO)(CN)_5]$	634	–
2.	$[Ru(NH_3)_4(NO)(OH)]Cl_2$	628	–
3.	$[RuNH_3)_4(NO)(Cl)]Cl_2$	608	–
4.	$[RuNH_3)_4(Br)(NO)(]Br_2$	591	–
5.	$[Ru(NH_3)_4(I)(NO)(]I_2$	572	–
Cis effect			
6.	$Na_2[Ru(NO_2)_4(OH) NO)] \cdot 2H_2O$	638	568
7.	$[Ru(NH_3)_4(OH) (NO)]Cl_2$	628	565
8.	$Ag_2[Ru(Cl)_4(OH) (NO)]$	600	519

The M–NO mode of stretching vibrations in $[M(NO)X_5]^{2-}$ form of complexes (where M refers to Ru or Os) has been allocated by several research groups. The allocation for ν(M–NO) was ranging from 552 to 613 cm^{-1} when M = Ru and from 585 to 623 cm^{-1} when M is Os. In $[Co(NO)(NH_3)_5]^{2+}$ complexes, designations of ν(Co–NO) ranging from 573 to 644 cm^{-1} and δ(Co–N–O) at 578 cm^{-1} were given by Durig et al. [31] in 1967. Owing to ^{15}N substitution, the shifting of the ν(Co–NO) stretching vibrations towards lower frequency was noticed in these complexes. Table 2.8 displays results for $[M(NO)(CN)_5]^{n-}$ (M = Fe, Mn or Cr) reported by two research groups [32, 33].

Table 2.8: Vibrational assignment (cm^{-1}) for some cyanonitrosyl complexes.

S. no.	Compound	ν(NO)	ν(M–NO)	δ(M–N–O)
1.	$Na_2[Fe(NO)(CN)_5]\cdot 2H_2O$	1,935	658, 647	–
2.	$Cs_2[Fe(NO)(CN)_5]$	1,929, 1,910	663, 658, 651	–
3.	$[(CH_3)_4N]_2\,[Fe(NO)(CN)_5]$	1,908, 1,883	660, 650	–
4.	$K_3[Cr(NO)(CN)_5]$	1,630	616	–
5.	$K_3[Mn(NO)(CN)_5]$	1,700	655	–
6.	$[(CH_3)_4N]_3\,[Cr(NO)(CN)_5]$	1,625	616	
7.	$K_3[Cr(^{14}NO)(CN)_5]\cdot H_2O$	1,643	620	610
8.	$K_3[Cr(^{15}NO)(CN)_5]\cdot H_2O$	1,610	617	600

2.9.2 Electronic spectral studies

Data of electronic spectral studies are sparse. However, the electronic spectra of some metal nitrosyls have been reported, which are being presented in this section.

Based on the energy-level scheme (Figure 2.29) derived by Manoharan and Gray [34] for $[Fe(NO)(CN)_5]^{2-}$, the assignments of observed electronic transitions for a series of cyanonitrosyl complexes are shown in Table 2.9. The computed theoretical values of orbital energies in several forms of $[M(NO)(CN)_5]^{n-}$ share similarity with $[Fe(NO)(CN)_5]^{2-}$ compounds displaying identical levels, namely, 6e < $2b_2$ < 7e < $3b_1$ < $5a_1$ < 8e. However, in $[Cr(NO)(CN)_5]^{3-}$, $3b_1(x^2–y^2)$ and $5a_1(z^2)$ fall under the same energy, whereas for $[V(NO)(CN)_5]^{5-}$, $5a_1(z^2)$ level is less than $3b_1$ $(x^2–y^2)$.

Additional interesting dissimilarity in the complexes is found in the composition of the 6e and 7e energy levels. The population analysis data given in Table 2.10 disclose this behaviour. In general, the metal d_{xz}, d_{yz} character in the 6e level and the π*NO character in the 7e level reduces in going from $[Fe(NO)(CN)_5]^{2-}$ to $[V(NO)(CN)_5]^{5-}$. Particularly, for $[Mn(NO)\text{-}(CN)_5]^{3-,2-}$ and $[Fe(NO)(CN)_5]^{2-}$, the 6e level is mainly d_{xz}, d_{yz}, whereas it has mainly π*NO character in the $[V(NO)(CN)_5]^{5-}$ and $[Cr(NO)(CN)_5]^{3-}$ complexes. Similarly, the 7e level is mainly π*NO for $[Mn(NO)(CN)_5]^{3-,2-}$ and $[Fe(NO)(CN)_5]^{2-}$, and mainly d_{xz}, d_{yz} for $[V(NO)(CN)_5]^{5-}$. The $[Cr(NO)(CN)_5]^{3-}$ complex represents an intermediate case in which 7e level has nearly equal π*NO and d_{xz}, d_{yz} character. However, the former character is slightly predominating.

Raynor and co-workers [35] carried out the electronic spectral studies of the complexes, namely, $K_3[Cr(NO)(CN)_5]$, $[Cr(NO)(H_2O)_5]Cl_2$ and $[Cr(NO)(NH_3)_5]Cl_2$ in dimethylformamide (DMF). The results of their studies are also given in Table 2.9. Following the energy-level scheme worked out by these authors (Figure 2.30), the appropriate assignments of the observed maxima are also included in this table. A

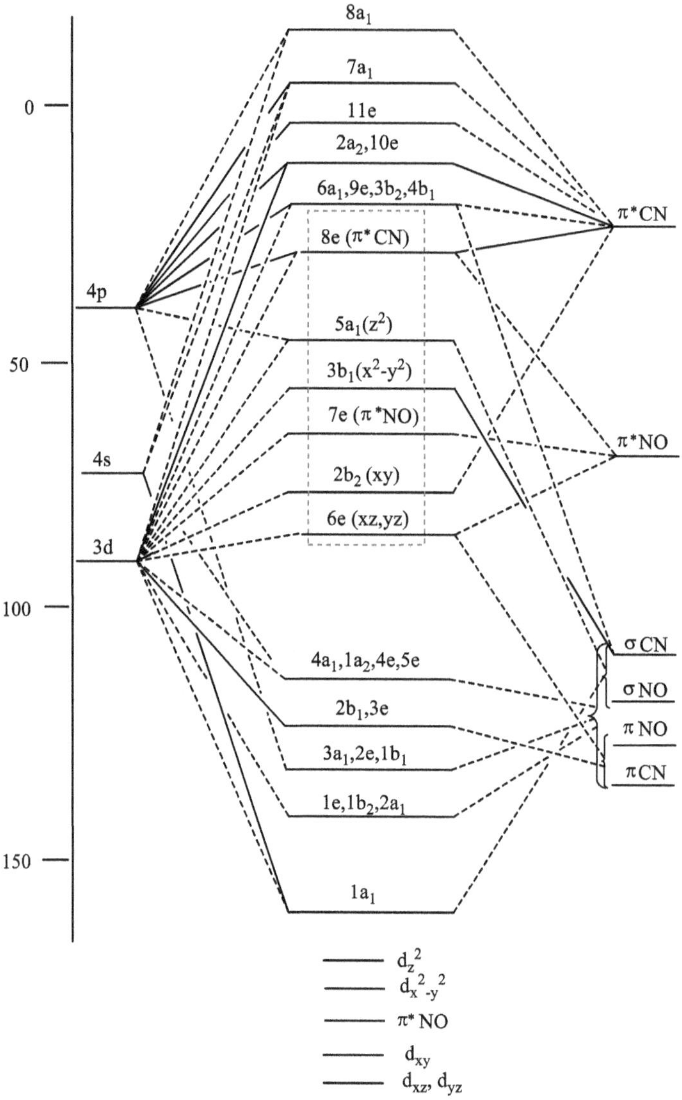

Figure 2.29: Proposed ordering of the d-levels in $[M(NO)(CN)_5]^{n-}$ (Manoharan and Gray's scheme).

reasonable energy-level scheme for the family of $[Cr(NO)(L)_5]^{2+}$ complexes seems to be 6e, $2b_2$, 7e (π*NO), $5a_1$, $3b_1$ in C_{4v} symmetry. This is analogous to that suggested by Manoharan and Gray for $[Cr(NO)(CN)_5]^{3-}$. However, the energy levels of $3b_1$ and $5a_1$ are reversed.

Table 2.9: Electronic spectral data of $[M(NO)(CN)_5]^{n-}$ and $[Cr(NO)(L)_5]Cl_2/ClO_4)_2$ (L = H_2O, C_2H_5OH or NH_3).

S. no.	Complex	Observed maxima (cm^{-1})	Band assignments
1.	$[Fe(NO)(CN)_5]^{2-}$	20,080; 25,380; 30,300; 37,800; 42,000; 50,000	$2b_2$→7e; 6e→7e; $2b_2$→$3b_1$; 6e→$5a_1$; 6e→$3b_1$; $2b_2$→8e
2.	$[Mn(NO)(CN)_5]^{3-}$	18,520; 24,690; 28,980; 37,850; 42, 550; 45,450	$2b_2$→7e; 6e→7e; $2b_2$→$3b_1$; 6e→$5a_1$; 6e→$3b_1$; $2b_2$→8e
3.	$[V(NO)(CN)_5]^{5-}$	12,900; 21,160; 32,470; 37,410	$2b_2$→7e; $2b_2$→$3b_1$; 6e→7e; $2b_2$→8e
4.	$[Cr(NO)(CN)_5]^{3-}$	13,700; 15,380; 22,200; 27,320; 37,300; 43,480	6e→$2b_2$; $2b_2$→7e; 6e→7e; $2b_2$→$3b_1$; 5e→$2b_2$; $2b_2$→8e
5.	$[Mn(NO)(CN)_5]^{2-}$	12,650; 18,600; 25,960; 28,570; 32,280; 37,630; 48,540	6e→$2b_2$; $2b_2$→7e; 6e→7e; $2b_2$→$3b_1$; 5e→$2b_2$; 6e→$5a_1$; $2b_2$→8e
6.	$[Cr(NO)(CN)_5]^{3-}$	13,700; 22,200; 27,320; 37,300; 43,480	6e→$2b_2$; 6e→7e; $2b_2$→$3b_1$; 5e→$2b_2$; $2b_2$→8e
7.	$[Cr(NO)(H_2O)_5]Cl_2$	17,700; 22,400; 30,300; 26,000; 40,500	6e→$2b_2$; 6e→7e; $2b_2$→$3b_1$; 5e→$2b_2$; $2b_2$→8e
8.	$[Cr(NO)(NH_3)_5]Cl_2$	17,300; 22,150; 33,200; 28,700; 41,000	6e→$2b_2$; 6e→7e; $2b_2$→$3b_1$; 5e→$2b_2$; $2b_2$→8e
9.	$[Cr(NO)(H_2O)_5](ClO_4)_2$	17,636; 22,371; 25,641; 30,864	No band assignments
10.	$[Cr(NO)(C_2H_5OH)_5]Cl_2$	17,543; 21,881; 30,769;	No band assignments
11.	$[Cr(NO)(NH_3)_5]Cl_2$	21,881; 28,735; 33,112; 38,910	No band assignments

Table 2.10: Population analysis of $2b_2$, 6e and 7e energy levels of $[M(NO)(CN)_5]^{n-}$ complexes [34].

M.O. level	Orbital	V $n = 5$	Cr $n = 3$	Mn $n = 2$	Mn $n = 3$	Fe $n = 2$
$2b_2$	π*CN	21.00	8.24	3.64	4.7	1.58
6e	*xz,yz*	20.48	38.05	52.19	46.63	60.71
	π*NO	73.68	50.18	32.80	42.22	24.79
7e	*xz,yz*	52.77	40.75	27.23	37.09	22.90
	π*NO	21.71	45.97	64.96	54.84	72.53

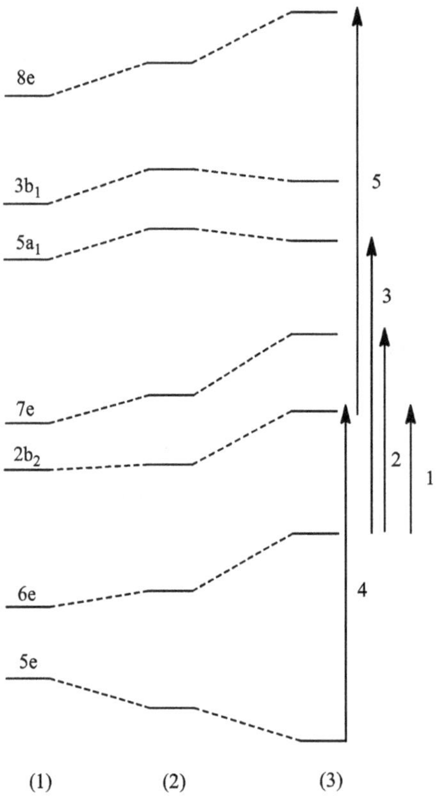

Figure 2.30: The relative order of energy levels in $[Cr(NO)(H_2O)_5]^{2+}$ (1) $[Cr(NO)(NH_3)_5]^{2+}$ (2) and $[Cr(NO)(CN)_5]^{3-}$ (3).

Table 2.9 also includes electronic spectral data for $[Cr(NO)(L)_5]^{2+}$ (L = H_2O, C_2H_5OH or NH_3) complexes reported by Griffith [36] in aqueous solution.

The solution formed by dissolving Fe^{2+} in 0.2 M acetate buffer at 25 °C and thereafter saturating with NO gas displays three characteristic absorption bands in its UV–Vis spectrum centred at 336 (ε = 440), 451 (ε = 265) and 585 nm (ε = 85 M^{-1} cm^{-1}) when recorded at room temperature anaerobically [37] (Figure 2.31). The appearance of three absorption bands was considered owing to the formation of the pentaaquamononitrosyliron(II) complex ion, $[Fe^{II}(NO)(H_2O)_5]^{2+}$. When an inert gas was passed in the solution slowly, the spectrum was completely reversed. The reversal of the spectrum shows the characteristic of the pentaaquairon(II) ion, $[Fe^{II}(H_2O)_5]^{2+}$. This happening demonstrates the excessive lability of the $[Fe^{II}(NO)(H_2O)_5]^{2+}$ complex.

The UV–Vis characteristic spectral bands of some nitrosyl complexes of different classes are given in Table 2.11.

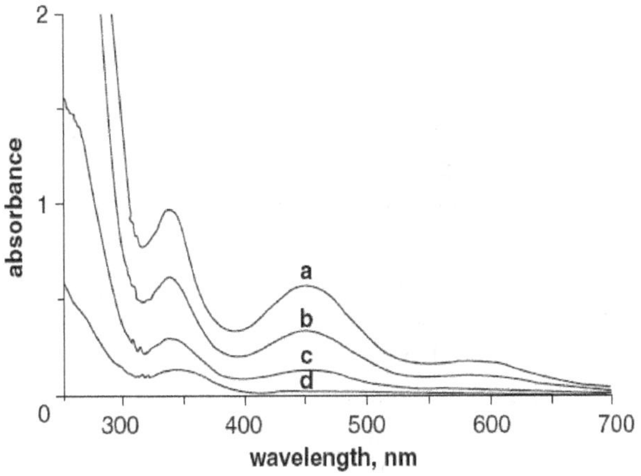

Figure 2.31: UV–Vis spectra of 3 × 10^{-3} M $[Fe^{II}(H_2O)_6]^{2+}$solution with NO under experimental conditions: 0.2 M acetate buffer, pH = 5.0, 25 °C; (a) $[Fe^{II}(H_2O)_6]^{2+}$ solution saturated with NO; (b) a + Ar (10s); (c) b + Ar (10s); (d) c + Ar (10s).

Table 2.11: UV–Vis spectral data of some nitrosyl complexes of different classes.

Type of the complex	Formula of metal nitrosyl in question	Absorption bands: nm/ε (M^{-1} cm^{-1})[a]
Mononitrosyl	$[(NO)Fe(H_2O)_5]^{2+}$	336/440, 451/265, 585/85 (water)
	$[(NO)Ru(NH_3)_5]^{3+}$	300/67.2, 460/14.4
	$[(NO)Fe(SC_9H_6N)_2]$	432/3,571, 512/2,967, 800/428 (THF)
	$[(NO)Fe(SPh)_3]^-$	500[b] (THF)
	$[(NO)Fe(SEt)_3]^-$	459[b] (THF)
	$[Et_4N][Fe(S_tBu)_3(NO)]^-$	370, 475 (THF)
Dinitrosyl	$[Fe(CS)_2(NO)_2]^-$	392/3,580, 603/299, 772/312 (water)
	$[(PhS)_2Fe(NO)_2]^-$	479, 798 (THF)
	$[(EtS)_2Fe(NO)_2]^-$	436, 802 (THF)
	$[PPN][(NO)_2Fe(S(CH_2)_3S)]$	430, 578, 807 (THF)
Dinitrosyl dimer	$[(NO)_4Fe_2(CS)_2]$	305, 362, 440 (sh), 755 (water)
Pseudo-haeme	[(NO)Fe(L)]	303/17,000, 403/4,000, 613/600
Haeme, 6C	[(NO)Fe(OEP)(SR)]	430, 536, 567
Haeme, 5C	$[(NO)Fe(To\text{-}F_2PP)]$	403, 475, 550

[a]Extinction coefficients (given where available);
[b]spectra recorded in the range 390–1,200 nm.

OEP, octaethylporphyrin; To-F_2PP, tetrakis-5, 10, 15, 20-(*o*-difluorophenyl) porphyrin; PPN, bis-triphenylphosphineiminium cation; CS, cysteamine; L, Schiff-base-tetradentate macrocyclic ligands.

Mir and Maurya [25] recently reported the preparation of three new molybdenumdinitrosyl complexes of the general composition $[Mo(NO)_2(L)(OH)]$, where L is *N*-(dehydroacetic acid)-4-aminoantipyrene (dha-aapH) (I), *N*-(4-acetylidene-3-methyl-1-phenyl-2-pyrazolin-5-one)-4-aminoantipyrine (amphp-aapH) (II) or *N*-(3-methyl-1-phenyl-4-propionylidene-2-pyrazolin-5-one)-4-aminoantipyrine (mphpp-aapH) (III) (Figure 2.32). Electronic spectra of these complexes were recorded taking 10^{-3} molar DMF solutions of these compounds. The detailed electronic spectral parameters of these compounds are given in Table 2.12. All the complexes under study display five transitions. The naming of spectral bands given in the table are based on M.O. diagram (Figure 2.33) applicable to hexa-coordinated dinitrosyl complexes reported by Gwost and Caulton [38] in 1974.

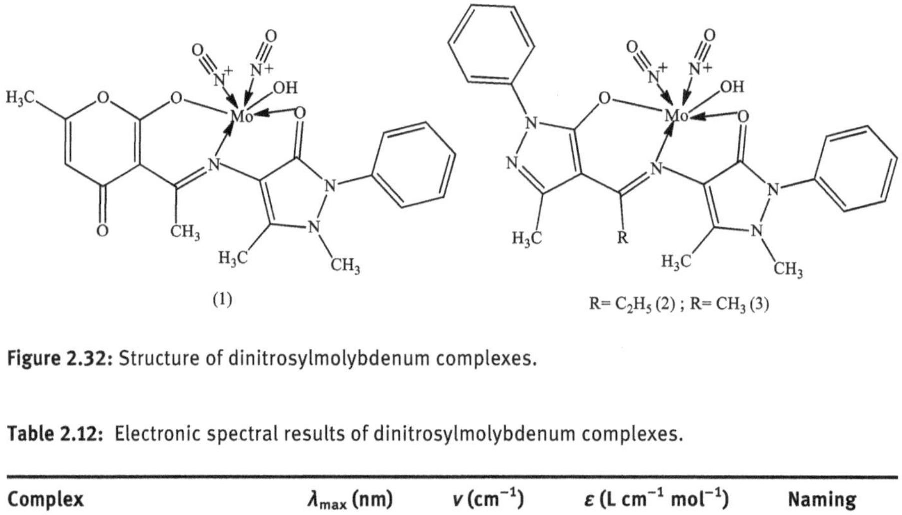

Figure 2.32: Structure of dinitrosylmolybdenum complexes.

Table 2.12: Electronic spectral results of dinitrosylmolybdenum complexes.

Complex	λ_{max} (nm)	ν (cm^{-1})	ε (L cm^{-1} mol^{-1})	Naming of the peaks
$[Mo(NO)_2$(dha-aap)(OH)]	274	36,496	3,722	$1b_2 \rightarrow 2a_2$
	296	33,783	4,312	$1b_2 \rightarrow 3a_1$
	325	30,769	4,582	$1b_2 \rightarrow 2b_2$
	349	28,653	4,821	$1b_2 \rightarrow 1b_1$
	440	22,797	1,942	$1b_2 \rightarrow 2a_1$
$[Mo(NO)_2$(amphp-aap)(OH)]	293	34,129	4,110	$1b_2 \rightarrow 2a_2$
	329	30,395	5,255	$1b_2 \rightarrow 3a_1$
	356	28,089	5,046	$1b_2 \rightarrow 2b_2$
	387	25,839	3,021	$1b_2 \rightarrow 1b_1$
	444	22,522	2,722	$1b_2 \rightarrow 2a_1$

Table 2.12 (continued)

Complex	λ_{max} (nm)	v (cm^{-1})	ε (L cm^{-1} mol^{-1})	Naming of the peaks
[Mo(NO)$_2$(mphpp-aap)(OH)]	289	33,898	4,411	$1b_2 \rightarrow 2a_2$
	313	31,645	4,666	$1b_2 \rightarrow 3a_1$
	339	28,490	5,141	$1b_2 \rightarrow 2b_2$
	357	27,397	5,054	$1b_2 \rightarrow 1b_1$
	416	21 881	3,480	$1b_2 \rightarrow 2a_1$

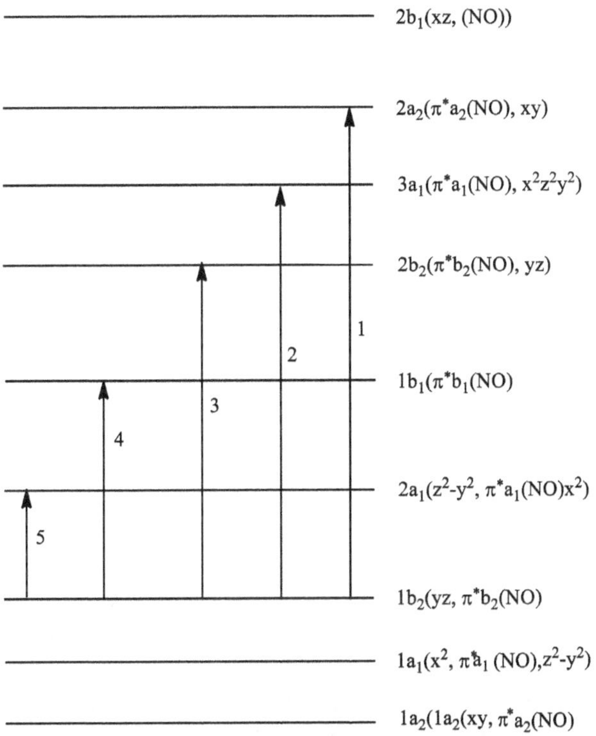

Figure 2.33: Representation of M.O.s of *cis*-dinitrosyl complexes molybdenum complex (O_h).

For exploring the theoretical electronic spectral analysis, the TD-DFT LANL2DZ/RB3LYP and 6-311G/RB3LYP were used by the authors for a representative ligand amphp-aap (II) and the complex[Mo(NO)$_2$(amphp-aap)(OH)]. The characteristics of electronic spectra of all the complexes carried out experimentally suggest the presence of an octahedral geometry around molybdenum that was correlated theoretically as well.

Maurya et al. [26] reported the preparation and analyses of three molybdenum dinitrosyl complexes, namely, [Mo(NO)$_2$(tsc-dha)(H$_2$O] (tsc-dhaH$_2$ = Schiff base derived from thiosemicarbazide and dehydroacetic acid (I), $[Mo(NO)_2(mtsc\text{-}dha)(H_2O]$ (mtsc-dhaH$_2$ = 4-methyl-3-thiosemi-carbazide and dehydroacetic acid (II) and $[Mo(NO)_2(ptsc\text{-}dha)(H_2O]$ (ptsc-dhaH$_2$ = 4-phenyl-3-thiosemicarbazide and dehydroacetic acid (III) (Figure 2.34).

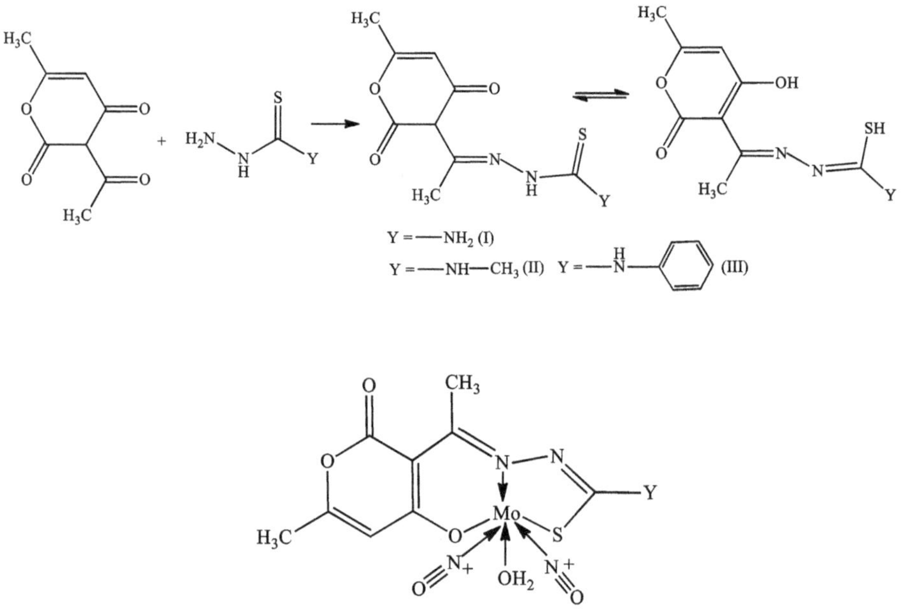

Figure 2.34: Structure of dinitrosylmolybdenum(0) complexes with Schiff bases.

Electronic spectra of the two representative compounds were recorded in 10^{-3} molar DMF solutions displaying five spectral peaks. The assignments of these spectral peaks are done using the M.O. diagram appropriate to hexa-coordinated dinitrosyl complexes as given in Figure 2.33.

Likewise, the preparation and analysis of a novel dicyanodinitrosyl molybdenum complex of the composition, $[Mo(NO)_2(CN)_2(nic)(H_2O)] \cdot 2H_2O$ (Figure 2.35), wherein "nic" is nicotine has been carried out by Maurya and Mir [39].

UV–Vis spectrum of the complex was recorded in 10^{-3} molar DMF solution (Figure 2.36). This spectrum displays five electronic transitions. The designations of these transitions have been made and are based on M.O. diagram pertinent to hexa-coordinated dinitrosyl complexes given in Figure 2.33.

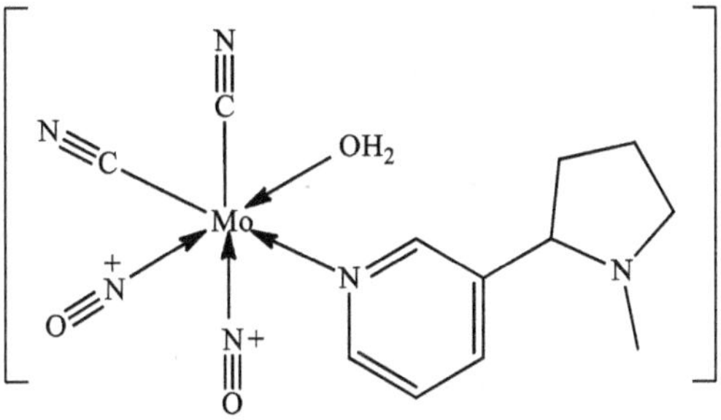

Figure 2.35: Structure of dinitrosylmolybdenum(0) complex with nicotine.

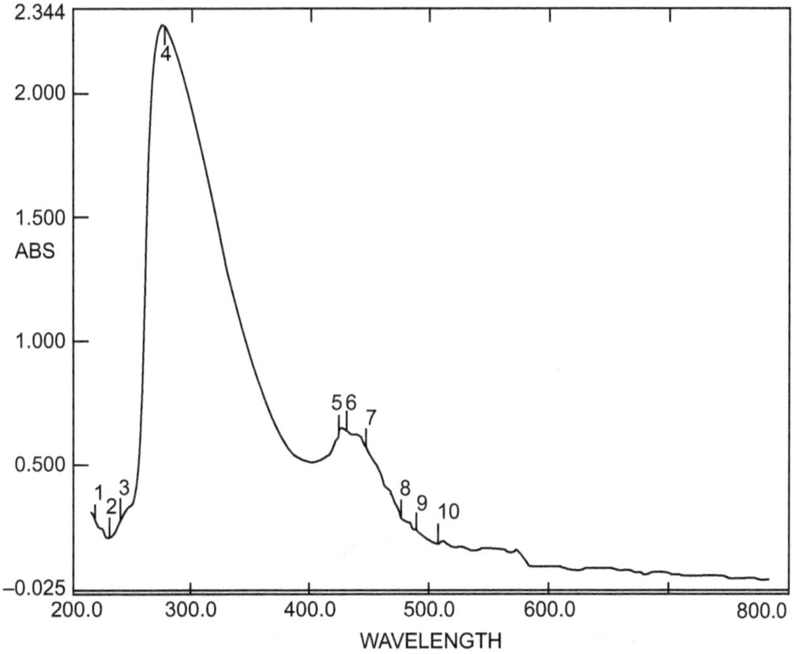

Figure 2.36: UV–Vis spectrum of $[Mo(NO)_2(CN)_2(nic)(H_2O)] \cdot 2H_2O$.

2.9.3 Magnetic properties

It is well known that studies on magnetic behaviour of nitrosyl complexes involve the measurement of magnetic moments. This opens the way of knowing the number of unpaired electron(s) and stereochemistry of complexes under investigations. Although a lot of papers can be seen in the existing literature, wherein the magnetic studies have been reported along with others. However, only few of them are chosen and the magnetic data reported therein, are presented in Table 2.13 in order to discuss the results in the light of bonding schemes developed by Manoharan and Gray [34] (Figure 2.29) and Enemark and Feltham [21] (Figure 2.37).

Table 2.13: Magnetic data of some metal nitrosyl complexes along with M.O. electronic configurations.

S. no.	Complex	Electron configuration $\{M(NO)_x\}^n$	M.O. configuration Enemark	Gray et al.	Expected μ(B.M.)**	Observed μ(B.M.)
1.	$K_3[V(NO)(CN)_5]\cdot 2H_2O$	$\{VNO\}^4$	$(2e)^4(1b_2)^0$	$(6e)^4(2b_2)^0$	0	D
2.	$K_4[V(NO)(CN)_6]\cdot 2H_2O$	$\{VNO\}^4$	–	–	0	D
3.	$[V(NO)_2(CN)_2(dipy)]\cdot H_2O$	$\{V(NO)\}^5$	$(1a_2)^2(1a_1)^2(1b_2)^{1***}$	–	1.73	1.5
4.	$[V(NO)_2(CN)_2(o\text{-phen})]\cdot H_2O$	$\{V(NO)\}^5$	$(1a_2)^2(1a_1)^2(1b_2)^{1***}$	–	1.73	1.6
5.	$K_3[Cr(NO)(CN)_5]\cdot H_2O$	$\{CrNO\}^5$	$(2e)^4(1b_2)^1$	$(6e)^4(2b_2)^1$	1.73	1.87
6.	$K_4[Cr(NO)(CN)_5]\cdot 2H_2O$	$\{CrNO\}^6$	$(2e)^4(1b_2)^2$	$(6e)^4(2b_2)^2$	0	D
7.	$(Me_4N)_2[Cr(NO)(CN)_4]\cdot H_2O$	$\{CrNO\}^5$	$(2e)^4(1b_2)^1$	$(6e)^4(2b_2)^1$	1.73	1.7
8.	$[Cr(NO)(CN)_2(dipy)]$	$\{CrNO\}^5$	$(2e)^4(1b_2)^1$	–	1.73	1.60
9.	$[Cr(NO)(CN)_2(o\text{-phen})]$	$\{CrNO\}^5$	$(2e)^4(1b_2)^1$	–	1.73	1.67
10.	$K_3[Mn(NO)(CN)_5]\cdot H_2O$	$\{MnNO\}^6$	$(2e)^4(1b_2)^2$	$(6e)^4(2b_2)^2$	0	D
11.	$K_2[Mn(NO)(CN)_5]$	$\{MnNO\}^5$	$(2e)^4(1b_2)^1$	$(6e)^4(2b_2)^1$	1.73	1.73
12.	$[Mn(NO)(CN)_3(dipy)]^*$	$\{MnNO\}^5$	$(2e)^4(1b_2)^1$	$(6e)^4(2b_2)^1$	1.73	3.9
13.	$(PPh_4)_2[Mn(NO)_2(CN)_4]\cdot H_2O$	$\{Mn(NO)_2\}^7$	$(1a_2)^2(1a_1)^2(1b_2)^2 1(b_1)$ 1***	–	1.73	1.9
14.	$Na_2[Fe(NO)(CN)_5]\cdot 2H_2O$	$\{FeNO\}^6$	$(2e)^4(1b_2)^2$	$(6e)^4(2b_2)^2$	0	D
15.	$(Et_4N)_2[Fe(NO)(CN)_4]$	$\{FeNO\}^7$	–	–	1.73	1.75
16.	$K_4[Mo(NO)(CN)_5]$	$\{MoNO\}^6$	$(2e)^4(1b_2)^2$	$(6e)^4(2b_2)^2$	0	D
17.	$(PPh_4)_3[Mo(NO)_2(CN)_4]$	$\{MoNO\}^5$	$(2e)^4(1b_2)^1$	$(6e)^4(2b_2)^1$	1.73	1.96
18.	$Cs_2[Mo(NO)(Cl)_4(H_2O)]$	$\{MoNO\}^5$	$(2e)^4(1b_2)^1$	$(6e)^4(2b_2)^1$	1.73	1.77

Table 2.13 (continued)

S. no.	Complex	Electron configuration $\{M(NO)_x\}^n$	M.O. configuration Enemark	Gray et al.	Expected μ(B.M.)**	Observed μ(B.M.)
19.	$(PPh_4)_3[Mo(NO)(Cl)_4]$	$\{MoNO\}^6$	$(2e)^4(1b_2)^2$	$(6e)^4(2b_2)^2$	0	D
20.	$(PPh_4)_2[Mo(NO)(CN)_5]\cdot 2H_2O$	$\{MoNO\}^4$	$(2e)^4(1b_2)^0$	$(6e)^4(2b_2)^0$	0	D
21.	$[Mo(NO)(CN)_3(dipy)]\cdot H_2O$	$\{MoNO\}^4$	$(2e)^4(1b_2)^0$	$(6e)^4(2b_2)^0$	0	D
22.	$(PPh_4)_2[Mo(NO)_2(CN)_4]\cdot 2H_2O$	$\{Mo(NO)_2\}^6$	$(1a_2)^2(1a_1)^2(1b_2)^2$***	–	0	D
23.	$[[Mo(NO)_2(CN)_2(o\text{-phen})]]\cdot$	$\{Mo(NO)_2\}^6$	$(1a_2)^2(1a_1)^2(1b_2)^2$***	–	0	D
24.	$(PPh_4)_3[Cr(NO)(NCS)_5]$	$\{CrNO\}^5$	$(2e)^4(1b_2)^1$	$(6e)^4(2b_2)^1$	1.73	2.23
25.	$(PPh_4)_2[Os(NO)(NCS)_5]$	$\{OsNO\}^6$	$(2e)^4(1b_2)^2$	$(6e)^4(2b_2)^2$	0	D

* Ref. [41].

** Spin-only value; D, diamagnetic

*** Based on M.O. diagram reported by J. H. Enemark and R. D. Feltham, in *Topics in Inorganic and Organometallic Stereochemistry*, ed. G. Geoffroy, Wiley, New York, 1981, vol. 12, pp. 152–215.

In the light of the energy-level diagrams developed by Gray et al. (Figure 2.29) and Enemark and Feltham (Figure 2.37), the experimental magnetic moments are fairly expected. Considering the metal and nitrosyl groups together (because NO being known paramagnetic have an unpaired electron), the total number of M.O. configuration-based electrons referred to this group (based on Enemark and Gray's schemes) has been given in Table 2.13. As a result of various physicochemical analyses, a strong covalent nature of M–NO is established. The binding fashion adopted thus has been schematically shown in Figures 2.29 and 2.37 reflecting the nature of this interaction.

Although cyanide group in the cyanonitrosyl complexes mostly behave as a good π-bonding ligand, the strong interaction between the metal ion and the nitrosyl group is quite enough to influence the separation of bonding, non-bonding and antibonding orbitals to a large extent. We are aware that in the hexa-coordinated complexes, the maximum symmetry expected is C_{4v}. This leads to a strong tetragonal distortion. Thus, the sequence of these orbitals according to Figures 2.29 and 2.37 is $6e < 2b_2 < 7e$, and $2e < 1b_2 < 3e$, respectively. Obviously, the bonding orbitals below 6e/2e in the respective M.O. diagram will be occupied by the ligating

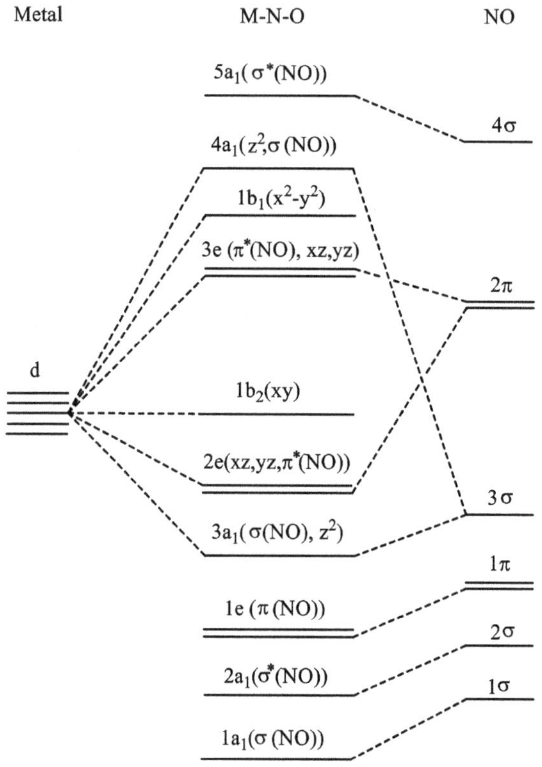

Figure 2.37: The molecular orbital diagram for six-coordinate complexes involving M–N–O grouping linear. The M–N–O bond passes through the *z*-axis.

electrons. Thus, electronic fillings start from the metal d-orbitals using 6e/2e level. Such a rationalization has been followed in the discussion to adopt bonding scheme to consider NO^+ as a nitrosylating agent, rather a ligand.

Based on the concept of "Inorganic Functional Group" introduced by Enemark et al. (already discussed in previous section), the nitrosyl complexes may have the $\{M(NO)_x\}^n$ core formula, where n is 4, 5, 6 or 7, counting directly the involved number of electrons belonging to d-orbitals. Therefore, $n = 4$ or 6 pertaining complexes are expected as diamagnetic materials and whereas for $n = 5$ or 7 paramagnetic behaviour for unpaired electron gets established. The determined magnetic moment measurements of some model complexes connected with these assumptions are given in Table 2.13.

It is notable here that for the complexes having d^5/d^7 configurations, the resultant spin–orbit coupling mainly contributes in the determination of magnetic moment values. The spin–orbit coupling constants for Mo(I) core is −450, for Mn(0)/Mn(II) is −300 and for Cr(I) −190 cm^{-1}. The magnetic moment mainly depends on these coupling constants as compared to the respective spin-only value at ambient temperature. The experimental results for the complexes 5(Cr), 13 (Mn), 17 (Mo) and 24 (Cr) given in Table 2.13 follow this trend. However, the bulk susceptibility measurement on a solid sample of the complex 11, $K_2[Mn(NO)(CN)_5]$, gives μ_{eff} = 1.73 B.M. This indicates $S = ½$ with no abnormal behaviour. The low magnetic moments, 1.60 and 1.67 B.M. for compounds 8 and 9, respectively, have been explained by Maurya [40] on the basis of the polymeric nature of the complexes.

The magnetic moments (3.9 B.M.) of compound 12 is quite interesting and looks unusual for $\{MnNO\}^5$ electron configuration. This is expected insofar as the molecular symmetry species in both the structures (a) and (b) shown will not be greater than C_s and hence its expected M.O. diagram [41] with the valance electron occupancy is as shown in Figure 2.38.

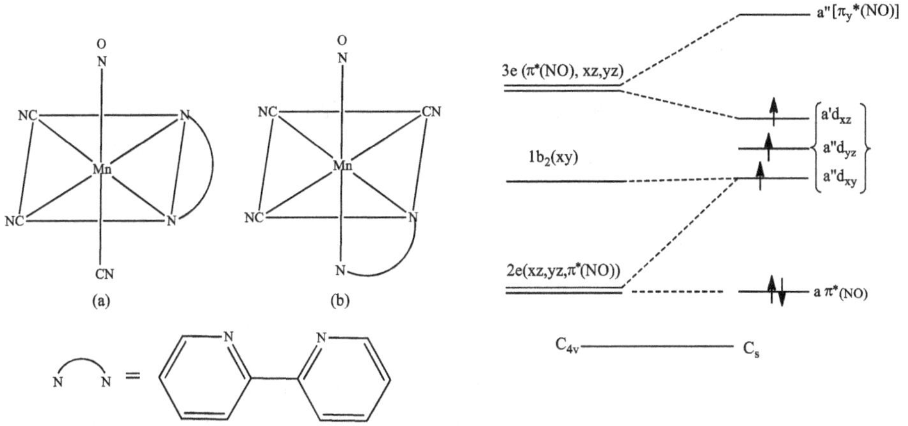

Figure 2.38: Expected M.O. diagram for $[Mn(NO)(CN)_3(dipy)]$ in C_s symmetry.

Synthesis and characterization of some dinitrosylmolybdenum(0) complexes have been recently reported by Maurya et al. [25, 26, 39]. Their magnetic properties along with other details are given in Table 2.14.

2.9.4 Electron spin resonance (ESR) studies

Among nitrosyls complexes, the general quest to identify coordinated NO groups on preliminary functional identification tests mainly involves the use of some colour

Table 2.14: Magnetic properties of some dinitrosylmolybdenum(0) complexes.

Complex	Electronic configuration	Electronic structure	Expected μ (B.M.)	Observed magnetic property
$[Mo(NO)_2(CN)_2(nic)(H_2O)]\cdot 2H_2O$[a]	$\{Mo(NO)_2\}^6$	$(1a_2)^2(1a_1)^2(1b_2)^2$	0	Diamagnetic
$[Mo(NO)_2(tsc\text{-}dha)(H_2O]$[b]	$\{Mo(NO)_2\}^6$	$(1a_2)^2(1a_1)^2(1b_2)^2$	0	Diamagnetic
$[Mo(NO)_2(mtsc\text{-}dha)(H_2O]$[b]	$\{Mo(NO)_2\}^6$	$(1a_2)^2(1a_1)^2(1b_2)^2$	0	Diamagnetic
$[Mo(NO)_2(ptsc\text{-}dha)(H_2O]$[b]	$\{Mo(NO)_2\}^6$	$(1a_2)^2(1a_1)^2(1b_2)^2$	0	Diamagnetic
$[Mo(NO)_2(dha\text{-}aap)(OH)]$[c]	$\{Mo(NO)_2\}^6$	$(1a_2)^2(1a_1)^2(1b_2)^2$	0	Diamagnetic
$[Mo(NO)_2(amphp\text{-}aap)(OH)]$[c]	$\{Mo(NO)_2\}^6$	$(1a_2)^2(1a_1)^2(1b_2)^2$	0	Diamagnetic
$[Mo(NO)_2(mphpp\text{-}aap)(OH)]$[c]	$\{Mo(NO)_2\}^6$	$(1a_2)^2(1a_1)^2(1b_2)^2$	0	Diamagnetic

[a]Ref [39].
[b]Ref [26].
[c]Ref. [25].

changing reagents. But as per the main question of identifying paramagnetic behaviour of ESR active core, the respective ESR spectroscopy can be used for gaining deeper insights in metal nitrosyls. Irrespective of metallic core whether ESR active or inactive, but the coordinated NO group/s due to having unpaired electron could be detected by this technique. Notably, frozen solution of free NO generates a signal having $g \approx 1.95$. It can be detected only at very low temperatures below 20 K. It also requires high concentrations of NO [42].

The ESR data for some reported mononitrosyl complexes are given in Table 2.15.

Several research groups have examined the ESR spectrum of pentacyanonitrosylchromate(I) anion, $[Cr(NO)(CN)_5]^{3-}$. There is no complete parity on the interpretation of the ESR data for this complex. However, the important experimental results can be presented as follows: (i) Both the g tensor and chromium hyperfine tensor have axial symmetry. (ii) The nitrogen hyperfine tensor has near-axial symmetry. The maximum deviation of 9° from the axial symmetry is analogous to the observed deviation of 4.4° from linearity of the CrNO group. (iii) $g_{\perp} > g_{\parallel}$ (iv) The nitrogen hyperfine tensor is highly anisotropic.

In octahedral complexes of the type $[Cr(NO)(L)_5]$ [43, 44], ESR data are found as per expectation. Moreover, these data depend on the electronegativity of the co-ligand L.

Table 2.15: ESR data of some nitrosyl complexes.

Complex	g_{av}	$g_{\perp}$	$g_{\parallel}$	$A_{\perp}$ (^{14}N) (G)	$A_{\parallel}$ (^{14}N) (G)	A^{*}_{iso}/ A_{avg} (^{14}N) (G)	A_{iso}/ (^{15}N) (G)	$A_{\perp}$ (^{53}Cr or ^{55}Mn) (G)	$A_{\parallel}$ (^{53}Cr or ^{55}Mn) (G)
1. $[Cr(NO)(H_2O)_5]^{2+}$ (aqu. solution)	1.966	–	–	–	–	6.44	–	–	–
2. $[Cr(NO)(H_2O)_5]^{2+}$ (rgd. aq. ol., 113°K)	1.966	1.991	1.916	–	–	–	–	–	–
3. $[Cr(NO)(NH_3)_5]^{2+}$ (aqu. solution)	1.980	–	–	–	–	–	–	–	–
4. $[Cr(^{14}NO)(CN)_5]^{3-}$ (aqu. solution)	1.9945	–	–	–	–	5.26	–	–	–
5. $[Cr(^{14}NO)(CN)_5]^{3-}$ (dil. single crystal)	1.9945	2.0052	1.9475	–	–	5.26	–	–	–
6. $[Cr(^{15}NO)(CN)_5]^{3-}$ (10^{-3} M aq. KCN sol.)	1.994	–	–	–	–	–	7.35	–	–
7. $[Fe(NO)(CN)_5]^{3-}$ (rigid sol., 77°K)	2.0231	2.0313	2.0059	14.75	17.10	15.53*	–	–	–
8. $[Mn(NO)(CN)_5]^{2-}$ (rigid sol., 77°K)	2.0144	2.0279	1.9873	<1.9	<1.9	<1.9*	–	34.0	164.2
9. $[Mn(NO)(CN)_5]^{2-}$ (single crystal)	2.0181	2.0311	1.9922	4.75	1.91	3.80	–	36.6	159.98
10. $Cs_2[Mo(NO)(Cl)_4$-$(H_2O)]$ (in solid phase and 6 M HCl sol.)	1.95	1.99	1.97	–	–	–	–	–	–
11. $[Mo(NO)(o\text{-phen})_2Cl]$ Cl (in solid phase and 6 M HCl sol.)	1.97	1.98	1.97	–	–	–	–	–	–
12. $(PPh_4)_3[Cr(NO)$-$(NCS)_5]$	1.970	–	–	–	–	–	–	–	–
13. $(PPh_4)_3[Mo(NO)$-$(CN)_5]$	2.004	–	–	–	–	–	–	–	–
14. $[Cr(NO)(CN)_2$-$(BUMPP)_2(H_2O)$ (powdered sample)	1.984	–	–	–	–	–	–	–	–
15. $[Cr(NO)(CN)_2$-$(CAMPP)_2(H_2O)$ (powdered sample)	1.982	–	–	–	–	–	–	–	–

Table 2.15 (continued)

Complex	g_{av}	$g_{\perp}$	$g_{\parallel}$	$A_{\perp}$ (^{14}N) (G)	$A_{\parallel}$ (^{14}N) (G)	A^{*}_{iso}/ A_{avg} (^{14}N) (G)	A_{iso}/ (^{15}N) (G)	$A_{\perp}$ (^{53}Cr or ^{55}Mn) (G)	$A_{\parallel}$ (^{53}Cr or ^{55}Mn) (G)
16. $[Cr(NO)(CN)_2$-$(CBMPP)_2(H_2O)$ (powdered sample)	1.985	–	–	–	–	–	–	–	–
17. $[Cr(NO)(CN)_2$-$(ACPMP)_2(H_2O)$ (powdered sample)	1.987	–	–	–	–	–	–	–	–

BUMPP, 4-butyryl-3-methyl-l-phenyl-2-pyrazoline-5-one; CAMPP, 4-chloroacetyl-3-methyl-l-phenyl-2-pyrazoline-5-one; CBMPP, 4-*p*-chlorobenzoyl-3-methyl-l-phenyl-2-pyrazoline-5-one; ACPMP, 4-acetyl-1-(3′-chlorophenyl)-3-methyl-2-pyrazoline-5-one taken from R. C. Maurya and D. D. Mishra, *Synth. React. Inorg. Met.-Org. Chem.*, 21 (1991)845–857.

Therefore, the value of g_{av} reaches close to the assigned value of a free electron in relevance to the delocalization phenomenon. This reflects in ν(NO) too.

Making use of ESR method, Raynor and co-workers [45] have demonstrated that acid hydrolysis of pentacyanonitrosylchromate(I) anion displays successive replacement of cyanide group by water.

Sometimes, ESR spectroscopic investigation can be used to envisage the stereochemical aspects of a complex undoubtedly. The conformational applications in this context was illustrated for $[Mo(NO)(CN)_5]^{3-}$ at times when the isolated form of this complex was not available [46]. The ESR parameters acquired for the oxidized species $[Mo(NO)(CN)_5]^{3-}$ in the host of $K_3[Co(CN)_6]$ substantiate the formation of the parent diamagnetic (d^6) complex as $[Mo(NO)(CN)_5]^{4-}$ instead of $[Mo(NO)(CN)_5(OH)_2]^{4-}$. It was later on confirmed by X-ray studies.

Manoharan and Gray [47] and Burgess et al. [45] carried out the detailed ESR studies on $[Mn(NO)(CN)_5]^{2-}$. The interest behind this study was to highlight the ordering of energy levels in the corresponding Cr(I) and Fe(II) complexes. There is some disparity about the spin density in the NO group. When one group accepts this due to spin–orbit coupling, the other group feels that the spin polarization mechanism [48] is more probable for this.

For complexes 10 and 11, the g-tensor values observed are in the order $g_{\perp}>g_{\parallel}$. This suggests that the unpaired electron must be in the d_{xy} orbital. The observed g_{av} values for the complexes 12–17 given in Table 2.15 are coherent with low-spin d^5 configuration of Mo(I) and Cr(I).

Paramagnetic nitrosyl complexes of iron have been shown to be easily studied using ESR technique. Therefore, for biological sample analysis to identify and quantify the presence of NO, the technique can be used. Biological iron samples in NO environment are generally believed as low-spin complexes with $S = 1/2$. These types of complexes show ESR activity in the form of a narrow and discrete signals collected at ambient temperature having g values falling in the range between 2.012 and 2.04 [49].

In several experimentally approachable exemplary compounds, the total numbers of d electrons available in iron with addition to π*NO electrons are designated by notations $\{FeNO\}^{6-8}$. The EPR activity is shown by the complexes following $\{FeNO\}^7$ and $\{Fe(NO)_2\}^9$ notations, and the known system $S = 3/2$ along with $S = ½$ are the commonly known for $\{Fe{-}NO\}^7$ cores. Fe–NO complexes having $S = 1/2$ with haeme iron are well known and reported by several research groups. On the other hand, the low-spin non-haeme iron nitrosyl complexes are infrequently available. Importantly, the ground state with $S = 1/2$ is characteristic for relatively strong ligand field systems.

It has been observed that in some of the non-haeme metalloproteins involving Fe(II) centres interact with NO reversibly, and form nitrosyl complexes of iron having $S = 3/2$. These iron nitrosyl complexes are characterized by EPR with the g values around [50] 4.0 and 2.0.

Joannou and Cui [51] reported that the EPR signal at $g_{\perp} = 1.999$ and $g_{||} = 1.927$ with resolved ^{14}N hyperfine splitting has been observed for an intermediate $Fe^{I}NO^{+}$ complex during the reduction of nitroprusside by thiols. Reaction of nitroprusside with a thiol such as cysteine (Figure 2.39) forming single-electron reduction resulting in the formation of $[Fe(CN)_5NO]^{3-}$ as a paramagnetic form due to NO-centred unpaired electron, and with the respective EPR results showing $g_x = 1.9993$, $g_z = 2.008$ and $g_y = 1.9282$.

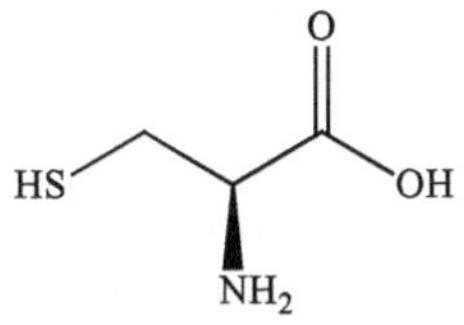

Figure 2.39: Chemical structure of cysteine.

Additional reaction in $[Fe(CN)_5NO]^{3-}$ involves internal electron transfer and the formation of an $Fe^{I}NO^{+}$ complex $[Fe(CN)_4NO]^{2-}$. The experimentally observed EPR parameters for the new complex so obtained are $g_z = 2.0054$, $g_x = 2.036$ and $g_y = 2.0325$.

2.9.5 Nuclear magnetic resonance (NMR) spectral studies

NMR spectroscopy principally rests on the use of ^{14}N and ^{15}N NMR studies of diamagnetic complexes. This is why the technique is quite limited in its use. It is noteworthy that ^{14}N nucleus is quadrupolar (I = 1) and has a high natural abundance but small relaxation times. Contrary to this, ^{15}N (I = 1/2) has a very low natural abundance but long relaxation times. Having dissimilar properties, both nuclei have their own importance in NMR spectroscopy. Notably, most commonly available data on NMR are based on ^{15}N studies on metal nitrosyl complexes [52]. In view of the above, most of N-NMR studies have been carried out on both synthetic and natural metalloporphyrins. Additionally, N-NMR studies have also been made on cobalt and ruthenium nitrosyl complexes. Certain ^{59}Co measurements have also been reported in the literature on NMR studies.

In general, the ^{15}N chemical shifts (δ_N) vary over a wide range in ppm unit. This also happens in metal nitrosyls that have linear M–N–O groupings. Ordinarily, it has increasing trend across the transition metal series, and also down in a particular group. In metal nitrosyl complexes having bent M–N–O species, normally ^{15}N chemical shifts are significantly more deshielded (with respect to nitromethane) compared to the linear nitrosyl complexes. As per the available reports, δ_N varies from around 950 to 350 ppm in metal nitrosyls with bent M–N–O groupings. On the other hand, chemical shift varies from about 200 to 10 ppm in nitrosyl complexes involving bent M–N–O groupings.

Also, shield tensors are known to give some substantial information for metal nitrosyl complexes. For instance, a technique called cross-polarized magic angle spinning provides information on fluxionality between linear and bent M–N–O groupings in solid metal nitrosyls. Using the same technique, one can get information on fluxional behaviour of metal nitrosyls in solution also [52]. Moreover, shielding tensor analysis in transition metal nitrosyl complexes having bent M–N–O arrangement, namely, [Co(NO)(tpp)], provides information on the swinging or spinning of the NO group over the face of the porphyrin ring.

^{15}N NMR studies in the explanation of linear versus bent dinitrosyls of $[Ru(NO)_2Cl(PPh_3)_2]^+$ form do not show exchange effect in the respective solid state. However, in solution phase, it was found that the complex under square-pyramidal geometry isomer with linear and bent M-NO attains an equilibrium with respect to trigonal-bipyramidal conformation bearing linear M–NO functionality as indicated by Figure 2.40. The corresponding osmium nitrosyl exists as a solid-state stable isomer with $\angle$M–N–O of 169° and 171°. Both forms actually show a rapid fluxionality making linear versus bent identity indistinctive in NMR scale [53].

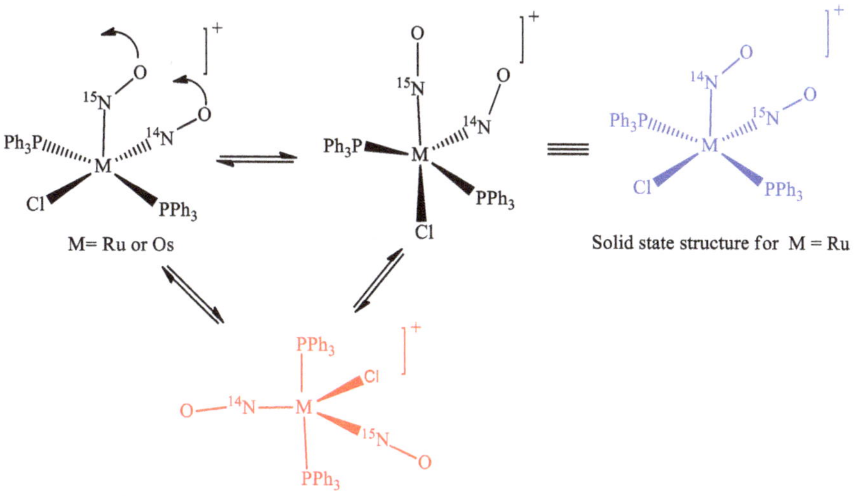

Figure 2.40: Suggested dynamic processes to explain the ^{15}N NMR recorded results of dinitrosyl complexes $[MCl(NO)_2(PPh_3)_2]^+$ (M = Ru or Os).

2.9.6 X-ray photoelectron spectroscopy or ESCA studies

The electronic distribution in the M–N–O group in nitrosyl complexes lies in between two extreme labels NO^+ and NO^-. The X-ray photoelectron spectroscopy may be used to throw light in distinguishing these two extreme labels of bonding in nitrosyl complexes by studying the binding energies of the electrons of the atoms in these complexes. The difference in the binding energies for a particular atom present in a functional group when free and coordinated may reflect on the change of electron distribution of the atom on coordination. A positive change conforms to the effective +ve charge of an atom in the complex compound, whereas the negative move conforms to the effective –ve charge.

The N_{1s} binding energy for coordinated cyano groups in compounds 4–10 and coordinated nitrosyl group of the nitrosyl complexes along with N_{1s} value for NO^+ in $NO^+ClO_4^-$ and for neutral NO molecule is presented in Table 2.16. Two valuable information can be drawn from the ESCA data:

(i) While the N_{1s} electron binding energy (E_b) value for the NO ligand varies from one complex to another, the N_{1s} electron binding energy (E_b) value for the CN ligand is practically unchanged. This means that the degree of π-bonding within the M–C–N group is changed very little compared to the more pronounced change in the degree of π-bonding within the M–N–O group. This result also correlates a constant value of C–N bond distance and a pronounced variation in N–O bond distance in cyanonitrosyl complexes (vide section bonding in nitrosyl complexes under six coordinate complexes).

Table 2.16: ESCA data for $NO^+ClO_4^-$, NO, NO^- and some nitrosyl complexes.

S. no.	Species	E_b N_{1s} (NO) (eV)	E_b N_{1s} (CN) (eV)	Reference
1.	$NO^+ClO_4^-$	409.0	–	[53a]
2	NO (gas)	406.0	–	[53a]
3.	NO^-	402.0*	–	[53a]
4.	$Na_2[Fe(NO)(CN)_5]\cdot 2H_2O$	403.9	398.7	[53a, 53b]
5.	$Fe[Fe(NO)(CN)_5]$	402.6	–	[53a]
6.	$Zn[Mn(NO)(CN)_5]\cdot H_2O$	402.1	398.3	[53b]
7.	$K_3[Cr(NO)(CN)_5]\cdot H_2O$	401.4	399.0	[53b]
8.	$K_3[Mn(NO)(CN)_5]\cdot 2H_2O$	401.6	398.3	[53b]
9.	$K_4[Mo(NO)(CN)_5]$	401.1	398.8	[53b]
10.	$K_3[V(NO)(CN)_5]$	400.6	398.8	[53b]
11.	$[Mo(NO)_2(Cl)_2(PPh_3)_2]$	401.6	–	[53c]
12.	$[Mo(NO)_2(Et_2\text{-}DTC)_2]$	400.0	–	[53d]
13.	$[Mo(NO)_2(Cl)_2(diars)]$	401.6	–	[53d]
14.	$(Ph_4P)_2[Mo(NO)(Cl)_4]$	401.0	–	[53e]
15.	$Cs_2[Mo(NO)(Cl)_4(H_2O)]$	401.0	–	[53e]

* Extrapolated value.

(ii) With the possible exception of $[Fe(NO)(CN)_5]^{2-}$, the N_{1s} binding energies for all other nitrosyl complexes come lower to that of NO^- and even for the nitroprusside, the value is lower than that of neutral NO. This strongly suggests that all nitrosyl complexes mentioned in Table 2.16 may be considered as containing NO^-. However, for allocating the oxidation state of the metal atom in a nitrosyl complex, binding energies of a particular atom can be compared as a function of the different oxidation states. For a series of iron and molybdenum complexes, metal binding energies are given in Table 2.17.

From this table, it is clear that Mo $3d_{5/2}$ binding energy gradually increase with increase in oxidation state of molybdenum where ligand environments are non-controversial. Also whenever the complex is in anionic form, this binding energy is a bit higher than that of neutral complex having the same oxidation state of molybdenum. Thus, the overall charge on the complex has an obvious result on the binding energies. On the basis of this fact, the reported Mo $3d_{5/2}$ binding energies of some of the neutral $\{Mo(NO)_2\}^6$ derivatives, which are slightly lower than that in case of $Cs_2[Mo(NO)_2(Cl)_4]$, are justified. Feltham et al. comparing the binding energy Mo $3d_{5/2}$ for $[Mo(NO)_2(Cl)_2(diars)]$ (230.3 eV) with *trans*-$[MoOCl\text{-}(diars)_2]PF_6$ (230.1 eV) concluded that as this dinitrosyl derivative is having slightly higher value for Mo $3d_{5/2}$,

Table 2.17: Binding energies of metals in some Fe and Mo complexes along with their oxidation states.

S. no.		E_b Mo3d$_{5/2}$ or E_b Fe2p$_{3/2}$ (eV)	Oxidation state	Reference
1.	$K_3[Fe(CN)_6]$	710.3	III	[53a]
2.	$K_4[Fe(CN)_6]$	708.8	II	[53a]
3.	$Fe[Fe(NO)(CN)_5]$	711.3	–	[53a]
4.	$Na_2[Fe(NO)(CN)_5]$	711.0	–	[53a]
5.	$Cs_2[MoOCl_5]$	231.9	V	[53f]
6.	$Cs_2[MoOBr_5]$	231.4	V	[53f]
7.	$[Mo_2Cl_{10}]$	231.2	V	[53f]
8.	$[MoCl_4(PPh_3)_2]$	231.1	IV	[53c]
9.	$K_3[MoCl_6]$	230.0	III	[53g]
10.	$Rb_3[Mo_2Cl_8H]$	229.8	III	[53h]
11.	$[Mo_2Cl_4(py)_4]$	228.6	II	[53f]
12.	$(NH_4)_5[Mo_2Cl_9]\cdot H_2O$	229.3	II	[53f]
13.	*trans*-$[Mo(N_2)(Cl)(dppe)_2]$	228.2	I	[53d]
14.	*trans*-$[Mo(N_2)(Br)(dppe)_2]$	228.1	I	[53d]
15.	*trans*-$[Mo(N_2)_2(dppe)_2]$	227.2	0	[53d]
16.	$[Mo(NO)_2(Cl)_2(diars)]$	230.3	–	[53d]
17.	$[Mo(NO)_2(Cl)_2(PPh_3)_2]$	230.5	–	[53c]
18.	$[Mo(NO)_2(Et_2\text{-}DTC)_2]$	229.6	–	[53d]
19.	$Cs_2[Mo(NO)_2(Cl)_4]$	231.3	–	[53i]

the formal oxidation state of Mo in this complexes may be taken as (+V). However, the binding energy of Mo in the cationic complex should be viewed with caution, as the overall positive charge may affect the observed value.

For a series of neutral complexes containing $\{Mo(NO)_2\}^6$ moiety, the Mo 3d$_{5/2}$ binding energies are lower than those reported for the pentavalent molybdenum. Moreover, Walton [2] showed that the binding energy of Mo(V) may even be equal to that of Mo(IV). On these grounds, the oxidation state of the Mo in this $\{Mo(NO)_2\}^6$ series may be considered as (+IV). This is in conformity with the assignment of the nitrosyl group as NO^- in these complexes. However, the situation is not as simple as it appears from the above discussion. For example, in the series $K_4[Fe(CN)_6]$, $K_3[Fe(CN)_6]$ and $Na_2[Fe(NO)(CN)_5]$, the Fe2p$_{3/2}$ binding energies reported are 708.8, 710.3 and 711.0, respectively. Since the oxidation state of iron is (+II) and (+III) in ferrocyanide and ferricyanide, respectively, the oxidation state of iron should be more than (+III) in nitroprusside.

2.9.7 Mössbauer (MB) spectral studies

MB spectroscopy is an important experimental technique in bioinorganic chemistry. This is owing to the fact that the technique provides an extremely fine energy resolution and can identify even minute changes in the nuclear surroundings of the pertinent atoms. This technique can give information on the spin and oxidation states of the MB-active atom depending on the chemical analysis of surrounding effects on nuclear interactions. The method depends on the recoilless resonant absorption of γ-radiation by the MB-active nuclei from a source emitter isotope. This spectroscopy follows the fact that the excitation energy of the nucleus (caused by the absorption of gamma rays) is disturbed by its chemical surroundings. The disturbance is counted by measuring the Doppler shift essential to cause the resonant absorption between the sample and the reference.

Applications of the MB effect to solid-state research are principally based on hyperfine structure of the spectra. Usually, three hyperfine interactions are examined in this spectroscopy. In fact, these three interactions provide a link to the electronic structure in this technique. These are (i) isomer shift (or chemical shift), (ii) quadrupole splitting (QS) and (iii) hyperfine splitting/magnetic hyperfine splitting (also known as Zeeman splitting).

(i) Isomer shift

The isomer shifts (δ) give an idea about the metal–ligand bond covalency and also about the spin and oxidation states of the MB-active atom in question. The extent of the isomer shift depends on the electron shielding covalency influences and the changes in bond distances. The MB isomer shifts are proportional to the s-electron density at the nucleus. The s-electron density in turn is influenced by the d-electrons shielding. Accordingly, the isomer shift is a sensitive measure of the above-mentioned properties of an MB-active atom.

Isomer shifts in MB spectroscopy are usually interpreted taking into consideration of the following rules: (i) the isomer shifts of the spin-paired complexes are lower compared to those of the spin-free complexes, (ii) the isomer shift increases with the increase in coordination number. However, the related lengthening of bonds must also be taken into consideration. (iii) The isomer shift increases with the decrease in oxidation state, and (iv) the isomer shifts of metals coordinated by soft ligands (e.g. sulphur donor) are less than those coordinated by hard ligands (e.g. N, O donor). Mostly, higher the ligand field strength, the greater is the isomer shifts (δ) observed. Therefore, the important considerations are the σ-donor and π-acceptor features of ligands.

(ii) Quadrupole splitting

The interaction between the nuclear energy levels and the surrounding electric field gradient is revealed by the QS.

The majority of absorber MB nuclei are found to have non-zero spins and also most of them have half-integral spins rather than integral spins. Further according to a selection rule, the spins of the excited state must be different from that of the ground state. Thus, it follows that either or both of the nuclear states will have spin >1/2. It is well known that nuclei having spin I >1/2 keep an electric quadrupole moment because of the non-spherical charge distribution. Therefore, one or both nuclear states will have electric quadrupole moment. The electric quadrupole moment so produced will interact with electric field gradients at the nucleus produced by its surroundings.

A fairly common situation for the excited state nucleus is to have I = 3/2 (and thus have an electric quadrupole moment) and the ground state I = 1/2. This is found, for example, in ^{57}Fe, ^{119}Sn and ^{129}Xe. With I = 3/2, there are ($2I$ + 1) or 2 × 3/2 + 1 = 4 possible orientations for the excited nucleus in an electric field along the vertical (z) direction (Figure 2.41). These orientations are specified by the quantum number m_I being 3/2, 1/2, −1/2 and −3/2.

As the angle which the nucleus makes with the electric field gradient is the same in the m_I = ±1/2 states, these two states will have the same energy in the electric field. Similarly, the two states m_I = ±3/2 have the same energy, though different from that of the m_I = ±1/2 states.

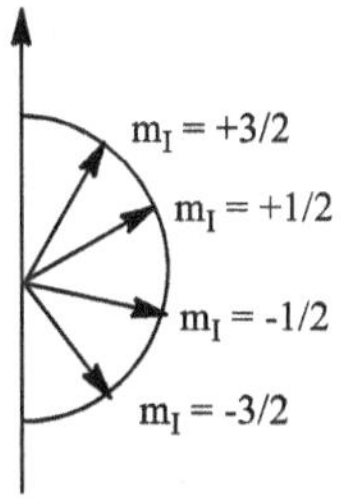

Figure 2.41: The four allowed spin orientations to an excited nucleus having I = 3/2 in an electric field.

The electric quadrupole moment of the excited nuclear state having I = 3/2 interacts with the electric field gradient. This results in the splitting of the excited nuclear energy into two levels. If the quadrupole moment is positive (cigar shape), the ±3/2 states are raised in energy and the ±1/2 states are lowered (Figure 2.42), while a reverse situation exists where the quadrupole moment is negative (tangerine-shaped nucleus). It is notable here that the ground state with I = 1/2 will not split because of having no electric quadrupole moment. Thus, the QS parameters are very sensitive to the changes in the population and geometry of MB-active-atom-derived M.O.s.

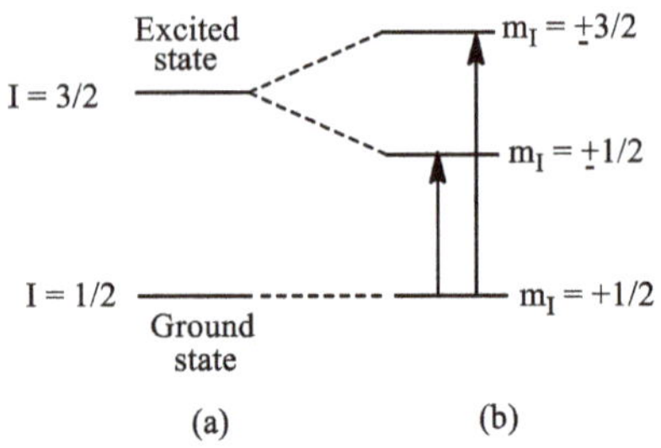

Figure 2.42: (a) The ground and excited state energy levels of a nucleus in the absence of an electric field. (b) Quadrupole interaction of a nucleus having $I = 3/2$ in the excited state in the presence of an electric field.

Example

Ferrocyanide ion {$[Fe(CN)_6]^{4-}$}and nitroprusside ion {$[Fe(CN)_5(NO)]^{2-}$}

Electrical field around the iron nucleus in ferrocyanide ion $[Fe(CN)_6]^{4-}$ is sufficiently symmetrical because of the spherical charge distribution ($t_{2g}^6e_g^0$). This causes no electric field gradient at the iron nucleus, and thus no QS. Thus, its MB spectrum has a single sharp peak without any splitting.

Replacement of one of the cyano groups in ferrocyanide ion by NO yields [Fe $(CN)_5$–$(NO)]^{2-}$wherein symmetrical electric field is lost. This produces sufficient electric field gradient to cause observable QS. Consequently, the spectrum shows two peaks. Both the MB spectra are shown in Figure 2.43.

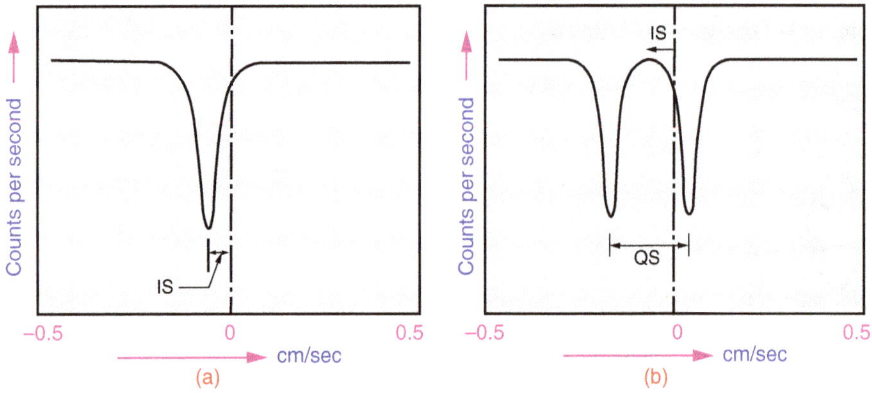

Figure 2.43: Mössbauer spectra of (a) ferrocyanide ion and (b) nitroprusside ion. Spectrum (b) displays quadrupole splitting.

QS is thus a sensitive and powerful means of monitoring the symmetry of a given compound.

(iii) Magnetic hyperfine splitting

Indeed, the nucleus and the any magnetic field surrounding it interacts, and this results in the happening of hyperfine splitting or magnetic hyperfine splitting. Notably, in the vicinity of magnetic field, a nucleus having spin I splits into $(2I + 1)$ sub-energy levels. This gives a hyperfine structure in the MB spectrum. A hyperfine structure in atoms happens because of the energy of the nuclear magnetic dipole moment in the magnetic field produced by the electrons, and the energy of the nuclear electric quadrupole moment in the electric field gradient as a result of the charge distribution within the atom. Owing to the domination of the two effects described earlier, the molecular hyperfine structure also involves the energy related with the interaction between the magnetic moments of different magnetic nuclei in a molecule. The hyperfine structure also includes the energy between the nuclear magnetic moments and the magnetic field produced by the rotation of the molecule.

MB studies on metal nitrosyls

Mostly, nitrosyl complexes of iron and ruthenium have been extensively studied by MB spectroscopy. However, some early reports on iridium nitrosyl complexes are also available in the literature. MB spectral studies on cobalt complexes are of special interest. This is due to the important role it plays in many biological paths as a protein cofactor.

It is well known that the M–N–O electronic arrangement in metal nitrosyls is very sensitive to its coordination environment. Therefore, MB spectroscopy is very much relevant in examining the structures of metal nitrosyls. When the measurements in MB spectroscopic are pooled with the results of DFT calculations, this study becomes more effective. Sandala and co-workers [54] reported that the uses of MB spectroscopic data in the calibration of isomer shift permit one to precisely assess the Fe and NO spin populations in iron nitrosyls. DFT [55] is gradually becoming a popular tool in calculating structures and MB properties of iron nitrosyl compounds.

In order to study the metalloenzymes, particularly, containing iron, MB study that conjoints with DFT [56] has been used. By using these two methods conjointly, these studies provide an atomic-level interpretation of the mechanisms of action of metalloenzymes. The interactive comparison of computational calculations with the experimental one permits a confirmation of the existing data and calibration of the MB parameters for a range of the frequently used DFT functionals. This type of methodology has been effectively used to test a group of non-haeme iron nitrosyls involving mono- and poly-nucleated low- (S = 1/2) and high-spin (S = 3/2) iron centres having $\{FeNO\}^7$ and $\{Fe(NO)_2\}^9$ electronic configurations. The ligand field and geometry of the nitrosyl complexes strongly influence their MB spectra. MB

Table 2.18: Mössbauer data for some iron and ruthenium nitrosyl compounds along with electronic configuration of {MNO) core.

Complex	Electron config./ metal oxidation state	Isomer shift δ (mm s^{-1})	Quadrupole splitting parameter, [ΔEQ, mm s^{-1}]	Reference
(a) Complexes: non-haeme				
$[Fe(NO)(H_2O)_5{}^{2+}$	$\{FeNO\}^7$	0.760	2.10	[57]
$[Fe(dtciPr_2)_2(NO)]$	$\{FeNO\}^7$	0.350	0.89	[58]
$[Fe(NO)_2(SPh)_2]^{1-}$	$\{Fe(NO)_2\}^9$	0.182	0.69	[69]
$[Fe(NO)_2(SC_2H_3N_3)(SC_2H_2N_3)]$	$\{Fe(NO)_2\}^9$	0.310	1.12	[60]
$[Fe_2(EtHPTB)(O_2CPh)\ (NO)_2]^{2+}$	2×$\{FeNO\}^7$	0.670	1.44	[61]
(b) Complexes: haeme				
[Fe(NO)(TPP)]	$\{FeNO\}^7$	0.35	1.24	[62]
[(Cytochrome cd$_l$)haeme-NO]	$\{FeNO\}^7$	0.34	0.8	[63]
(c) Metal nitrosyls: ruthenium				
$K_2[RuCl_5(NO)]$	$\{RuNO\}^6$	−0.43	0.11	[64]
$K_2[Ru(NO_2)_4(OH)(NO)]$	$\{RuNO\}^6$	−0.22	<0.07	[64]
$[Ru(NH_3)_5(NO)]Cl_3$	$\{RuNO\}^6$	−0.2	0.41	[64]
$[RuCl(py)_4(NO)](PF_6)_2$	$\{RuNO\}^6$	−0.2	0.37	[64]
$[RuBr(py)_4(NO)](PF_6)_2$	$\{RuNO\}^6$	−0.2	0.37	[64]

dtciPr$_2$, di-isopropyl dithiocarbamate; Ph, phenyl; Et-HPTB, *N*,*N*,*N*′,*N*′-tetrakis(*N*-ethyl-2-benzimidazolylmethyl)-2-hydroxy-1,3-diaminopropane; TPP, tetraphenylporphyrin.

parameters for some typical iron and ruthenium nitrosyl complexes along with the electron configuration of the {MNO} cores are given in Table 2.18.

The MB spectrum of the $[Fe(H_2O)_5(NO)]^{2+}$ solution enriched with ^{57}Fe nucleus measured in zero field at 80 K displays a peak at approximately 1.8 mm s^{-1} having δ = 0.76 mm s^{-1} and ΔEQ = 2.10 mm s^{-1}. Remarkably, the MB and EPR parameters of the pentaaquairon complex $[Fe(H_2O)_5(NO)]^{2+}$ in question are closely similar to that of the $\{FeNO\}^7$ cores of the other nitrosyl complexes characterized well. Based on the above observations, it was concluded that its electronic structure is best presented as a high-spin (S = 5/2) Fe^{3+} coupled to the NO^- (S = 1) antiferromagnetically. This gives the observed spin quartet ground state, S = 3/2. The calculated Mulliken iron spin population in the $[Fe(H_2O)_5(NO)]^{2+}$ complex with S = 3/2 was observed between 3.4 and 3.8. However, this opinion was revised by Cheng et al. [65] in 2003. On the basis of DFT calculations, these authors proposed that the spin-quartet ground state of the $[Fe(H_2O)_5(NO)]^{2+}$ is best designated somewhat as the Fe^{II} (S = 2) coupled to the NO (S = 1/2) [not to NO^- (S = 1)] antiferromagnetically. This gives the correct assignment of the pentaaquairon complex as $[Fe^{II}(H_2O)_5(NO)]^{2+}$, and not the $[Fe^{III}(H_2O)_5(NO^-)]^{2+}$ or the $[Fe^{I}(H_2O)_5(NO^+)]^{2+}$ given in the conventional textbooks. The Fe–N–O bond angle in the optimized structure of $[Fe(H_2O)_5(NO)]^{2+}$ was found as linear.

The isomer shifts (δ), in certain ruthenium(II) compounds, were found to be approximately −0.20 mm s^{-1}. These values in isomer shifts are nearly equal for most of the known nitrosyls and rational with a +2 charge of the central ion. However, the too much negative value of δ for $K_2[Ru(NO)Cl_5]$ (−0.43 mm s^{-1}) specifies a much stronger ligand field in this compound compared to the other ruthenium nitrosyl complexes. The two distinct MB peaks found in ruthenium nitrosyl complexes, such as $[Ru(NO)Cl(py)_4](PF_6)_2$, $[Ru(NO)Br\text{-}(py)_4](PF_6)_2$ and $[Ru(NO)(NH_3)_5]Cl_3$ (with $\Delta EQ \approx 0.40$ mm s^{-1}), exemplify the un-symmetric environment of the central ion. For each of the complexes under this investigation, the backdonation of the metal d-electrons to the π^* orbitals of the NO ligand is very strong, but almost the same for in each. This is substantiated by the relatively small variations in stretching frequency of NO [ν(NO)] in the IR spectra of these complexes. Thus, the observed variations in the isomer shifts are due to the contribution from the remaining ligands to the ligand field.

2.9.8 Kinetic studies

With an intention of extending the work of kinetic studies carried out on metal carbonyls, several research groups were extensively involved in the kinetics and mechanism of substitution reactions of metal nitrosylcarbonyls [66–71]. Remarkably, the existence of NO ligand with CO in metal nitrosylcarbonyls considerably affects the mechanism so that they undergo SN^2-type reactions. On the other hand, SN^1 mechanism is usual for metal carbonyls such as $[Ni(CO)_4]$. Most of the kinetic studies of substitution reactions on cobalt nitrosylcarbonyl $[Co(NO)(CO)_3]$ and its derivatives have been carried out. Additionally, kinetic studies of substitution reactions in $[Mn(NO)(CO)_4]$ and $[Fe(NO)_2(CO)_2]$ have also been explored with similar results.

In the oldest known cyanonitrosyl complex ion, the so-called nitroprusside ion, $[Fe(NO)(CN)_5]^{2-}$, kinetics of the substitution reactions with many substrates have been explored. One of the simplest reactions studied is that of the nitroprusside ion with OH^-. This has long been known to give rise to the formation of $[Fe(NO_2)(CN)_5]^{4-}$. Swinehart and Rock [72] in 1966 have carried out the detailed thermodynamics and kinetic investigation. They suggest the mechanism for the substitution reaction as follows:

$$[Fe(NO)(CN)_5]^{2-} + OH^- \rightarrow [Fe(NO_2H)(CN)_5]^{3-} \tag{2.1}$$

$$[Fe(NO)(CN)_5]^{2-} + 2OH^- \rightarrow [Fe(NO_2)(CN)_5]^{4-} + H_2O \tag{2.2}$$

In this substitution reaction, the rate-determining step is reaction (2.1). The p*K*a value was determined by the stopped-flow method for the equilibrium $[Fe(NO_2)(CN)_5]^{4-} \rightleftharpoons [Fe(NO_2H)(CN)_5]^{3-}$. This has been found to be about 6.4.

The reaction of nitroprusside ion with the substrate SH^- completes through a similar mechanism as that given with substrate OH^-. $[Fe(NOS)(CN)_5]^{4-}$, which is unstable in solution, is formed through an intermediate that most probably contains NOSH group [72].

When mixtures of nitroprusside ion $[Fe(NO)(CN)_5]^{2-}$ and NCS^- are irradiated, a deep blue colouration is obtained. This may also be prepared by acidifying a basic solution of the two reactants. The kinetic studies carried out by Dempir and Masck [73] in 1968 suggested that the species responsible for the deep blue colour is $[Fe(CNS)(CN)_5]^{4-}$. This blue colour species $[Fe(CNS)(CN)_5]^{4-}$ is formed by the reaction of $[Fe(NO_2)(CN)_5]^{4-}$ or $[Fe(H_2O)(CN)_5]^{3-}$ with NCS^-.

Swinehart and W. G. Schmidt [74] applied the spectrophotometric methods to explore the kinetics and mechanism of the reactions of the nitroprusside ion $[Fe(NO)(CN)_5]^{2-}$ with ketones, namely, acetone and acetophenone in basic medium. The deep red intermediates formed in these reactions were believed to be of the type shown in Figure 2.44.

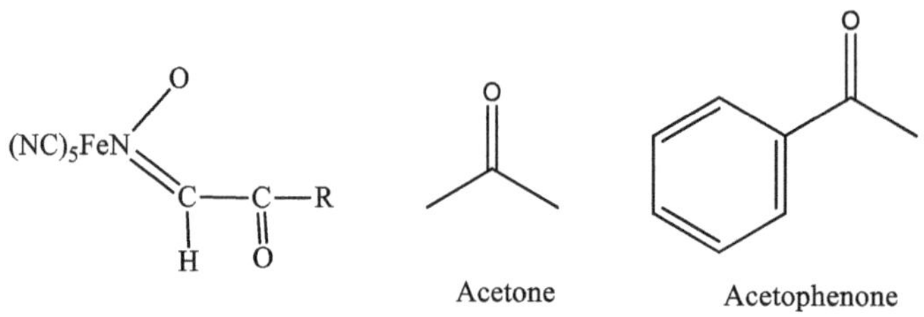

Figure 2.44: Deep red intermediates formed by $[Fe(NO)(CN)_5]^{2-}$ with acetone and acetophenone.

For the first time in 1970, West and Hassemer [75] explored the kinetics of the reactions of $[Fe(NO)(CN)_5]^{2-}$ with different thioureas.

The hydrolysis of pentacyanonitrosylchromate (I), $[Cr(NO)(CN)_5]^{3-}$ was studied by Burgess et al. [45] in 1968 using ESR method. Thereafter, Bottomley et al. [76] investigated the same hydrolysis reaction employing the electrochemical method. The ESR spectra of all the aquation products $[Cr(NO)(H_2O)_x(CN)_{5-x}]^{x-3}$ have been examined by the research group of J. Burgess and as per their findings the processes

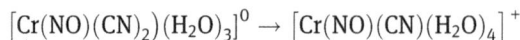

$$[Cr(NO)(CN)_2)(H_2O)_3]^0 \rightarrow [Cr(NO)(CN)(H_2O)_4]^+$$

and

$$[Cr(NO)(CN))(H_2O)_4]^+ \rightarrow [Cr(NO)(H_2O)_5]^{2+}$$

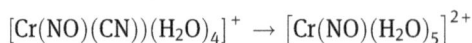

were shown to be of first order.

2.9.9 Nuclear resonance vibrational spectroscopic studies

NRVS is a synchrotron-based technique. This investigates vibrational energy levels [77]. It is notable that the technique is precise for those samples that contain MB-active [78] nuclei. NRVS is selective for vibrations involving displacement of MB-active nuclei. Most of the reported NRVS results are centred around ^{57}Fe isotope. The massive amount of the NRVS data for iron results, owing to its presence in the haeme and non-haeme centres of proteins. It is expected that the NRVS spectrum comprises an MB signal with vibrational sidebands.

Dissimilar from Raman and IR spectroscopies, NRVS follows no optical selection rules. Hence, all of the iron–ligand modes can be observed in the spectrum. For example, in haeme (Figure 2.45) molecule, these comprise the in-plane Fe–N (macrocyclic ring) vibrations and the Fe–N (imidazole) stretching for six-coordinated porphyrinates. These stretching vibrations were not reported in resonance Raman spectral studies. Other low-frequency vibrations, comprising the haeme doming, can also be explored by NRVS. Consequently, the NRVS experiment can provide the complete set of bands corresponding to the modes that involve motion of the iron atom. Besides these, those modes that do not follow the optical selection rules of Raman or IR spectroscopy are also observed in NRVS spectrum. Moreover, NRVS is an ideal tool for the identification of metal–ligand stretching vibrations. This is because it samples the kinetic energy distribution of the vibrational modes. The band intensities are, therefore, proportional to the amount and direction of the iron motion in a normal mode [79]. This makes the metal–ligand stretching vibrations very intense in NRVS.

The applied aspect of NRVS for metal–ligand stretching vibrational data in metal nitrosyls is exemplified here by taking an example of ferrous nitrosyl tetraphenylporphyrin [^{57}Fe(TPP)(NO)] (TPP=tetraphenylporphyrin). NRVS data for this five-coordinated compound, both in a powder sample and as an oriented single-crystal array, were obtained with all porphyrin planes at an angle of 6° to the incident X-ray beam. This allowed increasing the intensity of in-plane modes in the crystal relative to the powder sample, while the out-of-plane mode intensities were reduced comparative to the powder sample. The modes that are less intense than their equivalents in the powder spectrum represented out-of-plane vibration modes' character. As per rule, the remaining modes clearly had a strong in-plane character. The in-plane modes in D_{4h} symmetry are Raman inactive, and were previously unobserved. Although in the case of the [Fe(TPP)(NO)], the D_{4h} symmetry breaking caused by the axial NO ligand might allow Raman observation of these in-plane vibrations, they have not been reported. On the other hand, NRVS will always allow observation of these in-plane modes. The in-plane modes were found in the region 200–500 cm^{-1} for a series of five-coordinated iron porphyrin nitrosylates [80]. Vibrations connected with hindered rotation/tilting of the NO and haeme doming are expected at low frequencies, where Fe motion perpendicular to the haeme (out-of-plane vibrations) is recognized experimentally for [Fe(TPP)(NO)] at 73 and 128 cm^{-1}.

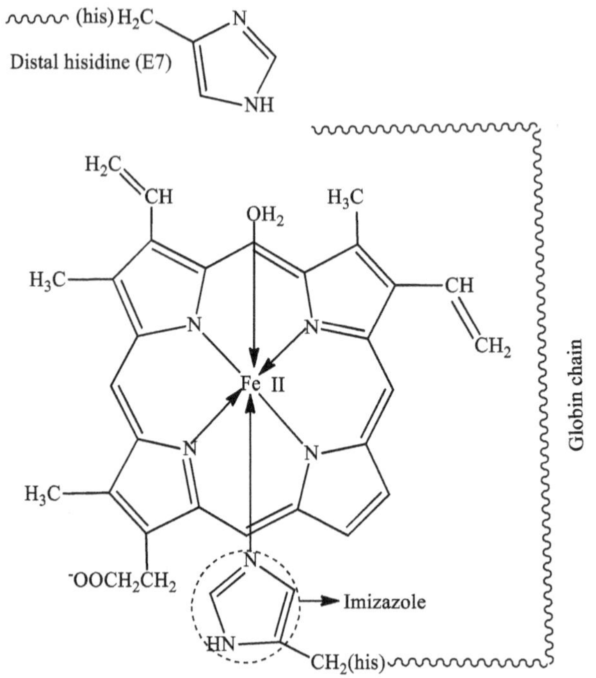

Figure 2.45: Structure of haeme.

Scheid et al. [81] reported that DFT calculations signify that no Fe–N–O stretching mode can be assigned independently. In fact, the Fe–NO stretch modes have a significant Fe–N–O bending character as well as a Fe–NO stretching character. The Fe–N–O bending mode was also calculated to contribute in the in-plane Fe vibrations. The ^{57}Fe excitation probabilities for a series of iron porphyrins (Figure 2.46) have been measured by the research group of Scheid. Accordingly, these measurements display that iron-nitrosyl porphyrinato complexes have Fe–NO stretch/bend modes in the range 520–540 cm^{-1}. Moreover, the bands found in the region 300–400 cm^{-1} were associated with stretching of the four in-plane $Fe–N_{pyrrole}$ bonds of the nitrosyl compound [Fe(TPP)(NO)].

2.9.10 Mass spectral studies

It is notable that the mass spectral studies on metal nitrosyl complexes encompass the measurement of certain physical quantities, namely ionization energies, metal–ligand dissociation energies as well as examination of fragmentation patterns.

Interestingly, in some carbonylnitrosyl complexes, such as $[Fe(NO)_2(CO)L]$ and $[Co(NO)(CO)_2L]$, mass spectral studies carried out by two research groups [82, 83]

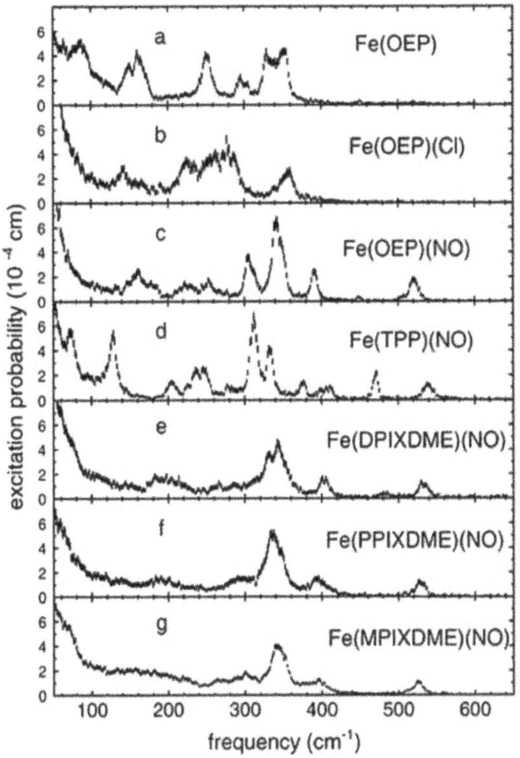

Figure 2.46: Nitrosyl complexes displaying an Fe–NO stretch/bend mode in the range 520–540 cm^{-1} and stretching of the four in-plane Fe–$N_{pyrrole}$ bonds in the range 300–400 cm^{-1}.

conclude that a linear correlation is found between the first ionization energy of the complex and the first ionization energy of the corresponding ligand L. It has been observed that a decrease in the σ-donating ability of ligand L is accountable for a decrease in the first ionization energy of the complex. Moreover, calculated first ionization energies of $[Fe(NO)_2(CO)L]$ and $[Co(NO)(CO)_2L]$ are found to be in good parity with the measured values from their mass spectra.

Johnson et al. [84] studied the fragmentation pathways of several dinitrosyl complexes of cobalt and iron of the general composition $[M(NO)_2(X)]_2$ (where X = Cl, Br and I for M = Co and X = Cl and Br for M = Fe). They observed that the nitrosyl groups are removed stepwise from the iron compounds. The final residue found in these iron complexes is $[Fe_2X_2]^+$. On the other hand, mass spectra of the cobalt complexes display a higher profusion of mononuclear ions containing nitrosyl groups. These results conclude that the iron compounds have metal–metal bonds. This increases the stability of the dinuclear fragments. In complexes $[Fe(NO)_2(SR)]_2$ and $[Co(NO)_2(SR)]_2$, the nitrosyl groups are only removed. This show that the thio-bridges are more than the halo-bridges in $[M(NO)_2(X)]_2$ (where X = Cl, Br and I for M = Co and X = Cl and Br for M = Fe).

The mass spectrum of dimeric mononitrosyl complex of molybdenum $[Mo(C_5H_5)(NO)(I)_2]_2$ (I) (Figure 2.47) is interesting. The highest molecular ion peak observed in this case is due to the formation of $[Mo(C_5H_5)(NO)(I)]_2^+$ (II) (Figure 2.47). This happens due to the loss of the two terminal iodide ligands from the parent compound [85] as shown. The complex $[Mo(C_5H_5)(NO)(I)]_2$ was then synthesized in the laboratory of the R. B. King.

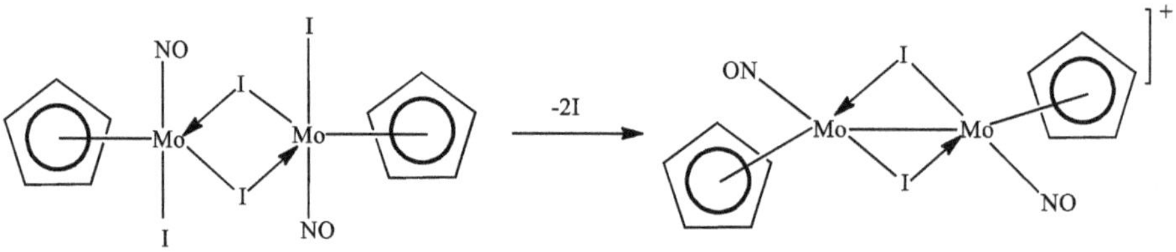

(I) (II)

Figure 2.47: Fragmentation pattern of $[Mo(C_5H_5)(NO)(I)_2]_2$ (I) forming $Mo(C_5H_5)(NO)(I)]_2^+$ (II).

A thrilling restructuring was observed by J. Müller [86] in the mass spectrum of several metal carbonylnitrosyls. For instance, in the fragmentation pathway of $[V(Cp)(NO)_2(CO)]$, the ion VO^+ is observed. On the other hand, in that of $[Fe(Cp)(NO)]_2$, the ion $CpFe_2O^+$ is demonstrated. Metastable peaks display that these ions are formed through the following ways:

$$[V(Cp)(NO)]^+ \rightarrow VO^+ + C_5H_5N$$

and

$$[Fe_2(Cp)_2NO)]^+ \rightarrow CpFe_2O^+ + C_5H_5N$$

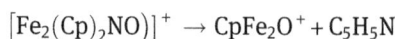

The restructurings are thought to happen as follows:

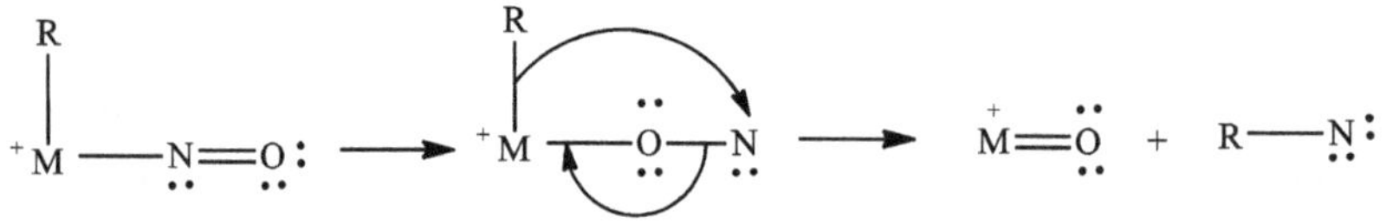

Interestingly, analogous restructurings do not take place in the mass spectrum of $[Ni(Cp)NO]$ or $[M(Cp)(NO)(CO)_2]$ (M = Cr, Mo or W). The dissimilar behaviour in restructurings is considered to occur in the spectrum of $[Ni(Cp)NO]$ or $[M(Cp)(NO)(CO)_2]$ because ions $[V(Cp)(NO)]^+$ and $[Fe_2(Cp)_2NO)]^+$ formed as the intermediate hold an even number of electrons. On the other hand, the resulting $[Ni(Cp)(NO)]^+$ and $[M(Cp)(NO)]^+$ have an odd number of electrons. The latter ions are unable to undertake the types of restructuring shown above.

Maurya et al. [87] have examined the mass spectrum of chromium(I) nitrosyl complex $[Cr(NO)(acac)_2]$ (acacH = acetylacetone) (Figure 2.36). In addition to the parent ion peak at m/z = 280, fragmented ions observed were NO^+, Cr^+, $Cr(acac)^+$, $Cr(NO)(acac)^+$, $Cr(acac)_2{}^+$ and $Cr(acac)_3{}^+$. Looking over the fragmentation patterns, the nitrosyl compound in question was established as a monomer.

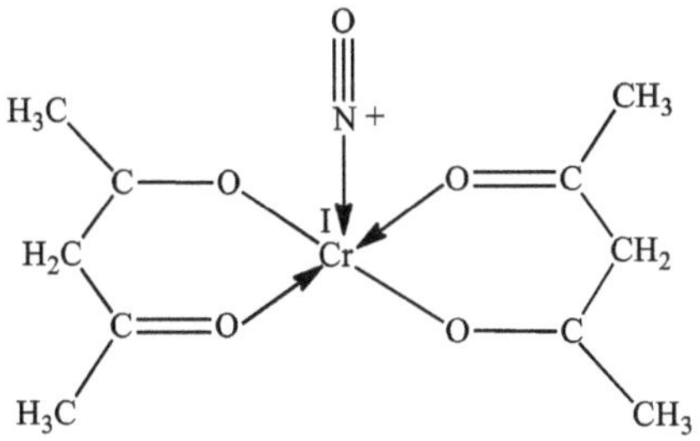

Figure 2.48: Structure of $[Cr(NO)(acac)_2]$.

Maurya et al. [1] recently reported the preparation and diagnosis of a ruthenium(II) nitrosyl complex (involving Schiff base ligand obtained from the condensation of dehydroacetic acid and thiosemicarbazide) of composition $[Ru^{II}(NO)(dha\text{-}tsc)(Cl)(H_2O)]$ (Figure 2.49). In addition to other characterization tools, namely elemental analysis, NMR, FT-IR, UV–Vis, cyclic voltammetry and thermogravimetric analysis, mass spectrometry was also carried out in support of characterization of this compound. The synthesis of the Schiff base ligand and its complex is shown in Figure 2.49.

The THERMO Finnigan LCQ advantage maxion trap mass spectrometer was used for recording the spectrum of the complex in question in the mass ranges of 100–700 (Figure 2.50). The different mass fragments of the complex show well-defined m/z peaks of ruthenium isotopes ranging from 422 to 430. This indicates the formula weight of the compound. The difference in heights of the respective peaks is as a result of isotopic profusion of the metal. In addition, m/z peak near 240 indicates the mass of Schiff base ligand. ESI-mass spectrometry in combination with other instrumental techniques provides the sufficient proof of molecular composition of the complex compound in question.

2.10 Reactivity of nitric oxide coordinated to transition metals

The use of pentacyanonitrosylferrate(II), $[Fe(NO)(CN)_5]^{2-}$, as the source of qualitative spot test determination of SH^-, $SO_3{}^{2-}$ and a range of organic compounds has a very extensive and remarkable history. However, in 1967 it was reviewed by J. H. Swinehart [88] that these tests involve reaction of the coordinated NO and that the typical colours produced are accompanying with the chromophoric group [Fe{N(O)R}],

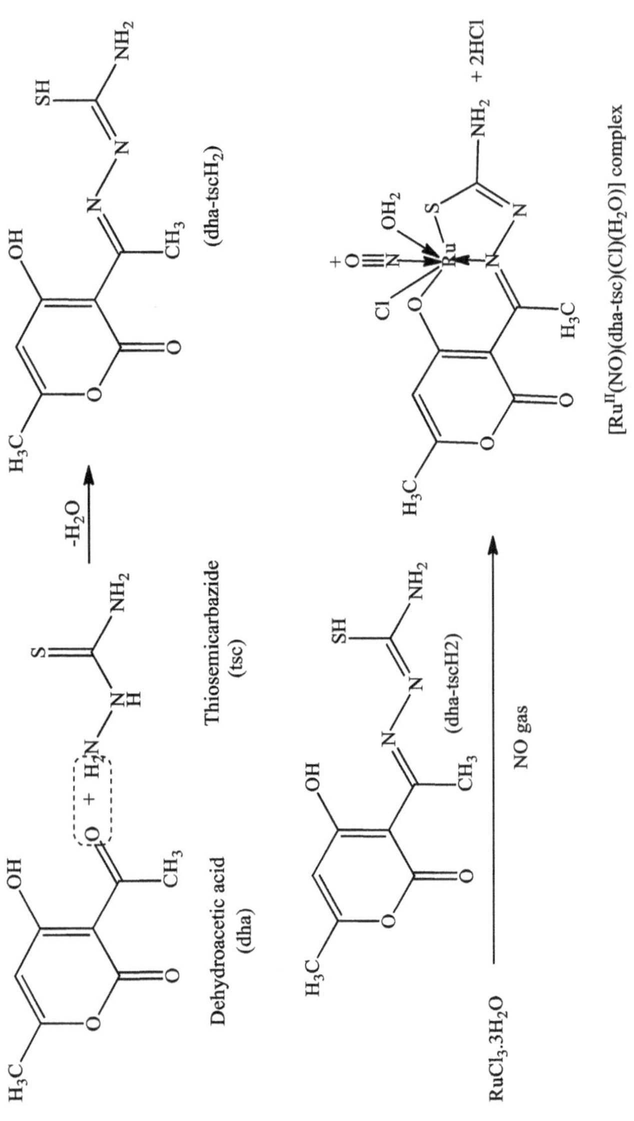

Figure 2.49: Synthesis of [Ru^{II}(NO)(dha-tsc)(Cl)(H_2O)].

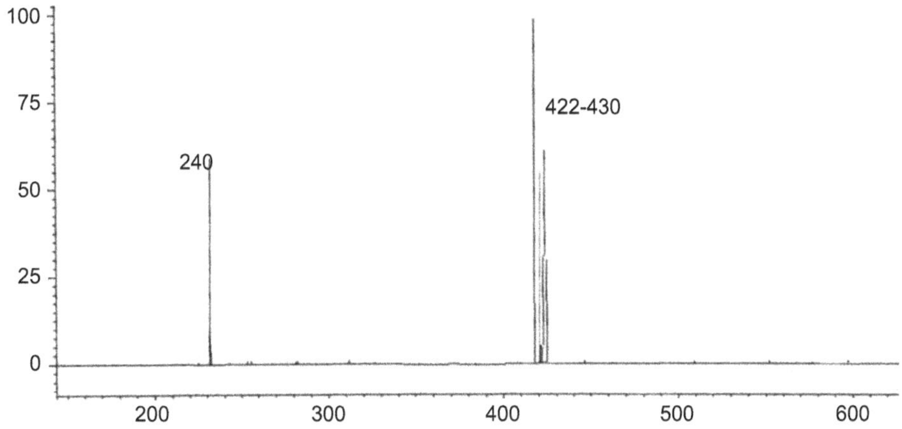

Figure 2.50: ESI MS of the complex.

where R is sulphide, sulphite and carbanion. Apart from the analytical use of this work, however, the attention paid to the reactivity of the metal nitrosyls is very inadequate compared to the metal carbonyls. This is because of the fact that the coordinated NO is strongly bonded with the metal in nitrosyl complexes. As a result, the ligand displacement reactions are not as frequent as those with the metal carbonyls. Due to the enormous studies on metal carbonyls having industrial significance, many aspects concerning the reactivity of nitrosyl complexes have come out. Two things are of key importance in these studies: (i) the first one is to deal with the environmental pollution triggered by NO and (ii) the second one is the development of newer catalytic system of industrial applicability.

The reactions of the coordinated NO fall under the following category:

2.10.1 Nucleophilic attack

It is noteworthy that most of the spot tests developing colours using pentacyano-nitrosylferrate(II), $[Fe(NO)(CN)_5]^{2-}$, for example, the test for detection of SH^-, so-called Gmelin reaction, the Boedeker test for SO_3^{2-} and so many others, are now considered as a nucleophilic attack upon the coordinated NO group of the pentacyanonitrosylferrate(II). This results in the production of an unusual ligand bound to Fe(II). Sometimes the displacement of NO by another species, usually H_2O, occurs. Some of the colour reactions occurring in such tests are presented in Table 2.19.

Table 2.19: Colour development in $[Fe(NO)(CN)_5]^{2-}$ with different substrates.

Substrate	Colour	Product
SH^-, S^{2-}	Red	$[Fe(NOS)(CN)_5]^{4-}$
Se^{2-}	Blue-green	$[Fe(NOSe)(CN)_5]^{4-}$
Te^{2-}	Black	$[Fe(NOTe)(CN)_5]^{4-}$
$SO_3{}^{2-}$	Deep-red	$[Fe\{N(O)OSO_2\}(CN)_5]^{4-}$
Mercaptans	Purple-red	$[Fe\{N(O)SR\}(CN)_5]^{3-}$
Indole	Violet	$[Fe(N_2OC_8H_6)]^{4-}$
$PhCH_2CN$	Red	$PhC(=NOH)CN$
m-$C_6H_4(NO_2)_2$	Red-violet	$[Fe(CN)_5(H_2O)]^{3-}$

Another well-known example of nucleophilic attack (already known since the middle of the nineteenth century) is the reaction of nitroprusside ion with nucleophile OH^-, as follows:

$$2OH^- + [Fe(NO)(CN)_5]^{2-} \rightleftharpoons [Fe(NO_2)(CN)_5]^{4-} + H_2O$$

This reaction (J. H. Swinehart, *Coord. Chem. Rev.*, 2 (1967) 385–402) has an equilibrium constant of $1.5 \times 10^6\ L^2\ mol^{-2}$. This may be compared to the value of 2.3×10^{31} for the parent reaction,

$$NO^+ + 2OH^- \rightleftharpoons NO_2{}^- + H_2O$$

Such a wide difference in magnitudes of equilibrium constant suggests that coordinated NO^+ in $[Fe(NO)(CN)_5]^{2-}$ is a considerably weaker electrophile [88] compared to free NO^+.

The well-known ruthenium dinitrogen complex $[Ru(NH_3)_5(N_2)]^{2+}$ is the result of the nucleophilic attack of OH^- over $[Ru(NH_3)_5(NO)]^{3+}$ in a strong alkaline solution. Bottomley et al. [76] proposed that the reaction occurs via the following reaction mechanism:

$$[Ru(NH_3)_5(NO)]^{3+} + OH^- \rightleftharpoons [Ru(NH_2)(NH_3)_4(NO)]^{2+} + H_2O$$

$$Ru(NH_3)_5(NO)]^{3+} + [Ru(NH_2)(NH_3)_4(NO)]^{2+} \rightarrow$$

$$\left([(ON)(NH_3)_4Ru(NH_2)(O)NRu(NH_3)_4(NO)]^{5+}\right)(X)$$

$$X \rightarrow cis-[Ru(OH)(NH_3)_4(NO)]^{2+} + [Ru(NH_3)_5(N_2)]^{2+} + H^+$$

The compound $[RuCl(A\text{-}A)_2(NO)]^{2+}$ [where A-A = 2,2′-dipyridyl or *o*-phenylene-bis (dimethylarsine)] undergoes a range of electrophilic reactions [89, 90]. On reaction with OH^-, the respective nitro complexes are formed as follows:

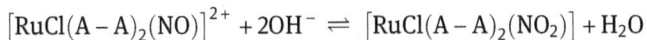

$$[RuCl(A-A)_2(NO)]^{2+} + 2OH^- \rightleftharpoons [RuCl(A-A)_2(NO_2)] + H_2O$$

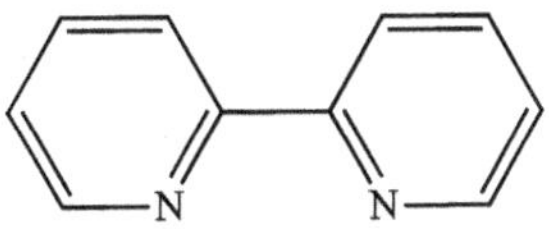

2,2'-Dipyridyl

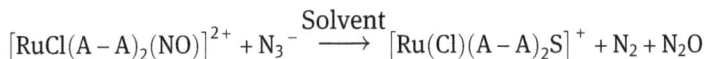

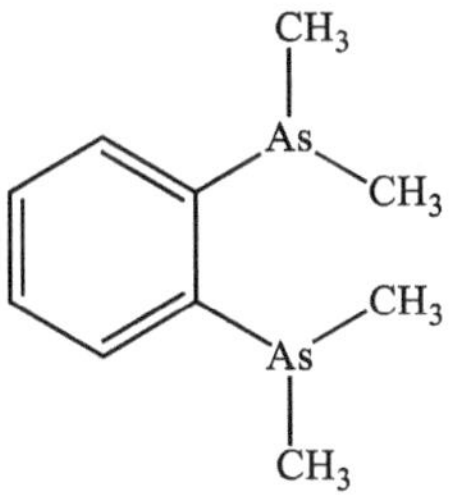

o-Phenylenebis(dimethylarsine)

Reaction of $[RuCl(A\text{-}A)_2(NO)]^{2+}$ with azide ion (N_3^-) [91] undertakes in the following way:

$$[RuCl(A-A)_2(NO)]^{2+} + N_3^- \xrightarrow{\text{Solvent}} [Ru(Cl)(A-A)_2S]^+ + N_2 + N_2O$$

where S = H_2O, MeCN, MeOH or Me_2CO.

2.10.2 Electrophilic attack of coordinated NO

The activity of the coordinated NO in the electrophilic attack would be just opposite to that happened in the nucleophilic attack. The electrophilic attack is, therefore, principally associated with the nitrosyl complexes where the M–N–O group is bent or, in other words, the nitrosyl group contains NO^- grouping.

Research group of Collman [92] and Roper [93] examined the electrophilic attack of coordinated NO in a nitrosyl complex of Os. The osmium dinitrosyl complex $[Os(NO)_2(PPh_3)_2]$ (wherein the ∠Os–N–O is a bit distorted from linearity) reacts in a reversible manner with 2MHCl to give $[Os(Cl)_2(NHOH)(NO)(PPh_3)_2]$. An analogous reaction of $[Os(Cl)(CO)(NO)(PPh_3)_2]$ leads to the formation of $[Os(Cl)_2(CO)(HNO)(PPh_3)_2]$. The former reaction is thought to occur in the following three steps:

$$[Os(NO)_2(PPh_3)_2] + H^+ \longrightarrow [Os(NO)(HNO)(PPh_3)_2]^+$$

$$\downarrow +Cl^-$$

$$[Os(Cl)\ (NO)(HNO)(PPh_3)_2] \xrightarrow{H^+}$$

$$[Os(Cl)\ (NO)(NHOH)(PPh_3)_2]^+ \xrightarrow{+Cl^-} [Os(Cl)_2(NHOH)(NO)(PPh_3)_2]$$

It is expected that continuous protonation of $[Os(NO)_2(PPh_3)_2]$ may give rise to the formation of $[Os(Cl)_3(NH_2OH)(NO)(PPh_3)_2]$. However, this has not been perceived with the osmium complex. However, the related Rh and Ir nitrosyl compounds can be reduced so.

The nucleophilic and electrophilic attacks of the coordinated NO discussed above, therefore, suggest that the metal nitrosyls containing linear M–N–O group (i.e. having NO^+ grouping) should be unstable in alkaline medium. On the other hand, metal nitrosyls containing bent M–N–O group (i.e. containing NO^- grouping) should be unstable in acidic medium. As far as the synthetic point of view is concerned, the cyanonitrosyl complexes synthesized in acidic medium should behave like nitroprusside anion, and this, in fact, has been found to be factual. On the other hand, the cyanonitrosyl complexes synthesized using hydroxylamine method, though rather unstable in acidic medium, do not contain bent M–N–O grouping. Surprisingly, it has been shown by Sarkar and Müller [94] that complexes containing the same nitrosyl grouping can be isolated in alkaline as well as in acidic media. This again reflects the lack of our knowledge regarding the difference of actual electron density in the linear M–N–O grouping. Attempts were made by different research groups to answer this uncertainty. However, a complete explanation is still awaited.

2.10.3 Reduction reactions

It is well known that nitrosonium ion (NO^+) is reduced by one electron to give neutral NO. Therefore, it appears rational to expect that certain metal nitrosyl complexes formally containing NO^+ grouping should behave in the same way. That is, after one electron reduction, they should result in compounds formally containing neutral NO.

The cyanonitrosyl complex ion of iron $[Fe(NO)(CN)_5]^{2-}$(nitroprusside ion) after one electron reduction accomplished by chemical, electrochemical or radiolytic technique affords, at least initially, the brown $[Fe(NO)(CN)_5]^{3-}$ species [95]. This compound was reported to formally contain Fe(II) and neutral NO. Because of the lability of this compound, it readily loses a CN^- (presumably *trans* to NO), and thus affording the blue colour $Fe(I)NO^+$ species, $[Fe(NO)(CN)_4]^{2-}$.

Certain other nitrosyl complexes [96–99] undergo a reversible one electron reduction process. This affords species containing formally coordinated neutral NO, namely,

$$[Cr(NO)(CN)_5]^{3-} + e^- \rightleftharpoons [Cr(NO)(CN)_5]^{4-}$$

$$[Ru(NO)(NH_3)_5]^{3+} + e^- \rightleftharpoons [Ru(NO)(NH_3)_5]^{2+}$$

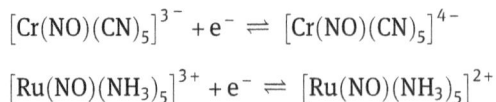

$$[Ru(NO)(dipy)(X)]^{n+} + e^- \rightleftharpoons [Ru(NO)(dipy)(X)]^{(n-1)+}$$

$$(X = Cl^-, N_3^- \text{ or } NO_2^-;\ n = 2)(X = NH_3, \text{Py or MeCN};\ n = 3)$$

$$[Fe(NO)(S_2C_2R_2)_2]^- + e^- \rightleftharpoons [Fe(NO)(S_2C_2R_2)_2]^{2-}\ (R = CN \text{ or } CF_3)$$

The cyanonitrosyl complex $[Mn(NO)(CN)_5]^{3-}$ is formally isoelectronic with $[Fe(NO)(CN)_5]^{2-}$. This cyanonitrosyl complex of manganese on polarographic reduction gives $[Mn(NO)(CN)_5]^{5-}$. This has been interpreted in terms of an initial two-electron process. Moreover, this compound finally decomposes to give Mn^{2+}, CN^- and either N_2 or NH_3. The electrochemical reduction process of $[Cr(NO)(CN)_5]^{3-}$ is found to be pH dependent. Therefore, a reversible one-electron process gives $[Cr(NO)CN)_5)]^{4-}$ in alkaline medium. This corresponds to the formal reduction of Cr(I) to Cr(0). This suggests the addition of an electron to a metal *d* orbital rather than to the NO group. Interestingly, as pH decreases, the electrode reaction becomes a two-electron process. Finally, at pH less than 3, a second three-electron transfer is observed. The final products of the reduction are found to be Cr^{2+}, CN^- and NH_3. A pathway encompassing sequential protonation of the NO group has been suggested as follows:

$$[Cr(NO)(CN)_5]^{3-} \underset{-e^-}{\overset{+e^-}{\rightleftharpoons}} [Cr(NO)(CN)_5]^{4-} \overset{H^+}{\rightleftharpoons} [Cr(NOH)(CN)_5]^{3-} \xrightarrow{H^+} [Cr(NOH_2)(CN)_5]^{2-}$$

$$[Cr(NOH_2)(CN)_5]^{2-} \xrightarrow{-e^-} [Cr(NOH_2)(CN)_5]^{3-} \xrightarrow{H^+} [Cr(NH_2OH)(CN)_5]^{2-} \xrightarrow{3e^-, 2H^+}$$

$$Cr^{2+}, CN^-, NH_3, H_2O$$

The chromium(I) nitrosyl complex $[Cr(NO)(H_2O)_5]^{2+}$ on treatment with Cr(II) in aqueous acid gives hydroxylamine. The suggested equation is given as follows:

$$2[Cr(NO)(H_2O)_5]^{2+} + 4[Cr(H_2O)_6]^{2+} + 2H_3O^+ \rightarrow 3[Cr(OH)(H_2O)_4]_2^{4+} + 2NH_3OH^+ + 6H_2O$$

The reaction seems to be a two-electron reduction comprising a binuclear intermediate.

2.10.4 Oxygenation reactions in metal nitrosyls

Remarkably, coordinated NO reacts with oxygen affords either a nitrato or nitro/nitrito species. It has been observed by Clarkson and Basolo [100] that cobalt nitrosyl Schiff base complexes (I or II), $[Co(NO)(en)_2]^{2+}$ or $[Co(NO)(S_2CNMe_2)_2]$ react with O_2 in the presence of bases (pyridines or MeCN) give nitro products. For example,

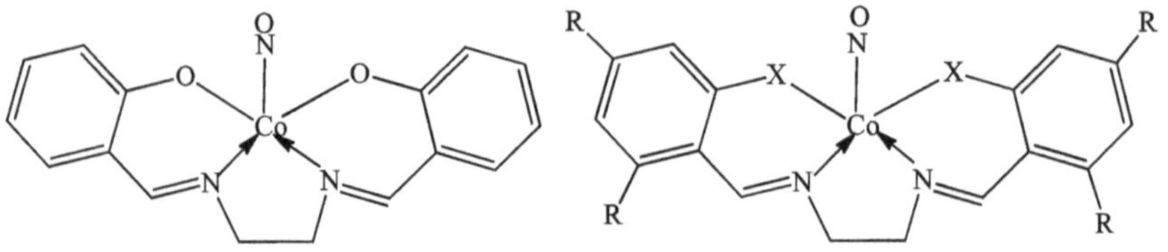

[Co(NO(salen) (I) (II)

X= O, R= Me, Ph; X=S, R=Me

$[Co(NO)(salen)] + Py + 1/2O_2 \longrightarrow [Co(NO_2)(salen)(Py)]$

$[Co(NO)(en)_2]^{2+} + MeCN + 1/2O_2 \longrightarrow [Co(NO_2)(en)_2(NCMe)]^{2+}$

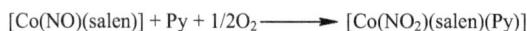

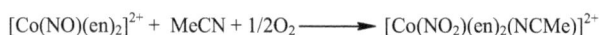

Analogous oxygenation of NO takes place when [NiCl(NO)(diphos)] reacts with O_2 either in boiling DMF or under UV light in dichloromethane (DCM). Moreover, quantitative yields of the nitro product, [NiCl(NO_2)(diphos)], were obtained upon irradiation, but no reaction at all took place in the dark.

Graham et al. [101] in 1972 reported the use of the oxygenation reaction in the catalytic conversion of triphenylphosphine to triphenylphosphine oxide by [Ru(CO)(NO)-(NCS)(PPh_3)$_2$]. The pathway proposed for this process is given below. The rate-determining step is believed to be O-atom transfer to the coordinated P atoms, giving the unstable phosphine oxide complex:

$[Ru(CO)(NO)(NCS)(PPh_3)_2] + O_2 \longrightarrow [Ru(NO)(NCS)(O_2)(PPh_3)_2] + CO$

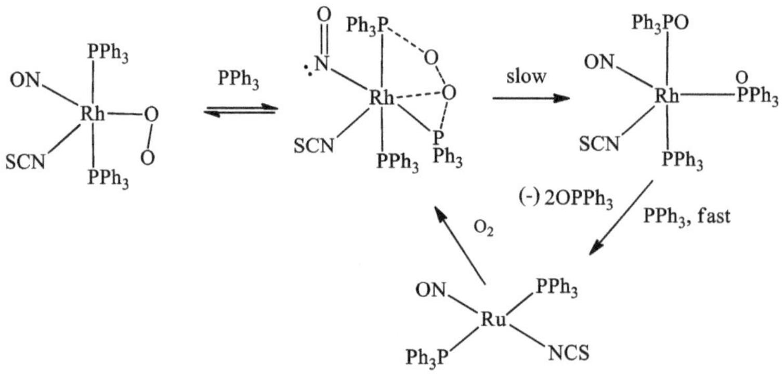

2.11 Other reactions

2.11.1 Nitric oxide (NO): insertion reactions

This class of reactions may not apparently appropriate within a discussion of coordinated NO. But there is a strong possibility that intermediates encompassing metal–NO bonds may involve in the formation of “inserted products.” One of the earliest recognized reactions of NO with a transition metal compound was reported by Frankland [102] in 1857. This was the insertion of NO into a zinc–carbon bond giving species of the type $[Zn\{ONN(R)O\}_2]$ (R = alkyl).

The above reaction appears to be general for main group metal alkyls, because similar compounds containing $[ONN(R)O]^-$ are obtained upon the reaction of NO with Grignard reagents, cadmium, aluminium and even some boron alkyls. Hydrolysis of some of these compounds, followed by treatment of the aqueous extract with Cu^{2+}, affords the copper(II) complexes of *N*-nitroso-*N*-alkylhydroxylaminate (I). These species are entirely analogous to the copper(II) complex of *N*-nitrosophenylhydroxylamine (cupferron). Moreover, reactions of group 2 alkyls with NO provide a convenient route to the alkyl analogues of cupferron [103].

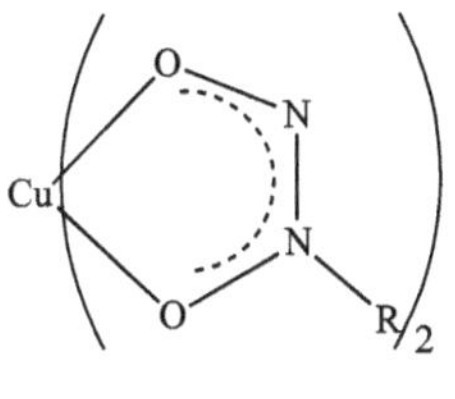

(I)

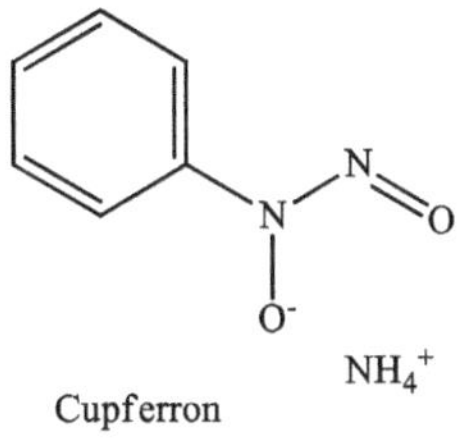

Cupferron

2.11.2 Transfer of coordinated NO to the other metals

It is one of the interesting procedures for the synthesis of metal nitrosyls involving reactions of coordinated NO, which has already been discussed in the previous section of synthetic procedures of nitrosyl complexes.

Some of the other reactions of coordinated NO, which in fact made the nitrosyl chemistry important in the present era, will be covered in the following section.

2.12 Transition metal nitrosyls: organic synthesis and in pollution control

2.12.1 Transition metal nitrosyls: organic synthesis

NO ligand is comparable to the carbonyl ligand in terms of steric and electronic properties. Nevertheless, the nitrosyl complexes have so far attracted much lesser attention than metal carbonyls. The purpose of this section is to highlight some of applications of transition metal nitrosyls in organic synthesis. The reaction of metal nitrosyls in this regard must be divided into two main heads: (a) catalytic reactions and (b) reactions of the coordinated NO group.

2.12.1.1 Catalytic reactions

(i) Oligomerization

Wide-ranging olefin self-condensation reactions happen through transition metal nitrosyls. This leads to the formation of oligomers. The oligomerization of butadiene, isoprene, norbornadiene and phenylacetylene has been successfully completed using a range of nitrosyl complexes [104, 105] of iron and cobalt.

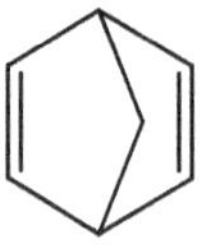
Norbornadiene

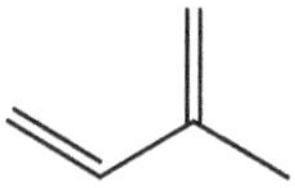
Isoprene

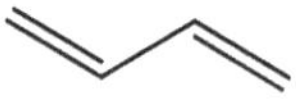
Butadiene

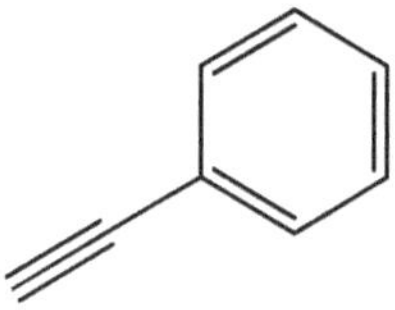
Phenylacetylene

The nitrosyl complex of rhodium, $[Rh(NO)(NCCH_3)_4](BF_4)_2$, is found to catalyse the oligomerization of 2-methylpropene. This reaction gives dimeric, trimeric and tetrameric oligomers. However, the formation of the trimeric oligomer is prevented in the presence of *cis*-butene [106].

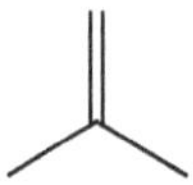
2-Methylpropene

(ii) Polymerization of olefins

Connelly et al. [106] reported the catalytic polymerization of 1,3-butadiene to *trans*-1,4-polybutadiene (95%) using the rhodium nitrosyl complex $[Rh(NO)(NCCH_3)_4](BF_4)_2$ as a catalyst. Reports [107, 108] are also available wherein some cationic nitrosyl complexes of cobalt and iron have been found to induce polymerization of olefins, namely, isoprene, acrylonitrile and styrene.

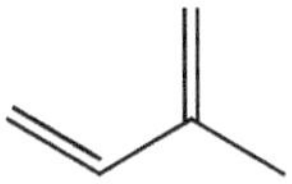

Isoprene

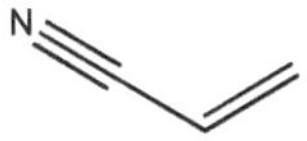

Acrylonitrile

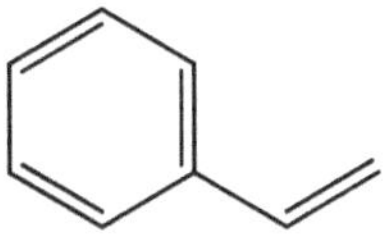

Styrene

(iii) Chemical oxidation

Catalysed oxidations of olefins using Pd(I1) and Pt(I1) as catalysts are reactions of much significance. An aqueous solution of $PdCl_2$ in the presence of NO gas oxidizes alkenes of the type $RCH = CH_2$ (R = H or alkyl) to ketones. For instance, propene is oxidized to acetone [109]. Relevant reactions showing the formation of acetone are as follows:

$$2PdCl_2 + 2NO + H_2O \rightarrow [Pd(NO)Cl] + [Pd(NO_2)Cl_3]^{2-} + 2H^+$$
$$[Pd(NO_2)Cl_3]^{2-} + CH_3CH = CH_2 \rightarrow [Pd(NO)Cl] + CH_3COCH_3 + 2Cl^-$$

Graham et al. [101, 110] reported the catalytic oxidation of PPh_3 to $OPPh_3$ using some Ru nitrosyl complexes, namely, $[Ru(NO)(X)(O_2)(PPh_3)_2]$ ($X^- = Cl^-$, OH^-, CN^- or NCS^-). Surprisingly, these authors observed that the catalytic activity depends on X^- and it decreases in the order $NCS^- > CN^- > Cl^- > OH^-$.

Fleischer et al. have shown that the molybdenum nitrosyl complexes $[Mo(NO)_2Cl_2L_2]$ and $[Mo(NO)Cl_3L_2]$, where L = hexamethylphosphortriamide (HMPT), dimethylformamide, 1/2(2,2′-bipyridine), C_2H_5CN, $OPPh_3$, PPh_3, $AsPh_3$ or $SbPh_3$, catalyse the epoxidation of cyclohexene with *tert*-butylhydroperoxide. On the other hand, the corresponding chromium and tungsten nitrosyl complexes are found to be not much active. However, the epoxidations of cyclohexene with cyclohexenylhydroperoxide, and of cyclopentene, cycloheptene, cyclooctene and octene-1 with *tert*-butylhydroperoxide are successfully catalysed by the molybdenum dinitrosyl complex $[Mo(NO)_2Cl_2(HMPT)_2]$ [111].

Torso et al. have found that the cobalt mononitrosyl complex [Co(NO)(saloph)] [saloph = *N*, *N′*-bis(salicylidene)-*o*-phenylenediamine] on reaction with O_2 in the presence of pyridine (a Lewis base) gives the nitro complex $[(Py)Co(NO_2)(saloph)]$. The nitro complex so obtained catalyses the oxidation [112] of triphenylphosphine to triphenylphosphine oxide. The pertinent reactions are as follows:

$$[Co(NO)(saloph)] + Py + 1/2O_2 \rightarrow [(Py)Co(NO_2)(saloph)]$$
$$[(Py)Co(NO_2)(saloph)] + PPh_3 \rightarrow [(Py)Co(NO)(saloph)] + OPPh_3$$

The catalytic cycle is said to be completed by re-oxidation of the nitrosyl ligand by O_2. Moreover, it has been observed that the catalytic activity of the whole system decreases on decreasing the amount of Lewis base pyridine owing to the formation of the less active $[Co(NO_2)(saloph)(PPh_3)]$.

Interestingly, the cobalt thionitrosyl complexes [Co(NS)Cl$_2$L$_2$] [L = P(OPh)$_3$ or PPh$_3$] are easily oxidized by O$_2$ to the thionitro complexes [Co(NSO)Cl$_2$L$_2$]. These complexes catalyse the oxidation of triphenylphosphine to triphenylphosphine oxide in the absence of Lewis base [113].

It has been observed that the complexes [(Py)Co(NO)(saloph)] and [(Py)Co(NO)(TPP)] (TPP = tetraphenylporphyrin) in the presence of Lewis acids A (A = BF$_3$·C$_2$H$_5$O or LiPF$_6$) react with oxygen giving rise to the formation of nitro complexes [(Py)Co(NO$_2$)L.A] (L = saloph or TPP). The complexes so obtained stoichiometrically oxidize primary alcohols to aldehydes and secondary alcohols to ketones [114]. A possible pathway has been suggested as follows:

$$[Py(Co)(NO)L] + 1/2O_2 + A \longrightarrow \left[(Py)LCo{-}N \overset{\displaystyle O}{\underset{\displaystyle O{-}{-}{-}{-} A}{}} \right]$$

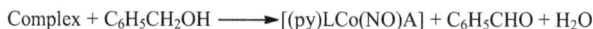

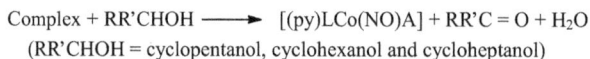

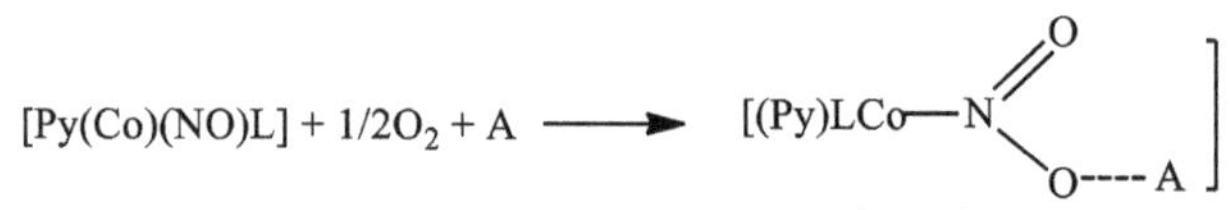

Complex

Complex + C$_6$H$_5$CH$_2$OH \longrightarrow [(py)LCo(NO)A] + C$_6$H$_5$CHO + H$_2$O

Complex + RR'CHOH \longrightarrow [(py)LCo(NO)A] + RR'C = O + H$_2$O
(RR'CHOH = cyclopentanol, cyclohexanol and cycloheptanol)

In fact, Lewis acids when added increase the electrophilicity of the nitro group of the complex by combining with [PyCo(NO)L]. This increases the oxidizing capacity of the resulting cobalt(III) nitro complexes.

(iv) Significance in saturating unsaturated compounds by hydrogen addition

The first report claiming the saturation of unsaturated compound, that is, benzoquinone using the quinoline solvent was put forth by Calvin [115] by intervening Cu(CH$_3$COO)$_2$ as a catalyst. Similar reactions for several unsaturated systems have been studied under the catalytic role of metal nitrosyls [116]. For instance, several RuNOH-based complexes have been found to show hydrogenation in the same fashion targeting propionaldehyde, styrene, crotonaldehyde and acetone (Table 2.20). From the tabulated data, the differences in the respective catalytic efficiencies among the complexes could be attributed to the dissimilarity in the binding mode of NO [117].

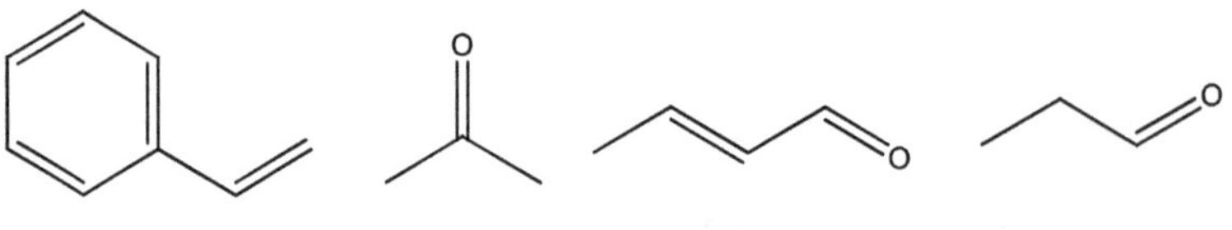

Styrene Acetone Crotonaldehyde Propionaldehyde

In further studies regarding the examination of factors influencing the catalytic activities, one of the examples, [RuH(NO)(PPh$_3$)$_3$], experimented in the presence of water was found to enhance the catalysis. This observation led to accept the utility of water for such reactive hydrogenations [118]. Similarly, using Os in place of Ru in the above complex expresses the activity of the same fashion in preparing ethylbenzene from styrene. In case of hydrogenation of unsaturated hydrocarbons, rhodium(–I) NO complexes have been reported to behave as catalysts of considerable efficiency [120, 121].

Table 2.20: Hydrogenation of some unsaturated organic compounds using some ruthenium hydridonitrosyls as catalysts.

Substrate	Catalyst	Product
Styrene	$[RuH(NO)(PPh_3)_3]$	Ethylbenzene
	$[RuH(NO)(PPr^iPh_2)_3]$	Ethylbenzene
	$[RuH(NO)\{P(C_6H_{11})Ph_2\}_3]$	Ethylbenzene
Acetone	$[RuH(NO)(PPh_3)_3]$	2-Methoxyethanol (22%)
	$[RuH(NO)(PPh_3)_3]$ +2.5% H_2O	2-Methoxyethanol (97%)
Crotonaldehyde	$[RuH(NO)(PPh_3)_3]$	Butanol
Propionaldehyde	$[RuH(NO)(PPh_3)_3]$	Propanol

When the catalytic activity of the similar manner using Ir–NO and Rh–NO was compared, it was found that the reduction of the selected compounds, namely, styrene, heptene and hexene derivatives is slower in former than latter nitrosyl complex [121]. Like aforementioned, here the difference in the catalytic efficiency is again creditable to the lower dissociation of phosphine in the Ir–NO complex, and dissimilar binding fashion of metal–NO (α_D = 0.25 in Ir complex with M–N–O 180° and (α_D = 2.0 for Rh–NO complex with M–N–O 153°). Dichloro bis(triphenylphosphine) complex of Rh–NO has been shown to catalyse several hexanone derivatives for homogenous hydrogenation.

(v) Nitrosyl complex–mediated isomerization

Isomerization phenomenon brought up by several catalysts represents an area of huge interest. In dealing with the isomerization of 1-pentene, Zuech and co-workers suggest the use of dinitrosyl complexes of dichloro-bistriphenylphosphinemolybdenum and ethylaluminium dichloride as co-catalyst [123]. In the application of PPh_3-coordinated Ru–NO- and OS–NO-based catalytic systems have shown promising results during the isomerization between 1-hexene and internal olefins [124]. Among several factors affecting the rate of such reaction, the nature of phosphine derivative in the coordination core has been found to be quite effective, and hence the reaction rate under [RuH(NO)] form of $PMePh_2$ catalysis is 1/1,000 times slower as compared to the catalysis operational with [RuH(NO)] form of PPh_3, and here also M–N–O bonding mode has been the reason behind the observed dissimilarity [125].

Also, it has been found that the catalysis of terminal olefins fails using [Rh(NO)$(PPh_3)_3$] as a catalyst while H_2 and O_2 are not present. On the other hand, the traces of air complex become catalytically active [126, 127] for 1-hexene isomerization (10%) using DCM. In case of presence of hydrogen for the same reaction using the same catalytic class of nitrosyls does not show the formation of isomerized product. However, if benzene or THF (tetrahydrofuran) are used in place of DCM, the presence of H_2 induces the respective catalysis.

2.12.1.2 Metathesis

Metathesis reactions are industrially very significant targeting olefins (acyclic and cyclic), but in organic preparations the metathesis is very scarcely used. The applied interest of metal nitrosyls in metathesis has been widely accepted as the homogenous catalysts especially employing Mo–NO or W–NO complexes, and using alkylaluminium as a co-catalyst in the respective reactions. Several examples of this class of reactions have been illustrated in Table 2.21.

Table 2.21: Metathesis reaction of olefins in the presence of transition metal nitrosyl as a catalyst.

Olefin	Catalyst	Products
Pent-1-ene	$[Mo(NO)_2Cl_2(PPh_3)_2] + (CH_3)_3Al_2Cl_3$	4-Octene (48%), ethylene
	$[Mo(NO)_2Cl_2(PPh_3)_2] + C_2H_5AlCl_2$	Butene (18%), pentene-2 (25%), hexene (28%), heptene (11%)
	$[Mo(NO)_2Cl_2] + (CH_3)_3Al_2Cl_3$	4-Octene (24%)
Pent-2-ene	$[Mo(NO)_2Cl_2(PPh_3)_2] + (CH_3)_3Al_2Cl_3$	2-Butene and 3-hexene
	$[Mo(NO)_2Cl_2(PPh_3)_2] + C_2H_5AlCl_2$	2-Butene and 3-hexene
	$[W(NO)_2Cl_2(PPh_3)_2] + (CH_3)_3Al_2Cl_3$	2-Butene(9%), 3-hexene (l5%) and heptene(l%)
Hexa-1,5-diene	$[Mo(NO)_2Cl_2(PPh_3)_2] + (CH_3)_3Al_2Cl_3$	1,5,9-Decatriene (24%), 1,5,9,13-tetradecatetraene (7%), 1,5,9,13,17-octadeca-pentaene and ethylene

W. B. Hughes [128] while studying kinetics of pent-2-ene metathesis under dinitrosylmolybdenum complexes, $[Mo(NO)_2Cl_2L_2]$ (L = C_5H_5N, PPh_3 or $OPPh_3$) concluded the first order with respect to the rate of the reaction. The organo-aluminium co-catalysts served for the usefulness in the order $(CH_3)_3Al_2Cl_3 > C_2H_5AlCl_2 > (C_2H_5)_3Al_2Cl_3$. Quaternary ammonium group containing olefins give (α–ω)-bifunctional olefins on metathesis under the catalytic action of $[Mo(NO)_2Cl_2(PPh_3)_2]–C_2H_5AlCl_2$. As an example, let us consider the formation of 2-butene and $[Me_3N(CH_2)_2CH{=}CH(CH_2)_2NMe_3]^{2+}$ using iodo salt of *trans*-$[Me_3NCH{=}CH(CH_2)_3NMe_3]^+$, indicating that metathesis of such olefins might be because of the possibility of weakening in basic strength among the constituent nitrogens [129] and may be suggested as a fundamental reason for all other similar systems studied for metathesis.

2.12.2 Coordinated NO group: some reactions

2.12.2.1 Oxime (R = NOH) formation

(a) Pentacyanonitrosylferrate(II) shows the following anion production on hydrolysis giving 2-oxopropanol-1-oxime [130]:

$$CH_3COCH_3 + \left[Fe(NO)(CN)_5\right]^{2-} \xrightarrow{OH^-/H_2O} \left[Fe(H_2O)(CN)_5\right]^{3-} + H_3CC(O)CH{=}NOH$$

(b) McCleverty et al. [131] have demonstrated the formation of benzaldoximes by conducting the reaction of benzyl bromide ($C_6H_5CH_2Br$) with ruthenium dinitrosyl, $[Ru(NO)_2(PPh_3)_2]$, under constant refluxing method in toluene ($C_6H_5CH_3$) fitted with carbon monoxide mediation. The mechanism reveals that the reaction mainly involves the electrophilic $PhCH_2^+$ addition to NO group giving rise to the formation of $RuN(O)CH_2Ph$.

(c) Similarly, Foâ and Cassar [132] reported aldoxime formation from a series of derivatives of benzyl halides using cobalt carbonylnitrosyls of general formula $N(C_4H_9)_4[Co(CO)_2(NO)X]$ with the respective yield of 30–50%. Despite the unknown reaction path for these transformations, it has been suggested that primary stage of the reaction mainly involves the oxidative addition of the respective benzyl halide towards the cobalt centre followed by change in the position of the benzyl group by occupying the nitrogen binding site of NO functionality to form the final desirable aldoxime. The pathway described thus may be shown as follows:

$$C_6H_5CH_2Cl + \left[Co(CO)_2(NO)X\right]^- \rightarrow C_6H_5CH_2-Co(CO)_2(NO)X + Cl^-$$

$$C_6H_5CH_2-Co(CO)_2(NO)X \rightarrow \underset{\substack{\downarrow \\ C_6H_5CH=NOH}}{C_6H_5CH_2-N=O} + \left[Co(CO)_2X\right]$$

2.12.2.2 Diazotization of aromatic amines

Diazotization reaction which serves as one of the most significant industrially relevant transformations carried out catalytically can be conducted using [Ru(NO)Cl(bipy)$_2$](PF$_6$)$_2$ [133] as catalyst as shown in the following illustration:

$$\left[(\mathrm{bipy})_2(\mathrm{Cl})\mathrm{Ru(NO)}\right]^{2+} + \mathrm{H_2NAr} \xrightarrow{\mathrm{CH_3CN}} \left[(\mathrm{bipy})_2(\mathrm{Cl})\mathrm{Ru}-\overset{\mathrm{O}}{\mathrm{N}}-\mathrm{NH_2Ar}\right]^{2+}$$

$$\left[(\mathrm{bipy})_2(\mathrm{Cl})\mathrm{Ru}-\overset{\mathrm{O}}{\mathrm{N}}-\mathrm{NH_2Ar}\right]^{2+} + \mathrm{B(Base)} \rightarrow \left[(\mathrm{bipy})_2(\mathrm{Cl})\mathrm{Ru}-\overset{\mathrm{O}}{\mathrm{N}}-\mathrm{NHAr}\right] + \mathrm{BH^+}$$

$$\downarrow \mathrm{BH^+}$$

$$(\mathrm{bipy})_2(\mathrm{Cl})\mathrm{Ru}-\mathrm{N}=\mathrm{NAr} + \mathrm{B} + \mathrm{H_2O}$$

2.12.3 Metal nitrosyls as depolluting agents

As the composition of atmosphere indicates that among different gases oxygen and nitrogen are 99% abundant present in several forms of oxides in addition to the molecular existence. All the forms of NO$_x$ vary in their concentration, and N_2O is considered as the most important one. This oxide form of nitrogen mainly confines to the lower regions among different atmospheric layers. Volumetric concentration of the nitrogen oxides other than N_2O is having abundance of the order of 10^{-8}.

The combustion of fuels by the internal combustion engine is linked with the production of air pollutants such as NO, CO, incompletely oxidized hydrocarbons and SO_2. Notably, CO and NO are the key air pollutants. However, partial oxidization (burning) of hydrocarbons results in the production of smog forming harmful compounds. Similarly, the other oxide pollutants of sulphur, halogens and lead are also remarked pollutants (though bearing meagre importance in this context).

Motivational measures linked with reducing the harmful exhausts from combustion phenomenon, in general, and the gaseous substances emitted by vehicle engines in special triggered the respective studies for so many years brought up by several governments and EPA (Environmental Protection Agency) for the interest towards air depollution. For example, N_2 and O_2 with the assistance of CO_2 to change into NO. This reaction has been shown to get initiated and completed when the combustion is at post-flame stage. The sequential sub-stages of the reaction may be illustrated as follows:

$$\mathrm{CO} + \mathrm{OH} \rightarrow \mathrm{CO_2} + \mathrm{H};\ \mathrm{O_2} + \mathrm{H} \rightarrow \mathrm{OH} + \mathrm{O}$$

$$\mathrm{N_2} + \mathrm{O} \rightarrow \mathrm{NO} + \mathrm{N};\ \mathrm{O_2} + \mathrm{N} \rightarrow \mathrm{NO} + \mathrm{O}$$

The thermokinetic studies reveal that the NO generation via these sequential steps is directly proportional to the temperature change in the presence of gases like CO_2 [134].

2.12.3.1 Scientific temper towards CO-free vehicle exhausts

For each harmful gas produced during engine combustions, separate methodologies are in practice to capture them before getting into air or technical advancements are in demand to construct vehicles having least or minimized harmful gaseous emissions. Meanwhile, following the EPA (USA-1975) directions in removing CO from such exhausts using sufficient air (i.e. with air–fuel ratio >14.7 weight-wise), and also by incorporating secondary air inlets to completely change the CO to CO_2 by the help of catalysts like Pb, P and S gained considerable attention.

2.12.3.2 Transformation of nitric oxide (NO) to less harmful products

Harmful gases have triggered scientists to develop strategies to convert the toxic into nontoxic form or to capture these pollutants in way to purify the environment or to make their concentration to a non-lethal level. Among these toxic gases, these transformation aspects have also been applied in case of the NO. Efforts concerned with converting NO have been appreciated by conducting this form of oxide to products like ammonia, nitrogen gas and N_2O. Therefore, this interest has principally come up from efforts to remove, or at least reduce, the concentration of pollutant NO in exhaust gases emanating from internal combustion engines. Meanwhile, consistent investigations [135–137] have been reported describing NO transformation by employing heterogeneous catalysis under the activity of transition metals. Several successful catalytic works of this form in special attention to Pd-, Pt- and Ru-based systems reported the conversion of harmful NO to NH_3 restricted under reducing conditions. In addition to ammonia production, Ru-based catalysis was proven to act as an effective N_2 formation catalyst from NO under CO environment.

Due to the catalytic role of transition metal nitrosyls in this specific conversion, the purpose of the following section is to present the role of homogeneous processes regarding this transformation strategy applicable for NO.

2.12.3.3 Formation of N_2

Hieber and Pigenot [138] observed that the cobalt-based nitrosyl complexes having general representation $[Co(NO)_2X]_2$ (where X is Cl or Br) when allowed to react with PPh_3 or PPh_3 derivatives of As and Sb form N_2 as follows:

$$[Co(NO)_2X]_2 + 6PPh_3 \rightarrow [Co(NO)(PPh_3)_3] + [Co(X)_2(OPPh_3)_2] + OPPh_3 + 3/2N_2$$

Similarly, ammonia reacts with $[Ir(NO)X_5]^-$, it converts the parent nitrosyl complex [139] to $[Ir(NH_3)X_5]^{2-}$ with formation of N_2. Iron dinitrosyl $[Fe(NO)_2Br]_2$ dissolves in molten triphenylphosphine, which gives rise to the formation of $[Fe(NO)_2(PPh_3)_2]$, $[Fe(NO)Br(PPh_3)_2]$, $OPPh_3$ and N_2. Interestingly, iron dinitrosyl $[Fe(NO)_2I]_2$ in a similar fashion reacts with PPh_3, but on the other hand, $[Fe(NO)_3I]$ while interacting with pyridine forms $[Fe(py)_6](NO_2)I$ and nitrogen gas [140].

Trithiazyltrichloride [141] reacts with $[Rh(NO)(PPh_3)_3]$ and $[Rh(NO)_2X_2\ (PPh_3)_2]$ (X = Cl, Br) in THF gives $[Rh(\mu\text{-}NS)Cl_2(PPh_3)]_2$, $OPPh_3$, $SPPh_3$ and N_2.

2.12.3.4 Formation of N_2O

The second important strategy employed in changing NO to N_2O using metal nitrosyls can be explained by taking the example of NO-tagged rhodium complex [142] of general formula $[Rh(NO)_2Cl]_4$. The compound has been observed to react with PPh_3 or As/Sb forms of PPh_3, to give N_2O at RT. This transformation has been interpreted in terms of disproportionation reaction. The similar reaction in case of Co/Rh dinitrosyls has been reported [143]. Both these reactions can be illustrated as follows:

$$[Rh(NO)_2Cl]_4 + 12PPh_3 \rightarrow 2[Rh(NO)(PPh_3)_3] + 2[Rh(NO)Cl_2(PPh_3)_2] + 2OPPh_3 + 2N_2O$$

$$[M(NO)_2Cl]_n + 4PPh_3 \xrightarrow[THF]{Na/Hg} [M(NO)(PPh_3)_3] + OPPh_3 + N_2O$$

Broomhead et al. [144] demonstrated the production of N_2O in the similar way using $[Mo(NO)_2(S_2CNEt_2)_2]$. Two research groups [145, 146] while targeting Ir/Fe dinitrosyls have demonstrated nucleophilic attack on NO using hydroxylamine to produce N_2O. The reactions may be shown as follows:

$$[Ir(NO)X_5]^- + NH_2OH + H_2O \rightarrow [Ir(H_2O)X_5]^{2-} + N_2O + H_3O^+$$
$$[Fe(NO)(CN)_5]^{2-} + + NH_2OH + H_2O \rightarrow [Fe(H_2O)(CN)_5]^{3-} + N_2O + H_3O^+$$

In several examples, free NO when allowed to react with metal-coordinated NO gives NO_2^- or $N_2O_2^{2-}$ form of the respective complexes in association with free N_2O. In general, hyponitrito-based complexes also release N_2O under acidic conditions [147]. For instance, these reactions may be presented as follows:

$$[Co(NO)(en)_2Cl]Cl + 2NO \xrightarrow{CH_3OH} [Co(NO_2)(en)_2Cl]Cl + N_2O$$

$$[Co(NO)(en)_2(MeOH)]Cl_2 + 2NO \xrightarrow{CH_3OH} [Co(NO_2)(en)_2Cl]Cl + N_2O$$

$$[Co(NO)(DMGH)_2(MeOH)] + 2NO \xrightarrow{CH_3OH/py} [Co(NO_2)(DMGH)_2(py)] + N_2O$$

$$[Ir(NO)_2Br(PPh_3)_2] + 2NO \xrightarrow{2-Butano} [Ir(NO)(NO_2)Br(PPh_3)_2] + N_2O$$

$$[Co(NH_3)_5(NO)]^{2+} + NO \rightarrow [Co(NH_3)_5(N_2O_2)]^{2+} \text{ (hyponitrito complex)}$$

$$[Co(NH_3)_5(N_2O_2)]^{2+} + NO + H_3O^+ \rightarrow [Co(NH_3)_5(OH)]^{2+} + NO_2^- + N_2O$$

$$2[Rh(NO)_2Cl(PPh_3)_2] \rightarrow [\{Rh(NO)(PPh_3)_2Cl\}_2N_2O_2] \text{ (hyponitrito complex)}$$

$$[\{Rh(NO)(PPh_3)_2Cl\}_2N_2O_2] \xrightarrow{HCl} [Rh(NO)Cl_2(PPh_3)_2] + N_2O + H_2O$$

2.12.3.5 Formation of N_2O and CO_2 from NO and CO

Transformation technique (highly harmful to less harmful/no harmful) is of fundamental interest for being significant in decreasing the environmental deterioration:

$$2NO + CO \rightarrow N_2O + CO_2$$

It may be noted that the above transformation reaction does not take place even at 450 °C without any suitable catalyst. Under heterogeneous catalysis, the reaction gets completed at 400 °C. Several reports do exist related to the application of homogeneous catalysis in case of these reactions at low temperature. Only few of them are presented here.

Johnson and co-workers [148–150] reported the reaction given below catalysed by the help of Ir or Rh dinitrosyl complexes containing co-ligand PPh_3 in the following way (where M represents Ir or Rh). It may be noted that the catalyst gets finally reproduced by replacing CO with NO:

$$\left[M(NO)_2(PPh_3)_2\right]^+ + 4CO \rightarrow \left[M(CO)_3(PPh_3)_2\right]^+ + N_2O + CO_2$$

Dissimilar to the above two complexes, Ru and Os dinitrosyl complexes [150] undergo the reaction with CO at a very slow rate at RT. However, $[Pt(N_2O_2)(PPh_3)_2]$ [149] shows stereospecific reaction while converting to respective carbonyl forming *cis*-dicarbonyl form with the evolution of N_2O and CO_2 as follows:

$$\left[Pt(N_2O_2)(PPh_3)_2\right] + 3CO \rightarrow cis - \left[Pt(CO)_2(PPh_3)_2\right] + N_2O + CO_2$$

Therefore, based on the above findings, the general reaction representing the catalytic conversion suggests the primary step involving CO attack to transform $[M(NO)_2(PPh_3)_2]^+$ into $[M(N_2O_2)(CO)(PPh_3)_2]^+$ followed by to continue reacting with CO resulting in the formation of $[M(CO)_3(PPh_3)_2]^+$ along with the evolution of CO_2 and N_2O. Hence, the initial form of catalyst is finally regained by letting NO to react with $[M(CO)_3(PPh_3)_2]^+$ keeping 80 °C as the temperature of the reaction. The overall transformatory equations may be summarized [150] as follows:

$$[M(N_2O_2)(CO)(PPh_3)_2]^+ + 3CO \longrightarrow [M(CO)_3(PPh_3)_2]^+ + N_2O + CO_2$$

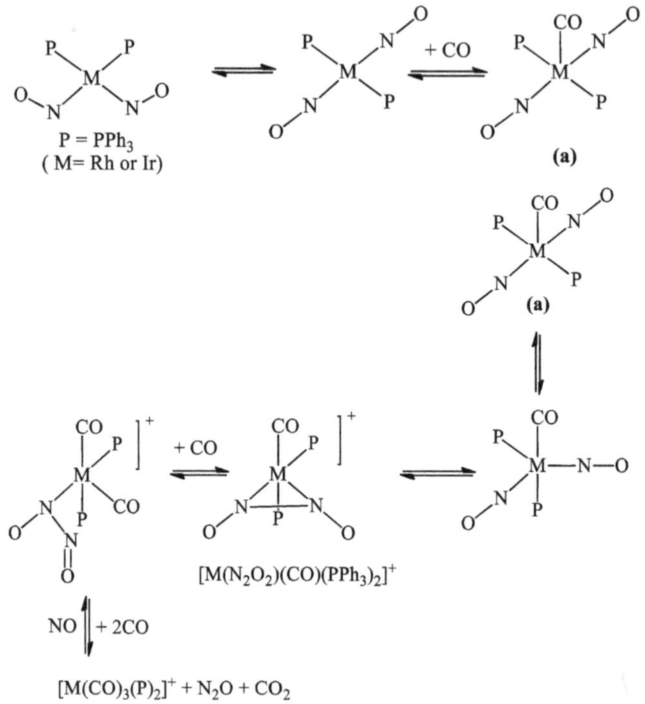

Since the involvement of such catalysts mainly confines with the backdonation towards NO, the respective efficiency follows the order showing maximum tendency for Fe and lowest for Rh in the series Rh, Ir, Ru, Os, Co and Fe indicating other metal centres in between Fe and Rh. In addition to backdonation tendencies towards NO solvents used in these types of reactions also matter. Haymore and Ibers [151] in 1974 also demonstrated the catalysis of reaction, 2NO + CO N_2O + CO_2 using some rhodium and iridium nitrosyl complexes. The effect of butanol and DMF studied by Kaduk et al. [152] suggests that rhodium dinitrosyl complex in butanol at 20 °C shows the above reaction in a slow manner as compared to the same reaction when carried out in DMF at 62 °C. The reported work also found that when water was added to this system, the reaction rate was noticed to increase, but less than $[Rh(CO)(NO)_2Cl_2]^-$ when employed as a catalyst [153]. This increasing trend may be because of the coordination effect of water over DMF [154].

2.12.3.6 Formation of NH_3

Followed by other possible interconversions laid for NO, the transformation into NH_3 has also been appreciated and highly fascinated as denoted by the work reported by Ellermann and Hieber [155], carrying out the reaction of carbonylnitrosyl compound of cobalt showing decomposition to N_2 and NH_3 as illustrated below:

$$3[Co(NO)(CO)_3] + 9OH^- \rightarrow [Co(CO)_4] + 2Co(OH)_2 + HCO_3^- + 3CO_3^{2-} + N_2 + NH_3$$

The mechanism of such reactions has been found to be a nucleophilic attack on NO ligand by any suitable nucleophile like H^- exemplified by two research groups [156, 157] by allowing $[Co(NO)_2(PPh_3)_2]^+$ and $[Rh(NO)_2Cl(PPh_3)_2]$ to react with sodium borohydride to produce $[Co(NO)(PPh_3)_3]$ and $[Rh(NO)(PPh_3)_3]$, respectively (generating NH_3), signifying the intervention of H^- as a nucleophile. In the current era, scientists are anxious to bring forth an efficient way to deal with nitrogen storage and interconversions. Taking NO as an intermediatory form of such transformations like modelling of nitrogenase action and other salt conversions, these reactions not only gesture the decontaminating stage but also represent highly economical ways to interpret nitrogen-based fertilizer formulations as well. Whether it is forward or backward direction of ammonia conversions, these sets of investigations represent highly fascinating zones of research.

2.12.3.7 Formation of N_2 and N_2O

The conceptual hindrances in finding the intermediatory products of several reactions connected with the production of N_2O and N_2 from cyanonitrosyl complexes, Swinhart and co-workers [158] studied the respective kinetics with the help of isotope labelling using the reaction between nitroprusside and azide ions. The results obtained by the group indicated the formation of reaction intermediate in the form of $[Fe(NO)N_3(CN)_5]^{3-}$, which directs decomposition to release N_2 and N_2O. Similar observations have been reported for some Ru- and Ir-based nitrosyl complexes [159, 160]. However, for $[Co(NO)(PPh_3)_3]$ [161] complex on direct NO reaction shows the following transformation. This transformation while in aqueous medium has been found a catalytic effect by geometrical isomerization as well [162]:

$$[Co(NO)(PPh_3)_3] + 7NO \rightarrow [Co(NO)_2(NO_2)(PPh_3)] + 2OPPh_3 + 2N_2O + 1/2N_2$$

2.13 Applications of metal nitrosyls

NO-based treatment therapies and industrial applications for systematic nitroso, nitrosyls and nitrite salts have been found interestingly showing upsurge in the context of applied aspects. Meanwhile, coordination compounds of NO with respect to carriage centre as a metallic moiety has furnished fascinating results shown by good record of investigations published so far. The potentiality of metal nitrosyls in biomedical field and several other areas in chemical industry together with a catalytic role as depolluting agents and the significant contribution of this molecule in almost every kingdom of living things have built up a huge field of its form called as NO chemistry. This is supported by the existence of several journals and societies framed in the name of NO. Therefore, the potential applications of NO-bonded metallic compounds (metal nitrosyls) have been addressed in this section.

2.13.1 Biomedical science

The biological recognition of NO revealed in 1992 as a vital signalling molecule persuaded the scientists to declare NO as "Molecule of the year" published in *Science* – a famous journal of the AAAS. The evaluation of NO in further areas wondered the whole world when platelet adhesion/aggregation, neurotransmitting physiology, vascular regulation, several liver metabolic pathways and normal kidney functioning were also found to be related with this molecule. The ethereal lived NO with 2–300 nM concentration was found to be an optimal level for normal physiological phenomena. At higher μM concentrations, the biological role as a significant immunomodulator and tumour suppressor expanded the interest. Such amazing revelations convinced scientists to declare NO as a 1998 Nobel Prized molecule (medicine) shared by Robert F. Furchgott, Louis J. Ignarro and Ferid Murad.

In the past four decades, NO is found to be a significant signalling molecule in a wide range of physiological processes. These comprise blood pressure control, neurotransmission, immune response and cell death. Just after these findings, efforts are being made by researchers of different disciplines constantly towards the development of exogenous synthetic NO donors that can deliver NO to biological targets to elicit desired responses. Several NO donors have been developed in this regard, including organic nitrites and nitrates, nitrosothiol, diazeniumdiolates (NONOates) and lastly transition metal–based NO donors such as SNP including other nitrosyl complexes. The search for new storage release systems, capable of delivering NO to desired targets, has stimulated the chemistry of metal nitrosyl complexes. As of now to develop NO-releasing molecules (NORMs) for a desirable target and transition metal nitrosyls are continuously examined in special reference to Fe, Ru and Mn nitrosyls for bearing NO lability under low-energy electromagnetic radiations (EMR; visible light, excitation). The various factors that are said to be concerned with NO donors include the specificity in terms of biological target, the structural framing of a usable compound, reactive behaviour in biological media, pH, purpose of NO release and many other redox concepts. In most of the cases, the available NO donors show non-specificity, and release NO spontaneously without controlled manner and optimal concentration. In due course, some cases have shown that the rate of NO release within a biological target can be modified by change in temperature, pH or other biological catalysts. The current success in managing tumour or other forms of cancers using NO therapy favourably triggered the design of controlled releasers of NO at a selected target.

With the beginning of photodynamic therapy (PDT) [163] as a general medication for certain (especially skin) cancers, NO-donating molecules activated by light have got too much interest. Moreover, the laser treatment permits for more specific targeting than systemic drugs alone. At the beginning, it was documented that NO complexes of transition metals could release NO when exposed to light. For instance, quite a few iron-based nitrosyls comprising SNP ($Na_2[Fe(NO)(CN)_5]$) and Roussin's salts were observed to release NO when exposed to light. On the contrary, these complexes also release NO spontaneously (i.e. in the dark), and frequently changes in pH and temperature also. These factors persuade loss of NO, making them non-specific for PDT. Moreover, side effects of labile subordinate cyanide ligands often confine the use of SNP.

2.13.1.1 Endogenous NO generation/biosynthesis of nitric oxide

Present in mammal's cells, arginine is one of the most multipurpose amino acids, serving as a precursor for the synthesis of not only proteins but also of **NO**, urea, polyamines, proline, glutamate, creatine and agmatine [(4-aminobutyl)guanidine]. Physiologically, the NOS enzyme is responsible to generate NO from L-arginine (targeting guanidino nitrogen) (Figure 2.51). Enzyme NOS (NO synthase) converts L-arginine (an amino acid available in living organism) to L-citrulline and NO. The co-substrates for the reaction encompass NADPH and O_2. The redox insights of the overall reaction indicates that this process of NO generation is accompanied by five-electron oxidation reaction, wherein NADPH contributes three electrons and further two are sourced by O_2. From the fully addressed reports on the isoforms of NOS, endothelial NOS denoted as eNOS accounts for maintenance of vasodilation, inducible NOS assigned as iNOS is immune responsive form of NOS, and the form of NOS labelled as nNOS is found to be responsible for the maintaining of normal neurotransmission.

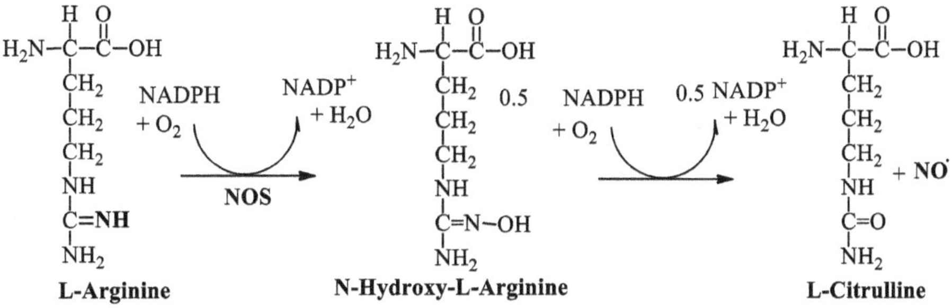

Figure 2.51: Generation of NO from L-arginine catalysed by NOS in the presence of NADPH and O_2.

The regeneration of the substrate L-arginine from L-citrulline (biosynthesis) is executed by the cytosolic enzymes argininosuccinate synthetase (ASS) and argininosuccinate lyase (ASL) (Figure 2.52). The reaction catalysed by ASS has the need of L-aspartate in the presence of ATP as a co-substrate and is the rate-limiting step. ASS and ASL appear to be distributed in many cells, although the degree of distribution and the efficiency of this pathway appear to differ substantially between different cells. Since during NO synthesis, L-arginine, via *N*-hydroxy-L-arginine as an intermediate product, is converted to L-citrulline, the immediate use of L-citrulline for the re-synthesis of L-arginine may be an effective way to maintain sufficient substrate supply for a prolonged NO synthesis and has therefore been described as citrulline/NO cycle.

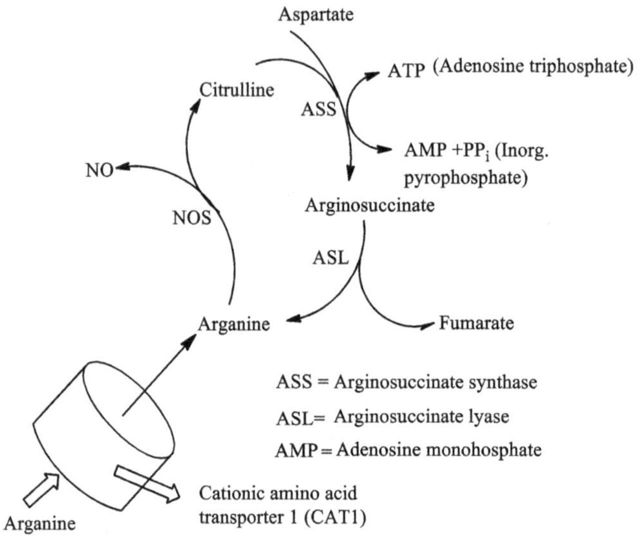

Figure 2.52: A model showing regeneration of L-arginine from L-citrulline under eNOS operation maintained by the cationic amino acid transporter 1 (CAT1).

2.13.1.2 Exogenous NO donating molecules

While glyceryl trinitrate (GTN) and nitroprusside (SNP) have been used for many decades to regulate hypertension, methodical research on exogenous NO donors has gained significant momentum because of the potent and significant role of NO in maintaining mammalian homeostasis. Both organic and inorganic chemists have played crucial roles in such strive and selected examples of compounds of such endeavour are presented below.

2.13.1.2.1 Organic nitrates

Organic nitrates ($RONO_2$) are nitric acid esters of mono- and polyhydric alcohols. These are the oldest class of NO-donating compounds, which have been clinically used. Illustrative organic nitrates encompass GTN, pentaerythrityl tetranitrate, isosorbide dinitrate (ISDN), isosorbide 5-mononitrate and nicorandil (Figure 2.53). The partially denitrated metabolites of GTN, glyceryl dinitrate and glyceryl mononitrate are still pharmacologically active but considerably less effective than GTN. The key biological effects of nitrates are owing to the formation of NO. The NO release from organic nitrates needs either enzymatic or non-enzymatic bioactivation, where a three-electron reduction [164] is comprised. When GTN is exposed to mammalian tissues, NO is released generally from the C-3 carbon [165]. As per the Ignarro hypothesis, GTN enters the smooth muscle cell, where it is changed to NO_2^- by reaction with cysteine. The nitrite then releases NO via nitrous acid (HNO_2). The NO so released combines with thiol to generate nitrosothiol, which activates soluble GC for vasodilation of blood vessels.

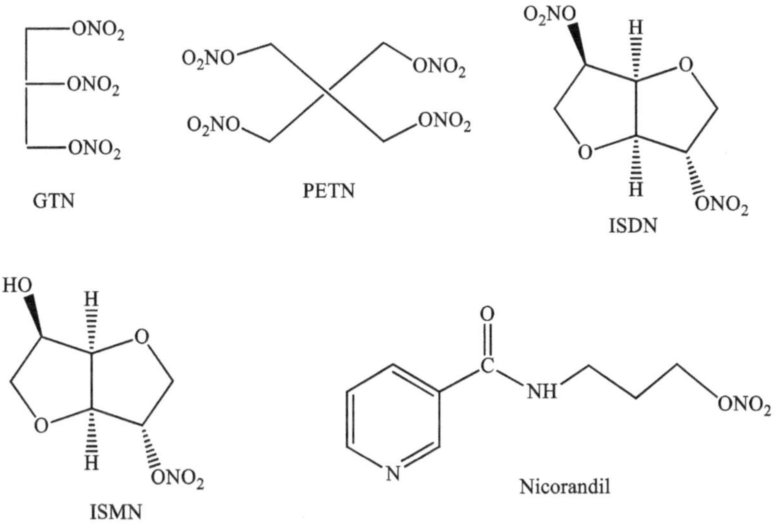

Figure 2.53: Structure of some organic nitrates.

GTN is believed to help in treating chest pain (angina pectoris) and is one of the most significant organic nitrates applicable for vasodilation. Figure 2.53 displays the potent therapeutic agents administered via sublingual and/or transdermal pathways. The mechanistic studies of evaluating the reasons lying behind such vasodilatory effects prove that vasodilation responsiveness is because of activation developed within the GC. This in further perspective indicates that biotransformatory process of a nitrate with respect to sulphhydryl or other ferrous functionalities within biological media finally generates NO.

In fact, all types of blood vessels are relaxed using GTN. Coronary arteries are found to be more susceptible to GTN than peripheral arteries. GTN has also been used as medication of critical myocardial infarction [166] (it is a permanent damage to the heart muscle. "Myo" means muscle, "cardial" refers to the heart and "infarction" means death of tissue due to lack of blood supply). Moreover, GTN is used in congestive heart failure [167]. The smooth muscle relaxation produced by the use of GTN medication for anal fissures [168] treatment and as erectile functioning agent [169] gesture significant use of this NO-based drug.

2.13.1.2.2 Organic nitrites

Organic nitrites shown in Figure 2.54 are clinically administered vasodilators and are famous nitrites having so long history of medical use. IAMN (isoamyl nitrite) is not only usable for vasorelaxation but also renders inhalant antidote action for cyanide poisoning, and the mechanism is simple to combat CN^- at haeme site with NO generated from IAMN in association with cytochrome c oxidase to develop the respective inhibitory effects.

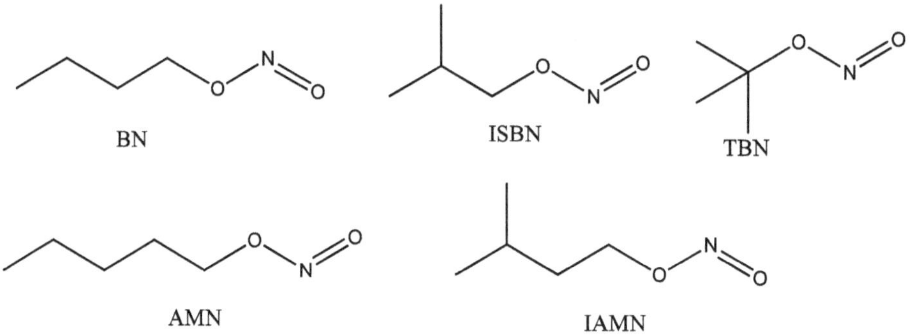

Figure 2.54: Structure of some organic nitrites.

Organic nitrites can produce NO [170] in vivo. The conversion of nitrites to NO needs one-electron reduction. They also undergo rapid hydrolysis to give nitrite ion and the corresponding alcohol. The nitrite ion is not an active intermediate in vascular metabolism of nitrite esters but can be reduced to NO. Since the nitrosyl moiety of nitrites can be readily transferred to a sulphhydryl group [171] (group consisting of sulphur bonded to a hydrogen atom), *S*-nitrosothiols may be formed from organic nitrites in vivo. This shows that NO release from nitrites, an enzymatic process in vascular smooth muscle, was primarily associated with the cytosolic [172] (as opposed to a microsomal fraction in the case of the organic nitrates). The metabolism of alkyl nitrites studied by several research groups reveal glutathione transferase–mediated reaction to generate *S*-nitrosoglutathione [173] or *S*-nitrosothiols compatible with biological membranes [174], and recently Doel et al. [175] reported that the involvement of xanthine oxidase could catalyse organic nitrites anaerobically to produce NO.

Alkyl nitrites used as inhalers result in vasodilation, increased heart rate and decreased systolic blood pressure [176]. The vasodilatory effect after inhalation of organic nitrites has been used to ameliorate angina pectoris [177] since 1867. They are more effective than nitrate esters. Amyl nitrite and *n*-butyl nitrite have been commonly used in ameliorating angina pectoris. Dissimilar to nitrates, such as GTN, continuous infusion of organic nitrites [178] in the same animal model does not produce tolerance. In fact, Fung and Bauer [179] claimed the long-term, continuous administration of organic nitrites as an effective vasodilator therapy.

2.13.1.2.3 Other organic donors

S-Nitrosothiol analogue like *S*-nitroso-*N*-acetyl-pencillamine (SNAP) (Figure 2.55) has been employed as an NO donor drug in animals for safety against haemorrhagic shock. SNAP-containing creams have been found to be successful in treating cutaneous leishmaniasis (skin infection caused by protozoans). The diazeniumdiolates, namely, diethylamino-NONOate (DEA-NO) (commonly known as NONOates) produce two equivalents of NO via a pH-dependent pathway (in aqueous environment at physiological pH). Moreover, decomposition of this type of NO donor releases two molecules of NO (Figure 2.56).

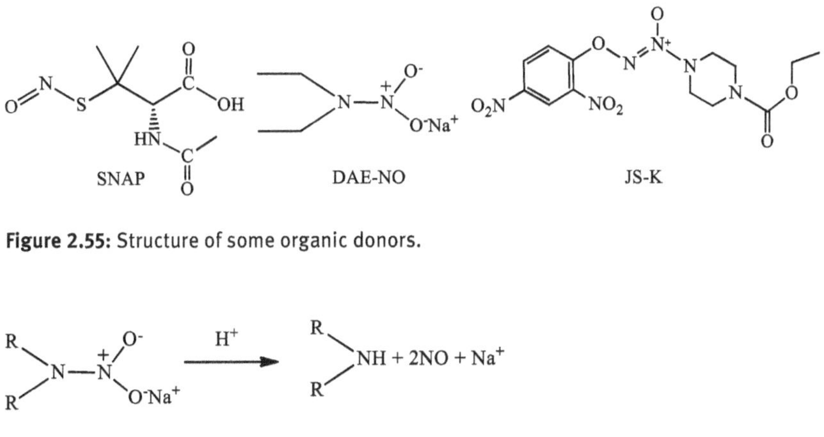

Figure 2.55: Structure of some organic donors.

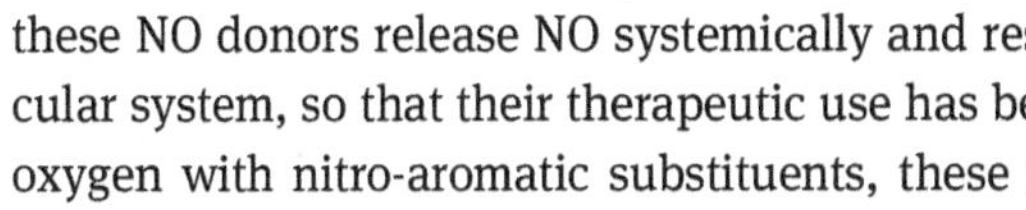

Figure 2.56: Decomposition of diazeniumdiolate, a nitric oxide donor.

Diazeniumdiolate [180] compounds have shown anti-leukaemia activity. However, these NO donors release NO systemically and result in severe side effects on the vascular system, so that their therapeutic use has been restricted. On modification at the oxygen with nitro-aromatic substituents, these derivatives can release NO in target cells after a hydrolytic or enzymatic action. JS-K, an example of O-protected diazeniumdiolate developed by the US National Cancer Institute, has attracted much attention as promising anticancer drugs [181]. Moreover, $NaNO_2$, a combined ascorbic acid treatment of skin inflammation, has been found considerably beneficial [182].

2.13.1.2.4 NO donors incorporated in polymeric matrices

Different concerns in specific use of NO donors have triggered investigations in finding the link of a suitable compound with various regulatory factors, protective approach and side effects. Hence, various NO donors have been reported so far, and based on the efficacy and ease of NO release polymer-based biocompatible matrices have been keenly eyed. In practical sense, such incorporated/immobilized materials have been reported to have a well-remarkable use as patches, wound dressings and are also served in the form of coatings over blood-containing devices in medical use. The main NO donors mentionable in that context are basically light-sensitive NO releasers.

The major problem accompanying with the conventional NO donors, namely, GTN, SNAP and SNP, is the absence of control of NO release. The drug goes everywhere in the body and NO is released via various enzymatic and non-enzymatic (e.g. heat or pH) pathways. This unstoppability of NO release and the unpredictable rates of its release become a major concern in many biomedical applications. For example, NO release from *N*-diazeniumdiolates is initiated by pH and it is relatively difficult to control the NO release inside the body via pH manipulation. If NO release is based on enzymatic reaction (e.g. in case of GTN and isoamyl nitrite), then it is more difficult to use the NO donors at selected targets (e.g. malignant sites). On the other hand, control of NO release by light is quite reasonable. As the developments in fibre optic technology allow one to provide light at almost all cellular targets, exogenous NO donors that could be activated by light are very desirable NO drugs (photoactive NO donors) for site-specific NO delivery.

In the recent past, NO donors have been amalgamated in various polymeric matrices and such composites have been utilized to deliver NO to selected targets under controlled conditions [183]. As the presence of NO in the vascular system prevents blood platelet adhesion and clot formation, exogenous NO donors [184] have been used to prevent blood clotting on surgical equipment, vascular stents and pacemaker contacts. Since both NO(g) and NO-donating drugs have been shown to drastically reduce microbial (bacterial, fungal and parasitic) loads in human and animal models [185], agents like SNAP and DEA-NO have been incorporated in different polymeric materials to treat chronic infections, wounds and diabetic leg ulcers via transdermal (through skin) delivery of NO. These formulations are particularly effective against different antibiotic-resistant strains of common pathogens such as *E. coli* and *Staphylococcus aureus*. NO has also been mainly found as an assisting agent in the synthesis of collagen, and regeneration of a wounded site of a tissue in addition to microbicidal effects.

Besides preventing platelet aggregation, NO donors, namely diazeniumdiolates and *S*-nitrosothiols, have also been shown to persuade apoptosis in a range of cell types comprising cancer cells. In order to use NO donors as prospective anticancer agents, micro- to milli-molar level of NO must be generated at a selected location for a short duration of time, mimicking the action of iNOS in apoptotic cells. As mentioned previously, these systematic NO-releasing materials cannot be used to deliver NO at malignant sites on demand and, therefore, NO donors that release NO photochemically have triggered much attention. Of the photosensitive NO donors, only *S*-nitrosothiols have been amalgamated into polymeric materials. Olabe and coworkers [186] modified a gold surface with dithiothreitol, which upon reaction with a solution of nitrosothiol formed S–NO bonds on the surface. Revelation of the S–NO monolayer to visible light (30 W) resulted in the photorelease of NO (Figure 2.57). This resembles the behaviour of nitrosothiols in aqueous solution. *S*-Nitrosothiols act as NO donors via homolytic breaking of the S–NO bond by light of 330–350 or 550–600 nm wavelengths, releasing NO and producing disulphide bonds between parent thiols.

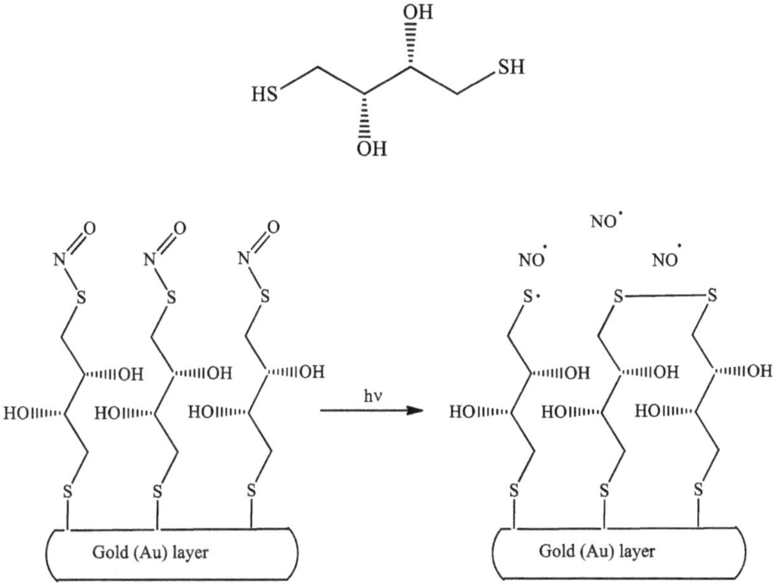

Figure 2.57: Gold-coated-S-nitrosylated dithiothreitol stimulates NO photorelease on exposure to visible light irradiation.

2.13.1.3 Transition metal nitrosyl complexes

The chemistry of metal nitrosyl complexes has got additional significance in the recent past because of the important role transition metals play in the biological process of NO, as well as the possibility of producing thermodynamically stable and kinetically labile species. Such approaches have focused on the development of pharmacological substances able to release NO at specific rates in tissues, in order to conquer NO deficiency. The great affinity of d^5 and d^6 low spin and in low oxidation state metal complexes for NO and the versatility of NO as a ligand make the nitrosyl complexes a good alternative for such a proposal. Iron, ruthenium and manganese are good candidates as a model for NO carriers. Indeed, several iron-based nitrosyls comprising SNP ($Na_2[Fe(NO)(CN)_5]$) and Roussin's salts [187] (Figure 2.58) were found to release NO when illuminated with light. In spite of its well-known toxicity owing to the presence of five labile cyano groups, SNP is largely used as an NO releaser in clinical practices for blood pressure regulation in cases of severe hypertension [188].

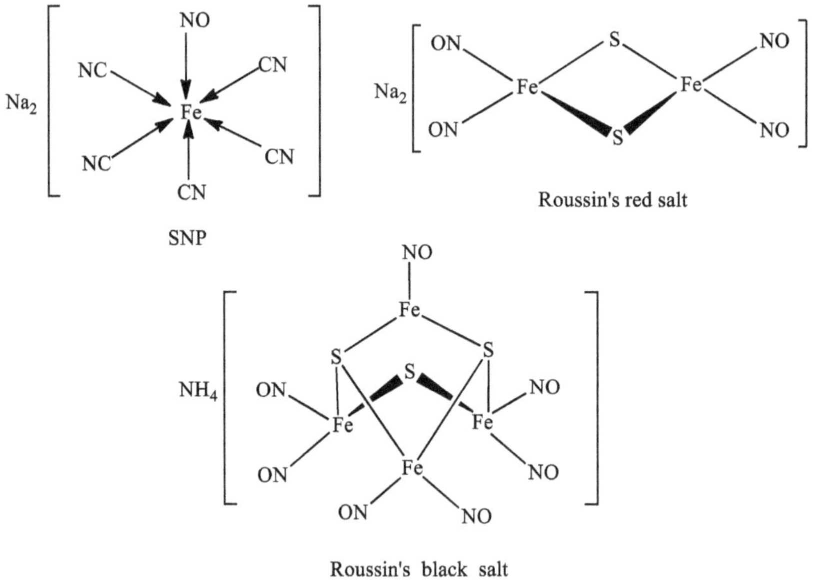

Figure 2.58: Structure of some iron-based nitrosyls.

In recent years, the formulation and medical applications encompassing the feasibility of the complex form of metals involving NO as a co-ligand stretched this area of research when SNP was successfully found to stabilize hypertension. With the enhanced fascination driven towards other forms of SNP-like compounds involving Roussin's salts (Figure 2.58) [189], the light-induced NO-donoring capability using 300–500 nm energetic light radiation is really among admirable findings related to NO chemistry. For example, the iron-based nitrosyl complex $[(PaPy_3)Fe(NO)](ClO_4)_2$ studied by Mascharak and co-workers [190, 191] was shown to behave as NO releaser under visible light irradiation. Similar approach was reflected from several nitrosyl complexes of chromium [192] and manganese [193]. The stability parameter in biological medium set some challenges for NORM designers. The only reported exception to this case has been found in the form of a manganese nitrosyl $[(PaPy_3)Mn(NO)](BF_4)$ [194] having the potentiality to deliver NO specifically at biological targets involving myoglobin, cytochrome c oxidase and papain [195]. Therefore, under such a limited availability of light-active (visible light) nitrosyls, compounds displaying physiological stability in O_2 and pH ~7 have triggered to search for more efficacious compounds of the desirable potentiality.

2.13.1.4 Photodynamic aspects of metal nitrosyls as NORMS: medicinal perspective

2.13.1.4.1 Photodynamic therapy: a brief introduction and photosensitization mechanism

Photoinduced medical therapy is one of the significant biomedical applications usable for both the photodiagnosis and photodynamics having broad scope in cancer biology, biomarking, decontamination, tissue-specific light treatment and so on. Photodynamic form of this therapy (PDT), hence, represents a relatively new therapeutic treatment currently approved for palliative and curative medication of some forms of cancers, precancerous injuries and also for age-related macular degeneration (an eye disease leading to vision loss). PDT involves the oxygen-mediated manifestation of two main (cellular interacting and non-toxic) components operational in manner that one component is a photosensitive molecule (photosensitizer: PS), and the second part of the therapy is the light administration of a specific energy that results in the sensitizer activation. The light-assisted activation mainly causes the molecule to produce highly reactive oxygen species (ROS) including singlet oxygen triggering a series of complex photochemical reactions to occur through a cascade mechanism under several photobiological events. This ultimately leads to the death of targeted cells in a specific manner without harming the healthy cells (Figure 2.59). Thus, PDT is a photochemical reaction and a photobiological event used to selectively destroy the tissue.

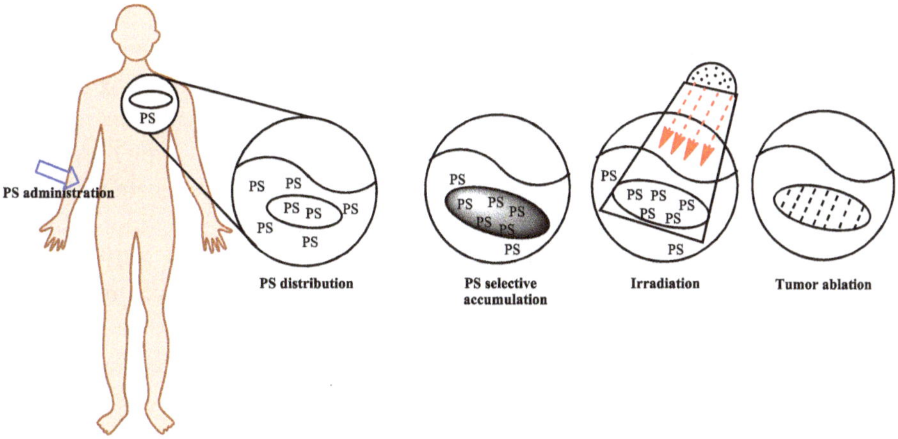

Figure 2.59: The functioning of PDT. (i) Administration of PS systemically (or topically), (ii) PS distribution, (iii) PS-selective accumulation in the target tissue, (iv) irradiation of PS for activation and (v) PS-mediated production of reactive oxygen.

At the initial stage, photosensitization had been performed with the use of conventional gas discharge lamps. The advent of lasers equipped with optical fibres modernized photosensitization and expanded its applicability in medicine, enabling the endoscopic (medical instrument for viewing interior of body) delivery of light to almost every site of the human body. Photodynamic medication in dermatology is simplified by the approachability of the skin to light application and allows using any light source with the appropriate spectrum. For instance, metal halogen lamp, which emits 600–800 nm radiations at high power density, short-arc xenon lamp, tuneable over a bandwidth between 400 and 1,200 nm, can be used. However, for the medication of large wounds, the broad light beam produced by incoherent lamps is useful.

Contrary to traditional incandescent lamps, lasers provide the exact selection of wavelengths and the precise application of light. Pulsed lasers such as the gold vapour laser and copper vapour laser-pumped dye laser produce brief light pulses of millisecond to nanosecond duration. The comparison of continuous-wave and pulsed lasers in practice has shown no difference. Tuneable solid-state lasers, such as the neodymium:YAG laser, are particularly useful for PDT. The above-listed laser systems are expensive, reasonably immobile and require frequent repair. The development of semiconductor diode lasers is a novel approach to avoid these disadvantages. Portable diode lasers, such as the gallium–aluminium–arsenide laser, produce light in the range between 770 and 850 nm, which corresponds to the absorption peaks of many new PSs.

The fundamental routes [196] by which the amalgamation of a PS, light and O_2 resulting in photosensitized cell death are shown in Figure 2.60. PSs that are in the ground state (S_0) can absorb light (photons) transforming first to an electronically excited singlet state (S_1) and, thereafter, via intersystem crossing, to the excited triplet state (T_1), a somewhat long-lived excited state. The energy of the excited PS can be dispersed either by thermal decay, emission of fluorescence (from S_1) or phosphorescence (from T_1). The excited triplet state(T_1) can also transfer a hydrogen atom or an electron to biomolecules (lipids, proteins, nucleic acids, etc.) or generate radicals that interact with O_2 to form ROS products such as the superoxide anion ($O_2{\bullet}^-$), the hydroxyl radical ($OH{\bullet}$) and hydrogen peroxide (H_2O_2) (type I reaction). Additionally, the T_1 state can transfer energy directly to the ground-state molecular oxygen in its usual triplet ($^3O_2$) state, which then produces the non-radical but highly reactive singlet oxygen, $^1O_2$ (type II reaction). The ROS ($^1O_2$, H_2O_2, $O_2{\bullet}^-$, $OH{\bullet}$) so generated via type I and II reactions result in cellular damage, which can be repaired or can kill tumour cells mainly by two routes: either necrosis or apoptosis.

Though the exact reaction paths of PDT are not fully known by now, it is commonly believed that the singlet oxygen, $^1O_2$ formed through type II reaction is principally responsible for cell death. The impact of both type I and type II reactions to cell death depends on several factors comprising the PS, the subcellular localisation, the substrate and the presence of O_2. In hypoxic conditions (inadequate oxygen in body tissues), radicals generated from type I photoreaction of some PSs are

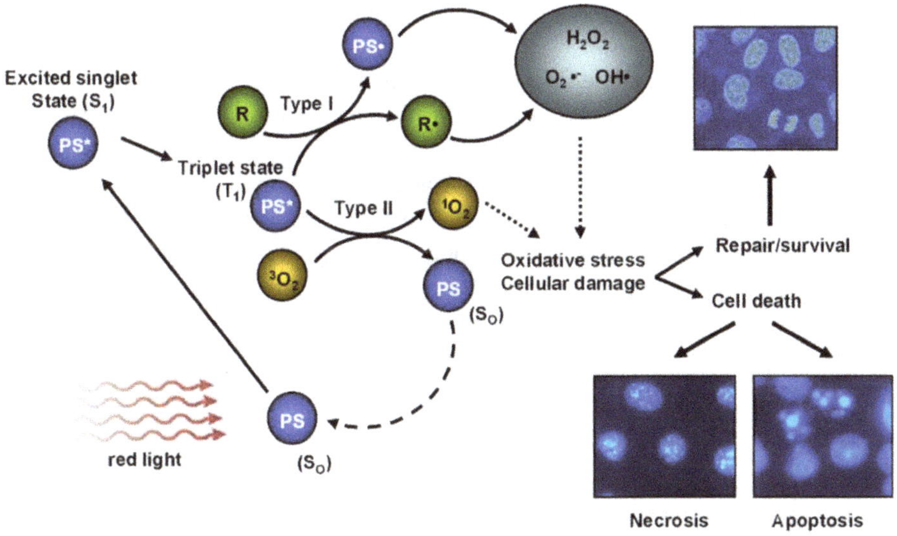

Figure 2.60: The simple pathways by which the amalgamation of a PS, light and O_2 results in photosensitized cell death. Transfer of energy from T_1 to biological substrates (R) and molecular oxygen, via type I and II reactions, generates ROS ($^1O_2$, H_2O_2, $O_2{\bullet}^-$, OH•). They produce cellular damage, which can be repaired or can kill tumour cells mainly by necrosis or apoptosis.

mainly responsible for sensitization. In an oxygenated medium, $^1O_2$ largely facilitates photosensitization, but the additional role of H_2O_2, OH• and $O_2^{\bullet -}$ is also significant. It is notable here that ROS are highly reactive and have a very short half-life. Therefore, only substrates located very close to the places of ROS production will be firstly influenced by the photodynamic treatment.

2.13.1.4.2 Photoactive metal nitrosyls as NO donors

Light-driven NO release from NO-containing compounds represents a highly fascinating area of medicinal research. In due course of designing systems that could behave in a perfect manner for PDT, the implications of bonded system and the specific target matter. Ultimately infusing direct NO for the treatment options could be replaced by controlled releasers, and more strictly under light effect beneficial accounts have been updated. In general, PDT, the binary componental theory of NO brings in an easy route for ROS generations and light absorptive phenomenon. Therefore, with the PDT linked NORMs serve as highly attentive option for cancer treatment with site and dose-specific context [197]. LASER-based site specificity precise the target cancerous cells than the usual drug therapy. Also, the expression of NO by LASER-exposed tissues/cells add interest to connect PDT with NO-releasing/NO-scavenging aspects [198]. In cancerous locus, the reactive intermediates in the form of ROS can also combine with NO to decrease the respective load, and hence

PDT-NO relation is an effective way of medication in the modern world to treat challenging diseases [199].

Since complexes of ruthenium are generally more stable, a range of ruthenium nitrosyl complexes [200] have been synthesized and studied in detail in terms of their structures and NO-donating capacities for a specific cellular target. Altogether, the non-porphyrin nitrosyls readily behave as NO donors upon light illumination by the intervention of Ru(III) photoproducts. On the other hand, photorelease of NO from ruthenium nitrosyls derived from porphyrins remains restricted due to rapid recombination again forming ruthenium nitrosyl complexes (Figure 2.61).

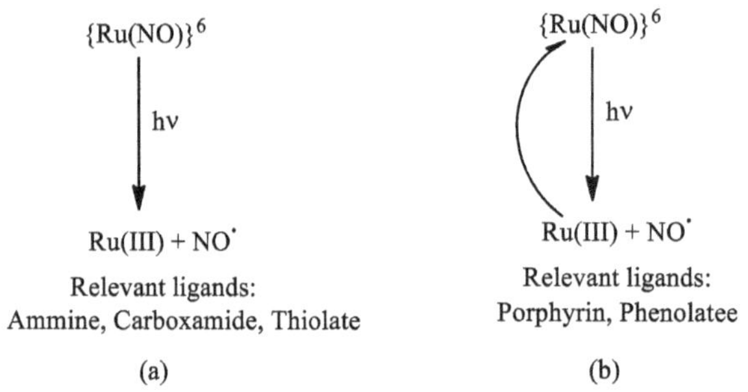

Figure 2.61: (a) Ru–NO-based reaction for NO and Ru(III) photoproduction and (b) photoactivated NO release.

Despite numerous reports of ruthenium nitrosyl complexes do exist displaying release of NO upon high energy irradiation (UV light irradiation) [201], but data on visible light [202] NO labilizations of ruthenium nitrosyls are few in numbers. Furthermore, the Ru^{III} solvent complexes that result after irradiation unavoidably undergo unwanted side reactions in solution. One persistent challenge is, therefore, to find ruthenium nitrosyl complexes that could release NO reversibly under very mild conditions without any metal-bound side reactions for biological applications.

Study carried out by Franco et al. [203] have shown that the NO release in polyamidoamine functionalized with ruthenium nitrosyl compounds is made through light irradiation ($\lambda = 355$ nm) and one-electron reduction.

NO release in some cases has been recognized to be possible only if they are activated by reduction [204] focussed on the nitrosyl ligand. Additional important recent advancement in the nitrosyl lability is from a trinuclear [205] ruthenium nitrosyl complex and the relevant in vitro cytotoxicity against melanoma cells. Certain ruthenium nitrosyl clusters have been designated pro-drug, NO releasers. It has also been observed that chronic corticosterone administration facilitates aversive memory recovery and increases glucocorticoid receptors/NOS immune reactivity [206]. As retrograde messenger, NO influences the formation of long-term potentiation and memory consolidation. Cell penetrating ability and cytotoxicity of NO-donating ruthenium complexes [207] are also quite interesting. Some concerns regarding the NO release with anti-HIV and anti-cancer activities have also been investigated.

In a recent report, Roveda et al. [208] reported the synthesis of ruthenium(II) nitrosylsulphito complex of composition *trans*-$[Ru(NH_3)_4(isn)\text{-}(N(O)SO_3)]^+$ (isn = isonicotinamide) by the reaction of sulphite ions (SO_3^{2-}) with the nitrosyl complex *trans*-$[Ru(NH_3)_4(isn)(NO)]^{3+}$ in aqueous solution:

$$\textit{trans}-\left[Ru(NH_3)_4(isn)(NO)\right]^{3+} + SO_3^{2-} \rightarrow \textit{trans}-\left[Ru(NH_3)_4(isn)(N(O)SO_3)\right]^+$$

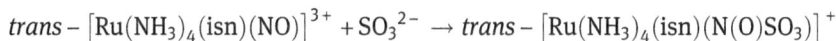

This compound on light irradiation of wavelength of 355 or 410 nm generates $NO^\bullet$, $SO_3^{\bullet-}$ and isonicotinamide in solution. The probable mechanism of this complex is given in Figure 2.62. This is based on detection of the aquo complex, *trans*-$[Ru(NH_3)_4\text{-}(H_2O)(isn)]^{2+}$ by laser flash photolysis, of $NO^\bullet$ by the reaction with catalase (by UV–Vis), of sulphite radical ($SO^{3\bullet-}$) (by EPR using DMPO as a spin trap) and on DFT results.

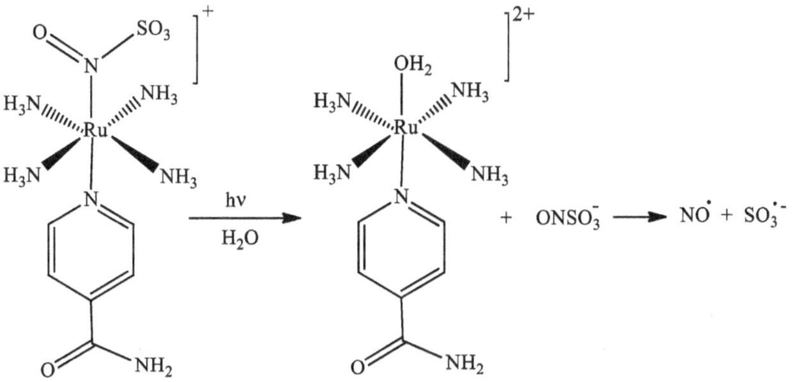

Figure 2.62: Light irradiation on *trans*-$[Ru(NH_3)_4(isn)(N(O)SO_3)]^+$ and generation of $NO^\bullet$, $SO_3^{\bullet-}$ and isonicotinamide in solution.

The photolability of NO in most of the metal nitrosyl complexes occurs from dπ (M)→π*(NO) transitions [209]. In ruthenium nitrosyls, such transitions occur in the range between 300 and 450 nm. Consequently, ruthenium nitrosyls display NO photolability upon exposure to UV light, which is often harmful to biological targets. Ruthenium nitrosyl complexes attached with appropriate dye molecules as light harvesters [210] (Figure 2.63) have recently created much interest. Such molecules rapidly release NO upon exposure to visible light. This type of sensitization permits the use of ruthenium nitrosyls in NO delivery to biological target without the need of harmful UV light. Some of these dye-nitrosyl conjugates have recently been used to inspire NO-induced apoptosis in cancerous cells [211] and destroy bacterial colonies [212] under the total control of light:

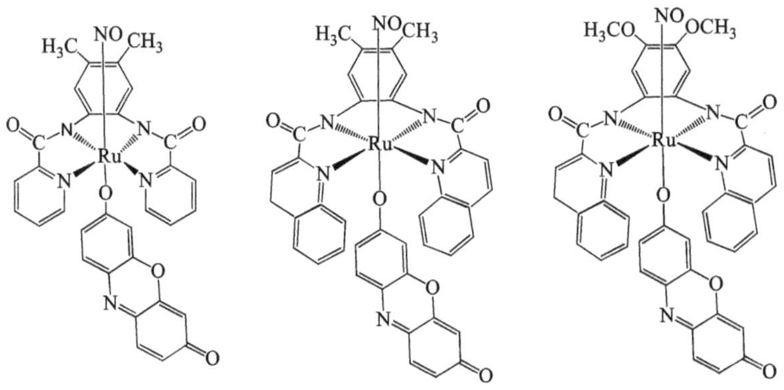

[(Me$_2$bpb)Ru(NO)(Resf)] [(Me$_2$bQb)Ru(NO)(Resf)] [(Me$_2$bQb)Ru(NO)(Resf)]
Me$_2$bpb =1,2-bis(pyridine-2-carboxamido)dimethybenzene
Me$_2$bQb =1,2-bis(quinoline-2-carboxamido)dimethylbenzene
Me$_2$bQb =1,2-bis(quinoline-2-carboxamido)dimethoybenzene
Resf = Resorufin dye

Figure 2.63: NO-releasing ruthenium nitrosyl complexes with dye attachment in visible light.

In our recent past, we have developed similar models of Ru-based systems capable of releasing NO at lower energy activation in special reference to salen Schiff base complex containing Cl as a co-ligand [213]. Theoretical-based DFT-supported study and in vitro anticancer evaluation (MTT) for targeting COLO-205 cell line show that the complex bears potent medicinal properties in special reference to NO-tagged implications (Figure 2.64). These observations indicate that in designing PDT agents, the Ru–Salen–NO complexes are beneficial in extra way as compared to the general two-componental PDT requirements. On the one hand, the Ru–NO compounds could act as PS and also due to potent anticancer properties are not dependants on oxygen for this bioactivity to kill cancerous cells. Release of NO can serve best to target cellular death. Also, the co-ligation associated with Ru–NO core is important in this context [214, 215]. For instance, carbohydrate ligation has been reported to enhance the PDT behaviour [216]. Therefore, coordination sphere represents a carriage of several components useful in PDT especially when Ru–NO core is the central part of the zone.

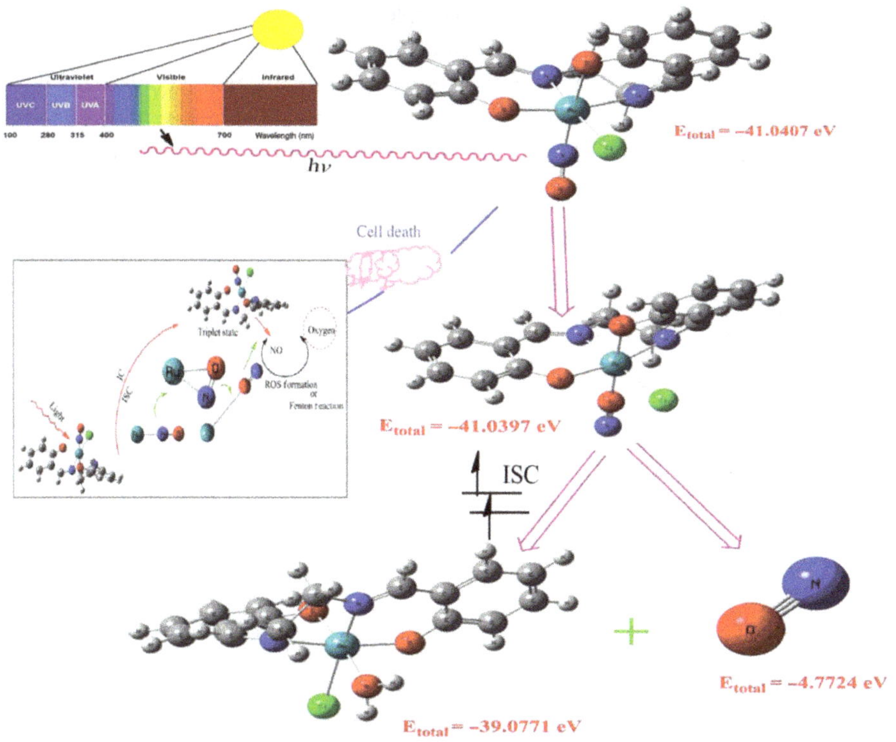

Figure 2.64: Energy profiling of NO release resulting from Ru–NO salen spin-activated system.

2.13.1.5 NO delivery from nanoplatforms

Among the several factors that show considerable effect upon a molecular system to behave as NO releaser, the size dependence of such systems have been significantly proven under nano-frameworks. Hence, in specific way nano-drug delivery approach for NO-donoring potential is keenly eyed by the NO chemists. Therapeutic-driven factors like NO payload, half-life and physiologically mediated are interestingly sought from nano-systems [217]. Tremendous fascination has been shown in this context for incorporating polymer-loaded NO, nanoparticles (NP; silica, TiO_2), several forms of quantum dots (QDs), NPs (bottom-up), carbon nanotubes, and so on.

2.13.1.5.1 Liposomes: microscopic artificial sacs whose walls are a double layer of phospholipids

Among several biological cellular components targeting spherical vesicle, liposome results in reaching highly compatibility site for drug delivery confined to long duration potentiality and modulation of immune responsiveness. *trans*-$[Cr(cyclam)(ONO)_2]^+$-based system investigated by Ford et al. [218] for light sensitivity was found in an encapsulated form by liposomes. Also, the results indicate liposome conjugation with light-sensitive CrONO as compared to free complex, and 470 nm illuminations resulted in a controlled NO release using aerated solutions.

Devoted to the similar form of work reported by Nakanishi and co-workers [219] targeting Ru–NO complex [Ru(L)Cl(NO)] (shown in Figure 2.65) when allowed to interact with liposome surface showed lipophilic behaviour of the Ru–NO system, and began to release NO continuously for 1.5 h, indicating high efficiency.

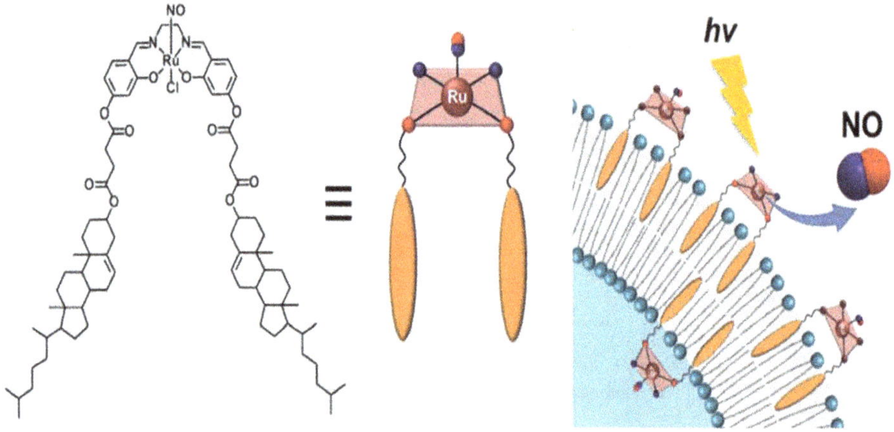

Figure 2.65: Illustration of [Ru(L)Cl(NO)] interaction with liposomes.

2.13.1.5.2 Polymers

While interrogating polymeric-based NP systems loaded with NO several fruitful results have been bagged. By such scaffolds, the respective half-lives and controlled NO release with respect to biological milieu have shown fascinating behaviour. Moreover, the additional effect in terms of quantity or quality of drugs generates a synergistic or combinatory treatmental effect, and hence can halt drug resistance also, for instance, *o*-phenylvinyl-anchored [Ru(salen)(NO)(Cl)] complex worked out by Borovik and co-workers [220] by linking methacrylate-polymered matrix. The resulting porous composite showed the NO-releasing behaviour when exposed to UV light generating photoproduct with the central metal as Ru^{III} state.

Similarly, Tfouni and co-workers [221] formulated a silicate (sol–gel approach) capsulated form of *cis*-, *trans*-[(cyclam)Ru(NO)(Cl)]$(PF_6)_2$ [222]. The selected model complexes were further illuminated in the UV range of EMR and resulted in their NORM-like behaviour. This significant approach involving sol–gel conjugation can be tamed in a way to revert back the form of NO-tagged complex when photoproduct is treated with NO. In the similar type of sol–gel-associated NO-releasing evaluation, Mascharak and co-workers [223] in 2010 characterized a photosensitive manganese nitrosyls of the general form [Mn(PaPy$_3$)(NO)]ClO_4. The observed releasing fashion of

NO and the quantum efficiency using visible light was highly appreciated for quick NO generation, site specificity and minimal harmful effects of radiation exposure.

Concisely speaking, the releasing molecules sought from incorporating polymeric material having porosity under metal-organic networks have thus reached an interesting relation within the context of NO chemistry. Interestingly, Furukawa et al. [224] reported several porous coordination materials incorporating polymers of biocompatible tone wherein nitro-containing imidazolate functionality or *N*-nitrosamine functional attachment [225] (Figure 2.66) were found to act as light-induced NO donors. The designed materials were found to show the drug release at the cost of energy falling in the beneficial range of EMR, that is, 300–600 nm.

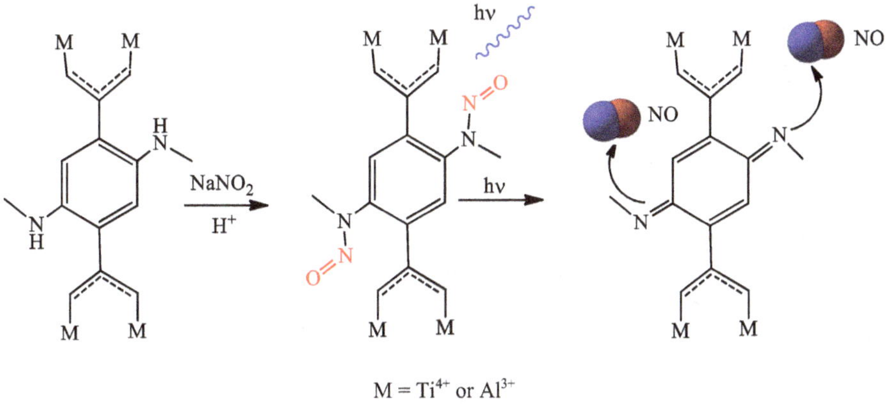

Figure 2.66: Scheme showing transformation of post-synthetic nitrosation of methylamine group and light-induced NO releasing.

In the same manner, Ostrowski et al. [226] while studying another example of photochemical reaction of a polymer-anchored CrONO core showed the visible light sensitivity in this context, and the NO-discharge timed for more than 30 h under restricted concentration (20–100 pmol^{-1}) of the sample (Figure 2.67).

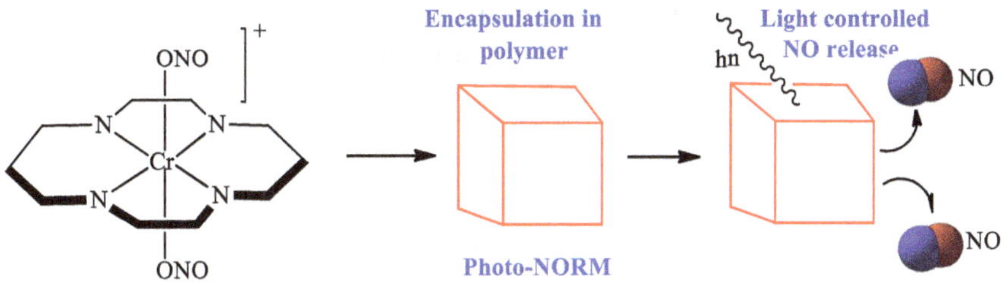

Figure 2.67: Encapsulation of CrONO compound in polymeric matrix and release of NO on irradiation of visible light.

In nanomedicine relevance of these types of composites, the target of hepatic stellate was aimed by Wang et al. [227]. The concluding remarks of this study suggest that these nanopolymers are potent medication advancements for liver fibrosis and hypertension. In the study, the primary carriage of NO in the form of *S*-nitrosoglutathione in association with vitamin A was the main specific means for the feasible delivery of the drug.

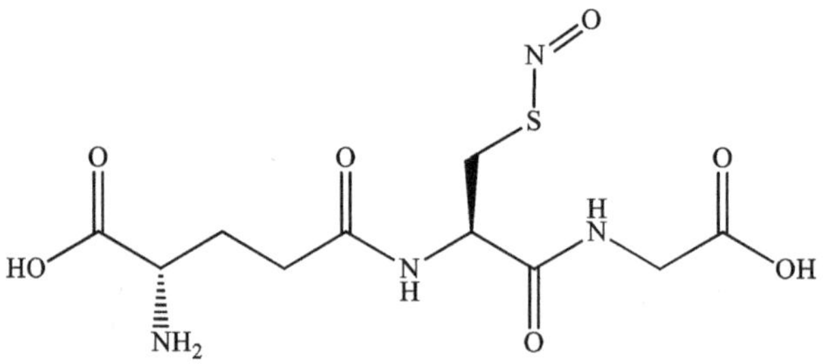

Figure 2.68: Chemical structure of *S*-nitrosoglutathione.

2.13.1.5.3 Nanoparticles: inorganic silica (SiO_2)

The considerable fascinating physicochemical behaviour of silica-based biocompatible functional models represent versatile scaffolds of desirable size, and stability, and are easy to synthesize. Silane forms diversified products by acting as a binding agent with so many functionalities (e.g. amines, carboxylates, thiols, olefins, halides and epoxides), and hence proves significant in grafting of surface-active drugs including other therapeutics [228]. In connection with the photoinduced NO release by a tetraaminenitrosyl complex of Ru, Tfouni et al. [229] observed the effect of nanoparticulate form of silica assembled with the complex for the suitability as a NO carrier capable of photoactive NO release as is shown in Figure 2.69.

In fact, silica NPs of mesoporous nature increases the ratio of surface to volume, and hence the enlarged pore spacing, which in turn is an increased storage (payload) NO capacity. In this context, Mascharak et al. [230] studied $[Mn(PaPy_3)(NO)](ClO_4)$ complex in association with electrostatically attached mesoporous silica (MCM-41), and found highly fascinating results against nosocomial infecting *Acinetobacter baumannii* demonstrating the effect of NO released by the light-activated system in visible range of EMR (Figure 2.70).

2.13.1.5.4 Titanium dioxide (TiO_2)

The significant and outstanding photochemical applications of TiO_2 have motivated researchers to use TiO_2 as a PS in treating cancer through PDT technique [231]. This in turn has gained consideration for employing as nanocomposites to serve for the purpose of cell imaging, analysing biological samples and more importantly in designing drug delivery systems. In association with NO-delivery system of a selected

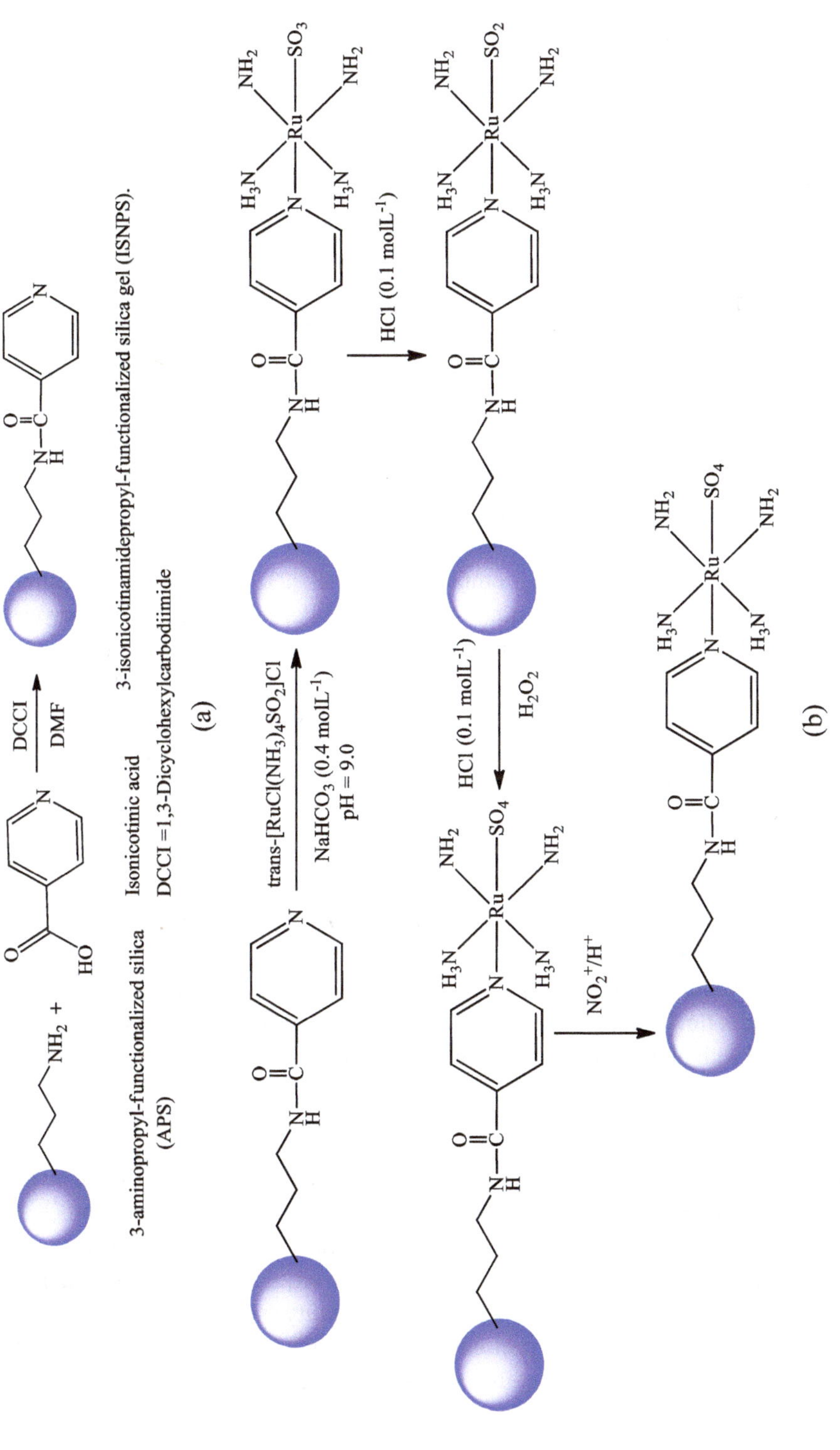

Figure 2.69: Preparation of isonicotinamide-functionalized silica gel and a photoactive ruthenium complex on it.

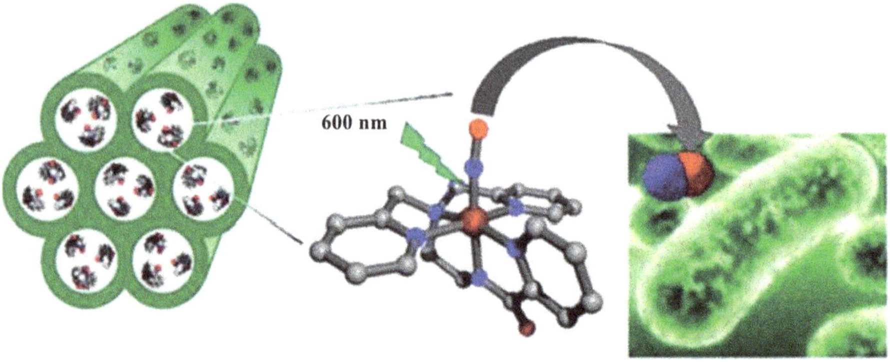

Figure 2.70: MCM-41-loaded $[Mn(PaPy_3)(NO)](ClO_4)$ (photoactivated) NO-releasing action against *Acinetobacter baumannii.*

Ru–NO complex tailored with TiO_2 composite, Liu et al. [232] developed a multifunctional nanoplatform of its outstanding form. Meanwhile, folic acid (FA) moiety was selected as the target wherein covalent-linked TiO_2 NPs were subjected to the light illumination in the visible range (light pulsed from 5 to 20 s) generated enhanced rate of NO release (150–300 nM) as compared to the free Ru–NO form. Overall, the study reveals that the combined effect of all the selected systems produced synergistic effects in inhibiting the respective cancer cell line.

Similarly, in another example of targeting dual targets of NO-delivery system for subcellular level, Xiang and co-workers [233] developed NO delivery nanoplatform {Lyso-Ru-NO@FA@C-TiO_2}(C-TiO_2 = carbon-doped TiO_2 NPs) using a lysosome-targeting Ru–NO and a folate receptor (FR). The platform was shown to exhibit optical density absorption in the range of 600–1200 nm, targeting FRs only. Therefore, a localized effect was achieved by using nanoplatform within the lysosomal area to deliver NO and ROS at the same time for the same purpose. Hence, again a combinatory cytotoxic effect was observed with high anticancer efficiency using this approach.

2.13.1.5.5 Semiconductor quantum dots (QDs)

QDs generally denote to very small NPs of only few nanometres in size. Semiconductor QDs are one of the most important QDs. Their size and shape can be precisely controlled by the duration, temperature and ligand molecules used in the synthesis. Owing to the high optical regards in the form of absorption, controlled models of optical properties and the surface reforming capability, semiconductor inorganic QDs [234] have gained significant attention. Using this approach, Ford et al. [235] conducted the congregational experiment of water-soluble CdSe/ZnS core/shell QDs associated with the cationic *trans*-$[Cr(cyclam)(ONO)_2]^{2+}$ complex employing coulombic interactions. The study observed QDs to serve as an efficient attenuator for light-harvesting process to trigger the drug (NO) release. The observations made in the experiment indicated high efficacy, light-controlled transfer from quantum particles to the NO carriage and stood as a first example to trigger NO release under nanoshield using hydrophilic NO-carrying scaffold.

Similarly, Tan and co-workers [236] used nanospheres of Ag_2S QDs, targeting *S*-nitrosothiols or Roussin's black salt (RBS) [237] as NO loading. The photoemission as well as photoabsorption indicated NO-releasing activation at visible light, and emitted radiation of near-IR (NIR) fluorescence range (stimulated by a NIR laser). This study suggests that such a system is applicable and fruitful for in vitro as well as in vivo cell imaging (Figure 2.71). For similar experiments carried out using Mn^{2+}-doped ZnS QDs in place of Ag_2S QD core resulted in activated NO release at an energy cost of NIR-II light (1,160 nm) [238] (Figure 2.72), wherein Mn^{2+}-doped capsulation of ZnS QDs served as a unique antenna for producing activation signals to let NO release in a controlled and feasible manner.

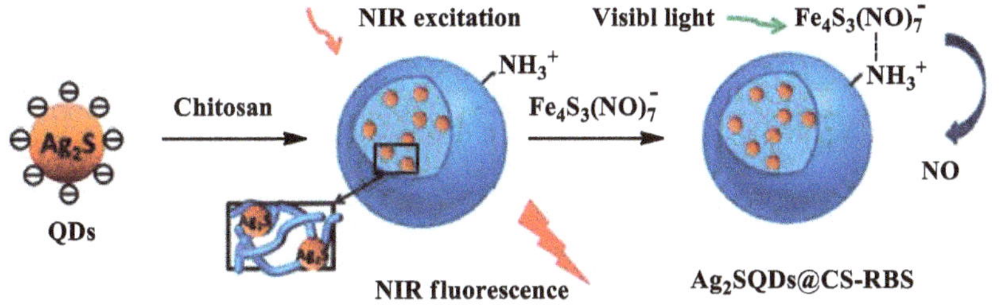

Figure 2.71: Diagram for synthesis and light-activated NO release pathway of NIR- fluorescent Ag_2S QDs@CS-RBS nanospheres (adapted from Ref. [237]).

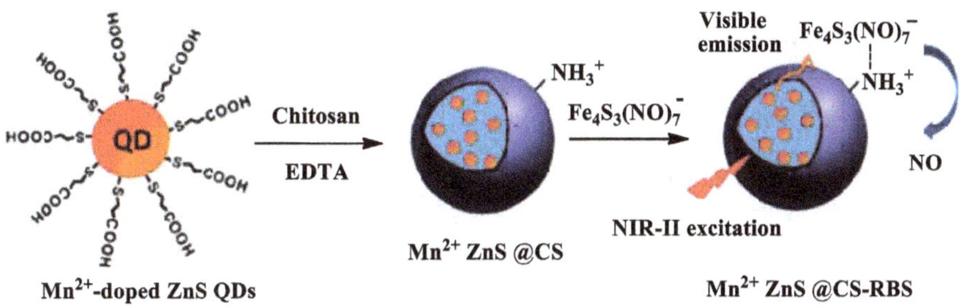

Figure 2.72: Diagram for synthesis and the NIR-II-activated NO-release pathway of Mn^{2+}–ZnS@CS-RBS (adapted from Ref. [238]).

2.13.1.5.6 Upconverting nanoparticles (UCNPs)

Another NIR-activated drug-release technique using upconverting NPs (UCNPs) employing lanthanide-doped nanosystem represents fascinating form in designing NO-releasing nanosystems [239]. The studies conducted in this context have shown such systems capable of serving as efficient modulators in changing low-energy NIR to high-energy UV–Vis form. This switching technique has been suggested to be useful in recording real-time cell imaging that too even if the target is lying within the deep tissues. In combination with PSs for carrying out photodynamic or photothermal cancer therapy UCNP-based drug-releasing phenomenon could be very significant [240]. For instance, Zhang et al. [241] formulated silica-coated $NaYF_4$:Yb/Er@$NaYF_4$ UCNPs (hexagonal) system with surface-appended RBS resulted in releasing NO when subjected to irradiation of NIR at 980 nm. Similarly, Burks and co-workers [242] set another example of NO-releasing polymer discs (UCNP–RBS–PDs) triggering the drug release at the same 980 nm NIR range (Figure 2.73) releasing NO in a concentration range of picomols to nanomols.

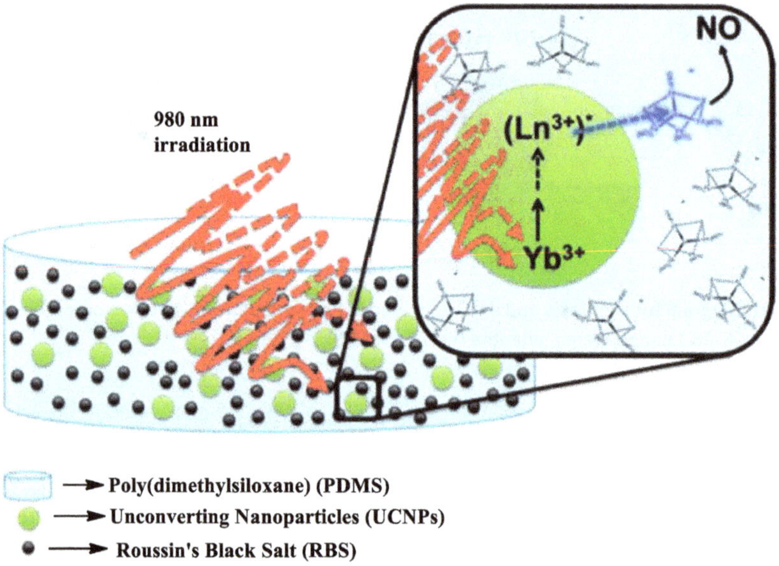

Figure 2.73: Diagrammatic illustration of photochemically active NO-releasing polymer discs (PDs) under NIR light excitation for NO release (taken from Ref. [242]).

2.13.1.5.7 Carbon-based nanostructures

In association with nanosizing of nanodrug-delivery investigations, carbon nanostructures like fullerenes, carbon dots (CDs), graphene QDs, graphene oxide (GO) and carbon nanotubes (CNTs) have been preferably studied for having outstanding properties desirable for drug-delivery assemblies. Having physicochemical properties like unique electrical and mechanical properties, CNTs have been found to be very useful in biomedical scientific applications involving their profitable role in biosensing, cell imaging and drug release. The planar forms of CNTs [243] represent high S/A values enabling such systems to act as a carriage for various drug molecules involving physisorption interactions. Chen et al. [244] reputedly carried out GO-based assemblage of BNN6 (*N*,*N'*-di-*sec*-butyl-*N*,*N'*-dinitroso-1,4-phenylene-diamine) congregated via expected π–π stacking interaction found that NIR absorption triggers the respective NO release controlled by NIR laser. This nanosystem-based intracellular NIR response remarked as a notable therapy for cancer treatment. Among several parameters that are generally eyed while designing drug-delivery systems like water solubility, nature of biocompatibility, cell permeability, photobehaviour and surface adjustability, carbon nanodots (CDs) represent much fascinating consideration for such criteria [245]. For example, Xu et al. [246] reported the CD system of NO releaser using *S*-nitrosothiol to deliver the drug at mitochondrial site under white light emitting diode irradiation. The intense fluorescence recorded in the nanosystem helped to track the release easily. This technical advancement made subcellular system damage like that of mitochondria in a more efficient way, and hence resulted in an enhanced pro-apoptotic implication as compared with the undesirable target in a system. In another example of {Ru-NO@FA@CDs}, while targeting FA reportedly conducted by Liu and co-workers [247] (Figure 2.74) indicate self-trackable cellular recognition of cancerous cells on binding FRs. This system studied the amount of NO release by the manipulation of visible light, thereby allowing NO release at optimal space and desirable duration applicable for highly demanded NO-mediated therapy for cancer treatment. Therefore, the nanoscience involvement in NO releasers has brought in improvement in the proficiency at the subcellular NO release. Liu and co-workers [248] while studying {Lyso-Ru-NO@FA@CDs} incorporated with another CD-based system targeting lysosome subcellular region (Figure 2.75) found that the respective nanoplatform {Lyso-Ru-NO@FA@CDs} indicated immediate release of NO by the activation energy of visible light or with 808 nm NIR light. In addition to the observed NO-swift release, an increase in temperature by 808 nm irradiation depicted the role in photothermal therapy. The overall observation again confirms the fruitful and purposeful use of nanocontrolled platform of NO-caged systems for a multitude of cancer cell lines.

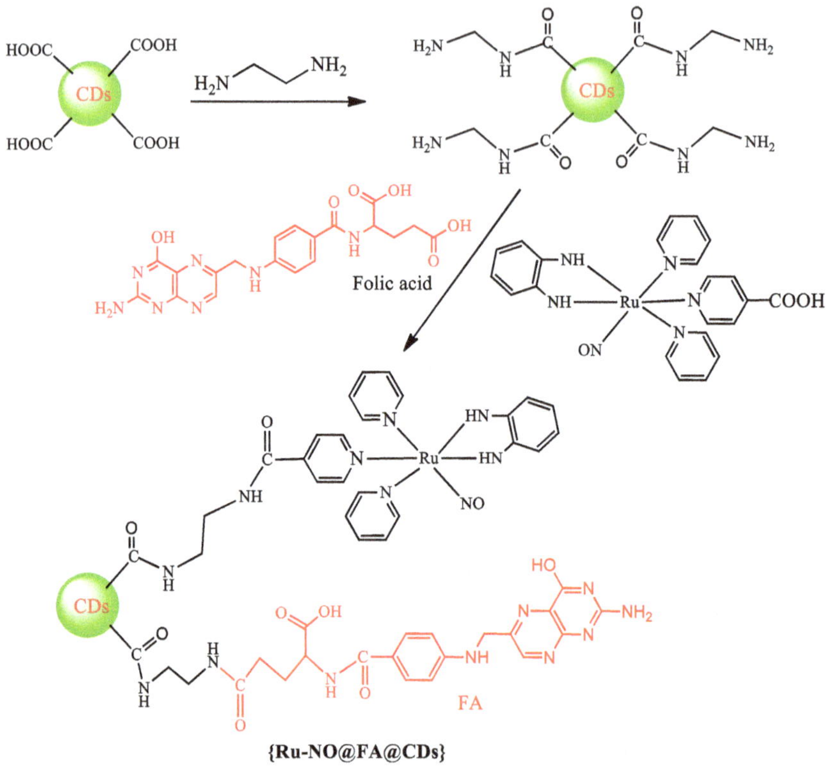

Figure 2.74: Reactions showing the formation of the {Ru-NO@FA@CDs} nanoplatform (taken from Ref. [247]).

2.14 NO news is good news for eyes: NO donors for the treatment of eye diseases

Since the dawn of NO recognition as a key signalling molecule, the diversified biological role of this free radical met with its extended role in so many areas of physiological investigations. Keeping in view that the combinatory functional status of NO and cGMP (cyclic guanosine monophosphate) is entailed with a range of biological actions, there are numerous evidences supporting the fact that the NO-metabolic pathways are also involved in the normal functioning of an eye.

Before coming to know the key role of exogenous NO donors in the medication of eye diseases, it is must to know the endogenous production of NO inside the eyes.

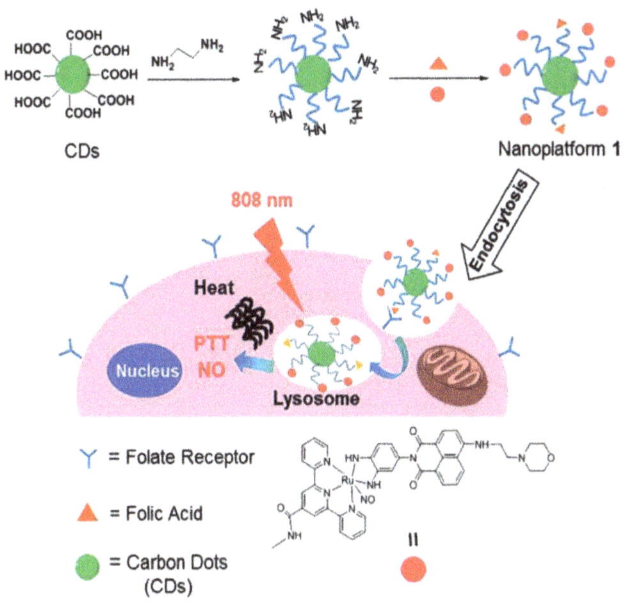

Figure 2.75: Reaction showing the formation of {Lyso-Ru-NO@FA@CDs} nanoplatform and its directed attack against cancer cell (adapted from Ref. [248]).

2.14.1 Biosynthesis of nitric oxide

NO, a biologically synthesized molecule, is now widely accepted to be produced via metabolic pathway of L-arginine to L-citrulline under NOS action (oxidation of the nitrogen-coloured red in Figure 2.76).

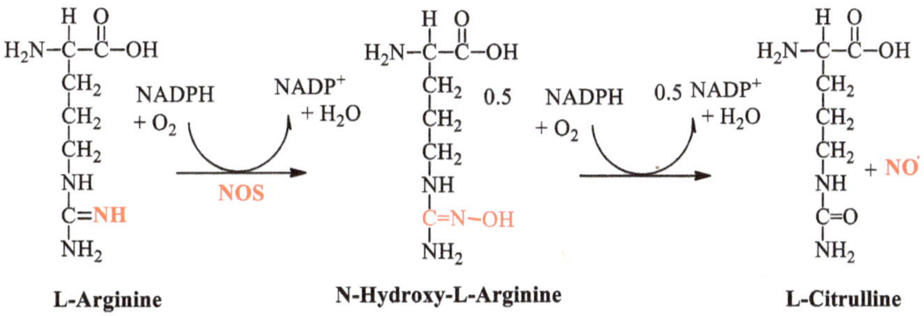

Figure 2.76: Biosynthesis of nitric oxide.

The bioregulatory pathways of NO are involved almost in every mammalian system, particularly under the action of NOS in association of oxygen, the reduced form of NADPH and several flavins. The three well-known NOSs show distinctive variation in the site of action and nature of requirement controlled by human physiological action. NOS-I or nNOS is neuron-related NOS, and the second NOS "NOS-II" (iNOS) refers to immunomodulation, and the third very important one is NOS-III (eNOS) denotes the vascular endothelium NOS. First and third isoforms produce NO in a very low concentration, mediated by the calcium/calmodulin activation (Figure 2.77). Whereas NOS-II is immunoresponsive (Figure 2.77) and produces NO in large quantities for a longer duration in a stimulus-dependant way [249].

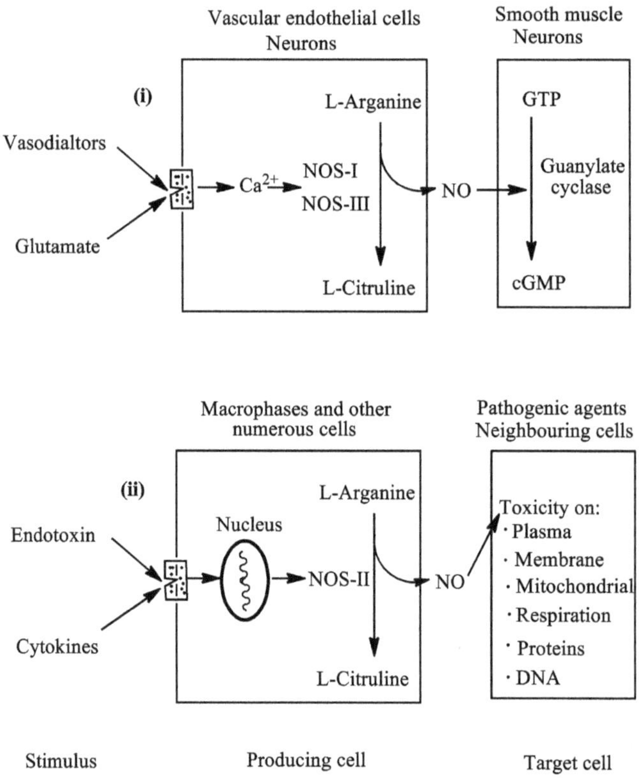

Figure 2.77: Nitric oxide interactions with cells after its generation by three isoforms of NOS enzyme.

2.14.2 Roles of nitric oxide in eyes

NO is found to have a precious role in maintaining homeostasis among humans. The role that it plays in sensory neurons activates the search for its possible concern as homeostatic mediator in the eye. These functional regulations include maintenance of aqueous humour (AqH) dynamics, neurotransmission of retina and photo-transduction. Any halt in its generation or malfunctioning of the respective metalloprotein actions could result in uveitis, retinitis, glaucoma or retinal degeneration [249].

It is now well known that eNOS and nNOS are activated in normal tissues to produce NO for physiological functioning of the eye [250]. The expressions/indications of neuronal and immunologic NOS have been detected in the retina. Diverse studies disclose that neuronal NOS may be responsible for producing NO in photoreceptors and bipolar cells. This is followed by stimulation of GC of photoreceptor rod cells and increases calcium channel currents. It has been established that NOS action if halted/stopped in the retina of cats results in phototransduction impairment [251]. Moreover, Müller cells and in retinal pigment epithelium contain iNOS responsible for keeping normal phagocytosis of the retinal outer segment. NO also caters retinal circulation [253]. Glaucoma is a progressive optic neuropathy caused by death/degeneration of the retinal ganglion cells (RGCs) and is the leading cause of irreversible blindness worldwide. The mechanism by which this progressive RGC death occurs is not fully understood. It is clear that multiple causes may give rise to the common effect of ganglion cell death.

The clinical investigations reveal that glaucoma is associated with increased IOP. In case of open-angle glaucoma (OAG), the finding of high IOP suggests imbalance between AqH generation and outflow. It is estimated that more than 60 million people suffer from primary OAG worldly, showing a possible graphic projection of about 79 million by the end of 2020 and more than 100 million by 2040. The general form of glaucoma is indicative of high IOP and hence is known as ocular hypertension. From the available data it is clear that one-third of glaucomatous patients (vision loss) show normotensive IOP (normotensive glaucoma and this disease has a major impact on age factor, that is, increases with age, independent of IOP [254]. This shows that this mechanistic approach is not the sole explanation of the cause of this defect.

Reducing IOP is not always a preventive measure of glaucomatous neurodegeneration, and many patients go with this disease in spite of having IOP in the normal range. It has been found that NO is a direct regulator of IOP and that dysfunction of the NO–GC pathway is associated with glaucoma incidence. Studies conclude that NO has potential as a novel therapeutic that decreases IOP, elevate ocular blood flow and bestow neuroprotection. Ideally, novel glaucoma therapeutics would take care of both IOP-dependent and -independent mechanisms [255] of the disease.

The complete roles of endogenous NO in eyes are depicted in Figures 2.78–2.81 with details given below each figure (adapted from Ref. [254]).

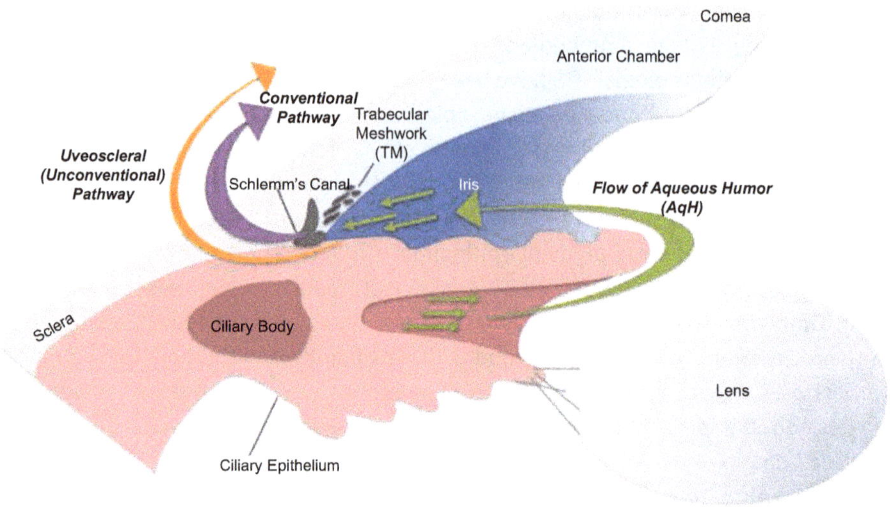

Figure 2.78: Diagrammatic representation of aqueous humour (AqH) equality. (a) AqH at the ciliary body in the eye flows (green arrows) through two routes independently that control AqH dynamics: (i) via the trabecular meshwork (TM) and Schlemm's canal (purple arrow) (conventional route) and (ii) via the uveoscleral tract (orange arrow) (non-conventional route). (b) The balance of (AqH) production at the ciliary body and elimination in the anterior chamber establishes IOP in the eye.

2.14.3 Use of NORMs in the treatment of eye defects

The results of NO donors on IOP have been observed in several species comprising mouse, rabbit, dog and monkey. NO donors such as nitroglycerin (GTN), ISDN, sodium nitrite and SNP when administered topically (Figure 2.82), they rapidly lowered IOP with a highest effect at 1–2 h in a normotensive (having normal blood pressure) rabbit model.

Interestingly, SNP and nitroglycerin have shown the result on IOP peaked at 0.1% and 0.03%, respectively. Remarkably, higher doses were found to be less effective [256].

Kotikoski et al. [257] have shown that topical or intravitreal administration of SNP, *S*-nitrosothiol and spermine NONOate in normotensive rabbits resulted in highest IOP lowering after 2–5 h.

It has been observed by Behar-Cohen and co-workers [258] that intravitreal or intracameral injection of the NO donors, namely, 3-morpholinosydnonimine (SIN-1) or SNAP (Figure 2.83), when given to a normotensive rabbit persuaded a dramatic decrease in IOP depending on the amount of NO delivered to the eye.

Sugiyama et al. [259] studied the hypotensive action of nipradiol [3,4-dihydro-8-(2-hydroxy-3-isopropyl-amino)-propoxy-3-nitroxy-2*H*-1-benzopyran (Figure 2.84), a β-adrenoceptor blocker with an NO-donating nitroxy group in normotensive rabbits. This study demonstrates that the hypotensive role of the drug nipradiol is dependent

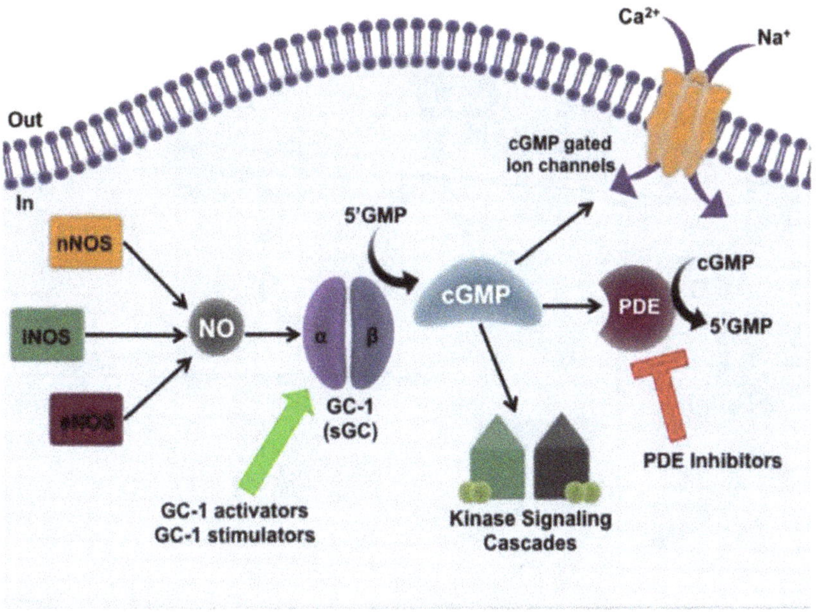

Figure 2.79: The NO-GC-1-cGMP pathway, steps in the way of IOP lowering. (a) NO is generated from L-arginine by nitric oxide synthase (NOS) available in three isoforms: (i) neuronal NOS-I (nNOS), (ii) endothelial NOS-III (eNOS) and (iii) inducible NOS-II (iNOS). (b) NO binds guanylate cyclase-1 (GC-1), a heterodimeric protein capable of converting guanosine 5′-monophosphate (5′-GMP) to cyclic guanosine monophosphate (cGMP). The cGMP so produced can target cGMP-gated ion channels and activate kinase signalling cascades. (c) Phosphodiesterase enzymes (PDEs) bind to cGMP and catalyse the decomposition of cGMP into 5′-GMP. PDEs act as important regulators of signal transduction mediated by cGMP. (d) The cGMP bioavailability in the cell can be increased in two ways: (i) by the use of GC-1 stimulators and activators, which increase production of cGMP, or (ii) by the use of PDE inhibitors that prevent the decomposition of cGMP in the cell.

on NO donation by pre-treating it with the NO trapping agent carboxy-PTIO [2-(4-carboxyphenyl)-4, 4, 5, 5,-tetramethylimidazoline-l-oxyl 3-oxide].

Studies conducted by Orihashi and co-workers [260] in hypertensive and normotensive rabbits have shown that nipradilol and SNP exhibited IOP lowering in both types of rabbits. Contrary to this, latanoprost (Figure 2.85) was without significant effect in lowering of IOP. When either nipradilol or SNP was given in combination with latanoprost, the IOP reduction was significantly greater than that of either of the NO donors administered alone. This suggests a synergistic effect of latanoprost and NO in lowering IOP.

Non-arteritic anterior ischemic optic neuropathy (NAION) [261], a common eye problem generally found in middle-aged group (though no age group is safe), is linked with phosphodiesterase (PDE) inhibitors (such as Sildenafil) presumably due

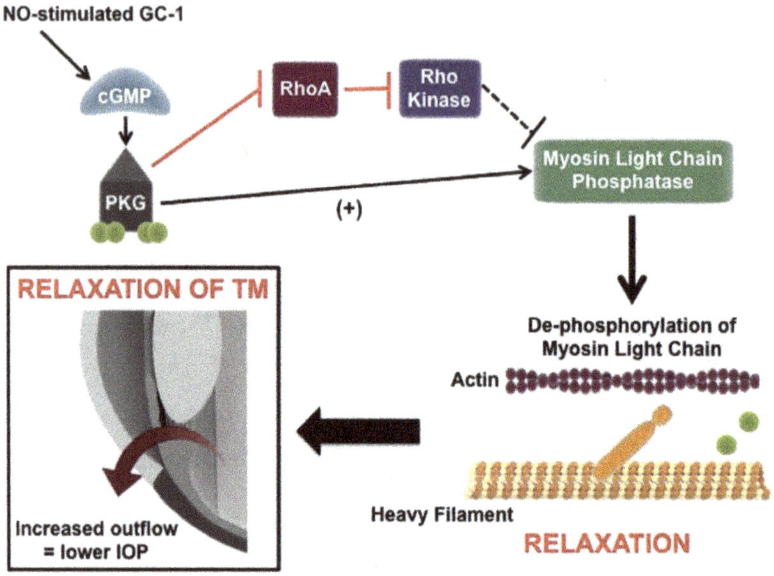

Figure 2.80: cGMP-assisted modification of IOP via increase in aqueous humour (AqH) outflow. (a) Nitric oxide activates generation of cGMP by GC-1. The cGMP activates protein kinase G (PKG). The PKG so activated can phosphorylate numerous targets with various downstream effects, including inhibition of Ras homolog family member A (RhoA). This prevents inhibition of myosin phosphatase by Rho Kinase. (b) Besides inhibition of RhoA, activated PKG can directly trigger myosin light chain phosphatase (MLCP). Thereafter, dephosphorylation of the regulatory light chain of myosin by MLCP prevents actin–myosin interaction, promoting cell relaxation. (c) This then leads to a widening of the intercellular spaces in the juxtacanalicular trabecular meshwork (TM) and Schlemm's canal. This facilitates conventional AqH outflow and thereby lowering IOP.

to the hypotensive effect and vasorelaxation. Hence, sildenafil (a well-known NORM) finds the application in lowering the blood pressure [262].

Several recent reports describe the use of erectile dysfunction (ED) drugs (tadalafil, avanafil and vardenafil) (Figure 2.86) questioning these drugs as responsible agents for NAION [263]. Many factors have been elaborated to set this belief of contribution towards NAION. Therefore, among warning factors, such possibilities of side effects must be highlighted. The same vision defects have been found among patients after sildenafil consumption [264].

It is established that PDE 5 (PDE in the corpus cavernosum) along gets inhibited by using the ED drugs, escaping degradation of 3′-5′-cGMP to guanosine 5′-monophosphate (5′-GMP). The NO linkage with guanylyl cyclase creates conformational modification in this enzyme, followed by catalytic cGMP generation from guanosine 5′-triphosphate (GTP), stimulating penis towards erection as has been displayed in Figure 2.87.

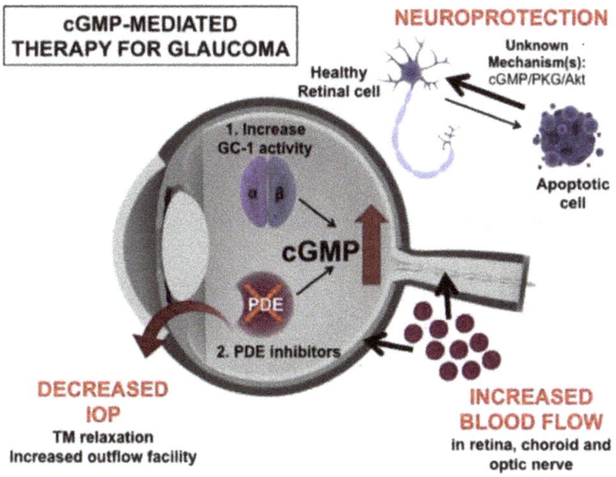

Figure 2.81: GC-1-directed therapy for glaucoma is pleiotropic in its action. Increased levels of cGMP are shown to have pleiotropic targets that are beneficial in the treatment of glaucoma. These are (a) relaxation of the trabecular meshwork to increase AqH outflow facility, which leads to lowering in IOP; (b) increasing blood flow to the retina, choroid and optic nerve head; (c) prevention of degeneration of retinal ganglion cells through mechanisms that may involve downstream kinase pathways. As shown in Figure 2.79, the cGMP concentrations in the eye can be increased in two ways: (i) by the use of GC-1 stimulators and activators, which aim to increase production of cGMP; or (ii) by the use of PDE inhibitors that prevent the decomposition of cGMP into 5′-GMP in the cell to increase its bioavailability.

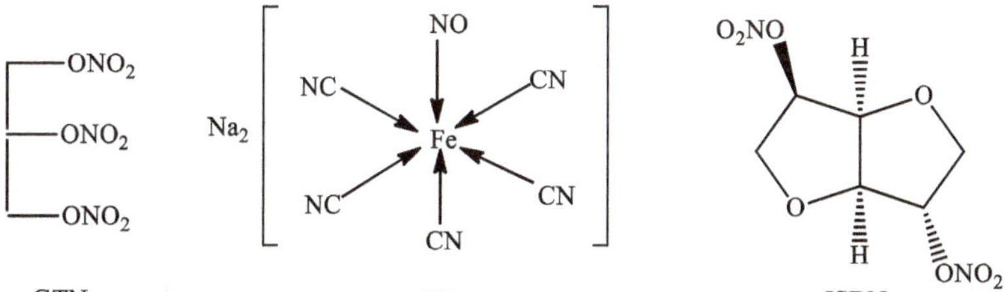

GTN SNP ISDN

Figure 2.82: Structure of some nitric oxide donors.

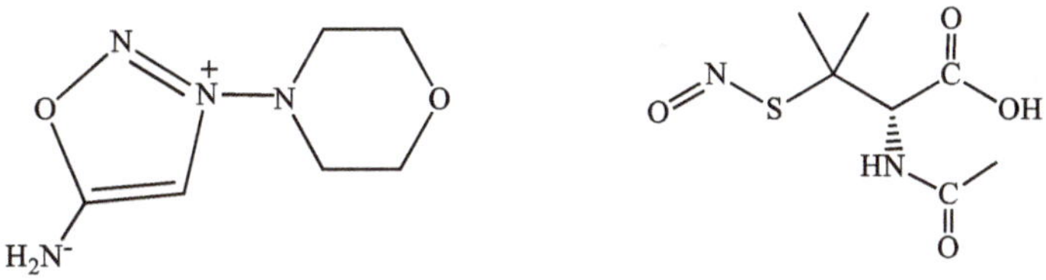

3-morpholinosydnonimide (SIN-1) S-nitroso-N-acetylpenillamine (SNAP)

Figure 2.83: Structure of SIN-1 and SNAP.

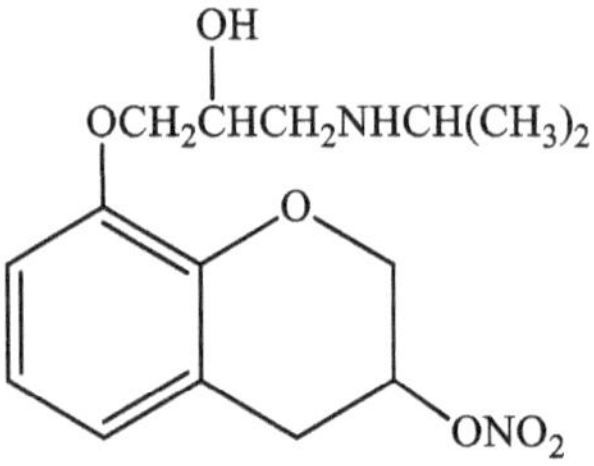

Figure 2.84: Chemical structure of nipradiol.

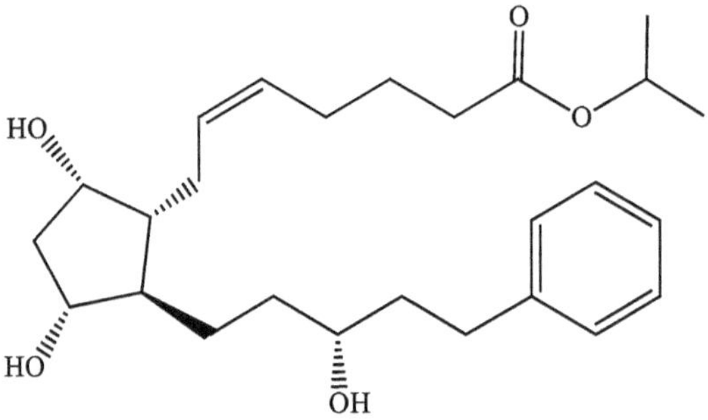

Figure 2.85: Structure of latanoprost.

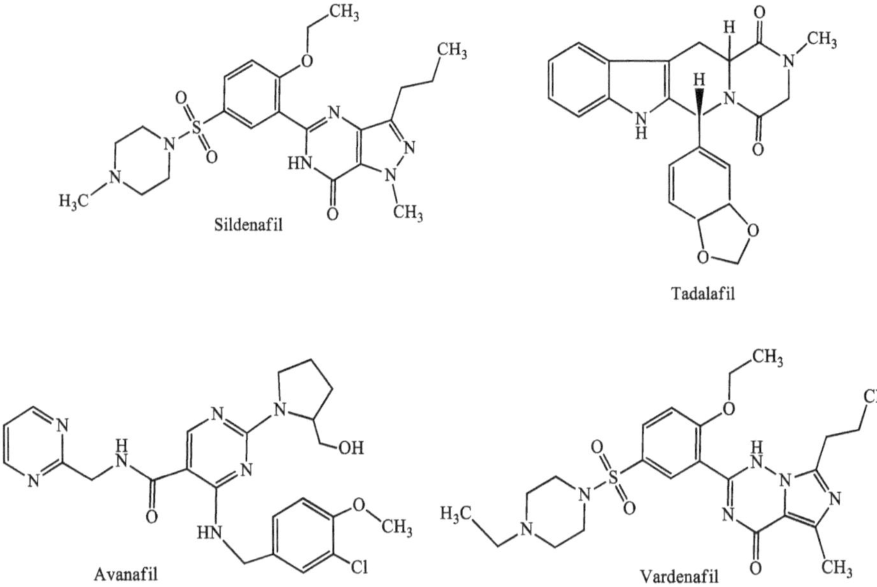

Figure 2.86: Structure of sildenafil and other similar ED drugs.

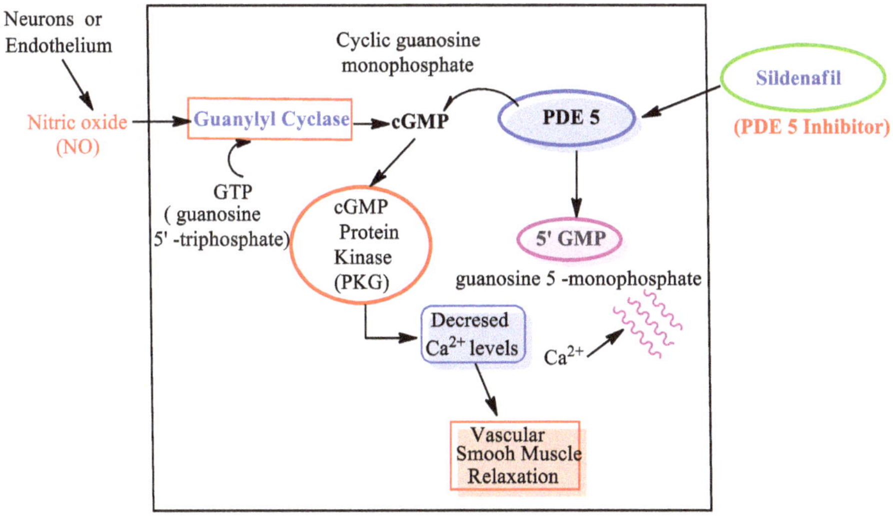

Figure 2.87: NO–cGMP routes for relaxation of arterial and trabecular smooth muscle.

Another familiar example of NO donor usable in lowering IOP is NO-bonded latanoprost acid called as latanoprostene bunod (LBN) and is generally referred for topical treatment, and its action of releasing NO is prostaglandin equivalent. The role of this compound in outflow of AqH has been described in Figure 2.78 and the mechanism of NO release is given in Figure 2.88.

Galassi et al. [265] investigated the outcome of a topical NO-releasing compound known as dexamethasone (NCX1021) involving NO-donating group (Figure 2.89) on IOP and ocular haemodynamics in a rabbit glaucoma model. This study concludes that NCX1021 may prevent the IOP increase, impairment of ocular blood flow and the morphological changes in the ciliary bodies possibly induced by corticosteroid treatment.

Moreover, numerous other NO donor compounds are known that exhibit preclinical effectiveness for IOP reduction in animal models, including the prostaglandin equivalent NCX 470 (a novel NO-donating bimatoprost) and two novel carbonic anhydrase inhibitors: NO dorzolamide and NO brinzolamide [254]. Altogether, these studies demonstrate that NO-donoring compounds are feasible and possibly effective agents for medication towards IOP lowering in glaucoma patients as well as those with ocular hypertension.

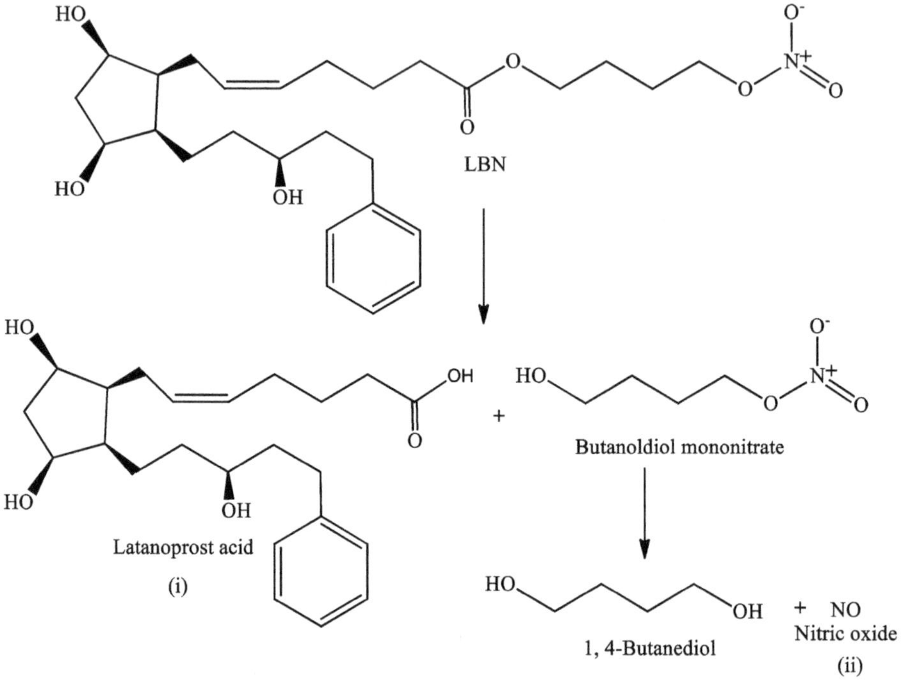

Figure 2.88: The release of nitric oxide from LBN (adapted from *J. Ocul. Pharma. Therap.*, 34 (2018) 52–60).

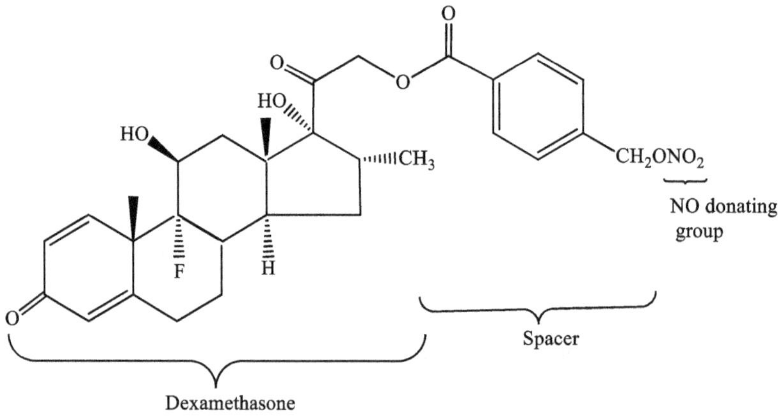

Figure 2.89: Structure of NCX1021 (dexamethasone) involving NO-releasing group.

2.14.4 Mechanism of action of NO in IOP lowering

Contemporary therapies for reduction in IOP comprise β-adrenergic receptor blockers, α-adrenergic receptor agonists and carbonic anhydrase inhibitors. They target

AqH formation. Moreover, prostaglandin equivalents are supposed to lower IOP through effects on the uveoscleral pathway [266]. There is evidence that NO stimulation may mediate IOP lowering effects through a predominant increase in the aqueous outflow, mainly with effects on the conventional pathway.

2.15 Role of NO and exogenous NO donors in plants

While the history of studies on NO in animals is substantially much more advanced, considerable attention has been given to the mechanism of NO synthesis and its roles in plants in the last decades. NO emission from soybean plants treated with herbicides was first observed by Klepper [267] in 1975, much earlier than in animals.

For the first time in 2005, Yamasaki [268] specified that plant systems are more open to the environment and to NO than those of vertebrates. Thus, Arasimowicz and Floryszak-Wieczorek [269] highlighted that plant NO signalling network should be more sensitive to exogenous NO emission. For example, soil bacteria nitrification/denitrification, soil fertilization or air pollutants than closed animal systems localized in specific tissues. With regard to the physiological roles of NO in plants, several works testified its association in the inhibition of plant leaves expansion, cell wall lignification, root organogenesis, sexual reproduction, germination and breaking of seed dormancy [270].

In view of the above, the coming section will, therefore, focus on different biological effects/roles of NO in plants in details.

2.15.1 Biosynthesis of nitric oxide

The synthesis of NO fluctuates with the type of plants and tissues. It also depends on their growing conditions. NO is generated at different centres, such as the cytosol, nucleus, peroxisome matrix and chloroplasts of the plant cells. Wojtaszek [271] in year 2000 reported that there are various biosynthetic sources of NO in plants based on enzymatic and non-enzymatic processes. Figure 2.90 illustrates these sources.

(a) Non-enzymatic processes

Quite a few non-enzymatic reactions are identified for the production of NO. These are being given here: (i) reaction of ascorbate and nitrous acid (HNO_2) in acidic conditions gives rise the formation of dehydroascorbic acid and NO [272]. (ii) NO could be generated by chemical reduction of NO_2^- at acidic pH. (iii) Carotenoids are found to convert NO_2 to NO in the presence of light [273].

(iv) NO is also produced by nitrification and denitrification routes [273]. (v) Non-enzymatically, NO can also be produced from NO donors, namely, SNP and SNAP (Figure 2.91) when applied to plants externally [274].

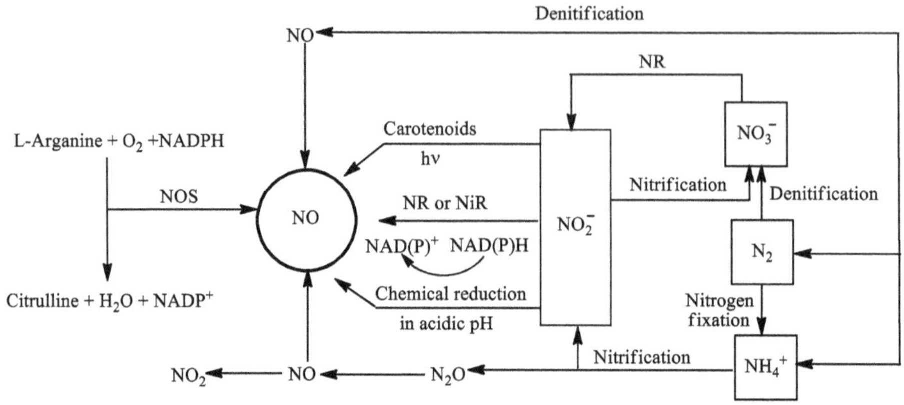

Figure 2.90: Possible biosynthetic sources of nitric oxide (NO). Here, NR, nitrate reductase; NiR, nitrite reductase; NOS, nitric oxide synthase (adapted from Ref. [270]).

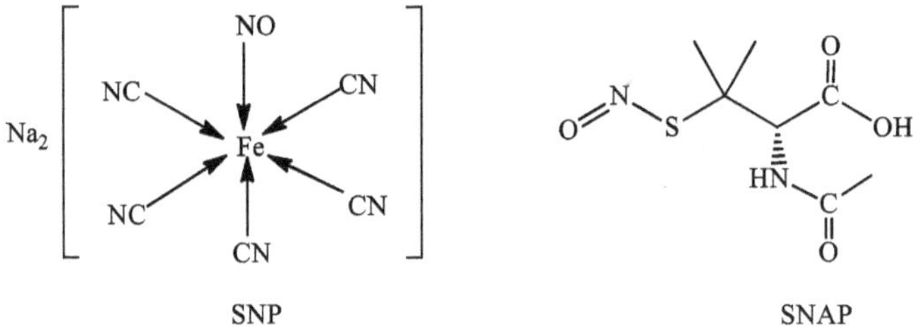

Figure 2.91: Structure of sodium nitroprusside (SNP) and *S*-nitroso-*N*-acetyl-penicillamine (SNAP).

(b) Enzymatic processes

Cevahir and co-workers [275] reported some enzymatic routes that generate NO. For instance, (i) in plant cells, NO can be generated enzymatically via the formation of L-citrulline from L-arginine by an unidentified enzyme, which resembles mammalian NOS in properties. (ii) NOS-like activity has been testified in chloroplasts and peroxysomes of plants. (iii) In peas, wheat and corn, a molecule similar to NOS has also been documented. (iv) Enzyme xanthine oxidoreductase [276] is known to generate NO in hypoxic conditions.

In higher plants, Ninneman and Maier [277] observed prominent concentrations of L-citrulline formation. Using inhibitors, namely, NG-nitro-L-arginine and NW-nitro-L-arginine methyl ester, both the researchers confirmed NOS activity in plants.

Besides the accepted NOS pathway of NO production, nitrate reductase (NR) may produce NO in plants. In fact, NR is a crucial enzyme for nitrate assimilation in higher plants. This enzyme catalyses the transformation of nitrate to nitrite. The resulting nitrite has been found to generate NO in the presence of NADPH at an optimum pH of 6.75 under in vitro and in vivo environments [278] both.

2.15.2 Nitric oxide action in plants

Metabolism in plants is very much affected by NO emission. Consequently, a new field of research has emerged, wherein the effects of NO in plant physiology are being focussed. This has been principally achieved by external/synthetic uses of NO donors in a varied range of concentrations to a great variety of plant systems. The main subjects of attention here are the possible roles of NO in plant growth and development, in signalling mechanism, and during plant stress situation.

2.15.2.1 NO in plant growth and development

(a) We are aware that plants under normal growth conditions release NO. It also accumulates in the atmosphere from several other sources, comprising industrial pollution. It is reported long back that photosynthesis in plants is badly affected by NO, and thereafter the effects of NO on plant growth were observed to be concentration dependent. It has been observed that the development in tomato, lettuce and pea plants is much inhibited, especially at high concentrations of NO (40–80 pphm) (pphm =parts per hundred million), while its low concentrations (0–20 pphm) stimulated overall growth in these plants. Recent data reported by two researchers [279], Takahashi and Yamasaki, suggested that NO can reversibly suppress electron transport and ATP synthesis in chloroplasts.

    As nitrite produces NO, it was proposed that when nitrite reduction by nitrite reductase (NR) is inadequate to produce NO, it could inhibit photosynthesis. As per this suggestion, antisense-nitrite reductase tobacco plants have been found to store increased NO and display reduced growth [280].

(b) Lengthening of hypocotyl and internode in plants is affected by NO. It has been observed by Beligni et al. [281] that the NO donors inhibit hypocotyl elongation and stimulate de-etiolation and increase in chlorophyll contents in potato and lettuce. NO is also found to enhance the chlorophyll continuum in the leaves of pea [282]. Also, the role of NO in increasing iron bioavailability reported by Graziano et al. [283] highlights the significance of NO in plants.

(c) Anti-senescence properties of NO have been reported by several research groups. Leshem and Haramaty [284] found that application of an NO donor like SNAP (Figure 2.91) increases the age of pea plant, incorporating anti-ageing influence [285]. Fruit ripening is another senescence-related process promoted by ethylene that can be delayed by NO [286]. Moreover, application of NO to vegetables and fruits also retarded senescence and prolonged their postharvest life.

(d) In several plant species, germination of seeds is found to be motivated by NO [287]. Use of NO donors, namely, SNP and SNAP, terminated dark-imposed seed dormancy in lettuce. Remarkably, it was inverted in the presence of the NO scavenger, 2-phenyl-4,4,5,5-tetramethylimidazoline-1-oxyl 3-oxide (PTIO). Such effects of NO might also explain the germination of dormant seeds of California chaparral plants stimulated by smoke containing nitrogen oxides [288]. Giba et al. [289] demonstrated that phytochrome-controlled germination of empress tree seeds is facilitated by NO using NO donors, namely, SNP, SNAP and SIN-1 (Figure 2.92). Nitrite promotes seed germination at low pH of 2.5 and 3. However, seeds are not germinated only in acidic conditions in the absence of nitrite.

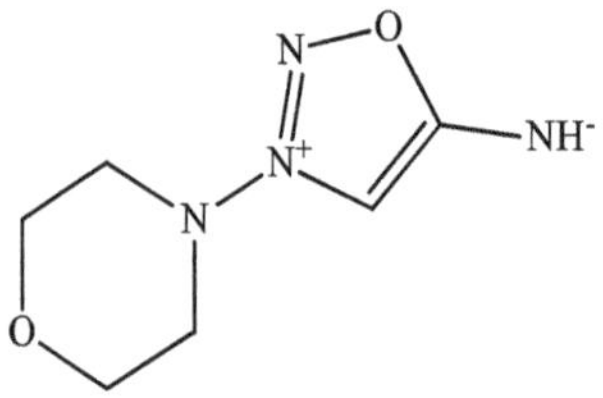

Figure 2.92: Structure of 3-morpholinosydnonimine (SIN-1).

In fact, NO are generated from nitrite under low pH conditions, and nitrite-induced NO synthesis via a non-enzymatic route can occur in plants (see under biosynthesis of NO above). Thus, earlier reports on the promotion of seed germination by nitrite might be explained by NO generation. These roles of nitrite (and thus NO) on seed germination are fascinating and suggest that the concentration of soil nitrite is an important factor for seed germination.

2.15.2.2 Interactions between nitric oxide and hormones

It is known that NO serves as a facilitator in the biological functioning of primary messenger molecules such as hormones. As an illustration, a plant hormone cytokinin was found to motivate synthesis of NO in some plants like tobacco, parsley and *Arabidopsis* cells. Interestingly, NO was shown to duplicate some roles of cytokinin [290] via stimulation of betalain accumulation in *Amaranthus seedlings*. NO may also participate in abscisic acid (ABA)-affected stoma closure. It is recently demonstrated that ABA motivates NO synthesis in the guardian cells [291] of peas, but had no effect in *Arabidopsis* suspension cultures. Pagnussat et al. [292] stated that SNP and SNAP used as NO donors in cucumber stimulated adventive root development. But auxin (Figure 2.93), a plant hormone, persuaded root growth and lateral root formation were hindered by the use of the NO inhibitor, 2-[4-carboxyphenyl]-4,4,5,5-tetramethylimidazoline-1-oxy-3-oxide (c-PTIO).

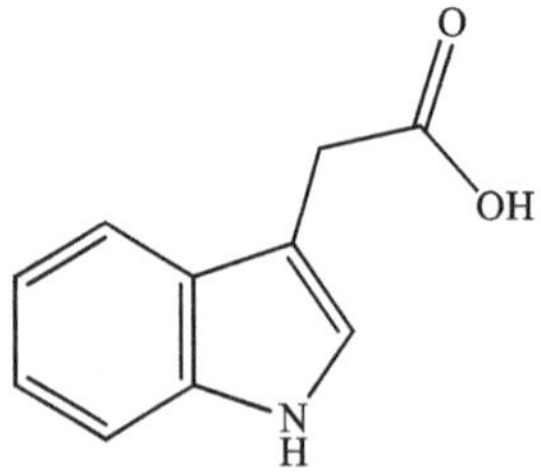

Figure 2.93: Chemical structure of auxin (indole-3-acetic acid).

2.15.2.3 Correlation of NO with abiotic stress plants

By abiotic stress in plants, it refers to exposure of plants to water scarcity, non-optimal temperatures, high energetic radiations and ozone. Such stresses result in the generation of ROS by plants in a responsive way, triggering different signalling stages, and meanwhile NO serves as an antioxidant agent to interact with these ROS. Thus, modifying superoxide and preventing lipid peroxidation defines this correlation. Overall, an optimum balance between ROS/NO is essential for normal functioning and to cater the abiotic stress because in excess NO productive nitrosative stress could harm the plant.

(a) Drought stress

Remarkably, drought stress is a foremost environmental limitation on crop productivity and performance. In view of this, knowledge of the cellular processes that improve the consequences of drought stress and conservation of water are obviously important. Importantly, ABA is synthesized subsequent to turgor loss of the plant cells, and this arouses guard cell (stress tolerance cell) NO synthesis. However, the effects of dehydration on NO production have not been determined by now. Leshem and Haramaty [293] reported that wilting (excess water loss from plant called wilting) augmented NO release from pea plants. However, in *Arabidopsis* (a small flowering plant), Magalhaes et al. [294] observed the opposite result. This may be due to species differences. Hence, it will be noteworthy to determine the results of water stresses at a range of reduced water potentials and over variable time periods. On the other hand, NO can assist in reducing transpiration by halting stomatal opening, and NO donors, like SNP and *S*-nitroso-*N*-acetylpenicillamine can serve for this purpose to facilitate NO availability in plants [295].

Accumulating evidences are available that ROS and NO interact to persuade ABA biosynthesis. Zhao and co-workers [296] reported that in reaction to drought stress, an increase in NOS-like activity was noticed in wheat seedlings. In addition, growth of ABA was slowed down by NOS inhibitors. Jiang and Zhang [297] reported that NADPH oxidase activity was induced by ABA during drought stress. This leads to increased ROS levels in maize seedlings, indicating a close interaction between ABA, ROS and NO levels.

(b) Temperature/heat and cold stress

Abiotic stresses in the form of too hot or too cold also influences the crop production to a considerable extent. Similar to other stresses, as per the relevant study reported by Suzuki and Mittler [298] suggests that oxidative stress is induced in plants exposed to high temperature resulting in lipid peroxidation that causes injury to cellular membrane and degrades several proteins, and other cellular organelles. In a similar approach concerned with the effect of cold stress indicates in the disturbance of normal ROS equilibrium in plants [299]. Again, the facilitation of NO through the administration of SNP or SNAP can prove to be beneficial in lowering this stress. Zhang et al. [300] selected a common reed (*Phragmites communis*) subjected to a temperature of 45 °C and found ion leakage and growth inhibiting factors at an increased amount in the selected temperature. Bavita and co-workers [301] used 50 and 100 μM of SNP as NO donors for two varieties of wheat as models [wheat-C306 (heat tolerant) and PBW550 (heat sensitive)] at 33 °C, and found that all antioxidant pathways of the respective enzymes showed increased active defence against the stress. Similar observations have been reported by Yang et al. [302], using SNP as exogenous NO source in mung bean. In several reports of the same interest, positive results of lowering the stress among plants have been reported using NO donors [303, 304].

(c) Ultraviolet radiation stress

Sunlight is the major energy provider for plants. Additionally, morphological, biochemical, physiological and molecular effects in plants have been suggested to occur because of UV-B radiation (among three classes) as reported by Kataria et al. [305]. In exposure of high energetic radiations, plants undergo oxidative stress resulting in adverse effects, increasing the responsiveness of plants in the form of AA, chloroplast and superoxide dismutase (SOD) activities [306, 307]. Meanwhile, here too NO can serve as a motivational antioxidant tool as has been reported by a study on soybean leaves [308]. It was found that SNP can alone be used to tackle with the oxidative stress caused by UV exposure.

(d) Salt stress

Another prolonged adverse condition disturbing plant growth and development occur due to salinity stress. High salt level in plants results in osmotic and ionic stress [309], which confines growth and development of plants owing to several key metabolic processes being affected in the prevailing condition. Moreover, salinity modifies the activities of several enzymes [310] comprising nitrate and sulphate assimilation pathway in plants, lowers their energy status and increases the demand for nitrogen and sulphur. Much of the injury at cellular level resulted because the salinity stress is associated with oxidative damage due to ROS. Plants appear to possess a wide range of defence strategies to protect from oxidative damage. Nevertheless, little is known about the involvement of NO in tolerance to plants to salt stress.

In salt stress, the use of an exogenous NO donor [311], SNP, significantly lessened the oxidative damage of salinity to seedling of rice, lupine and cucumber. Moreover, SNP enhanced seedlings' growth and increased the dry weight of maize and *Kosteletzkya virginica* seedlings under salt stress. Zheng et al. [312] have shown that pre-treatment of SPN effectively played a part in better balance between carbon and nitrogen metabolism by increasing total soluble protein and by increasing the activities of endopeptidase and carboxypeptidase in plants under salt stress.

(e) Heavy metal stress

Heavy metal pollution changes the environment [313] in many places globally. There are two key mechanisms for most of the heavy metals to exert their toxicity. These are (i) redox-active mechanism and (ii) not-redox-active mechanisms. Autoxidation of redox-active metals, namely, Fe or Cu may result in O_2^- formation and successively in H_2O_2 and $OH^•$. The toxicity of non-redox-active metals is owing to their knack to bind to oxygen, nitrogen and sulphur atoms.

Numerous studies have been conducted in order to evaluate the causes and remedy of different heavy metal concentration on plants. NO plays a vital role in augmentation of antioxidant enzyme activities and lessens the toxicity of heavy metals. Use of SNP from outside reduces the toxicity of Cu and NH_4^+ accumulation in rice leaves. The defensive effect of SNP on the toxicity and NH_4^+ accumulation can be inverted by carboxy-2-(4-carboxyphenyl-4,4,5,5-tetramethylimidazoline-1-oxyl-3-oxide (c-PTIO) (Figure 2.94), an NO scavenger, suggesting that the defensive effect of SNP is

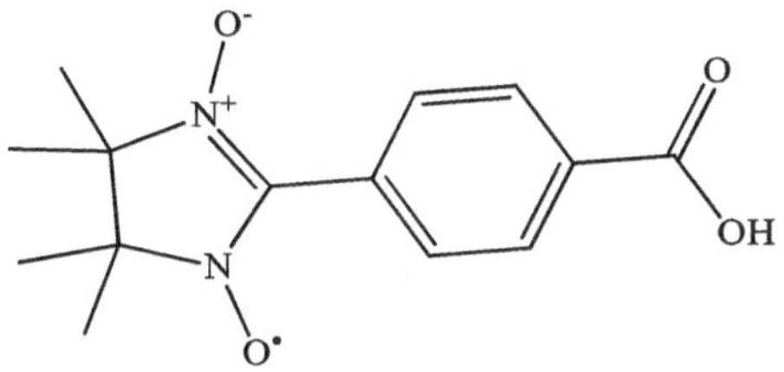

Figure 2.94: Chemical structure of c-PTIO.

because of the NO release. These results also suggest that lowering of Cu-induced toxicity [314] and NH_4^+ accumulation by SNP is most probably mediated through its ability to scavenge active oxygen species. Kopyra and Gwóźdź [315] also observed that SNP pre-treatment considerably reduces O_2^--persuaded specific fluorescence in *Lupinus luteus* roots under heavy metal treatment. The results this study indicate that antioxidant function of NO may be traced by scavenging of O_2^-, resulting in a decrease of superoxide anion. Kaur et al. [316] reported the exogenous NO interference with lead (Pb) persuaded toxicity by detoxifying ROS in hydroponically grown wheat (*Triticum aestivum*) roots. Furthermore, Tian and co-workers [317] established that NO decreases the aluminium (Al^{3+}) toxicity in root elongation of *Hibiscus moscheutos*.

It has been reported by Cui and co-workers [318] that the use of exogenous SNP promotes ROS-scavenging enzymes, reduces accumulation of H_2O_2, and persuades the activity of H^+-ATPase and H^+-PPase in plasma membrane or tonoplast. It also considerably improves the growth inhibition induced by $CuCl_2$ in tomato plants. These results suggest that exogenous NO could effectively encourage tomato seedlings to adjust physiological and biochemical mechanisms against Cu toxicity and maintain fundamentally metabolic capacity and normal growth under heavy metal stress.

Hu et al. [319] also found that pre-treatment of NO improves wheat seed germination and lessens oxidative stress against Cu toxicity by enhancing the activity of SOD and catalase and by decreasing the lipoxygenase activity and malondialdehyde synthesis. NO was reported to have the ability to reduce Cu-induced toxicity.

2.15.2.4 Correlation of biotic stress with NO in plants

Biotic stresses are a possible threat to global food security. The origin of new pathogens and insect races due to climatic and genetic factors is a major challenge for plant breeders in breeding biotic stress-resistant crops. Biotic stresses are the damage to plants instigated by other living organisms such as bacteria, fungi, nematodes, protists, insects, viruses and viroids. Fungi, bacteria, nematodes and viruses are the pathogens primarily responsible for plant diseases. Numerous biotic stresses (diseases) are of historical significance [320] unlike the abiotic stresses, for example, the potato blight in Ireland, coffee rust in Brazil, maize leaf blight caused by *Cochliobolus heterostrophus* in the United States and the great Bengal famine in 1943. These are some of the major events that devastated food production and led to millions of human deaths and migration to other countries in the past. Presently, the occurrence of new pathogen races and insect biotypes poses further threat to crop production.

The ability of NO to act as a stress-coping factor was studied for biotic stress in plants by several research groups. For example, treatment of potato tuber tissues with the NO donor NOC-18 (1–10 mM) [spontaneously release NO ($t_{1/2}$ = 20 h)] provoked an accumulation of the phytoalexin rishitin, an endogenous antibiotic compound. The effect of NOC-18 was neutralized by the specific NO-scavenger carboxy-PTIO and by the free radical scavenger Tiron [321]. Hyphal wall components (HWC) also stimulated the formation of rishitin and had an additive effect with NOC-18 (Figure 2.95). However, carboxy-PTIO did not affect HWC-induced rishitin increase, suggesting that NO and HWC signals are probably transduced through independent pathways.

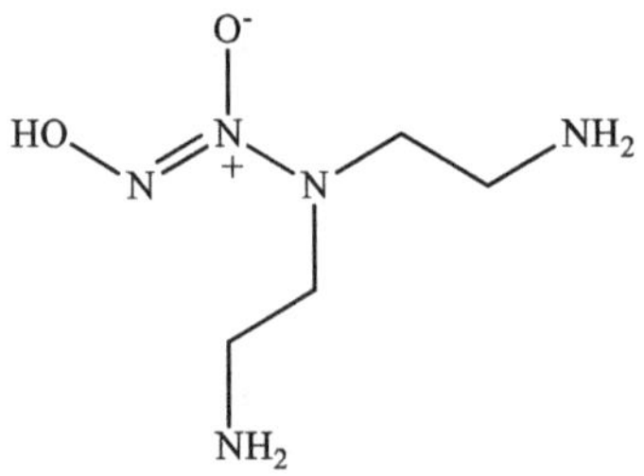

Figure 2.95: L-Hydroxy-2-oxo-3, 3-bis(2-aminoethyl)-1-triazene (NOC-18).

Laxalt et al. [322] testified that NO safeguards chlorophyll levels in potato leaves that were infected with the pathogen *Phytophthora infestans*. This was achieved with low NO concentrations (between 10 and 100 mm SNP). Lower concentrations were unproductive, whereas higher ones resulted in chlorophyll decay.

Huang and Knopp [323] observed that in the tobacco–*Pseudomonas solanacearum* pathosystem, NO donors were able to invoke a hypersensitive response (HR). (The HR is a mechanism, used by plants, to prevent the spread of infection by microbial pathogens.) HR was characterized by rapid host cell death at the pathogen penetration site. Consequently, the growth and development of the fungus is restricted and, thus, spreading to other parts of the plant is prevented.

Durner et al. [324] gave convincing evidence for the role of NO during plant defence responses. Infection of tobacco plants with HR-inducing ranges of tobacco mosaic virus–induced NOS activity that was inhibited by NOS inhibitors. NO also persuaded the synthesis of salicylic acid (SA) and expression of the defence-related gene *PR-1*. SA is a defence signalling molecule connected in the development of systemic acquired resistance (SAR) [325]. Thus, the data of Durner et al. suggested a role for NO in SAR. Since SA treatment leads to enhance NO production [326], a complex signalling relationship between H_2O_2, NO and SA during HR and SAR is likely [327].

2.15.2.5 Mechanism of nitric oxide signalling in plants

Even though research outputs clarifying NO's activity in plants are expanding progressively, only little facts regarding the processes of its signalling mechanisms have been accumulated by now. It seems improbable that specific receptors exist for NO as NO is such a simple, small and diffusible molecule. Nevertheless, cells are quite sensitive to NO and various cellular activities occur in its presence [328].

Significantly, NO signalling may encompass either direct activation of ion channel proteins or proteins that regulate their gene expression or indirect regulation of signal cascade proteins, such as protein kinases, ion channels or enzymes that produce second messengers [329]. NO signalling can be dependent or independent on the second messenger molecule cGMP [330]. In cGMP-dependent [331] signalling, NO activates the enzyme GC which catalyses the synthesis of cGMP from GTP. The resultant increase in cGMP levels raises cytosolic calcium levels by increasing levels of activated cyclic adenosine 5′-diphosphate ribose, another second messenger molecule. The cGMP also activates intracellular protein kinases. Alternatively, in cGMP-independent [332] NO signalling, proteins that contain thiol groups and metals, such as iron, copper and zinc, are among the primary targets of NO.

Besides GC, aconitases are a major NO target in animals, and NO directly affects aconitase functionality [333]. Aconitase is an iron–sulphur (4Fe–4S)-containing enzyme that catalyses the reversible isomerization of citrate to isocitrate. Navarre et al. [334] observed the effect of NO on tobacco (*Nicotiana tabacum*) aconitase. They recognized that the tobacco aconitases, like their animal counterparts, were inhibited by NO donors.

2.16 Conclusions

The applied aspects of NO in the entire set of living things ranging from microbial world to the mammals have brought up persuasion among scientists to expand the horizons of investigations of this molecule in the form of organic and inorganic frameworks. The catalytic role of nitrosyl complexes proven by different physicochemical analysis collaborated with theoretical front indicates the significance of this field. The coulombic interaction in the form of electron density functionalization helps in deciding the cellular or protein interaction of an NO compound. Therefore, though for a molecular system (whether neutral or having some surplus charge) the total charge is distributed among different atomic constituents, thereby creating a topographic divisions among different atoms.

Light-assisted decontamination of microbes is the current trend in microbial research. In connection with the phenomenon, NO could play a major role in disinfecting by a dual mechanism using photodiagnostic and photodynamic approach related to chemo- and phototherapy. Meanwhile, in countering several research problems, the main future directive aspects include the development of new molecular scaffolds showing varying coordination sphere with respect to metal and ligand. In the current times of pandemic COVID-19, antimicrobial studies should use metallic compounds as antibiotic substances to seek for an impressive antimicrobial agent. In addition to antimicrobial effects, anticancer and SOD behaviour, the complexes are suggested as antibiotics and anticancer compounds. The application of theoretical chemistry formalism should be stretched for all other basis sets and functionals for the selected compounds to bring forth the reliability evaluation of the selected theories. In the meantime, the use of the studied complexes in developing medication for viral strains like coronaviruses could pave the way for successful formulations against COVID-19 like pandemics. While looking at the literature support suggesting the key role of NO in biological world, the nitrosyl complexes serve as efficient tools to overcome oxidative stress in plants and animals. The use of such compounds for diversified applications is a gesture to conclude that NO is not only Noble Prize deserving in medicine, but has profoundly surprised all specialities ranging from the microorganism world to the industrial research.

References

[1] R. C. Maurya, N. Jan Mohammad Mir, P. S. Jain, W. K. Jaget, P. K. Vishwakarma, D. K. Rajak and B. A. Malik Urinary tract anti-infectious potential of DFT-experimental composite analyzed ruthenium nitrosyl complex of N-Dehydroacetic acid-thiosemicarbazide, *J. King Saud Univ. Agric. Sci.*, 31 (2019) 89–100.

[2] W. Hieber, R. Nast and G. Gehring, *Z. Anorg. Allgem. Chem.*, 256 (1948) 145, 169.

[3] W. P. Griffith, J. Lewis and G. Wilkinson, *J. Chem. Soc.*, 872 (1959).

[4] W. P. Griffith, J. Lewis and G. Wilkinson, *J. Chem. Soc.*, 1632 (1959).
[5] A. Müller, P. Werle, E. Diemann and P. J. Aymonino, *Chem. Ber.*, 105 (1972) 2419.
[6] R. G. Bhattacharya and P. S. Ray, *Transition Met. Chem.*, 9 (1984) 281.
[7] R. G. Bhattacharya, G. P. Bhatterjee and A. M. Saha, *Polyhedron*, 4 (1985) 583.
[8] W. Heiber, R. Nast and E. Proeschel, *Z. Anorg. Allgem. Chem.*, 256 (1948) 145.
[9] R. C. Maurya, A. Pandey, J. Chaurasia and H. Martin, J. *Mol. Struct.*, 798 (2006) 89–101.
[10] L. Malatesta and A. Sacco, *Z. Anorg. Allgem. Chem.*, 274 (1953) 341.
[11] S. Sarkar and A. Müller, *Z. Naturforsch*, 33B (1978) 1053.
[12] K. Weighardt, U. Quilitzsch, B. Nuberand and J. Weise, *Angew. Chem.*, 90 (1978) 381.
[13] K. Weighardt and W. Holzbach, *Angew. Chem.*, 91 (1979) 583.
[14] J. B. Godwin and T. J. Meyer, *Inorg. Chem.*, 10 (1971) 471.
[15] S. Sarkar and A. Müller, *Z. Naturforsch.*, 33b (1978) 1053–1055.
[16] S. Sarkar and R. C. Maurya, *Transition Met. Chem.*, 1 (1976) 49.
[17] R. C. Maurya, *Synthesis and Characterization of Some Novel Nitrosyl Compounds, Pioneer Publication*, Jabalpur, (2000).
[18] R. G. Bhattacharya and R. S. Ray, *Transition Met. Chem.*, 9 (1984) 280.
[19] W. Kaim, *Angew. Chem. Int. Ed.*, 50 (2011) 10498.
[20] C. K. Jørgenson, *Coord. Chem. Rev.*, 1 (1962) 164.
[21] J. H. Enemark and R. D. Feltham, *Coord. Chem. Rev.*, 13 (1974) 339–406.
[22] C. De La Cruz and N. Sheppard, *Spectrochim Acta: A Mol. Biomol. Spectrosc.*, 78 (2011) 7–28.
[22a] S. F. A. Kettle, *Inorg. Chem.*, 4 (1965) 1661.
[22b] J. H. Enemark, *Inorg. Chem.*, 10 (1971) 1952.
[23] R. Hoffmann, M. M. L. Chen, M. Elian, A. R. Rossi and D. M. P. Mingos, *Inorg. Chem.*, 13 (1974) 2666–2675.
[24] *Applications of Density Functional Theory to Chemical Reactivity*, Edited by M. V. Putz, D. Michael and P. Mingos, Springer, Heidelberg, New York, Dordrecht, London, (2012).
[25] R. C. Maurya and J. M. Mir, *RSC Adv.*, 8 (2018) 35102–35130.
[26] R. C. Maurya and J. M. Mir, *J. Chin. Chem. Soc.*, 66 (2019) 651–659; doi: doi.org/10.1002/jccs.-201800337.
[27] D. J. Sherman and D. M. P. Mingos, *Adv. Inorg. Radiochem.*, 34 (1989) 293.
[28] B. L. Haymore and J. A. Ibers, *Inorg. Chem.*, 14 (1975) 3060–3070.
[29] C. De La Cruz and N. Sheppard, *Spectrochim. Acta A: Mol. and Biomol. Spectrosc.*, 78 (2011) 7–28.
[30] G. Chottard and D. Mansuy, *Biochem. Biophys. Res. Commun.*, 77 (1977) 1333–1338.
[30a] E. E. Mercer, W. A. McAllister and J. R. Durig, *Inorg. Chem.*, 5 (1966) 1881–1886.
[31] E. E. Mercer, W. A. McAllister and J. R. Durig, *Inorg. Chem.*, 6 (1967) 1816–1821.
[32] P. Gans, A. Sabatini and L. Sacconi, *Inorg. Chem.*, 5 (1966) 1877–1881.
[33] N. M. Sinitsyn and K. I. Petrov, *Zh. Strukt. Khim.*, 9 (1968) 45.
[34] P. T. Manoharan and H. B. Gray, *Inorg. Chem.*, 5 (1966) 823, *J. Am. Chem. Soc.*, 87 (1965) 3340.
[35] B. A. Goodman, J. B. Raynor and M. C. R. Symons, *J. Chem. Soc.*, A (1968) 1973.
[36] W. P. Griffith, *J. Chem. Soc.*, A (1963) 3286.
[37] A. Wanat, T. Schneppensieper, G. Stochel, R. van Eldik, E. Bill and K. Wieghardt, *Inorg. Chem.*, 41 (2002) 4–10.
[38] D. Gwost and K. G. Caulton, *Inorg. Chem.*, 1974 13 (1974) 414–417.
[39] R. C. Maurya and J. M. Mir, *J. Chin. Adv. Mater. Soc. (JCAMS)*, 6 (2018) 620–639; doi: doi.org/10.1080/22243682.2018.1534608.
[40] R. C. Maurya, *Indian J. Chem.*, 22A (1983) 529.
[41] R. G. Bhattachaya, N. Ghosh and G. P. Bhattachajee, *J. Chem. Soc. Dalton Trans.*, 1963 (1989).

[42] G. W. Brudvig, T. H. Stevens and S. I. Chan Reactions of nitric oxide with cytochrome C oxidase, *Biochemistry*, 19 (1980) 5275–5285.
[43] L. S. Meriwether, S. D. Robinson and G. Wilkinson, *J. Chem. Soc. A*, (1966) 1488.
[44] B. A. Goodman, J. B. Raynor and M. C. R. Symons, *J. Chem. Soc. A*, 994 1966.
[45] J. Burgess, B. A. Goodman and J. B. Raynor, *J. Chem. Soc.*, A (1968) 501.
[46] S. Sarkar and A. Müller, *Z. Naturforsch.*, 33b (1978) 1053.
[47] P. T. Manoharan and H. B. Gray, *Inorg. Chem.*, 5 (1966) 823, *J. Chem Soc., Chem. Comm.* (1965).
[48] B. A. Goodman and J. B. Raynor, *Adv. Inorg. Radiochem.*, Edited by H. J. Emeleus and A. G. Sharp, 13 (1970) 135.
[49] A. F. Vanin, V. A. Serezhenkov, V. D. Mikoyan and M. V. Genkin, *Nitric Oxide*, 2 (1998) 224–234.
[50] B. D'Autreaux and N. Tucker, *Methods Enzymol.*, 437 (2008) 235–251.
[51] C. L. Joannou and X. Y. Cui, *Appl. Environ. Microbiol.*, 64 (1998) 3195–3201.
[52] J. Mason, L. F. Larkworthy and E. A. Moore Nitrogen NMR spectroscopy of metal nitrosyls and related compounds, *Chem. Rev*, 102 (2002) 913–934.
[53] D. M. P. Mingos and D. J. Sherman, *Trans. Met. Chem.*, 12 (1987) 400, 12 (1987)897.
[53a] N. I. Nefedov and W. Z. Karl, Marx. *Univ. Leipzig, Math. Naturwiss*, R. 25 Jg 383 (1976).
[53b] B. Folkesson, *Acta Chem. Scand. A*, 28 (1974) 491.
[53c] W. B. Hughes and B. A. Baldwin, *Inorg. Chem.*, 13 (1974) 1531.
[53d] P. Brant and R. D. Feltham, *Climax IInd International Conference on Chemistry and uses of Molybdenum*, p-39 (1976).
[53e] S. Sarkar and A. Müller, Z. *Naturforsch*, 33b (1978) 1053.
[53f] R. A. Walton, *Climax IInd International Conference on Chemistry and uses of Molybdenum*, p-34 (1976).
[53g] F. Lepage and P. B. Bardole, *Compt. Rend.*, 280C (1975) 1089.
[53h] J. R. Ebner, D. L. McFadden, D. R. Tyler and R. A. Walton, *Inorg. Chem.*, 15 (1976) 3014.
[53i] S. Sarkar and P. Subramanian, *Inorg. Chim. Acta*, 35 (1979) 357.
[54] G. M. Sandala, K. H. Hopmann, A. Ghosh and L. Noodleman, *J. Chem. Theory Comput.*, 7 (2011) 3232–3247.
[55] A. D. Bochevarov, R. A. Friesner and S. J. Lippard The prediction of ^{57}Fe Mössbauer parameters by the density functional theory: A benchmark study, *J. Chem. Theory Comput.*, 6 (2010) 3735.
[56] S. Friedle, E. Reisner and S. J. Lippard, *Chem. Soc. Rev.*, 39 (2010) 2768–2779.
[57] A. Wanat, T. Schneppensieper, G. Stochel, R. van Eldik, E. Bill and K. Wieghardt, *Inorg. Chem.*, 41 (2002) 4–10.
[58] R. J. Butcher and E. Sinn, *Inorg. Chem.*, 19 (1980) 3622–3626.
[59] T. C. Harrop and Z. J. Tonzetich, J. *Am. Chem. Soc.*, 130 (2008) 15602–15610.
[60] N. A. Sanina and O. A. Rakova, *Mendeleev Commun.*, 14 (2004) 7–8.
[61] A. L. Feig, M. T. Bautista and S. J. Lippard, *Inorg. Chem.*, 35 (1996) 6892–6898.
[62] H. Nasri and M. K. Ellison, *J. Am. Chem. Soc.*, 119 (1997) 6274–6283.
[63] M. C. Liu and B. H. Huynh, *Eur. J. Biochem.*, 169 (1987) 253–258.
[64] M. S. Goodman and M. J. Demarco, *J. Chem. Soc. Dalton Trans.*, (2002) 117–120.
[65] H. Y. Cheng, S. Chang and P. Y. Tsai, *J. Phys. Chem.* A, 108 (2003) 358–361.
[66] E. M. Thorsteinson and F. Basolo, *J. Am. Chem. Soc.*, 88 (1966) 3929.
[67] G. Cardaci, A. Foffani and G. Innorta, *Inorg. Chim. Acta*, 1 (1967) 340.
[68] J. P. Day, D. Diementc and F. Basolo, *Inorg. Chim. Acta*, 3 (1969) 363.
[69] G. Cardaci, S. M. Murgia and A. Foffani, *J. Organometal. Chem.*, 23 (1970) 275.
[70] H. Wawersik and F. Basalo, *J. Am. Chem. Soc.*, 89 (1967) 4626.
[71] D. E. Morris and F. Basolo, *J. Am. Chem. Soc.*, 90 (1968) 2531.
[72] J. H. Swinehart and P. A. Rock, Inorg. Chem., 5 (1966) 573–576.

[73] J. Dempir and J. Masck, *Inorg. Chim. Acta*, 2 (1968) 402, 443.
[74] J. H. Swinehart and W. G. Schmidt, *Inorg. Chem.*, 6 (1967) 232.
[75] D. X. West and D. J. Hassemer, *J. Inorg. Nucl. Chem.*, 32 (1970) 217.
[76] F. Bottomley, S. G. Clarkson and E. M. R. Kiremire, *J. Chem. Soc. Chem. Com.*, (1975) 91.
[77] C. Hu, A. Barabanschikov, M. K. Ellison, J. Zhao, E. E. Alp, W. Sturhahn, M. Z. Zgierski, J. T. Sage and W. R. Scheidt, *Inorg. Chem.*, 51 (2012) 1359–1370.
[78] W. R. Scheidt, S. M. Durbin and J. T. Sage, *J. Inorg. Biochem.*, 99 (2005) 60–71.
[79] J. T. Sage, et al., *J. Phys.: Condensed Matter*, 13 (2001) 7707.
[80] B. M. Leu, M. Z. Zgierski, et al., Quantitative vibrational dynamics of iron in nitrosyl porphyrins, *J. Am. Chem. Soc.*, 126 (2004) 4211–4227.
[81] W. R. Scheid, H. F. Duval, et al., Intrinsic structural distortions in five-coordinate (nitrosyl) iron (II) porphyrinate derivatives, *J. Am. Chem. Soc.*, 122 (2000) 4651–4659.
[82] A. Foffani, S. Pignataro, G. Distefano and G. Innorta Influence of the ligand donor ability on the ionization potentials and fragmentation patterns of transition metal nitrosyl complexes, *J. Organometal. Chem.*, 7 (1967) 473–479.
[83] G. Distefano, G. Innorta and A. Foffani Correlation between the ionization potentials of transition metal complexes and of the corresponding ligands, *J. Organometal. Chem.*, 14 (1968) 165–172.
[84] B. F. G. Johnson, J. Lewis and I. M. Wilson, *J. Chem. Soc.* A, (1967) 338–341.
[85] R. B. King, *Org. Mass Spectrom.*, 2 (1969) 401–412.
[86] J. Müller, *J. Organometal. Chem.*, 23 (1970) C38-C40.
[87] R. C. Maurya, R. K. Shukla and M. R. Maurya, *J. Indian Chem. Soc.*, 61 (1984) 799.
[88] J. H. Swinehart, *Coord. Chem. Rev.*, 2 (1967) 385–402.
[89] J. B. Godwin and T. J. Meyer, *Inorg. Chem.*, 10 (1971) 2150.
[90] P. G. Douglas and R. D. Feltham, *J. Am. Chem. Soc.*, 93 (1971) 84.
[91] P. G. Douglas and R. D. Feltham, *J. Am. Chem. Soc.*, 94 (1972) 5254.
[92] G. Dolcetti, N. W. Hoffman and J. P. Collman, *Inorg. Chim. Acta*, 6 (1972) 531.
[93] C. A. Reed and W. R. Roper, *J. Chem. Soc. A*, (1970) 3045.
[94] S. Sarkar and A. Müller, Z. Naturforsch., 33b (1978) 1053.
[95] R. P. Cheney, M. G. Simic and K. D. Asmus, *Inorg. Chem.*, 16 (1977) 2187.
[96] D. I. Bustin and A. A. Vlcek, *Coll. Czech. Chem. Comm.*, 31 (1965) 2374.
[97] J. N. Armor and M. Z. Hoffman, *Inorg. Chem.*, 14 (1975) 444.
[98] G. M. Brown and T. J. Meyer, *J. Am. Chem. Soc.*, 97 (1975) 894.
[99] J. A. McCleverty, N. M. Atherton, N. G. Connelly and C. J. Winscom., *J. Chem. Soc. A*, (1969) 2242.
[100] S. G. Clarkson and F. Basolo, *Inorg. Chem.*, 12 (1973) 1528.
[101] B. W. Graham, K. R. Laing, C. J. O'Connor and W. R. Roper, *J. Chem. Soc., Dalton Trans.*, (1972) 1237.
[102] E. Frankland, *Phil. Trans.*, 147 (1857) 73.
[103] J. A. McCleverty, *Chem. Rev.*, 79 (1979) 73.
[104] J. P. Candlin and W. H. Janes, *J. Chem. Soc. C*, (1968) 1856.
[105] D. Ballivet and I. Tkatchenko, *J. Mol. Catal.*, 1 (1975) 319.
[106] N. G. Connelly, P. T. Draggett and M. Green, *J. Organomet. Chem.*, 140 (1977) C10.
[107] D. Ballivet and I. Tkatchenko, *J. Mol. Catal.*, 1 (1975) 319.
[108] D. Ballivet, C. Billard and I. Tkatchenko, *J. Organomet. Chem.*, 124 (1977) C9.
[109] J. Smidt and R. Jira, *Chem. Ber.*, 93 (1960) 162.
[110] B. W. Graham, K. R. Laing, C. J. O'Connor and W. R. Roper, *J. Chem. Soc. Chem. Commun.*, (1970) 1272.
[111] J. Fleischer, D. Schnurpfeil, K. Seyferth and R. Taube, *J. Prakt. Chem.*, 319 (1977) 995.

[112] B. S. Tovrog, S. E. Diamond and F. Mares, *J. Am. Chem. Soc.*, 101 (1979) 270.
[113] R. D. Tiwari, K. K. Pandey and U. C. Agarwala, *Inorg. Chem.*, 21 (1982) 845.
[114] B. S. Tovrog, S. E. Diamond, F. Mares and A. Szalkiewicz, *J. Am. Chem. Soc.*, 103 (1981) 3522–3526.
[115] M. Calvin, *Trans. Faraday Soc.*, 34 (1938) 1181.
[116] C. W. Bird, *Transition Metal Intermediates in Organic Synthesis*, Academic Press, London, (1967), 248.
[117] R. A. Sanchez-Delgado, A. Andriollo, O. L. Deochoa, T. Suarez and N. Valencia, *J. Organomet. Chem.*, 209 (1981) 77.
[118] W. Strohmeier and K. Holke, *J. Organomet. Chem.*, 193 (1980) C63.
[119] S. T. Wilson and J. A. Osbom, *J. Am. Chem. Soc.*, 93 (1971) 3068.
[120] J. P. Collman, N. W. Hoffman and D. E. Morris, *J. Am. Chem. Soc.*, 91 (1969) 5659, G. Dolcetti, N. W. Hoffman and J. P. Collman, *Inorg. Chim. Acta*, 6 (1972) 531.
[121] W. Strohmeier and R. Endres, *Z. Naturforsch.*, 276 (1972) 1415.
[122] V. Z. Sharf, L. K. Freidlin, U. N. Krutii and I. S. Shekoyan, Izv. Akad. Nauk SSSR, Ser. Khim, (1974) 1330.
[123] C. A. Zuech, W. B. Hughes, D. H. Kubicek and E. T. Kittleman, *J. Am. Chem. Soc.*, 92 (1970) 528.
[124] S. T. Wilson and J. A. Osbom, *J. Am. Chem. Soc.*, 93 (1971) 3068.
[125] C. G. Pierpont and R. Eisenberg, *Inorg. Chem.*, 11 (1972) 1094.
[126] J. P. Collman, N. W. Hoffman and D. E. Morris, *J. Am. Chem. Soc.*, 91 (1969) 5659.
[127] G. Dolcetti, N. W. Hoffman and J. P. Collman, *Inorg. Chim. Acta*, 6 (1972) 531.
[128] W. B. Hughes, *J. Am. Chem. Soc.*, 92 (1970) 532.
[129] J. P. Lavel and A. Latters, J. *Chem. Soc. Chem. Commun.*, (1977) 502.
[130] J. H. Swinehart and W. G. Schmidt, *Inorg. Chem.*, 6 (1967) 232.
[131] J. A. McCleverty, C. W. Ninnes and I. Wolochowicz, *J. Chem. Soc. Chem. Commun.*, (1976) 1061.
[132] M. Foâ and L. Cassar, *J. Organomet. Chem.*, 30 (1971) 123.
[133] W. L. Bowden, W. F. Little and T. J. Meyer, *J. Am. Chem. Soc.*, 99 (1977) 4340.
[134] J. A. McCleverty, *Chem. Rev.*, 79 (1979) 65.
[135] R. J. H. Voorhoeve and L. E. Trimble, *J. Catal.*, 38 (1975) 80.
[136] A. Amirnazmi and M. Boudart, *J. Catal.*, 39 (1975) 383.
[137] R. J. H. Voorhoeve and L. E. Trimble, *J. Catal.*, 53 (1978) 251.
[138] W. Hieber and D. V. Pigenot, *Chem. Ber.*, 89 (1965) 610.
[139] F. Bottomley, S. G. Clarkson and S. B. Tong, *J. Chem. Soc. Dalton Trans.*, (1974) 2344.
[140] W. Hieber and R. Kramolowsky, *Z. Anorg. Allg. Chem.*, 321 (1963) 94.
[141] K. K. Pandey and U. C. Agarwala, *Inorg. Chem.*, 20 (1980) 1308, *Indian J. Chem.* A, 20 (1980) 74.
[142] W. Hieber and K. Heinicke, Z. *Anorg. Allg. Chem.*, 316 (1962) 321.
[143] W. Hieber and K. Heinicke, *Z. Anorg. Allg. Chem.*, 316 (1962) 305.
[144] J. A. Broomhead, J. Bridge, W. Grundy and T. Norman, *Inorg. Nucl. Chem. Lett.*, 11 (1975) 519.
[145] F. Bottomley, S. G. Clarkson and S. B. Tong, *J. Chem. Soc. Dalton Trans.*, (1974) 2344.
[146] S. K. Wolfe, C. Andrade and J. H. Swinhart, *Inorg. Chem.*, 13 (1974) 2567.
[147] K. K. Pandey, *Coord. Chem. Rev.*, 5 1 (1983) 69–98.
[148] B. F. G. Johnson and S. Bhaduri, *J. Chem. Soc. Chem. Commun.*, (1973) 6501.
[149] S. Bhaduri, B. F. G. Johnson, C. J. Savory, J. A. Segal and R. H. Walter, *J. Chem. Soc. Chem. Commun.*, (1974) 809.
[150] S. Bhaduri and B. F. G. Johnson, *Transition Met. Chem*, 3 (1978) 156.
[151] B. L. Haymore and J. A. Ibers, *J. Am. Chem. Soc.*, 96 (1974) 3325.
[152] J. A. Kaduk, T. H. Tulip, J. R. Budge and J. A. Ibers, *J. Mol. Catal.*, 12 (1981) 239.
[153] C. D. Meyer and R. Eisenberg, *J. Am. Chem. Soc.*, 98 (1976) 1364.
[154] D. E. Hendrickson, C. D. Meyer and R. Eisenberg, *Inorg. Chem.*, 16 (1977) 970.

[155] J. Ellermann and W. Hieber, *Chem. Ber.*, 96 (1963) 1667.
[156] K. K. Pandey and U. C. Agarwala, *J. Inorg. Nucl. Chem.*, 42 (1980) 293.
[157] B. F. G. Johnson, S. Bhaduri and N. G. Connelly, *J. Organomet. Chem.*, 40 (1972) C36.
[158] S. K. Wolfe, C. Andrade and J. H. Swinhart, *Inorg. Chem.*, 13 (1974) 2567.
[159] F. J. Miller and T. J. Meyer, *J. Am. Chem. Soc.*, 93 (1971) 1294.
[160] S. A. Adeyemi, F. J. Miller and T. J. Meyer, *Inorg. Chem.*, 11 (1972) 994.
[161] M. Rossi and A. Sacco, *J. Chem. Soc. Chem. Commun.*, (1971) 694.
[162] S. Naito, *J. Chem. Soc. Chem. Commun.*, (1978) 175.
[163] A. P. Castano, P. Mroz and M. R. Hamblin, *Nat. Rev. Cancer*, 6 (2006) 535–545.
[164] G. R. J. Thatcher and H. Weldon, *Chem. Soc. Rev.*, 27 (1998) 331–337.
[165] B. M. Bennett, D. C. Leitman, H. Schroder, J. H. Kawamoto, K. Nakatsu and F. Murad, *Pharmacol. Exp. Ther.*, 250 (1989) 316–323.
[166] B. I. Jugdutt Role of nitrates after acute myocardial infarction, *Am. J. Med.*, 70 (1992) 82–87.
[167] C. V. Leier, D. Bambach, M. J. Thompson, S. M. Cattaneo, R. J. Goldberg and D. V. Unverferth, *Am. J. Cardiol.*, 48 (1981) 1115–1123.
[168] B. Tander, A. Guven, S. Demirbag, Y. Ozkan, H. Ozturk and S. A. Cetinkursun A prospective, *J. Pediatr. Zurg.*, 34 (1999) 1810–1812.
[169] C. de May Opportunities for the treatment of erectile dysfunction by modulation of the NO axis-alternatives to Sildenafil Citrate, *Curr. Med. Res. Opin.*, 14 (1998) 187–202.
[170] B. Cederqvist, M. Persson and L. E. Gustafsson, *Biochem Pharmacol.*, 47 (1994) 1047–1053.
[171] B. A. Meloche and P. J. O'Brien, *Xenobiotica.*, 23 (1993) 863–871.
[172] S. L. Chung and H. L. Fung, J. *Pharmacol. Exp. Ther.*, 253 (1990) 614–619.
[173] D. J. Meyer, H. Kramer and B. Ketterer, *FEBS Lett.*, 351 (1994) 427–428.
[174] Y. Ji, T. P. M. Akerboom and H. Sies, *Biochem. J.*, 313 (1996) 377–380.
[175] J. J. Doel, B. L. J. Godber, T. A. Goult, R. Eisenthal and R. Harrison, *Biochem. Biophys. Res. Commun.*, 270 (2000) 880–885.
[176] S. Hadjimiltiades, I. P. Panidis, M. McAllister, J. Ross and G. S. Mintz, *Am. Heart J.*, 121 (1991) 1143–1148.
[177] T. L. Brunton, *Lancet* II, 5 (1867) 97–98.
[178] J. A. Bauer, T. Nolan and H. L. Fung, *J. Pharmacol. Exp. Ther.*, 280 (1997) 326–329.
[179] H. L. Fung and J. A. Bauer, *Cardiovasc. Drugs*, 8 (1994) 489–499.
[180] J. E. Saavedra, P. J. Shami, L. Y. Wang, K. M. Davies, M. N. Booth, M. L. Citro and L. K. Keefer, *J. Med. Chem.*, 43 (2000) 261–269.
[181] J. M. Mir, B. A. Malik and R. C. Maurya, *Inorg. Chem. Rev.*, 39 (2019) 91–112.
[182] G. M. Halpenny and P. K. Mascharak, *Anti-Infect. Agents Med. Chem*, 9 (2010) 187–197.
[183] A. Eroyreveles and P. K. Mascharak, *Future Med. Chem*, 8 (2009) 1497–1507.
[184] M. C. Frost, M. M. Reynolds and M. E. Meyerhoff., *Biomat.*, 26 (2005) 1685–1693, E. M. Hetrick and M. H. Schoenfisch, *Chem. Soc. Rev.*, 35 (2006) 780–789.
[185] F. C. Fang, *Nature Rev. Microbiol.*, 2 (2004) 820–832.
[186] R. Etchenique, M. Furman and J. A. Olabe, *J. Am. Chem. Soc.*, 122 (2000) 3967–3968.
[187] A. R. Butler and I. L. Megson, *Chem. Rev.*, 102 (2002) 1155–1166.
[188] A. S. Torsoni, B. F. Barros, J. C. Toledo Jr, M. Haun, M. H. Kriger, E. Tfouni and D. W. Franco, *Nitric oxide*, 2 (2002) 247–354.
[189] A. R. Butler and I. L. Megson, *Chem. Rev.*, 102 (2002) 1155–1166.
[190] A. K. Patra, J. M. Rowland, D. S. Marlin, E. Bill, M. M. Olmstead and P. K. Mascharak, *Inorg. Chem.*, 42 (2003) 6812.
[191] R. K. Afshar, A. K. Patra, M. M. Olmstead and P. K. Mascharak, *Inorg. Chem.*, 43 (2004) 5736.
[192] F. DeRosa, X. Bu and P. C. Ford, *Inorg. Chem.*, 44 (2005) 4157, M. A. De Leo, P. C. Ford, *J. Am. Chem. Soc.* 121 (1999) 1980.

[193] K. J. Franz and S. J. Lippard, *Inorg. Chem.*, 39 (2000) 3722, K. S. Suslick, R. A. Watson, *Inorg. Chem.* 30 (1991) 912.

[194] A. A. Eroy-Reveles, Y. Leung and P. K. Mascharak, *J. Am. Chem. Soc*, 128 (2006) 7166, K. Ghosh, A. A. Eroy-Reveles, M. M. Olmstead, P. K. Mascharak, *Inorg. Chem.* 44 (2005) 8469; K. Ghosh, A. A. Eroy-Reveles, B. Avila, T. R. Holman, M. M. Olmstead, P. K. Mascharak, *Inorg. Chem.* 43 (2004) 2988.

[195] I. Szundi, M. J. Rose, I. Sen, A. A. Eroy-Reveles, P. K. Mascharak and Ó. Einarsdóttir, *Photochem. Photobiol.*, 82 (2006) 1377.

[196] Á. Juarranz, P. Jaén, F. Sanz-Rodríguez, J. Cuevas and S. González Photodynamic therapy of cancer: Basic principles and applications, *Clin. Transl. Oncol.*, 10 (2008) 148–154.

[197] C. M. Pavlos, H. Xu and J. P. Toscano, *Curr. Top. Med. Chem.*, 5 (2005) 637–647.

[198] S. M. Ali and M. Olivo, *Int. J. Oncol.*, 22 (2003) 751–756.

[199] F. Yamamoto, Y. Ohgari, N. Yamaki, S. Kitajima, O. Shimokawa, H. Matsui and S. Taketani, *Biochem. Biophys. Res. Commun.*, 353 (2007) 541–546.

[200] M. J. Rose and P. K. Mascharak Photoactive ruthenium nitrosyls: effects of light and potential application as NO donors, Coord. Chem. Rev., 252 (2008) 2093–2114.

[201] G. E. Caramori, A. G. Kunitz, D. E. Coimbra, L. C. Garcia and D. E. P. Fonseca, *J. Braz.Chem. Soc.*, 24 (2013) 1487–1496.

[202] R. Prakash, A. U. Czaja, F. W. Heinemann and D. Sellmann, *J. Am. Chem.Soc.*, 127 (2005) 13758–13759.

[203] A. C. Roveda Jr, T. B. R. Papa, E. E. Castellano and D. W. Franco, *Inorg. Chim. Acta*, 409 (2014) 147–155.

[204] Z. A. Carneiro, J. C. B. Moraes, F. P. Rodrigues, R. C. Lima, C. Curti, Z. N. Rocha, M. Paulo, L. M. Bendhack, A. C. Tedesco, A. L. B. Formiga and R. S. Silva, *J. Inorg. Biochem.*, 105 (2011) 1035–1043.

[205] Z. A. Carneiro, J. C. Biazzotto, A. D. P. Alexiou and S. Nikolaou, *J. Inorg. Biochem.*, 134 (2014) 36–38.

[206] T. B. Santos, I. C. Cespedes and M. B. Viana, *Behav. Brain Res.*, 267 (2014) 46–54.

[207] L. E. Figueiredo, E. M. Cilli, R. A. S. Molina, E. M. Espreafico and E. Tfouni, *Inorg. Chem. Commun.*, 28 (2013) 60–63.

[208] A. C. Roveda Jr, W. G. Santos, M. L. Souza, C. N. Adelson, F. S. Gonçalves, E. E. Castellano, C. Garino, D. W. Franco and D. R. Cardoso Light-activated generation of nitric oxide (NO) and sulfite anion radicals ($SO_3^{\cdot -}$) from a ruthenium(II) nitrosylsulphito complex, *Dalton Trans.*, 48 (2019) 10812–10823.

[209] M. J. Rose and P. K. Mascharak, *Coord. Chem. Rev.*, 252 (2008) 2093–2114.

[210] N. L. Fry and P. K. Mascharak, *Acc. Chem. Res.*, 44 (2011) 289–298.

[211] M. J. Rose, N. L. Fry, R. Marlow, L. Hink and P. K. Mascharak, *J. Am. Chem. Soc.*, 130 (2008) 8834–8846.

[212] G. M. Halpenny, K. R. Gandhi and P. K. Mascharak, *ACS Med. Chem. Lett.*, 1 (2010) 180–183.

[213] J. M. Mir, N. Jain, P. S. Jaget and R. C. Maurya, *Photodiagn. Photodyn. Ther.*, 19 (2017) 363–374.

[214] L. K. Wareham, E. S. Buys and R. M. Sappington, *Nitric Oxide*, (2018); doi: 10.1016/j.niox.2018.04.010.

[215] J. M. Mir and R. C. Maurya, *RSC Adv*, 8 (2018) 35102–35130.

[216] J. T. Ferreira, J. Pina, C. A. F. Ribeiro, R. Fernandes, J. P. C. Tome, M. S. Rodríguez-Morgade and T. Torres, *ChemPhotoChem.*, 2 (2018) 640–654.

[217] J. Kim, G. Saravanakumar, H. W. Choi, D. Park and W. J. Kim, *J. Mater. Chem. B*, 2 (2014) 341–356.

[218] A. D. Ostrowski, B. F. Lin, M. V. Tirrell and P. C. Ford, *Mol. Pharm.*, 9 (2012) 2950–2955.

[219] K. Nakanishi, T. Koshiyama, S. Iba and M. Ohba, *Dalton Trans.*, 44 (2015) 14200–14203.

[220] J. T. Mitchell-Koch, T. M. Reed and A. S. Borovik, *Angew. Chem. Int. Ed.*, 43 (2004) 2806–2809, *Angew. Chem*, 116 (2004) 2866–2869.

[221] J. F. Bordini, P. C. Ford and E. Tfouni, *Chem. Commun.*, (2005) 4169–4171.

[222] K. Q. Ferreira, J. F. Schneider, P. A. P. Nascente, U. P. Rodrigues-Filho and E. Tfouni, *J. Colloid Interface Sci.*, 300 (2006) 543–552.

[223] G. M. Halpenny, K. R. Gandhi and P. K. Mascharak, *ACS Med. Chem. Lett.*, 1 (2010) 180–183.

[224] S. Diring, D. O. Wang, C. Kim, M. Kondo, Y. Chen, S. Kitagawa, K. Kamei and S. Furukawa, *Nat. Commun.*, 4 (2013) 2684–2691.

[225] C. Kim, S. Diring, S. Furukawa and S. Kitagawaa, *Dalton Trans.*, 44 (2015) 15324–15333.

[226] J. D. Mase, A. O. Razgoniaev, M. K. Tschirhartb and A. D. Ostrowski, *Photochem.-Photobiol. Sci.*, 14 (2015) 775–785.

[227] H. T. T. Duong, Z. Dong, L. Su, C. Boyer, J. George, T. P. Davis and J. Wang The use of nanoparticles to deliver nitric oxide to hepatic stellate cells for treating liver fibrosis and portal hypertension, *Small*, 11 (2015) 2291–2304.

[228] I. Roy, T. Y. Ohulchanskyy, H. E. Pudavar, E. J. Bergey, A. R. Oseroff, J. Morgan, T. J. Dougherty and P. N. Prasad Ceramic-based nanoparticles entrapping water-insoluble photosensitizing anticancer drugs: A novel drug-carrier system for photodynamic therapy, *J. Am. Chem. Soc.*, 125 (2003) 7860–7865.

[229] F. G. Doro, U. P. Rodrigues-Filho and E. Tfouni, *J. Colloid Interface Sci.*, 307 (2007) 405–410.

[230] B. J. Heilman, J. St. John, S. R. J. Oliver and P. K. Mascharak, *J. Am. Chem. Soc.*, 134 (2012) 11573–11582.

[231] Z. F. Yin, L. Wu, H. G. Yang and Y. H. Su, *Phys. Chem. Chem. Phys.*, 15 (2013) 4844–4858.

[232] H.-J. Xiang, L. An, -W.-W. Tang, S.-P. Yang and J.-G. Liu, *Chem. Commun.*, 51 (2015) 2555–2558.

[233] H.-J. Xiang, Q. Deng, L. An, M. Guo, S.-P. Yang and J.-G. Liu, *Chem. Commun.*, 52 (2016) 148–151.

[234] F.-G. Wu, X. Zhang, X. Chen, -W. Y.-W. Bao, X.-W. Hua, G. Gao and H.-R. Jia, Quantum dots for cancer therapy and bioimaging, nanooncology, *Nanomed. Nanotoxicol.*, doi: https://doi.org/10.1007/978-3-319-89878-0, Springer International Publishing AG, part of Springer Nature 2018.

[235] D. Neuman, A. D. Ostrowski, A. A. Mikhailovsky, R. O. Absalonson, G. F. Strouse and P. C. Ford, *J. Am. Chem. Soc.*, 129 (2007) 4146–4147.

[236] L. Tan, A. Wan and H. Li, *ACS Appl. Mater. Interfaces*, 5 (2013) 11163–11171.

[237] L. Tan, A. Wan, X. Zhu and H. Li, *Analyst*, 139 (2014) 3398–3406.

[238] L. Tan, A. Wan, X. Zhu and H. Li, *Chem. Commun.*, 50 (2014) 5725–5728.

[239] R. Singh, G. Dumlupinar, S. A. Engels and S. Melgar, *Internat. J. Nanomedicine*, 14 (2019) 1027–1038.

[240] J. Nam, N. Won, J. Bang, H. Jin, J. Park, S. Jung, S. Jung, Y. Park and S. Kim, *Adv. Drug Delivery Rev.*, 65 (2013) 622–648.

[241] J. V. Garcia, J. Yang, D. Shen, C. Yao, X. Li, R. Wang, G. D. Stucky, D. Zhao, P. C. Ford and F. Zhang, *Small*, 8 (2012) 3800–3805.

[242] P. T. Burks, J. V. Garcia, R. GonzalezIrias, J. T. Tillman, M. Niu, A. A. Mikhailovsky, J. Zhang, F. Zhang and P. C. Ford, *J. Am. Chem. Soc.*, 135 (2013) 18145–1815.

[243] G. Chen, I. Roy, C. Yang and P. N. Prasad, *Chem. Rev.*, 116 (2016) 2826–2885.

[244] J. Fan, N. Y. He, Q. J. He, Y. Liu, Y. Ma, X. Fu, Y. Liu, P. Huang and X. Y. Chen, *Nanoscale*, 7 (2015) 20055–20062.

[245] X.-D. Yang, H.-J. Xiang, L. An, S.-P. Yang and J.-G. Liu, *New J. Chem.*, 39 (2015) 800–804.

[246] J. S. Xu, F. Zeng, H. Wu, C. P. Hu, C. M. Yu and S. Z. Wu, *Small*, 10 (2014) 3750–3760.

[247] Q. Deng, H.-J. Xiang, -W.-W. Tang, L. An, S.-P. Yang, Q.-L. Zhang and J.-G. Liu, *J. Inorg. Biochem.*, 165 (2016) 152–158.

[248] H.-J. Xiang, M. Guo, L. An, S.-P. Yang, Q.-L. Zhang and J.-G. Liu, *J. Mater. Chem.B*, 4 (2016) 4667–4674.
[249] F. Becquet, Y. Courtois and O. Goureau Nitric oxide in the eye: multifaceted roles and diverse outcomes, *Surv. Ophthalmol*, 42 (1997) 71–82.
[250] G. C. Chiou, *J. Ocul. Pharmacol. Ther.*, 17 (2001) 189–98.
[251] I. M. Goldstein, P. Ostwald and S. Roth, *Vision Res.*, 36 (1996) 2979–2994.
[252] A. Yadav, R. Choudhary and S. H. Bodakhe, *Curr. Eye Res.*, 43 (2018) 1454–1464.
[253] R. Balasubramanian, A. Bui, X. Dong and L. Gan, *Mol. Neurobiol.*, 55 (2018) 2922–2933.
[254] L. K. Wareham, E. S. Buys and R. M. Sappington The nitric oxide-guanylate cyclase pathway and glaucoma, *Nitric Oxide*, 77 (2018) 75–87.
[255] J. M. Mir and R. C. Maurya, *Ann. Ophthalmol. Vis. Sci.*, 1 (2018) 1003.
[256] J. A. Nathanson Nitrovasodilators as a new class of ocular hypotensive agents, *J. Pharmacol. Exp. Ther.*, 260 (1992) 956–965.
[257] H. Kotikoski, P. Alajuuma, E. Moilanen, P. Salmenperä, O. Oksala, P. Laippala and H. Vapaatalo1 Comparison of nitric oxide donors in lowering intraocular pressure in rabbits: role of cyclic GMP, *J. Ocul. Pharmacol. Ther.*, 18 (2002) 11–23.
[258] F. F. Behar-Cohen, O. Goureau, F. D'Hermies, F. Y. Courtois and Y. Courtois Decreased intraocular pressure induced by nitric oxide donors is correlated to nitrite production in the rabbit eye, *Invest. Ophthalmol. Vis. Sci.*, 37 (1996) 1711–1715.
[259] T. Sugiyama, T. Kida, K. Mizuno, S. Kojima and T. Keda Involvement of nitric oxide in the ocular hypotensive action of nipradilol, *Curr. Eye Res.*, 23 (2001) 346–351.
[260] M. Orihashi, Y. Shima, H. Tsuneki and I. Kimura Potent reduction of intraocular pressure by nipradilol plus latanoprost in ocular hypertensive rabbits, *Biol. Pharm. Bull.*, 28 (2005) 65–68.
[261] J. Y. Hu and B. Katz Non-arteritic ischemic optic neuropathy and supplemental nitric oxide usage, *Am. J. Ophthalmol. Case Rep.*, 11 (2018) 26–27.
[262] A. Mahmud, M. Hennessy and J. Feely Effect of sildenafil on blood pressure and arterial wave reflection in treated hypertensive men, *J. Hum. Hypertens.*, 15 (2001) 707–713.
[263] S. S. Hayreh Erectile dysfunction drugs and non-arteritic anterior ischemic optic neuropathy: is there a cause and effect relationship?, *J. Neuro-Ophthalmol.*, 25 (2005) 295–298.
[264] S. S. Hayreh, P. A. Podhajsky and B. Zimmerman Non-arteritic anterior ischemic optic neuropathy: time of onset of visual loss, *Am. J. Ophthalmol.*, 124 (1997) 641–647.
[265] F. Galassi, E. Masini, B. Giambene, F. Fabrizi, C. Uliva, M. Bolla and E. Ongini A topical nitric oxide-releasing dexamethasone derivative: Effects on intraocular pressure and ocular haemodynamics in a rabbit glaucoma model, *Br. J. Ophthalmol.*, 90 (2006) 1414–1419.
[266] D. Sambhara and A. A. Aref Glaucoma management: Relative value and place in therapy of available drug treatments., *Ther. Adv.Chronic Dis.*, 5 (2014) 30–43.
[267] L. A. Klepper, Inhibition of nitrite reduction by photosynthetic inhibitors, *Weed Science*, 23 (1975) 188–190 Nitric oxide (NO) and nitrogen dioxide (NO_2) emissions from herbicide-treated soybean plants, *Atmospheric Environment*, 13 (1979)537–542.
[268] H. Yamasaki The NO world for plants: achieving balance in an open system, *Plant Cell Environ.*, 28 (2005) 78–84.
[269] M. Arasimowicz and J. Floryszak-Wieczorek Nitric oxide as a bioactive signalling molecule in plant stress responses, *Plant Science*, 172 (2007) 876–887.
[270] L. C. Ferreira1 and A. C. Cataneo Nitric oxide in plants: a brief discussion on this multifunctional molecule, *Sci. Agric. (Piracicaba, Braz.)*, 67 (2010) 236–243.
[271] P. Wojtaszek Nitric oxide in plants. To NO or not to NO., *Phytochem.*, 54 (2000) 1–4.
[272] G. Cevahir, E. Aytamka and C. Erol The role of nitric oxide in plants, *Biotechnol. Biotechnol. Equip.*, 21 (2007) 13–17.

[273] R. V. Cooney, P. J. Harwood, L. J. Custer and A. A. Franke Light-mediated conversion of nitrogen dioxide to nitric oxide by carotenoids, *Environ. Health Perspect.*, 102 (1994) 460–462.
[274] Y. Y. Leshem and F. Haramaty, E., The characterization and contrasting effects of the nitric oxide free radical in vegetative stress and senescence of *pisum sativum* linn. foliage, *J. Plant Physiol.*, 148 (1996) 258–263.
[275] G. Cevahir, E. Aytamka and Ç. Erol The role of nitric oxide in plants, *Biotechnol. Biotechnol. Equip.*, 21 (2007) 13–17.
[276] T. M. Millar, C. R. Stevens, N. Benjamin, R. Eisenthal, R. Harrison and D. R. Blake Xanthine oxidoreductase catalyses the reduction of nitrates and nitrite to nitric oxide under hypoxic conditions, *FEBS Lett.*, 427 (1998) 225–228.
[277] H. Ninnemann and J. Maier Indications for the occurrence of nitric oxide synthases in fungi and plants and the involvement in photoconidiation of *Neurospora crassa*, *Photochem. Photobiol.*, 64 (1996) 393–398.
[278] H. Yamasaki and Y. Sakihama, Simultaneous production of nitric oxide and peroxynitrite by plant nitrate reductase: *in vitro* evidence for the NR-dependent formation of active nitrogen species, FEBS Lett., 468 (2000) 89–92 N. M. Crawford, Mechanisms for nitric oxide synthesis in plants, J. *Exper. Botany*, 57 (2006) 471–478.
[279] S. Takahashi and H. Yamasaki Reversible inhibition of photophosphorylation in chloroplasts by nitric oxide, *FEBS Lett.*, 512 (2002) 145–148.
[280] Y. Morot-Gaudry-Talarmain, P. Rockel, T. Moureaux, I. Quileré, M. T. Leydecker, W. M. Kaiser and J. F. Morot-Gaudry Nitrite accumu-lation and nitric oxide emission in relation to cellular signalling in nitrite reductase antisense plants, *Planta*, 215 (2002) 708–715.
[281] M. V. Beligni and L. Lamattina Nitric oxide stimulates seed germination and de-etiolation, and inhibits hypocotyl elongation, three light-inducible responses in plants, *Planta*, 210 (2000) 215–221.
[282] Y. Y. Leshem, E. Haramaty, D. Iluz, Z. Malik, Y. Sofer, L. Roitman and Y. Leshem Effect of stress nitric oxide (NO): interaction between chlorophyll fluorescence, galactolipid fluidity and lipoxygenase activity, *Plant Physiol. Biochem.*, 35 (1997) 573–579.
[283] M. Graziano, M. V. Beligni and L. Lamattina, Nitric oxide improves iron availability in plants, *Plant Physiol.*, 130 (2002)1852–1859.
[284] Y. Y. Leshem and E. Haramaty The characterisation and contrasting effects of the nitric oxide free radical in vegetative stress and senescence of *Pisum sativum* Linn. foliage, *J. Plant Physiol.*, 148 (1996) 258–263.
[285] Y. Leshem, *Nitric oxide in plants*, Kluwer Academic Publishers, London, UK, (2001).
[286] Y. Y. Leshem and Y. Pinchasov Non-invasive photoacoustic spectroscopic determination of relative endogenous nitric oxide and ethylene content stoichiometry during the ripening of strawberries *Fragaria anannasa* (Duch.) and avocados *Persea americana* (Mill.), *J. Exp. Botany*, 51 (2000) 1471–1473.
[287] M. V. Beligni, Lamattina, Nitric oxide stimulates seed germination and de-etiolation, and inhibits hypocotyl elongation, three light-inducible responses in plants, *Planta*, 210 (2000), 215–221.
[288] J. E. Keeley and C. J. Fotheringham Trace gas emissions and smoke-induced seed germination, *Science*, 276 (1997) 1248–1250.
[289] Z. Giba, D. Grubisic, S. Todorovic, L. Sajc, D. Stojakovic and R. Konjevic Effect of nitric oxide-releasing compounds on phytochrome-controlled germination of Empress tree seeds, *Plant Growth Regul.*, 26 (1998) 175–181.
[290] G. F. E. Scherer and A. Holk, NO donors mimic and NO inhibitors inhibit cytokinin action in betalaine accumulation in *Amaranthus caudatus*, *Plant Growth Regul.*, 32 (2000) 345–350 N. N.Tun, A. Holk A and GF. E. Scherer, Rapid increase of NO release in plant cell cultures induced by cytokinin. *FEBS Letters* 509 (2001)174–176.

[291] S. J. Neill, R. Desikan, A. Clarke and J. T. Hancock Nitric oxide is a novel component of abscisic acid signalling in stomatal guard cells, *Plant Physiol.*, 128 (2002) 13–16.
[292] G. C. Pagnussat, M. Simontacchi, S. Puntarulo and L. Lamartine Nitric oxide is required for root organogenesis, *Plant Physiol.*, 129 (2002) 954–956.
[293] Y. Y. Leshem and E. Haramaty The characterisation and contrasting effects of the nitric oxide free radical in vegetative stress and senescence of *Pisum sativum* Linn. foliage, *J. Plant Physiol.*, 148 (1996) 258–263.
[294] J. R. Magalhaes, D. C. Monte and D. Durzan Nitric oxide and ethylene emission in Arabidopsis thaliana, *Physiology and Molecular Biology of Plants*, 6 (2000) 117–127.
[295] C. Garcia-Mata and L. Lamattina Nitric oxide induces stomatal closure and enhances the adaptive plant responses against drought stress, *Plant Physiol.*, 126 (2001) 1196–1204.
[296] Z. Zhao, G. Chen and C. Zhang Interaction between reactive oxygen species and nitric oxide in drought-induced abscisic acid synthesis in root tips of wheat seedlings, *Aust. J. Plant Physiol.*, 28 (2001) 1055–1061.
[297] M. Jiang and J. Zhang Involvement of plasma membrane NADPH oxidase in abscisic acid- and water stress-induced antioxidant defence in leaves of maize seedlings, *Planta*, 215 (2002) 1022–1030.
[298] N. Suzuki and R. Mittler Reactive oxygen species and temperature stresses: a delicate balance between signalling and destruction, *Physiol. Plant.*, 126 (2006) 45–51.
[299] M. G. Zhao, L. Chen, L. L. Zhang and W. H. Zhang Nitric reductase dependent nitric oxide production is involved in cold acclimation and freezing tolerance in Arabidopsis, *Plant. Physiol.*, 151 (2009) 755–767.
[300] L. Song, W. Ding, M. Zhao, B. Sun and L. Zhang Nitric oxide protects against oxidative stress under heat stress in the calluses from two ecotypes of reed, *Plant Science*, 171 (2006) 449–458.
[301] A. Bavita, B. Shashi and S. B. Navtej Nitric oxide alleviates oxidative damage induced by high temperature stress in wheat, *Indian J. Exp. Biol.*, 50 (2012) 372–378.
[302] J. D. Yang, J. Y. Yun, T. H. Zhang and H. L. Zhao Pre-soaking with nitric oxide donor SNP alleviates heat shock damages in mung bean leaf discs, *Botanical Studies*, 47 (2006) 129–136.
[303] M. Hasanuzzaman, M. A. Hossain, J. A. da Silva and M. Fujita, Plant responses and tolerance to abiotic oxidative stress: an antioxidant defence is a key factor, In: B. Venkateswarlu, A. K. Shanker, C. Shanker and M. Maheswari Eds., *Crop stress and its management: Perspectives and strategies*, Springer, Berlin, (2012) 261–315.
[304] M. Hasanuzzaman, K. Nahar, M. M. Alam, R. Roychowdhury and M. Fujita Physiological, biochemical, and molecular mechanisms of heat stress tolerance in plants, *Int J Mol Sci*, 14 (2013) 9643–9684.
[305] S. Kataria, A. Jajoo and K. N. Guruprasad Impact of increasing ultraviolet-B (UV-B) radiation on photosynthetic processes, *J. Photochem. Photobiol. B Biol*, 137 (2014) 55–66.
[306] A. Galatro and M. Simontacchi, Puntarulo, Free radical generation and antioxidant content in chloroplasts from soybean leaves exposed to ultraviolet-B, *Physiol. Plant*, 113 (2001), 564–570.
[307] S. Shi, G. Wang, Y. Wang, L. Zhang and L. Zhang Protective effect of nitric oxide against oxidative stress under ultraviolet-B radiation, *Nitric Oxide*, 13 (2005) 1–9.
[308] D. M. Santa-Cruz, N. A. Pacienza, C. G. Zilli, M. L. Tomaro, K. B. Balestrasse and G. G. Yannarelli Nitric oxide induces specific isoforms of antioxidant enzymes in soybean leaves subjected to enhanced ultraviolet-B radiation, *J. Photochem. Photobiol.*, 141 (2014) 202–209.
[309] M. N. Khan, M. H. Siddiqui, F. Mohammad, M. Naeem and M. M. N. Khan Calcium chloride and gibberellic acid protect Linseed (*Linum usitatissimum* L.) from NaCl stress by inducing antioxidative defence system and osmoprotectant accumulation, *Acta Physiol. Plant*, 32 (2010) 121–132.

[310] M. H. Siddiqui, F. Mohammad, M. N. Khan, M. Naeem and M. M. N. Khan Differential response of salt-sensitive and salt-tolerant Brassica juncea genotypes to N application: enhancement of N metabolism and anti-oxidative properties in the salt-tolerant type, *Plant Stress*, 3 (2009) 55–63.

[311] M. H. Siddiqui, M. H. Al-Whaibi and M. O. Basalah Role of nitric oxide in tolerance of plants to abiotic stress, *Protoplasma*, 248 (2011) 447–455.

[312] C. Zheng, D. Jiang, T. Dai, Q. Jing and W. Cao Effects nitroprusside, a nitric oxide donor, on carbon and nitrogen metabolism and the activity of the antioxidation system in wheat seedlings under salt stress, *Acta. Ecol. Sinica.*, 30 (2010) 1174–1183.

[313] A. Schützendübel and A. Polle Plant responses to abiotic stresses: heavy metal-induced oxidative stress and protection by mycorrhization, *J. Exp. Bot.*, 53 (2002) 1351–1365.

[314] C. C. Yu, K. T. Hung and C. H. Kao Nitric oxide reduces Cu toxicity and Cu-induced NH_4^+ accumulation in rice leaves, *J. Plant Physiol.*, 162 (2005) 1319–1330.

[315] M. Kopyra and E. A. Gwóźdź Nitric oxide stimulates seed germination and counteracts the inhibitory effect of heavy metals and salinity on root growth of *Lupinus luteus*, *Plant Physiol. Biochem.*, 41 (2003) 1011–1017.

[316] G. Kaur, H. P. Singh, R. D. Batish, P. Mahajan, R. K. Kohli and V. Rishi, Exogenous nitric oxide (NO) interferes with lead (Pb)-induced toxicity by detoxifying reactive oxygen species in hydroponically grown wheat (*Triticum aestivum*) roots, PLoS ONE, 10 (2015) 1–18; doi: https://doi.org/10.1371/journal.pone.0138713.

[317] Q. R. Tian, D. H. Sun, M. G. Zhao and W. H. Zhang Inhibition of nitric oxide synthase (NOS) underlies aluminum-induced inhibition of root elongation in Hibiscus moscheutous, *New Phytol.*, 174 (2006) 322–331.

[318] X. Cui, Y. Zhang, X. Chen, H. Jin and X. Wu, Effects of exogenous nitric oxide protects tomato plants under copper stress, Bioinformatics and Biomedical Engineering, 2009 (ICBBE 2009), 3rd International Conference on 11–13 June 2009, Beijing, p 1–7.

[319] K. D. Hu, L. Y. Hu, Y. H. Li, F. Q. Zhang and H. Zhang Protective roles of nitric oxide on germination and antioxidant metabolism in wheat seeds under copper stress, *Plant Growth Regul.*, 53 (2007) 173–183.

[320] B. Hussain Modernization in plant breeding approaches for improving biotic stress resistance in crop plants, *Turk. J. Agric. For.*, 39 (2015) 515–530.

[321] T. Noritake, K. Kawakita and N. Doke Nitric oxide induces phytoalexin accumulation in potato tuber tissues, *Plant Cell Physiol.*, 37 (1996) 113–116.

[322] A. Laxalt, M. V. Beligni and L. Lamartine Nitric oxide preserves the level of chlorophyll in potato leaves infected by *Phytophthora infestans*, *Eur. J. Plant Pathol.*, 73 (1997) 643–651.

[323] J. S. Huang and J. A. Knopp, Involvement of nitric oxide in *Ralstonia solanacearum* induced hypersensitive reaction in tobacco, In *Proceedings of the Second International Wilt Symposium* (Eds. P. Prior, J. Elphinstone and C. Allen), INRA, Versailles, France (1997).

[324] J. Durner, D. Wendehemme and D. F. Klessig Defence gene induction in tobacco by nitric oxide, cyclic GMP and cyclic ADP-ribose, *Proc. Natl. Acad. Sci. U.S.A*, 95 (1998) 10328–10333.

[325] J. Draper Salicylate, superoxide synthesis and cell suicide in plant defence, *Trends Plant Sci.*, 2 (1997) 163–165.

[326] L. Klepper NO_x evolution by soybean leaves treated with salicylic acid and selected derivatives, *Pestic Biochem Physiol*, 39 (1991) 43–48.

[327] W. Van Camp, D. Inze and M. V. Montagu, H_2O_2 and NO: redox signals in plant disease resistance, *Trends Plant Sci.*, 3 (1998) 330–334 F. Song and R. M. Goodman, Activity of nitric oxide is dependent on, but is partially required for function of, salicylic acid in the signalling pathway in tobacco systemic acquired resistance. *Molecular Plant-Microbe Interactions*, 14 (2001) 1458–1462.

[328] S. J. Neill, R. Desikan and J. T. Hancock Nitric oxide signalling in plants, *New Phytologist*, 159 (2003) 11–35.
[329] N. M. Crawford and F. Q. Guo, Trends Plant Sci., 10 (2005) 195–200.
[330] C. Bogdan, Trends Cell Biol., 11 (2001) 66–75.
[331] A. J. Trewavas, C. Rodrigues, C. Rato and R. Malho, *Curr. Opin. Plant Biol.*, 5 (2002) 425–429.
[332] D. Wendehenne, A. Plugin, D. F. Klessig and J. Durner, *Trends Plant Sci.*, 6 (2001) 177–183.
[333] P. R. Gardner, G. Costantino, C. Szabo and A. L. Salzman, *J. Biol. Chem.*, 272 (1997) 25071–25076.
[334] D. A. Navarre, D. Wendehenne, J. Durner, R. Noad and D. F. Klessig Nitric oxide modulates the activity of tobacco aconitase, *Plant Physiol.*, 122 (2000) 573–582.

Exercises

Multiple-choice questions/fill in the blanks

1. Nitric oxide was first characterized by
   (a) Johann Glauber
   (b) Joseph Priestley
   (c) Henry Cavendish
   (d) Sir Humphrey Davy

2. Who established the accepted chemical formula of NO by proving that it contains nitrogen and oxygen in equal proportions?
   (a) Joseph Priestley
   (b) Johann Glauber
   (c) Henry Cavendish and Sir Humphrey Davy
   (d) Any one of these

3. In the brown ring test of nitrate in the presence of concentrated H_2SO_4 and iron (II) sulphate, the composition of black nitrosyl complex cation formed is
   (a) $[Fe(NO)(H_2O)_5]^+$
   (b) $[Fe(NO)(H_2O)_5]^{2+}$
   (c) $[Fe(NO)(H_2O)_5]^{3+}$
   (d) None of these

4. Because of one of the most important physiological regulators, the nitric oxide molecule was voted as "Molecule of the Year" by the AAAS in year:
   (a) 1980 (b) 1990 (c) 1992 (d) 1998

5. Intraocular pressure (IOP) is related to
   (a) Heart (b) Lungs (c) Eye (d) None of these

6. Sort out a nitrosyl compound from the following:
   (a) 4-Nitrophenol
   (b) 4-Nitroaniline
   (c) *N,N*-Dimethy-4-nitroaniline
   (d) Sodium nitroprusside

7. In transition metal nitrosyl complexes, when M–N–O bond angle is ~180°, the mode of coordination of nitric oxide is
   (a) NO (b) NO^- (c) NO^+ (d) All of these

8. When M–N–O bond angle is ~120°, the mode of coordination of nitric oxide in nitrosyl complex is
   (a) NO^+ (b) NO (c) NO^- (d) All of these.

9. When the N–O bond order is 3, nitric oxide species will be
   (a) NO (b) NO^+ (c) NO^- (d) None of these

10. Sort out the correct species in which the N–O bond order is 1.5
    (a) NO (b) NO^+ (c) NO^- (d) None of these

11. Which one of the following is a non-innocent ligand?
    (a) H_2O (b) SCN^- (c) NH_3 (d) NO

12. Sort out the innocent ligand from the following:
    (a) H_2O (b) NO (c) SO_2 (d) Dithiolene

13. The concept of innocent and non-innocent ligands was first time given in 1960s by
    (a) F. A. Cotton
    (b) G. Wilkinson
    (c) D. M. P. Mingos
    (d) C. K. Jørgenson

14. The correct Enemark $\{M(NO)_x\}^n$ (M = metal) notation for $Na_2[Fe(NO)(CN)_5]$ is:
    (a) $\{FeNO\}^5$ (b) $\{FeNO\}^6$ (c) $\{FeNO\}^7$ (d) None of these

15. Sort out the appropriate nitrosyl compound having Enemark notation $\{M(NO)_2\}^6$ from the following:
    (a) $[Mn(NO)_2(CN)_2(AMPPHP)$
    (b) $[RuCl(NO)_2(PPh_3)_2]^+$
    (c) $[Mo(NO)_2(CN)_2(MPHP)_2]$
    (d) $[Os(NO)_2(PPh_3)_2(OH)]^+$

16. MB spectrum of $[Fe(CN)_5(NO)]^{2-}$ shows peak(s) in number:
    (a) 1 (b) 2 (c) 3 (d) None of these

17. Sort out a diamagnetic nitrosyl compound having the molecular orbital electronic configuration:
    (a) $(2e)^4(1b_2)^1$
    (b) $(2e)^4(1b_2)^2$
    (c) $(2e)^4(1b_2)^2(3e)^1$
    (d) None of these

18. Compounds containing the NO groups(s) are usually referred to as nitrosyl compounds when addendum is ______ in nature.

19. Which one is paramagnetic in the following nitrosyl compounds having Enemark notation?
    (a) $\{M(NO)\}^4$ (b) $\{M(NO)\}^5$ (c) $\{M(NO)\}^6$ (d) None of these

20. With regard to isomer shift (IS) in MB spectroscopy, which one of the following statements is correct?
    (a) IS of the low-spin complexes are higher than those of the high-spin complexes
    (b) IS of the low-spin complexes are lower than those of the high-spin complexes
    (c) IS of the low-spin complexes are equal to those of the high-spin complexes
    (d) IS of the low-spin complexes are double to those of the high-spin complexes

21. Nuclear resonance vibrational spectroscopy (NRVS) is a technique that probes
    (a) Rotational energy levels
    (b) Vibrational energy levels
    (c) Nuclear energy levels
    (d) All of these

22. In the Gmelin reaction between nitroprusside and S^{2-} giving red colouration is due to
    (a) Displacement of NO by NOS
    (b) Displacement of NO by H_2O
    (c) Displacement of NO by NS
    (d) Displacement of NO by S^{2-}

23. In the electrophilic attack of the coordinated NO in metal nitrosyls, the mode of coordination of nitric oxide is expected to be as
    (a) NO (b) NO^+ (c) NO^- (d) All of these

24. In the nucleophilic attack of the coordinated NO in metal nitrosyls, the mode of coordination of nitric oxide is expected to be as
    (a) NO (b) NO^+ (c) NO^- (d) All of these

25. Which one of the following nitrosyl compounds does not release NO when exposed to light?
    (a) $Na_2[Fe(NO)(CN)_5]$
    (b) $K[Fe_4S_3(NO)_7]$ (RBS)
    (c) $K_2[Fe_2S_2(NO)_4]$ (Roussin's red salt)
    (d) $K_3[Cr(NO)(CN)_5]$

26. Which of the following nitrosyl compounds is largely used as a NO donor drug in clinical practices for blood pressure control in cases of acute hypertension?
    (a) $Na_2[Fe(NO)(CN)_5]$
    (b) $[(PaPy_3)Fe(NO)](ClO_4)_2$
    (c) $K_2[Fe_2S_2(NO)_4]$ (Roussin's red salt)
    (d) $K[Fe_4S_3(NO)_7]$ (RBS)

27. Nitric oxide synthase (NOS) catalyses the conversion of L-arginine to L-citrulline in presence of NADPH and O_2 as co-substrates. The overall reaction is a five-electron oxidation process. Find the correct ratio of contribution of these five electrons by NADPH and O_2 from the following:
    (a) 3e from NADPH 2e from O_2
    (b) 4e from NADPH 1e from O_2
    (c) 5e from NADPH 0e from O_2
    (d) Ratio is not yet clear

28. The type of nitric oxide synthase (NOS) which releases large amounts of NO over longer periods that could diffuse and react with different cellular targets is
    (a) Constitutive NOS-I, found essentially in some neurons of the central and the peripheral nervous system
    (b) Constitutive NOS-III expressed by the vascular endothelium
    (c) Inducible NOS-II expressed in many cell types after challenge by immunological or inflammatory stimuli
    (d) All of these

29. The function of NO in eyes includes:
    (a) Maintenance of AqH dynamics
    (b) Neurotransmission of retina
    (c) Photo-transduction
    (d) All of these

30. NO emission from plants was first observed in 1975 by
    (a) H. Yamasaki
    (b) L. A. Klepper
    (c) L. C. Ferreira1
    (d) P. Wojtaszek

31. The isomer shifts of metals coordinated by soft ligands are _____ than those coordinated by hard ligands.

Short-answer-type questions

1. What are nitrosyl complexes? Explain with suitable examples.
2. Highlight the relevant reasons behind the little investigations on metal nitrosyls compared to metal carbonyls.
3. What are nitrosylating reagents? Give the name of at least five such reagents.
4. In the synthesis of nitrosyl complexes using nitric oxide (NO) gas as a nitrosylating agent, it is always recommended to use freshly prepared NO. Give the relevant reasons behind using freshly prepared NO.
5. In the synthesis of nitrosyl complexes using hydroxylamine (NH_2OH) as the nitrosylating agents, alkaline medium is required. Why?
6. Briefly describe the synthesis of nitrosyl complexes involving substitution of cyano group in parent cyanonitrosyl complexes.
7. Describe the synthesis of sodium nitroprusside and related compounds using nitric acid as the nitrosylating agent.
8. What is the $\{M(NO)_m\}^n$ formalism for metal nitrosyl complexes? Explain with suitable examples.
9. Briefly describe about the new notation for metal nitrosyls with the formal charges on the nitrosyl ligand and the formal metal oxidation state.
10. As per X-ray crystal structure studies on metal nitrosyls, NO^+ and NO^- bind to metals to give “linear” and “bent” nitrosyls, respectively. Justify these two types of bonding through hybridization in NO^+ and NO^-.
11. What are the valid reasons for deviation of M–N–O bond angles from 180° or 120°?
12. Correlate the ν(NO) with M–N–O bond angles.
13. Taking a suitable example, highlight the isomerism in metal nitrosyls, in which one isomer contains NO^+ and the other NO^-. How does IR spectroscopy differentiate these two isomers? Point out the utility of such type of metal nitrosyls.
14. What are the rules for corrections of ν(NO) proposed by C. De La Cruz and N. Sheppard in linear nitrosyl complexes for the charge on the complex and the electron donating/withdrawing effects of the ligands?
15. Giving the salient features of nuclear resonance vibrational spectroscopic studies, describe its use for metal–ligand stretching vibrational data in metal nitrosyls.
16. Briefly describe the reactivity of nitrosyl complexes as nucleophilic attack upon the coordinated NO group.
17. Taking suitable examples, highlight the reactivity of metal nitrosyls as electrophilic attack upon the coordinated NO group.
18. Describe the reactions of coordinated NO with oxygen taking suitable examples.

19. Present a brief note on homogeneous hydrogenation of unsaturated compounds using metal nitrosyl as catalyst.
20. Highlight the process of endogenous production of NO from L-arginine in living organisms. Also explain the way of regeneration of L-arginine from L-citrulline.
21. Present a brief account of organic nitrates and nitrites as exogenous NO donors for clinical uses.
22. Present an overview of endogenous production/biosynthesis of NO inside the eyes.
23. Present the mechanism of nitric oxide signalling in plants.

Long-answer-type questions

1. What are nitrosyl complexes? Present a detailed view of synthesis of nitrosyl complexes using different nitrosylating agents.
2. Explain Enemark $\{M(NO)_m\}^n$ notation for nitrosyl complexes of transition metals with suitable examples. Also highlight the limitations of the $\{M(NO)_m\}^n$ formalism.
3. Present a detailed view of an alternative notation for metal nitrosyl complexes that focuses on 18- and 16-electron rules in metal nitrosyls.
4. Describe in detail the simplified procedure for calculation of EAN of metal nitrosyls.
5. Based on X-ray crystal structure studies on metal nitrosyls, NO^+ and NO^- bind to metals to give "linear" and "bent" nitrosyls, respectively. Discuss the bonding in these two types of metal nitrosyls.
6. Present the detailed view of Walsh diagram methodology for linear to bent MNO bond angle transformation in hexa- and penta-coordinated nitric oxide complexes.
7. Present the Enemark–Feltham approach for the bent versus linear NO ligands in hexa- and penta-coordinated nitric oxide complexes.
8. Describe DFT-based molecular orbital calculations of bond lengths and bond angles in metal nitrosyls taking suitable examples.
9. Present the role of vibrational spectral studies in distinction of linear and bent NO ligands in metal nitrosyls.
10. Explain how the electronic spectral studies are helpful in characterization of nitrosyl complexes.
11. Describe the role of magnetic properties in characterization of nitrosyl complexes.
12. Describe in detail the utility of ESR spectroscopy in characterization of metal nitrosyl complexes. Also highlight its role in detection of NO in biological systems.
13. Present an explanatory note on ^{14}N and ^{15}N NMR studies of diamagnetic metal nitrosyl complexes.

14. The X-ray photoelectron spectroscopy is used to throw light in distinguishing NO^+ and NO^- labels of bonding in nitrosyl complexes by studying the binding energies of electrons of the atoms in these compounds. Justify this statement.
15. Highlight the basic principles and application of MB spectroscopy in metal nitrosyls with particular reference to Fe and Ru nitrosyls. How the characterized oxidation state (+III) of iron in $[Fe(H_2O)_5(NO)]^{2+}$ by MB spectroscopy has been modified as (+II) on the basis of DFT calculations?
16. What are the three hyperfine interactions that provide a link to the electronic structures in MB spectroscopy? Describe.
17. Present an overview of kinetic studies of substitution reactions in metal nitrosyls.
18. Present an account of mass spectral studies on metal nitrosyls.
19. Describe in detail the reactivity of nitric oxide coordinated to transition metals.
20. Discuss the role of transition metal nitrosyls in organic synthesis.
21. Present a detailed view of transition metal nitrosyls in pollution control.
22. Describe in detail the endogenous production of NO from L-arginine catalysed by different isoforms of nitric oxide synthase in living organisms.
23. Present a detailed view of exogenous NO-donating molecules, particularly, organic nitrates and nitrites along with their clinical uses.
24. Highlighting the fundamental problem associated with the conventional NO donors, describe the importance of NO donors incorporated in polymeric matrices.
25. What are the motivating factors behind the progress in metal nitrosyl chemistry in the last two decades? Describe in detail the journey of metal nitrosyls as NORMs in biomedical uses.
26. What is photodynamic therapy? Giving importance of this therapy, present a detailed view of photosensitization mechanism.
27. Present a detailed view of photoactive metal nitrosyls as NO donors for photodynamic therapy.
28. Present the latest development in nanoplatforms for controlled and effective NO delivery to biological targets.
29. Describe in detail the role of endogenous nitric oxide in eyes.
30. Describe the uses of NORMs in the treatment of eye defects.
31. Justify the statement that "NO news is good news for eyes."
32. Present an overview of biosynthesis of nitric oxide in plants enzymatically or non-enzymatically.
33. Discuss the role of nitric oxide donors in plant responses to abiotic and biotic stresses.
34. Present a detailed view of exogenous application of NO donors in plant growth and development.
35. Justify that "NO news is good news for plants."

Chapter III
Complexes containing carbon monoxide: synthesis, reactivity, structure, bonding and therapeutic aspects of carbon monoxide–releasing molecules (CORMs) in human beings and plants

3.1 Introduction

Since its toxicity was revealed by Claude Bernard in 1857, carbon monoxide (CO) has been recognized as a hazardous gas to mammals extensively. Definitely, owing to its toxicity, colourless, odourless and tasteless nature, the gas is generally known as "the silent killer". Its toxicity to some extent initiates from the high affinity of CO for the iron of haemoglobin. It strongly interacts with haemoglobin and forms *carboxyhaemoglobin* (*COHb*). The formation of COHb reduces the protein's ability to shuttle and transfer oxygen into tissues producing tissue hypoxia. CO poisoning results in more than 50,000 emergency department visits annually and is the second foremost basis of death from non-medicinal poisoning.

Two highly original and influential findings caused a new understanding of CO. Firstly, in 1949, Sjostrand reported that CO was produced endogenously and an oxidative metabolism of haeme was the source of CO in humans [1]. Secondly, the two CO-generating metabolic enzymes, HO-1 (haeme oxygenase-1) and HO-2 (haeme oxygenase-2) [2], were isolated and characterized in 1968. As a gasotransmitter, small amounts of endogenous CO are continuously produced and together with endogenously produced nitric oxide (NO) and dihydrogen sulphide (H_2S) are important for multiple physiologic functions. CO is now known to have a critical role in cellular functions including anti-inflammation, anti-apoptosis, anti-proliferative effects on smooth muscle, vasodilation as well as inhibition of platelet aggregation [3] under certain conditions and appropriate levels. Moreover, CO has been associated with the control of neuroendocrine functions such as inhibition of the release of hormones (corticotropin-releasing hormone, arginine vasopressin and oxytocin) involved in hypothalamo-pituitary-adrenal axis activation [4].

Deficiency in inside CO production results in systemic functional disorders. These include diabetes, inflammation or tissue cellular apoptosis. The finding of the unanticipated beneficial role of CO has led to its assessment as a therapeutic agent in clinics and hospitals. Moreover, the advantageous role of CO augmented interest in the chemistry of CO. The former has undoubtedly fuelled the latter as increasing biological and medicinal evidence indicates that CO can inhibit or lessen a variety of diseases. These include cardiovascular inflammation, hepatic ischemia and cell proliferation or have antiatherogenic or cytoprotective effects among others. In order to utilize its therapeutic

https://doi.org/10.1515/9783110727302-003

effects, an appropriate amount of CO must be delivered in a controlled manner to avoid tissue hypoxia and severe toxic side effects. However, in gaseous form, the molecule is difficult to handle and to administer in a precise level at a specific site within the organism.

As an approach to supply the gas by passing the respiratory system, CO-releasing molecules (CORMs) have emerged as an important research area. This bridge disciplines ranging from organometallic and organic chemistry to pharmacology and medicine. The growing interests in these molecules are owing to the fact that they, in principle, allow for the controlled and targeted delivery of the gas into wounded or injured tissues based on the nature and the specificity of the CORM used. Up to the present time, most CORMs are transition metal complexes having at least one carbonyl ligand, that is, metal carbonyls. Once released from the molecule with a defined stimulus, this ligand acts as the endogenously generated CO.

3.2 Metal carbonyls

Even though Paul Schützenberger was first to synthesize a metal carbonyl $[PtCl_2(CO)_2]$ in 1868, Chemistry Community considers Ludwig Mond as the father of carbonyl chemistry. The unexpected innovation of nickel tetracarbonyl $Ni(CO)_4$ in 1890 by Mond and co-workers was followed by discoveries of a few more metal carbonyls, namely, $Fe(CO)_5$ (1891), $Co_2(CO)_8$ and $Mo(CO)_6$ (1910) by the same group. Ludwig Mond, who was also the founder of the renowned Imperial Chemical Industry (England), was keen enough to notice the industrial potential of $Ni(CO)_4$ and obtained a British patent for making pure nickel by Mond process in 1890. His "Mond Nickel Company" was making over 3,000 tonnes of nickel in 1910 with a purity level of 99.9%. It is notable that $Ni(CO)_4$, a clear liquid (B.P. = 42 °C), decomposes at ~180 to pure nickel. The BASF company in Germany, by using a similar procedure, nowadays, manufactures more than 10,000 tonnes of $Fe(CO)_5$ per year for making iron oxide pigments used in magnetic tape manufacture.

It may be of interest to note that during this period of discovery of metal carbonyls that Alfred Werner's theory of bonding in metal complexes was being proposed (1893) and chemists had no idea about the actual shape of metal carbonyls. Indeed, the original formulae of $Ni(CO)_4$ and $Fe(CO)_5$ given in common chemistry textbooks (Figure 3.1) published around 1925 reflect the perception at that time.

The Lewis structure of CO molecule [:*C*: $:\ddot{O}:$] shows that it has lone pair (lp) of electrons on both carbon and oxygen atoms. The lower electronegative carbon lp (compared to oxygen) can be donated to transition metals to form complex compounds called metal carbonyls. Thus, metal carbonyls are complex compounds of transition metals with CO. The numbers of CO molecules that get chemically bonded to a metal through coordinate bonds vary with the nature of the metal. For example,

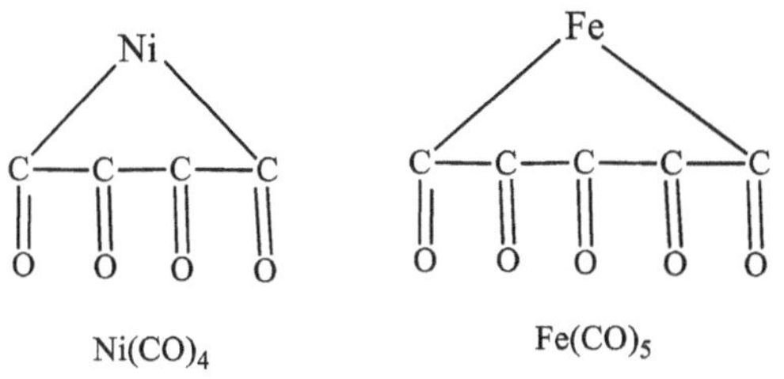

Figure 3.1: Original formula of $Ni(CO)_4$ and $Fe(CO)_5$.

nickel carbonyl contains four coordinated CO molecules while in chromium carbonyl-coordinated CO groups are six in number.

The bonding of the CO to the metal is highly sensitive to electronic effects on the metal centre. This is usually monitored by infrared (IR) spectroscopy. The followings are some well-known examples of metal carbonyls (Figure 3.2). Most of the metal carbonyls have metal in very low oxidation states, sometimes even zero or negative oxidation states.

Depending on the number of metal atoms in a given carbonyl, metal carbonyls are classified into the following two types:

(a) **Mononuclear carbonyls:** They contain only one metal atom per molecule and have the general composition $M(CO)_y$. Examples are $Ni(CO)_4$, $V(CO)_6$, $Cr(CO)_6$ and so on.
(b) **Polynuclear carbonyls**: They have more than one metal atom in each molecule and are of the general formula $M_n(CO)_y$. The common examples are $Co_2(CO)_8$, $Mn_2(CO)_{10}$, $Fe_2(CO)_9$ and $Fe_3(CO)_{12}$. Polynuclear carbonyls may be homonuclear, for example, $Mn_2(CO)_{10}$ and $Fe_3(CO)_{12}$ or containing more than one different atom (heteronuclear), for example, $MnCo(CO)_9$ and $MnRe(CO)_{10}$.

Numerous carbonyl compounds are of significant structural interest. Many of them are important industrially, and in catalytic and other reactions.

3.3 Synthesis of metal carbonyls

In 1868, Paul Schützenberger [5] prepared the first metal carbonyl compound, $PtCl_2(CO)_2]$, by the reaction of $PtCl_2$ with CO:

$$PtCl_2 + CO \longrightarrow [PtCl_2(CO)]_2 + cis-\left[PtCl_2(CO)_2\right]$$

Simple homoleptic or binary metal carbonyls are made by two basic methods: (i) direct reaction of metal with CO and (ii) reductive carbonylation in which a metal salt reacts

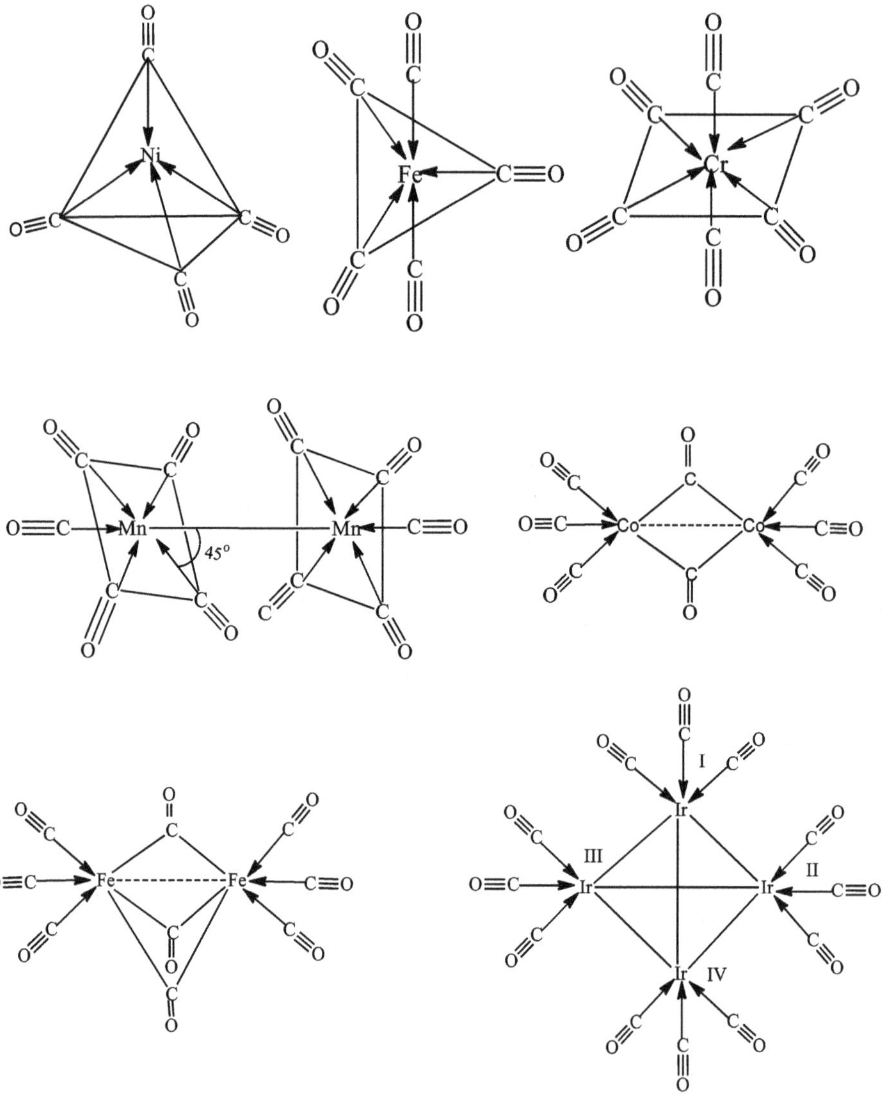

Figure 3.2: Some of the examples of homoleptic (containing only CO as ligands) metal carbonyls.

with CO in the presence of a reducing agent. Some other methods are also known for the synthesis of metal carbonyls.

(i) Direct carbonylation

By direct interaction of CO with finely divided transition metals under suitable conditions of temperature and pressure. This method is applicable to the synthesis of only a few metal carbonyls. $Ni(CO)_4$ is the classic example which was discovered by Ludwig Mond in 1890 and has since been utilized in the refining of nickel as well:

$$Ni + 4CO \xrightarrow[1\ atm.]{30\ ^\circ C} Ni(CO)_4$$

$$Fe + 5CO \xrightarrow[200\ atm.]{200\ ^\circ C} Fe(CO)_5$$

$$2Co + 8CO \xrightarrow[100\ atm.]{200\ ^\circ C} Co_2(CO)_8$$

(i) Reductive carbonylation:

When metal salts interact with CO-containing reducing agent in the reaction pot, metal carbonyls are obtained, for example,

$$CrCl_3 + 6CO \xrightarrow[LiAlH_4,\ ether]{200\ ^\circ C,\ 70\ atm.} Cr(CO)_6$$

$$VCl_3 + 3Na + 6CO \xrightarrow[High\ pressure]{120\ ^\circ C} V(CO)_6 + 3NaCl$$

$$2CoCO_3 + 2H_2 + 8CO \xrightarrow[250-300\ atm.]{120-200\ ^\circ C} Co_2(CO)_8 + 2H_2O + 2CO_2$$

$$2Mn(acac)_3 + 10CO \xrightarrow{(C_2H_5)_3Al} Mn_2CO_{10}$$

$$2Mn(OAc)_2 + 4Na + 10CO \xrightarrow[165\ ^\circ C/240\ bar]{DMF} NaOAc + Mn_2CO_{10}$$

$$MoCl_5 + 6CO + 5Na \xrightarrow[coordinating\ solvent]{Diglyme} Mo(CO)_6 + 5NaCl$$

$$WCl_6 + 3Zn + 6CO \xrightarrow[50\ atm]{Et_2O,\ 67\ ^\circ C} W(CO)_6 + 3ZnCl_2$$

$$RuCl_3 \cdot 3H_2O + CO + H_2 \xrightarrow[80-135\ ^\circ C/1\ atm;\ KOH,\ 75\ ^\circ C,\ Zn]{2-Ethoxyethanol} Ru_3(CO)_{12}$$

$$3Ru(acac)_3 + 12CO + 4.5H_2 \xrightarrow[150\ ^\circ C,\ 155\ atm]{MeOH} Ru_3(CO)_{12} + 9MeC(O)H_2C(O)Me$$

(ii) Synthesis using metal oxides:

In case of metal oxides, CO itself acts as a reducing agent and gets converted to CO_2 in the process. The reaction conditions are relatively more drastic for such reactions. For example,

$$Re_2O_7 + 17CO \xrightarrow[250\ ^\circ C]{200\text{–}300\ atm.} Re_2(CO)_{20} + 7CO_2$$

$$OsO_4 + CO \xrightarrow[125\,^\circ C/75\ atm]{MeOH} Os_3(CO)_{12} + 4CO_2$$

$$MoO_3 + CO \xrightarrow[300\,^\circ C]{200\ atm} Mo(CO)_6 + CO2$$

$$2CoCO_3 + 2H_2 + 8CO \xrightarrow[240\ bar]{100-150\,^\circ C} Co_2(CO)_8 + 2CO_2 + 2H_2O$$

$$CoO + CO \xrightarrow[80\,^\circ C]{1,900\ atm} Co_2(CO)_8 + CO_2$$

$$RuO_2 \cdot xH_2O + CO \xrightarrow[20atm,\ 160\,^\circ C]{Toluene} Ru_3(CO)_{12} + CO_2$$

$$Fe_2O_3 + CO \xrightarrow[225\,^\circ C]{2,000\ atm} Fe(CO)_5 + CO_2$$

(iii) Polynuclear carbonyls are generally prepared by thermal or photochemical decomposition of simple carbonyls, for example,

$$2Fe(CO)_5 \xrightarrow[Light]{UV} Fe_2(CO)_9 + CO$$

$$2Co_2(CO)_8 \xrightarrow{70\,^\circ C} Co_4(CO)_{12} + 4CO$$

$$2Os(CO)_5 \xrightarrow[Light]{UV} Os_2(CO)_9 + CO$$

(iv) Synthesis of $Mo(CO)_6$ and $W(CO)_6$ taking $Fe(CO)_5$ as a precursor

As CO groups present in $Fe(CO)_5$ are labile, they can be replaced by Cl^- ions by treating $Fe(CO)_5$ with $MoCl_5$ or WCl_6 and thus $Mo(CO)_6$ and $W(CO)_6$ are obtained.

$$3Fe(CO)_5 + MoCl_6 \xrightarrow[Ether]{100\,^\circ C} Mo(CO)_6 + 3FeCl_2 + 9CO$$

$$3Fe(CO)_5 + WCl_6 \xrightarrow[100\,^\circ C]{Ether} W(CO)_6 + 3FeCl_2 + 9CO$$

(v) There are also a variety of other methods for the synthesis of metal carbonyls. However, they are specific to a few metal carbonyls, clusters and industrially useful carbonyls:

$$RhCl_3 \cdot 3H_2O + CO \xrightarrow[60\,^\circ C,\ 40-50\ atm]{MeOH} Rh_6(CO)_{16}$$

$$RCHO + [RhCl(PPh_3)_3] \longrightarrow RH + [RhCl(CO)(PPh_3)_2]$$

$$WMe_6 + excess\ CO \xrightarrow{Deinsertion} W(CO)_6 + 3Me_2CO\ Na_3IrCl_6 + CO \xrightarrow[MeOH]{Reflux,\ 1\ atm} Ir_4(CO)_{12}$$

$$RhCl_3 \cdot 3H_2O + CO \xrightarrow[100\,^\circ C,\,1\,atm]{MeOH} [Rh(CO)_2Cl]_2 \xrightarrow[Heptane]{CO,\,H_2O,\,NaHCO_3} Rh_4(CO)_{12}$$

$$CO + [Ru(NH_3)_5H_2O]^{2+} \longrightarrow [Ru(NH_3)5(CO)]^{2+}$$

$$2Fe(CO)_5 \xrightarrow[Sunlight]{AcOH,\,hn} Fe_2(CO)_9 + CO$$

$$2Fe(CO)_5 + 6OH^- \xrightarrow{-3HCO_3^-} 3[HFe(CO)_4]^- \xrightarrow{3MnO_2} Fe_3(CO)_{12} + 3MnO$$

3.4 Physical properties

(i) With the exception of $Ni(CO)_4$, $Fe(CO)_5$, $Ru(CO)_5$ and $Os(CO)_5$, which are liquids at ordinary temperatures, all other carbonyls are crystalline solids. They melt or decompose at low temperatures.
(ii) All are typical covalent compounds, and hence they are soluble in non-polar solvents.
(iii) While $V(CO)_6$ is paramagnetic with respect to one paired electron, all the carbonyls are diamagnetic.

3.5 Chemical properties/reactivity of metal carbonyls

The reactions of metal carbonyls are quite varied. Some of the important reactions shown by these compounds are given further.

3.5.1 Displacement or substitution reactions

Displacement of one or more CO ligands from metal carbonyls may occur on interaction with electron donors like pyridine, C_5H_5N (py), phosphines (R_3P), isocyanides (RNC) and benzene. A molecule of CO is replaced by an electron pair donor. Consequently, a six-electron donor (like benzene) would replace 3CO molecules:

$$Fe(CO)_5 \xrightarrow[-CO]{+\,R_3P} R_3PFe(CO)_4 \xrightarrow[-CO]{+\,R_3P} (R_3P)_2Fe(CO)_3$$

$$Mo(CO)_6 \xrightarrow[-CO]{+py} (py)Mo(CO)_5 \xrightarrow[-CO]{+py} (py)_2Mo(CO)_4 \xrightarrow[-CO]{+py} (py)_3Mo(CO)_3$$

$$Cr(CO)_6 \xrightarrow[-3CO]{+\,C_6H_6} (C_6H_6)Cr(CO)_3$$

$$Fe(CO)_5 \xrightarrow[-CO]{+Me_3SiNC} (Me_3SiNC)Cr(CO)_3$$

$$Co_2(CO)_8 \xrightarrow[-2CO]{+RC \equiv CR} Co_2(CO)_6(RC \equiv CR)$$

$$2Fe(CO)_5 + 2C_5H_6 \longrightarrow [(h^5 - C_5H_5)Fe(CO)_2]_2$$

$$Mo(CO)_6 + 2C_5H_5N_a \longrightarrow Na(\eta^5 - C_5H_5)Mo(CO)_3$$

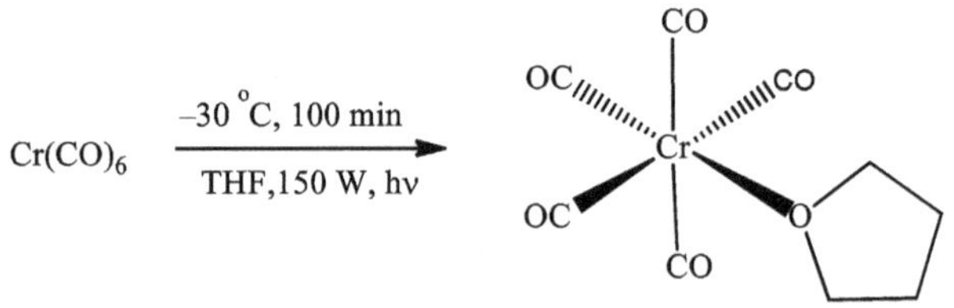
Cr(CO)6
-30 °C, 100 min
THF,150 W, hv
CO
OC
CO
Cr
OC
O
CO

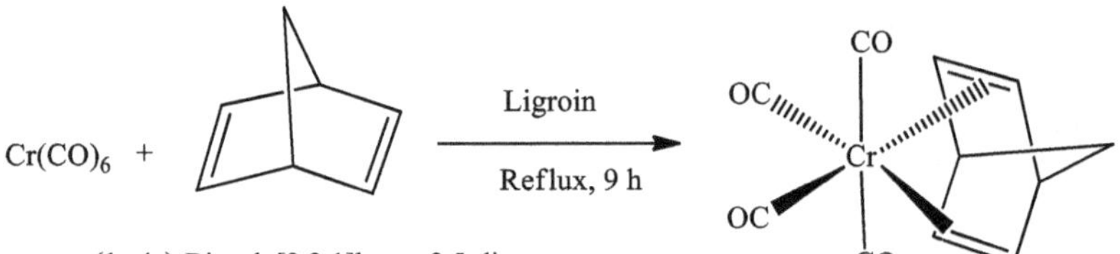
Cr(CO)6 +
Ligroin
Reflux, 9 h
CO
OC
Cr
OC
CO

(1*s*,4*s*)-Bicyclo[2.2.1]hepta-2,5-diene

It is notable here that RC ≡ CR, hepta-2,5-diene behaves as a four-electron donor, and cyclopentadiene acts as a five-electron donor. As per convention, five-electron donor cyclopentadiene is represented as η^5-C_5H_5.

3.5.2 Formation of cationic carbonyl complexes: carbonylate cations

Carbonylate cations from metal carbonyls may be prepared from the following methods:

(i) By protonation of carbonyls in strong acids:For example

$$[Fe(CO)_5] \xrightarrow[BCl_3]{HCl} [FeH(CO)_5]^+[BCl_4]^-$$

(ii) By the action of CO and a Lewis acid like $AlCl_3$ or BF_3 on carbonyl halide, for example,

$$Mn(CO)_5Cl + CO + AlCl_3 \longrightarrow [Mn(CO)_6]^+[AlCl_4]^-$$

3.5.3 Formation of anionic carbonyl complexes: carbonylate anions

Carbonylate anions from metal carbonyls may be prepared from the following methods:

(i) By the action of NaOH/KOH or nitrogenous bases (e.g. amines or pyridine) on metal complexes, for example,

$$[Fe(CO)_5] \xrightarrow{NaOH} Na_2{}^{+}[Fe(CO)_4]^{2-}$$

$$[Fe_2(CO)_9] \xrightarrow{KOH/MeOH} [Et_4N]_2{}^{+}[Fe_2(CO)_8]^{2-}$$

$$[Fe_3(CO)_{12}] \xrightarrow{en/H_2O} [Fe(en)_2]^{3+}[Fe_3(CO)_{11}]^{3-}$$

$$3[Co_2(CO)_8] + 12py \longrightarrow 2[Co(py)_6]^{2+}[Co(CO)_4]_2^{2-} + 8CO$$

(ii) By reduction of metal carbonyls with alkali metals, alkali metal amalgams or borohydrides, for example,

$$Cr(CO)_6 \xrightarrow[NH_3]{Na} Na_2[Cr(CO)_5]$$

$$Cr(CO)_6 \xrightarrow[THF]{NaBH_4} Na[HCr_2(CO)_{10}]$$

$$Cr(CO)_6 \xrightarrow[NH_3]{NaBH_4} Na_2[Cr_2(CO)_{10}]$$

$$Re_2(CO)_{10} \xrightarrow[THF]{NaBH_4} Na_2[Re_4(CO)_{16}]$$

$$Mn_2(CO)_{10} \xrightarrow[THF]{Na/Hg} Na[Mn(CO)_5]$$

$$Co_2(CO)_8 \xrightarrow[THF]{Na/Hg} Na[Co(CO)_4]$$

(iii) Carbonyl hydrides are synthesized from carbonyl anions, carbonyl halides and dimeric carbonyls. The term hydride is assigned based on the formal oxidation state. However, the properties of the organometallic hydrides vary from hydric to protic in character. The acidity of these complexes differs significantly depending on the nature of the metal and the ligands.

Synthesis of some carbonyl hydrido complexes are as follows:

Mode of synthesis	p*Ka*	Comparable acidity
$[Co(CO)_4]^- + H^+ \longrightarrow [HCo(CO)_4]$	1	HCl
$I_2Fe(CO)_4 + NaBH_4 \xrightarrow[150\,°C]{200\text{ bar}} H_2Fe(CO)_4$	4.7	CH_3COOH
$Mn_2(CO)_{10} + H_2 \longrightarrow 2HMn(CO)_5$	7	H_2S
$Fe(CO)_5 + OH^- \longrightarrow [HFe(CO)_4]^-$	14	H_2O

3.5.4 Redox reactions including the formation and cleavage of metal–metal bonds

$$2Na + \underset{O.S. = 0}{Co_2(CO)_8} \longrightarrow 2Na\underset{O.S. = -1}{[Co(CO)_4]}$$

$$\underset{O.S. = -1}{[Mn(CO)_5Br]} + \underset{O.S. = -1}{[Mn(CO)_5]^-} \longrightarrow \underset{O.S. = 0}{[Mn_2(CO)_{10}]} + Br^-$$

3.5.5 Reaction with NO

$$2NO + Fe(CO)_5 \xrightarrow{95\ °C} [Fe(CO)_2(NO)_2] + 3CO$$

$$4NO + 3Fe_2(CO)_9 \longrightarrow 2[Fe(CO)_2(NO)_2] + Fe(CO)_5 + Fe_3(CO)_{12} + 6CO$$

$$6NO + Fe_3(CO)_{12} \xrightarrow{85\ °C} 3[Fe(CO)_2(NO)_2] + 6CO$$

$$2NO + Co_2(CO)_8 \longrightarrow 2[Co(CO)_3(NO)_2] + 2CO$$

Moist NO gives a blue-coloured compound, Ni(NO)(OH) with $Ni(CO)_4$ while with dry NO a blue solution of the composition, $Ni(NO)(NO_2)$:

$$2Ni(CO)_4 + 2NO + 2H_2O \longrightarrow 2[Ni(NO)(OH)] + 8CO + H_2$$

$$Ni(CO)_4 + 4NO \longrightarrow 2[Ni(NO)(NO_2)] + 4CO + N_2O$$

3.5.6 Action of heat

Different metal carbonyls give different products on heating. For example,

$$Fe(CO)_5 \xrightarrow{200\,°C} Fe + 5CO$$

$$3Fe_2(CO)_9 \xrightarrow{70\,^\circ C,\ Cool} 3Fe(CO)_5 + Fe_3(CO)_{12}$$
(Intoluene)

$$Fe_3(CO)_{12} \xrightarrow{140\,^\circ C} 3Fe + 12CO$$

$$2Co_2(CO)_8 \xrightarrow[\text{Inert atmosphere}]{50\,^\circ C} Co_4(CO)_{12} + 4CO$$

$$3Fe_2(CO)_9 \xrightarrow{50\,^\circ C} 2Fe_3(CO)_{12} + 3CO$$

$$Ni(CO)_4 \xrightarrow{180\,^\circ C} Ni + 4CO$$

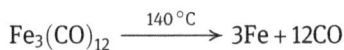

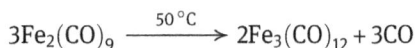

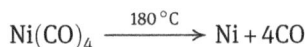

Because of wide reactivity, metal carbonyls are useful starting materials (precursors) for other organometallic carbonyls.

3.5.7 Insertion reactions

Mechanistic studies using ^{14}CO-labelled $CH_3Mn(CO)_5$ have shown that (i) the CO molecule that becomes the acyl ligand is not derived from the external CO, but from that already coordinated to the metal atom; (ii) the incoming CO is added *cis*- to the acyl group, as in the following reaction:

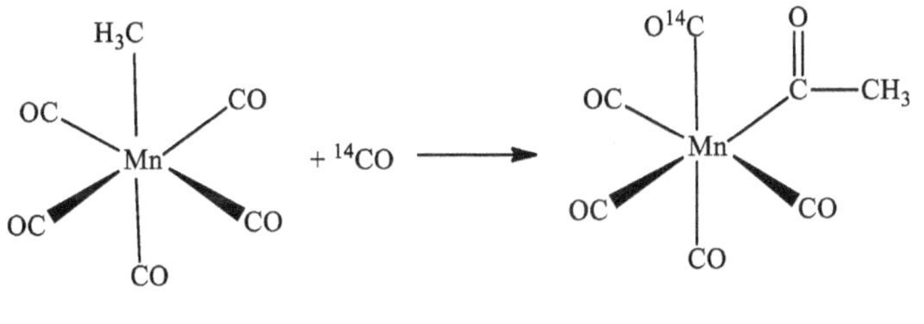

and (iii) the conversion of alkyl ligand into the acyl ligand can be promoted by addition of ligand other than CO, for instance, excess $P(C_6H_5)_3$ as in the following reaction:

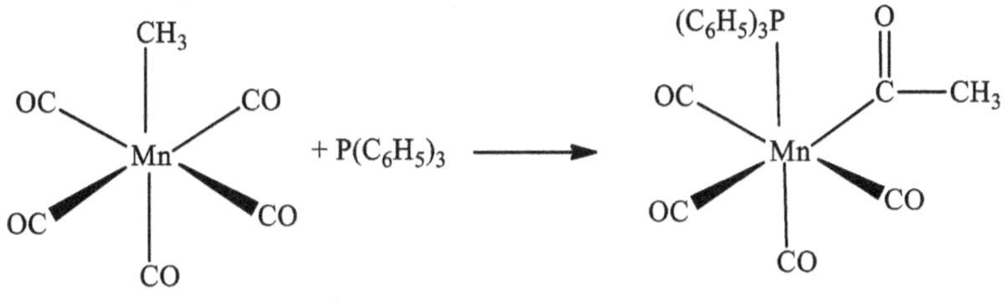

Kinetic studies of such insertion reactions show that the first step involves equilibrium between octahedral alkyl and a five-coordinate acyl intermediate

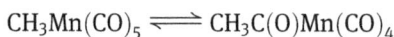

$$CH_3Mn(CO)_5 \rightleftharpoons CH_3C(O)Mn(CO)_4$$

The incoming ligand (whether CO, $P(C_6H_5)_3$ or the like) adds to the five-coordinate intermediate, as in the following reaction: $CH_3Mn(CO)$

$$CH_3C(O)Mn(CO)_4 + L \longrightarrow CH_3C(O)Mn(CO)_4L$$

3.5.8 Nucleophilic attack on coordinated CO

(i) Attack of nucleophiles such as OH^- on CO in $Fe(CO)_5$ gives $(CO)_4FeH^-$ and HCO_3^-:

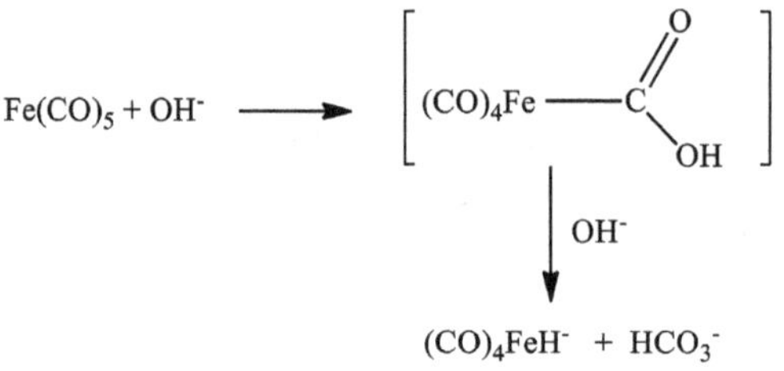

(ii) Attack of alkoxide ions RO^- on CO gives the M–C(O)OR group as in the following reaction:

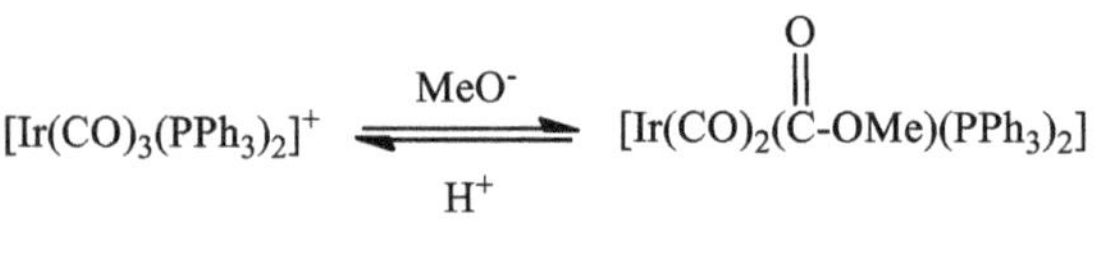

Such reactions are important in the catalysed synthesis of carboxylic acids and esters from alkenes, CO and water or alcohols.

(i) In organometallic chemistry, nucleophilic addition reaction to the electron-deficient carbon atom of a metal coordinated carbonyl group has become a powerful tool. It is commonly called the Hieber base reaction. Fischer carbenes can be synthesized by the addition of alkyl lithium to metal carbonyls followed by methylation. Carbenes are important reagents as well as catalysts for a host of organometallic transformation [6]:

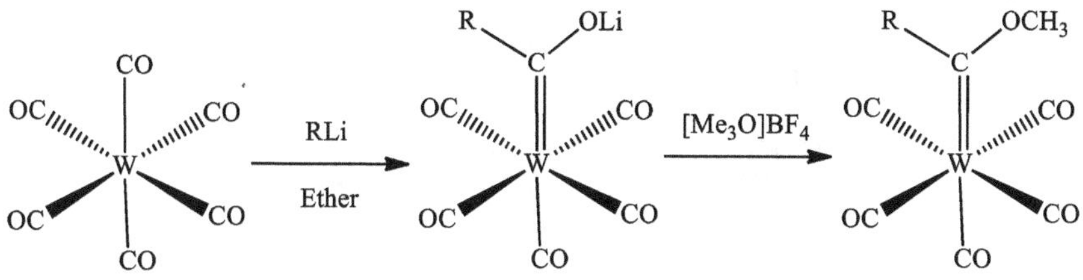

3.5.9 Electrophilic addition to the carbonyl oxygen

Lewis acid such as $AlEt_3$ can attach to the oxygen end of a coordinated carbonyl group. Only bridging carbonyl groups are nucleophilic enough to undergo this reaction as far as neutral metal carbonyls are considered. Addition of electrophiles becomes more facile with anionic carbonyl metallates:

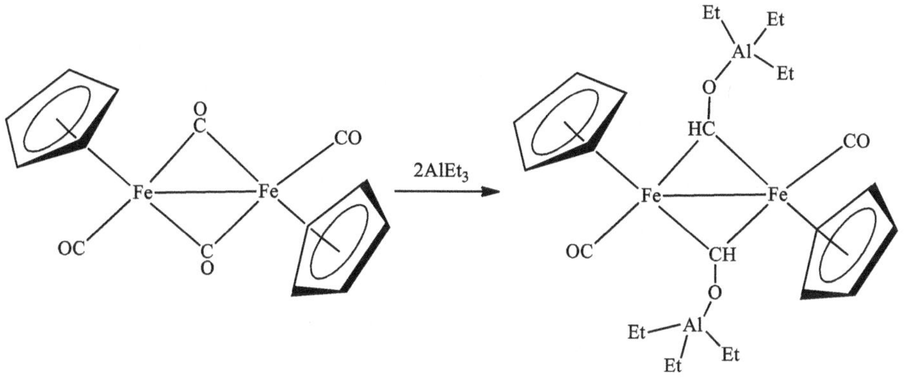

3.5.10 Disproportionation reaction

Multi-metal carbonyls may undergo disproportionation reaction in the presence of suitable ligands often leading to carbonyl anions. For example,

$$12Py + 3Mn_2(CO)_{10} \xrightarrow[-10\ CO]{120\,^{\circ}C} + 4[Mn(CO)_5]^- + 2[Mn(py)_6]^{2+}$$

3.5.11 Collman's reagent

$Na_2Fe(CO)_4$ called Collman's reagent has found many applications in organic synthesis. Organic halogen compounds can be functionalized in many ways via reaction with the Collman's reagent [7]:

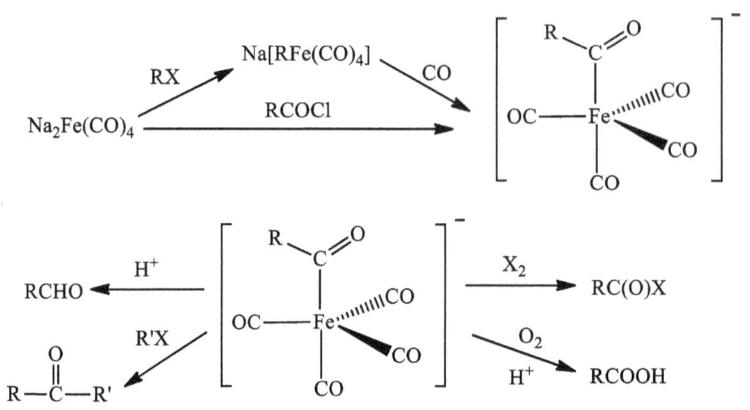

3.5.12 Oxidative decarbonylation

Carbonyl metal halides are prepared by oxidative decarbonylation and these are good reagents for reactions with metallated reagents. Moreover, substituted metal carbonyls can be converted to metal oxides by oxidative decarbonylation:

$$Mn_2(CO)_{10} + I_2 \longrightarrow 2Mn(CO)_5I \xrightarrow[-2CO]{NaCp} CpMn(CO)_3$$

$$Fe(CO)_5 + I_2 \longrightarrow Fe(CO)_4I_2$$

3.5.13 Photochemical substitution

Photochemical substitutions [8] are the most common photoreactions in organometallic chemistry. Kinetic studies have shown that the rate constant for the dissociation of one CO unit which precedes the entrance of a new ligand is increased upon photochemical excitation by a factor of 10^{16}:

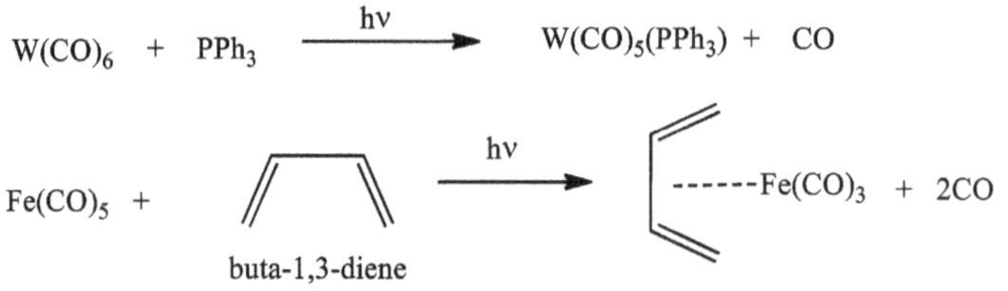

$$W(CO)_6 + PPh_3 \xrightarrow{h\nu} W(CO)_5(PPh_3) + CO$$

buta-1,3-diene

Photochemical substitution of CO by monodentate ligand often results only in the displacement of one carbonyl unit. This is because photochemical excitation of a metal carbonyl with a coordinated sphere of mixed substituent causes dissociation of the ligand which is mostly weak bonded in ground state. Hence, attempts to substitute a second carbonyl from an $M(CO)_5L$ species result in the displacement of L itself. If ligands of comparable ligand strength are used, one can displace all the carbonyl groups:

$$M(CO)_6 \xrightarrow[THF]{h\nu} M(CO)_5THF \xrightarrow{h\nu} M(CO)_5 + THF$$

$$Mo(CO)_6 \xrightarrow[P(OMe)_3]{h\nu} Mo[P(OMe)_3)]_6 + 6CO$$

Other photochemical reactions are

$$(arene)Cr(CO)_3 \xrightarrow[MeOH]{h\nu} Cr(OMe)_3 + arene + 3CO$$

$$(arene)Cr(CO)_3 \xrightarrow[R_2S_2]{h\nu} Cr(SR)_3 + arene + 3CO$$

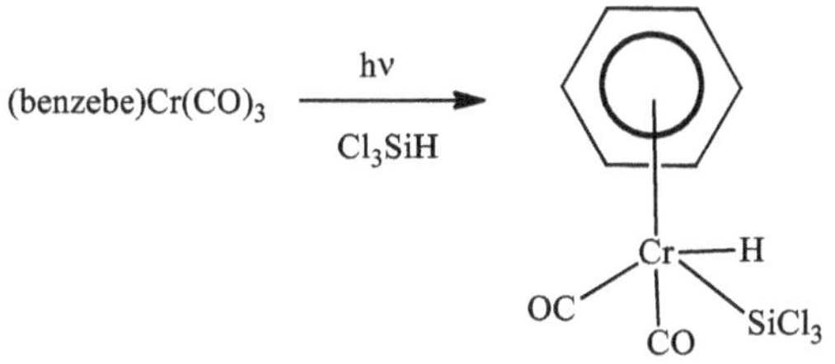

3.6 Catalytic aspect of metal carbonyls

Catalytic aspect of metal carbonyls may be understood by the following two reactions.

3.6.1 Hydrogenation of alkenes

Hydrogenation of olefins is a reaction of industrial importance, being used in petrochemical and pharmaceutical industries, where the preparation of chemical/drug often involves the hydrogenation of specific double bonds. For the hydrogenation of specific double bonds in compounds containing several double bonds, the search for new, more efficient and above all more selective catalysts is required.

One very successful catalyst for this process is chlorotris(triphenylphosphine) rhodium(I), $[RhCl(PPh_3)_3]$. As this compound was synthesized for the first time by Wilkinson and co-workers in 1965, it is now universally referred to as Wilkinson's catalyst. In addition to the hydrogenation reaction by $[RhCl(PPh_3)_3]$, the *trans*-hydridocarbonyl(triphenylphosphine)-rhodium(I), *trans*-$[RhH(CO)(PPh_3)_3]$, synthesized in Wilkinson's laboratory is another hydrogenation catalyst for alkenes:

$$H_2C=CHR + H_2 \xrightarrow[\text{Benzene}]{[RhH(CO)(PPh_3)_3]} H_3C-CH_2R$$

The chemistry of hydrogenation of an alkene with this catalyst can be understood with the following sequential steps:

(i) 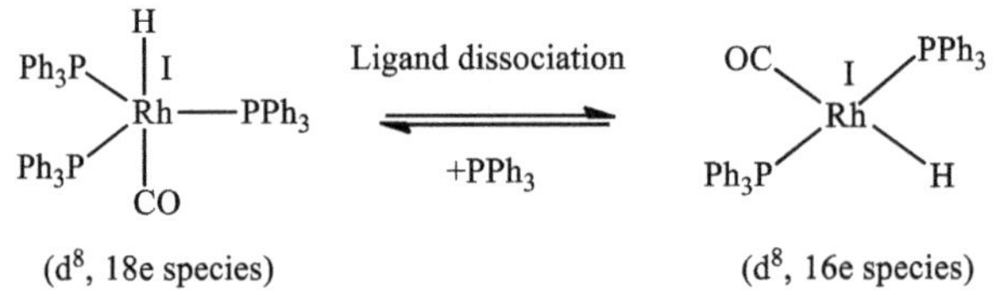

In spite of the higher stability of the 18-electron species in general, the tendency of the above dissociation may be understood on the basis of steric relief brought about by the dissociation of a bulky PPh_3 ligand.

(ii) 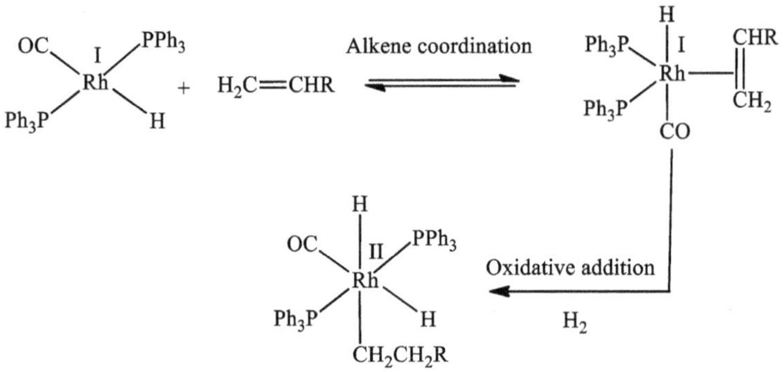

Coordination of alkene to the metal atom polarizes the C–C bond and allows facile migration of the alkyl group with the bonding electron pair along with oxidative addition.

(iii) 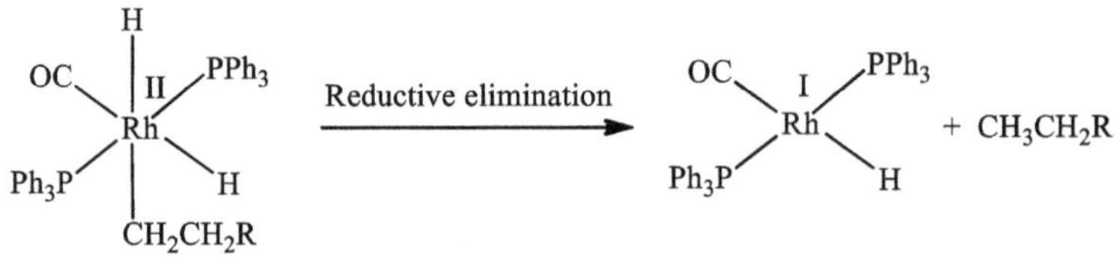

(iv) 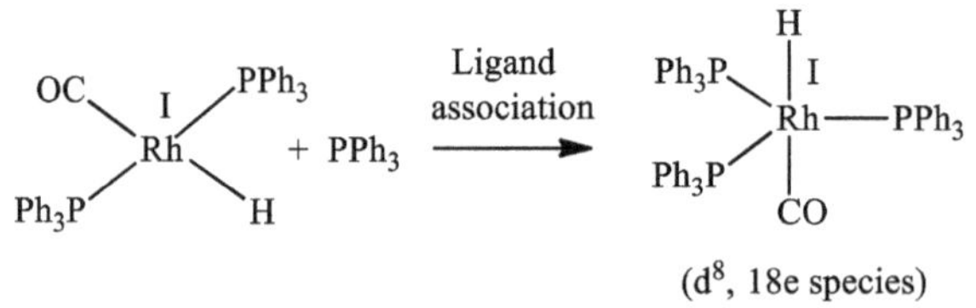

The unique features of the above catalytic steps are (i) hydrogenation of alkenes by this catalyst is unusual being highly selective for the hydrogenation of alk-1-ene compared toalk-2-ene, and (ii) hydrogenation follows anti-Markovnikov's rule.

3.6.2 Hydroformylation reaction

The hydroformylation reaction is the catalysed addition of H_2 and CO [or formally of H and the CHO (formyl) groups] to an alkene, usually a terminal or 1-alkene or alk-1-ene. The reaction is also known as OXO process because of inclusion of oxygen into the hydrocarbon:

$$\underset{1-\text{Alkene}}{RCH=CH_2} + H2 + CO \xrightarrow{\text{Catalyst}} \underset{\text{Aldehyde}}{RCH_2CH_2CHO}$$

Importance of OXO reaction or process

This reaction is of immense industrial importance. This is because the OXOreaction or process is used to convert alk-1-enes into aldehydes which can be further reduced to alcohols under the reaction conditions for the production of polyvinylchloride and polyalkenes. Moreover, in case of long-chain alcohols, in the production of detergents

$$\underset{\text{Aldehyde}}{RCH_2CH_2CHO} + H_2 \longrightarrow \underset{\text{Alcohol}}{RCH_2CH_2CH_2OH}$$

Further, resulting aldehydes give acid by oxidation of esters followed by alcohols to react with acids. Some of these secondary products are useful as plasticizers, lubricating oils and solvents:

$$\underset{\text{Aldehyde}}{RCH_2CH_2CHO} \xrightarrow[Na_2Cr_2O_7/H_2SO_4]{O} \underset{\text{Carboxylic acid}}{RCH_2CH_2COOH}$$

$$\underset{\text{Carboxylic acid}}{RCH_2CH_2COOH} + H-OR' \xrightarrow[HCl/H_2SO_4]{H^+ \text{(catalyst)}} \underset{\text{Ester}}{RCH_2CH_2COOR'}$$

3.6.2.1 Catalysts used for hydroformylation reaction

3.6.2.1.1 Co-based catalyst

Earlier, the reaction used to be catalysed by dicobalt octacarbonyl, Co_2CO_8, which was not so effective catalyst as the reaction had to be carried out at high temperature (150–180 °C) and pressure (~200 atmospheres). Moreover, the products were generally formed as a mixture of linear and branched-chain isomers approximately in 3:1 ratio. This is due to the anti-Markovnikov and Markovnikov additions, respectively:

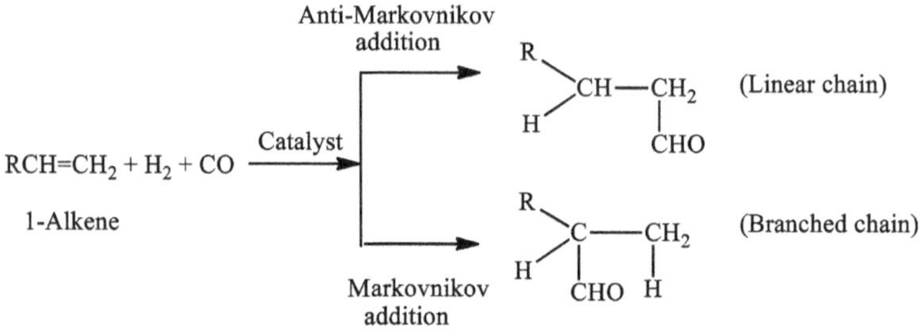

The role of cobalt catalyst, Co_2CO_8 can be illustrated by a catalytic cycle presented in Figure 3.3.

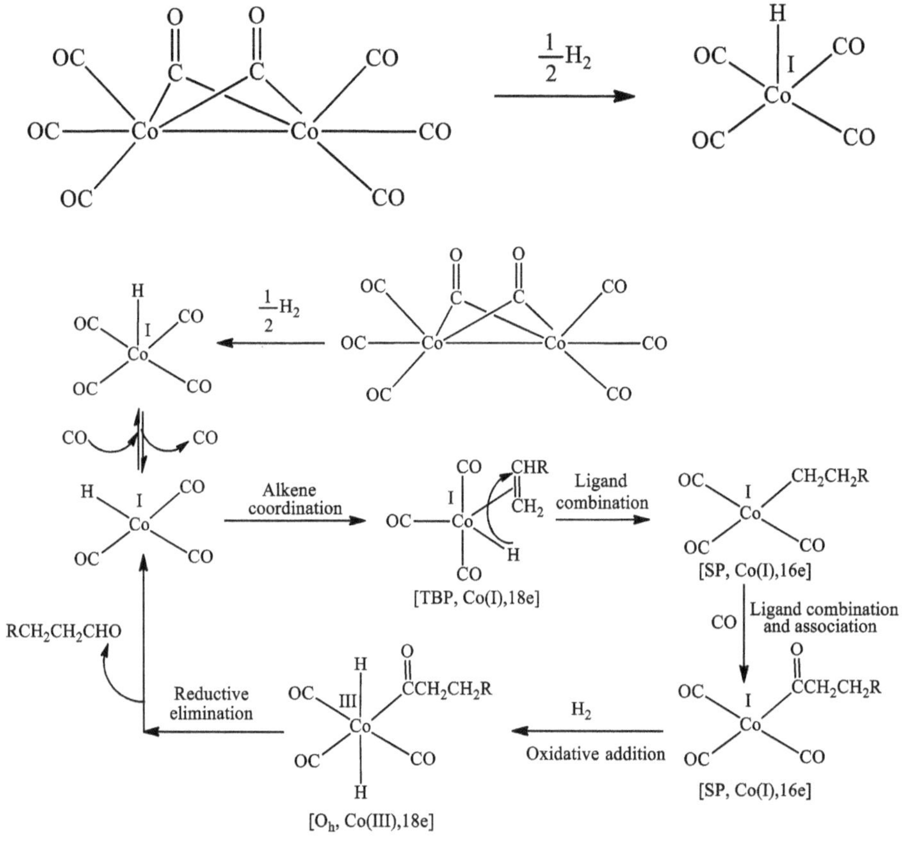

Figure 3.3: Hydroformylation of alkenes in the presence of cobalt catalyst, Co_2CO_8.

3.6.2.1.2 Rh-based catalyst

More recently, it has been found that the platinum group complexes such as [RhCl(CO)(PPh_3)$_3$] and [RhH(CO)(PPh_3)$_3$] are much more effective catalysts than the cobalt complexes. These complexes facilitate the hydroformylation reaction more effectively at lower temperature and pressure. For example, with [RhH(CO)(PPh_3)$_3$] as a catalyst, the hydroformylation reaction can proceed even at ambient conditions (25 °C and 1 atmospheric pressure) to yield aldehyde almost quantitatively.

In addition to more moderate working conditions, another distinct advantage for the rhodium-catalysed reactions is their high selectivity for the formation of only the straight-chain aldehydes particularly in the presence of excess phosphine (PPh_3). This excess phosphine tends to supress effectively the lower dissociative tendency of the intermediate rhodium bis complex, [RhH(CO)(PPh_3)$_2$], thus favouring the associative attack of the alkene on the bisphosphine complex, for which the specificity of the anti-Markovnikov addition is higher (vide the catalytic mechanism).

3.6.2.1.2.1 Catalytic mechanism

The probable mechanism for the rhodium-catalysed hydroformylation reaction is presented in Figure 3.4.

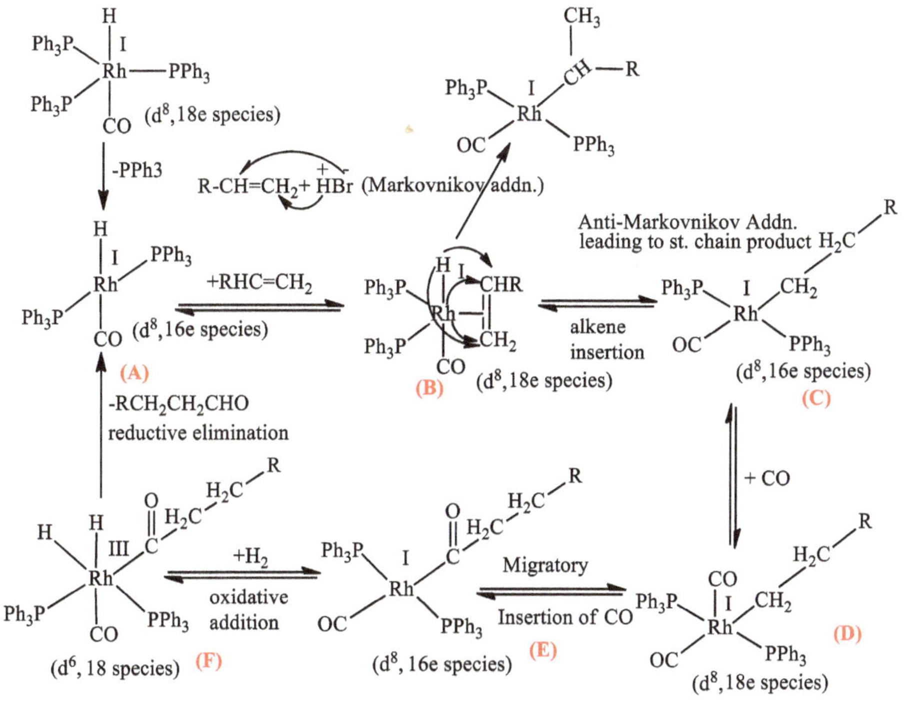

Figure 3.4: Hydroformylation of alkenes by rhodium catalyst, [RhH(CO)(PPh_3)$_3$].

The noteworthy reactions of this catalytic cycle are that **A to F** is reversible. Moreover, coordination numbers (C.N.) of the rhodium complexes range from 4 to 6, and the number of valence electrons changes between 16 and 18. The initial step is addition of alkene to compound **A**, followed by insertion of the olefin to give the alkyl complex, **C.** The latter then undergoes CO coordination and finally migratory inclusion of CO into Rh–C bond to give acyl derivative, **E.** Oxidative addition of H_2 then gives dihydrido acyl derivative, **F.** It is this last step that involves a change in oxidation state of Rh, which is the most likely rate-determining step in the cycle.

The final steps are the reductive elimination of the aldehyde to give **A.** It is notable that dissociation of PPh_3 from $[Rh(H)(CO)(PPh_3)_3]$ (18e species) is a required step for generating a free site to bind the olefin.

The catalyst $[Rh(H)(CO)(PPh_3)_3]$ has the great advantage over the conventional cobalt carbonyl catalyst, $Co_2CO_8/[HCo(CO)_4]$. It works competently at much lower temperature and provides straight-chain aldehydes as contrast to branched-chain aldehyde. The cause for its regioselectivity stands in the alkene insertion step of the cycle. Because of having two bulky PPh_3 groups, the attachment of $-CH_2CH_2R$ in the species **A,** via anti-Markovnikov addition resulting to a straight-chain product is easier that the attachment of a branched-chain product, via **Vladimir Vasilevich Markovnikov** addition.

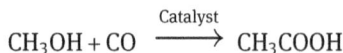

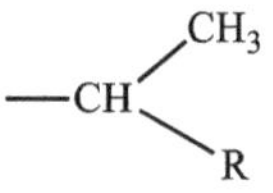

3.6.2.1.3 Ru-based catalyst

In addition to the above-mentioned rhodium catalysts, a ruthenium complex $[Ru(CO)_3$-$(PPh_3)_2]$ too has been found to be effective in hydroformylation of alkenes. Although the basic catalytic steps (Figure 3.5) are similar to those described in Figure 3.4, yet the mechanism of hydroformylation reaction is of a different pattern.

It is notable from the catalytic cycle that C.N. of the ruthenium complex ranges from 5 to 6 and the oxidation state cycles between 0 and 2.

3.6.3 Manufacturing of CH_3COOH by carbonylation of CH_3OH using metal carbonyl as catalyst

Carbonylation of CH_3OH (i.e. insertion of CO to methanol) to acetic acid has been detected by Reppe [9] in 1953. He used Co, Fe or Ni carbonyls as catalysts with halide promoters at high pressure of CO

$$CH_3OH + CO \xrightarrow{\text{Catalyst}} CH_3COOH$$

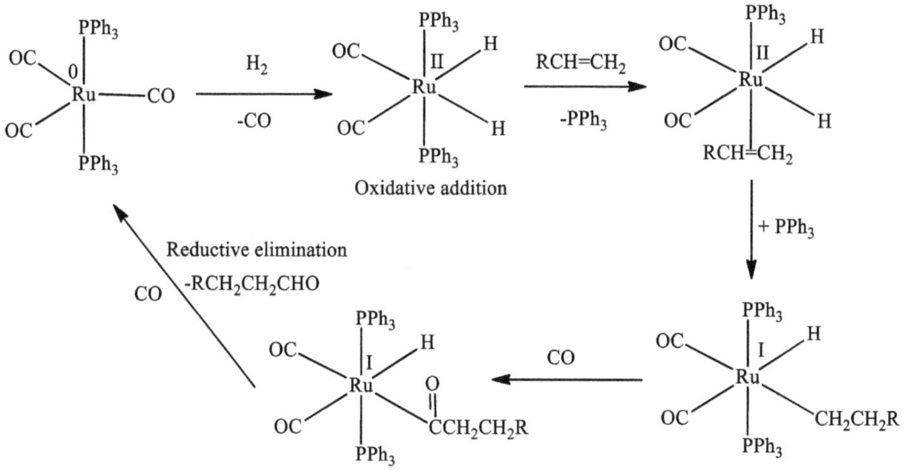

Figure 3.5: Catalytic cycle for hydroformylation of alkenes by ruthenium catalyst, $[Ru(CO)_3(PPh_3)_2]$.

This discovery led to a commercial process of CH_3COOH production by Badische Anilin and Soda Fabrik AG (BASF), using Co catalysts in the form of Cobalt(II) iodide under elevated temperature and pressure (200–250 °C and 500–700 atm) conditions with a selectivity to CH_3COOH with respect to methanol of ca. 90% [10].

The reactive components in the above high-pressure synthesis of CH_3COOH in the presence of CoI_2 used as a catalyst are probably $[HCo(CO)_4]$ and CH_3I, formed by the following sequence of reactions:

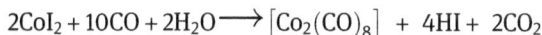

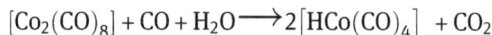

$$2CoI_2 + 10CO + 2H_2O \longrightarrow [Co_2(CO)_8] + 4HI + 2CO_2$$

$$[Co_2(CO)_8] + CO + H_2O \longrightarrow 2[HCo(CO)_4] + CO_2$$

$$CH_3OH + HI \longrightarrow CH_3I + H_2O$$

The role of these two catalytic species in the insertion of CO in CH_3OH leading to the production of acetic acid can be visualized as shown in the catalytic cycle given in Figure 3.6 in terms of the following three steps:

(i) Cobalt–alkyl bond formation,
(ii) insertion of CO in cobalt–alkyl bond and
(iii) cleavage of cobalt–acyl bond.

In the early 1970s Monsanto Company commercialized the use of Rh catalyst in the synthesis of CH_3COOH, and this appears to be technologically most successful example of homogeneous catalysis:

$$CH_3OH + CO \xrightarrow[180\ ^\circ C,\ 30-40\ atm]{RhI_2(CO)_2^-} CH_3COOH$$

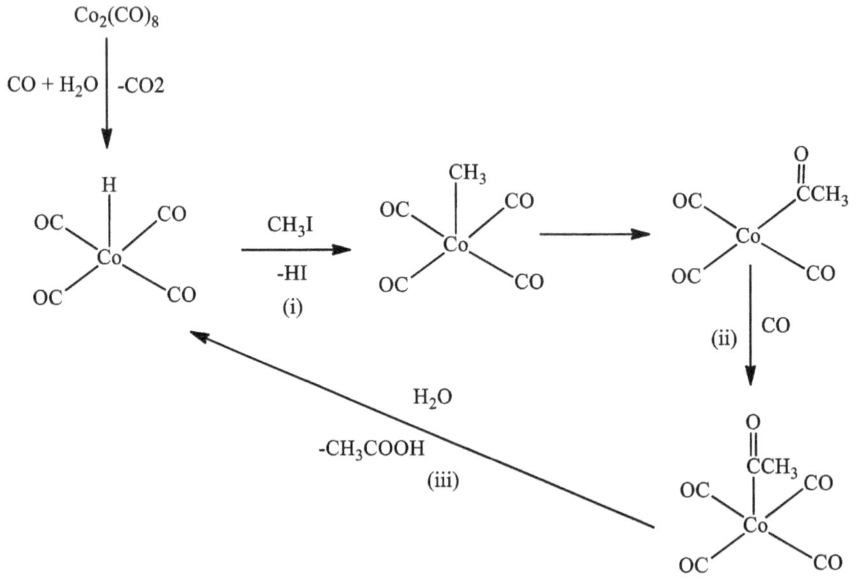

Figure 3.6: Synthesis of CH_3COOH in the presence of $[Co_2(CO)_8]$ (catalyst) and CH_3I (co-catalyst).

This is highly selective and extremely rapid reaction producing CH_3COOH in ~90% yields. The reaction path of the Monsanto acetic acid route has been studied in detail. The active species appears to be CH_3I formed from CH_3OH and HI,

$$CH_3OH + HI \longrightarrow CH_3I + H_2O$$

and diidodicarbonylrhodium anion formed by the interaction of Rh(III) iodide with CO:

$$RhI_3 + 3CO + H_2O \longrightarrow [RhI_2(CO)_2]^- + CO_2 + HI + H^+$$

The rate of reaction of CH_3COOH formation is first order with respect to Rh and CH_3I concentration, and above a certain CO pressure it is independent [11] of CO pressure. Asin the case of hydroformylation heterogenization of Rh was attempted for overcoming the problems of separation of the catalytic system, but no breakthrough has been reported. Efforts to exchange the extremely corrosive iodide acidic medium by other halides have been also without success yet.

The proposed catalytic mechanism involves a cycle of four basic steps shown in Figure 3.7.

(i) Oxidative addition,
(ii) migratory inclusion of alkyl to carbonyl,
(iii) coordination of CO and (iv) reductive elimination of acetyl iodide.

Finally water hydrolyses CH_3COI to CH_3COOH to complete the cycle:

$$CH_3COI + H_2O \longrightarrow CH_3COOH + HI$$

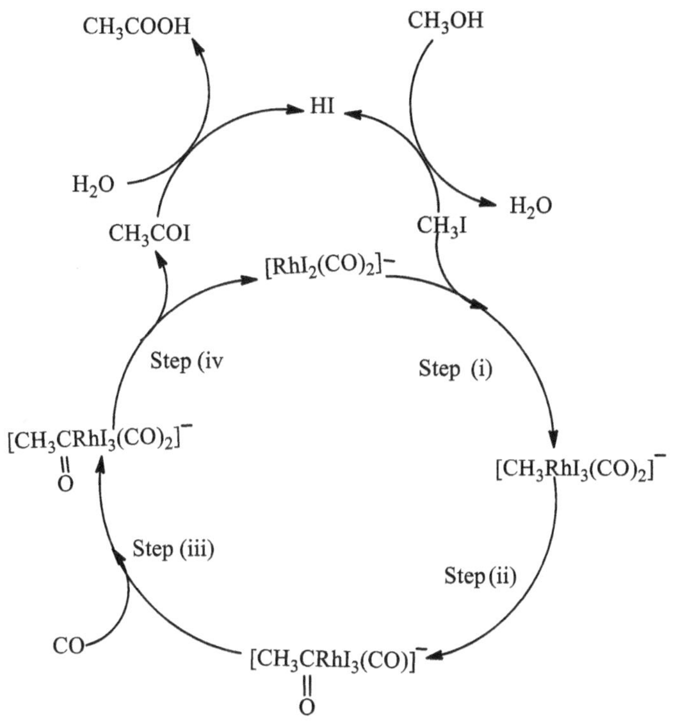

Figure 3.7: Catalytic cycle for the Rh-catalysed formation of CH_3COOH.

The rate-determining step is the oxidative addition of methyl iodide to the active catalyst $[RhI_2(CO)_2]^-$ forming the unstable, coordinatively saturated complex $[CH_3Rh1_3(CO)_2]^-$. CO insertion leads to the acetyl species $[CH_3CORh1_3(CO)_2]^-$, and thereafter the reductive elimination of CH_3COI and regeneration of the original active catalyst $[RhI_2(CO)_2]^-$. The completion of the catalytic cycle is obtained by the reaction of CH_3COI with H_2O to form CH_3COOH and HI, the latter reacting with CH_3OH to give the active CH_3I.

The use of platinum group carbonyls as homogeneous catalysts is increasing at a fast rate. Another example employing such catalysts is the homologation (i.e. increasing the chain length) of carboxylic acid as is shown in Figure 3.8, exemplifying the conversion of acetic acid using ruthenium carbonyl iodide as catalyst.

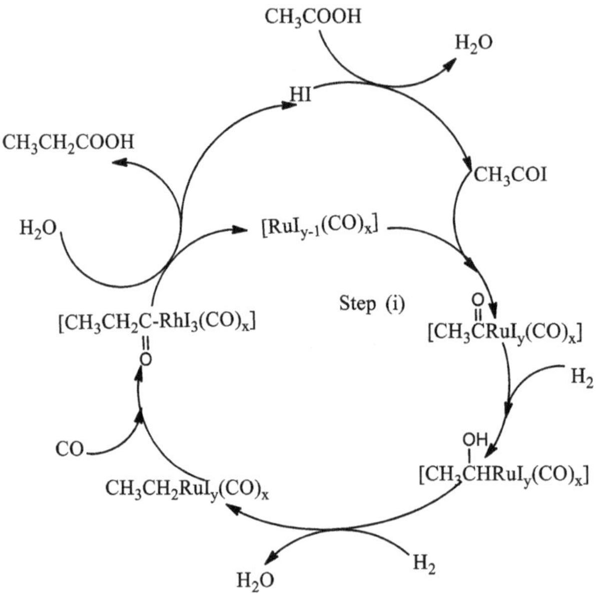

Figure 3.8: Homologation of acetic acid using ruthenium carbonyl iodide as a catalyst.

3.6.4 Manufacturing of acetic anhydride by carbonylation of CH_3COOCH_3

The carbonylation of carboxylic acid esters results in the formation of the respective acid anhydrides in the absence of water, for example, when using methyl acetate as feed, acetic anhydride is produced. This reaction was first described by Reppe (BASF) in 1951/52 using high partial pressures of CO with Co, Ni or Fe carbonyl complexes as catalysts [12].

After the discovery of the Rh-catalysed CH_3COOH production by Monsanto, new catalysts for the low-pressure carbonylation of esters have been described [13]. The process has been commercialized by Eastman Kodakwith Rh as a catalyst [14].

The process was first used in a plant, built on the site of coal mine, to convert synthesis gas to acetic anhydride. The latter is an important chemical used in acetylation reactions in the pharmaceutical industry. The Eastman process actually begins from coal which is gassified to get synthesis gas:

$$\text{Coal} \longrightarrow CO/H_2 \longrightarrow \text{methanol} \longrightarrow \text{methylacetate} \longrightarrow \text{aceticanhydride}$$

The catalytic cycle for the reaction has many similarities to the Monsanto process. The initial reactant has changed in the present process. Methanol (CH_3OH) is replaced by methyl acetate (CH_3COOCH_3), and acetate is the leaving group instead of hydroxyl group. It is notable that in the catalytic cycle, lithium iodide (LiI) replaces

HI (as water cannot be present in the synthesis of acetyl anhydride) which reacts with methyl acetate to yield methyl iodide.

After acquiring the Monsanto process, British Petroleum (BP) chemicals has also started coproduction of acetic acid and acetic anhydride from 1988. The process makes both these chemicals and generates methyl acetate in the process (Figure 3.9) using $[RhI_2(CO)_2]^-$ as a catalyst.

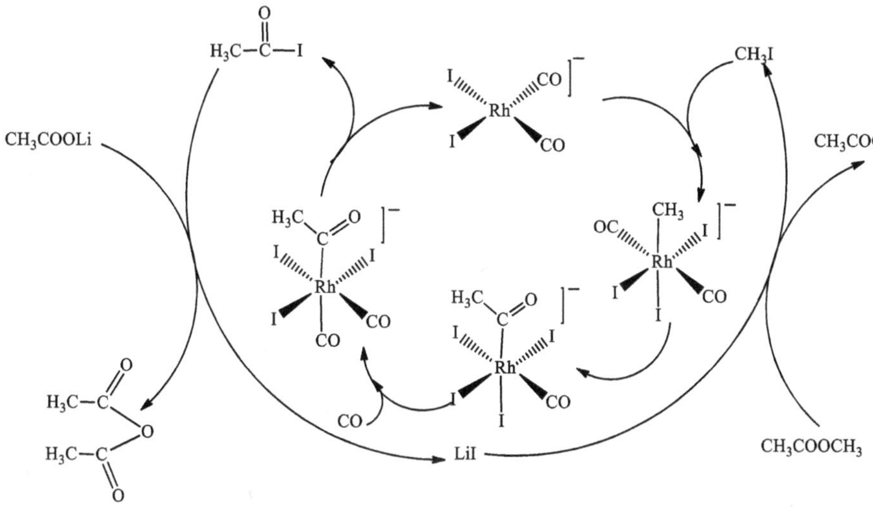

Figure 3.9: Catalytic cycle forEastman acetic anhydride process.

3.6.5 Importance of acetic anhydride

The main application of $(CH_3CO)_2O$ is in the large-scale production of cellulose acetate for photographic films and plastic materials. It is estimated that about 75% $(CH_3CO)_2O$ made per annum in the United States is used for this drive. About 1.5% of the yearly production is consumed in the synthesis of aspirin (acetylsalicylic acid) (Figure 3.10). Inter alia, uses of acetic anhydride are the manufacturing of industrial chemicals, pharmaceuticals, perfumes, plastics, synthetic fibres, explosives, weed killers and so on.

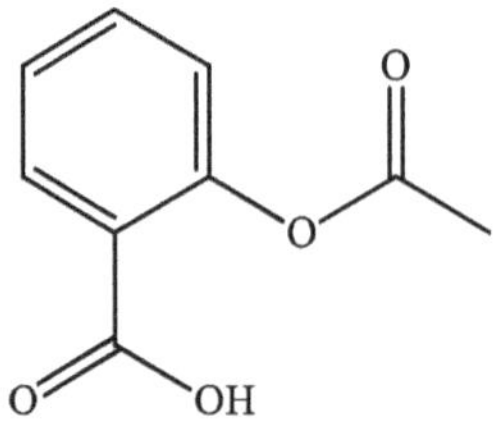

Figure 3.10: Structure ofaspirin.

3.6.6 Manufacturing of acetic acid by BP Cativa process using iridium carbonyl as a catalyst

A more cost-effective and efficient route called the Cativa method for the manufacturing of CH_3COOH came up from BP [15] in 1996. This process uses cheaper iridium carbonyl instead of rhodium carbonyl as a catalyst. The iridium carbonyl catalyst is found to remain stable under a wide range of conditions while the rhodium catalyst decomposes to inactive rhodium salt under such conditions. This process uses metal promoters such as InI_3 and $[Ru(CO)_3I_2]_2$. The mechanism of this process is shown in Figure 3.11.

The steps shown in catalytic cycle possibly occur in the presence of a promoter. An iodide ligand is transferred from the anionic iridium complex to the promoters. This creates a free site on the catalyst so that the third molecule of CO can bind to the catalyst. This species has been identified by IR spectral studies. It has been seen that the migratory insertion of CO occurs much more readily in the neutral complex than in the anionic precursor. Once this happens, an iodide can be transferred back to the iridium followed by reductive elimination. The promoter reduces the standing concentration of I^-, thus facilitating the loss of I^- from the catalyst. This is a slow step compared to the oxidative addition (step I). The addition of iodomethane to iridium, step I is ~150 times faster than the addition of iodomethane to rhodium. The promoter thus intervenes to speed up the slower step of the cycle.

Similar to the Monsanto method, the reaction is having 90% atom efficiency theoretically. In comparison of rhodium/iodide, using iridium/iodide as a catalyst has many benefits as follows:

(i) The cost of iridium is much less than rhodium.
(ii) The method is faster and more effective, demanding only small amount of catalyst.
(iii) This catalyst offers a very high turnover number, and thus reducing the frequency of catalyst replacement.
(iv) Because the catalyst iridium being more selective for methanol, it increases the overall produce and reduces by products. This results in lower purification costs and reduced waste.
(v) As compared to rhodium complexes, iridium complexes are more soluble in the reaction mixture. Consequently, the catalyst is not vanished by precipitation and does not require replacing so repeatedly.

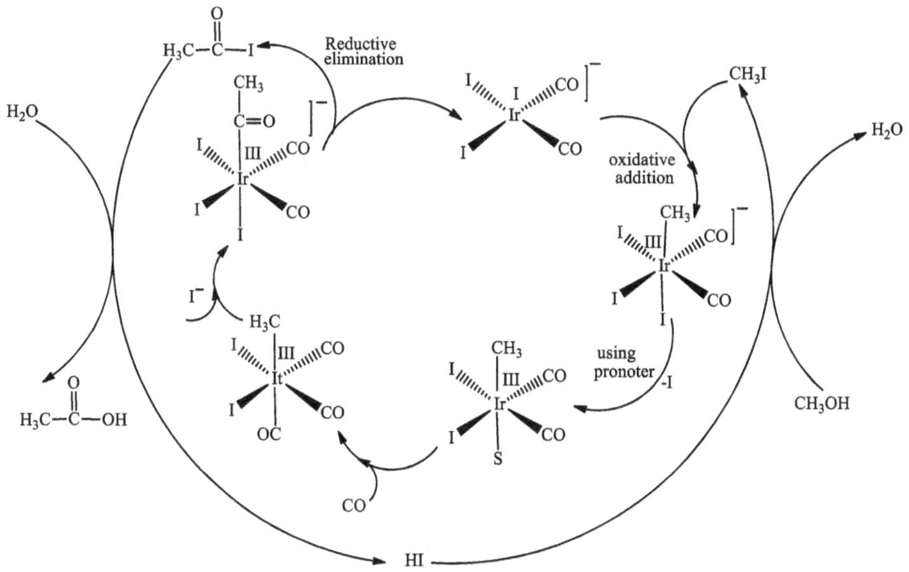

Figure 3.11: Catalytic cycle for BP's Cativa process using iridium carbonyl as a catalyst.

3.6.7 Carbonylation of olefins and acetylenes to carboxylic acids or esters or alcohols using metal carbonyls as a catalyst

Carbonyl compounds of Ni, Co, Rh, Pd, Pt, Ru and Fe are the most common catalysts, added either as metal carbonyls or produced in situ by reacting finely divided metals or metal salts with CO. The carbonyl complexes may be further modified by additional ligands, for example, trialkyl phosphines and tertiary amines in order to guide the reactions into a desired direction.

The most active catalyst for the carbonylation of acetylene is $Ni(CO)_4$ together with mineral acid. For the carbonylation of olefins, the carbonyls of Co, Rh and Ru are of similar activity as nickel carbonyl. In some cases, their activity may be even higher than that of $Ni(CO)_4$. The amount of linear or branched products formed during the reaction depends strongly on the catalyst composition. Starting from *n*-olefins mainly linear acids or derivatives are obtained when $Co_2(CO)_8$ or when $(R_3P)_2PdCl_2$ together with $SnC1_2$ as a co-catalyst is used as a catalyst, whereas mainly branched products are obtained in the presence of $Ni(CO)_4$ or $PdCl_2$ alone or $(R_3P)_2PdCl_2$ without a co-catalyst:

$$R-CH=CH_2+CO+H_2O \xrightarrow[(R_3P)_2PdCl_2+SnCl_2]{Co_2(CO)_8} R-CH_2-CH_2-COOH \text{ (linear product)}$$

$$R-CH=CH_2+CO+H_2O \xrightarrow[(R_3P)_2PdCl_2PdCl_2]{Ni(CO)_4} R-CH(CH_3)COOH \text{ (branch product)}$$

The possible variations in product distribution by changes in the catalyst system and/or in the reaction conditions are shown in the following reactions starting from butadiene [16–18] as an example for the carbonylation of conjugated dienes:

$$CH_2=CH-CH=CH_2+CO+CH_3OH \xrightarrow[135\ °C,\ 900\ bar]{Co_2(CO)_8} \underset{\text{Dimethyl adipate}}{H_3COOC-CH_2-CH_2-CH_2-CH_2-COOCH_3}$$

$$CH_2=CH-CH=CH_2+CO+H_2O \xrightarrow[200\ °C,\ 75\ bar]{RhCl_3+HI \text{ or } HBr} \underset{\text{Adipic acid}}{HOOC-CH_2-CH_2-CH_2-CH_2-COOH}$$

$$CH_2=CH-CH=CH_2+CO+ROH \xrightarrow[70\ °C,\ 300-700\ bar]{PdCl_2 \text{ or } (R_3P)_2PdCl_2} \underset{\text{3-Pentenoic acid ester}}{ROOC-CH_2-CH_2-CH_2-CH_2-COOR}$$

A further variant of the carbonylation reaction [19] is illustrated by the formation of alcohols from olefins with $Fe(CO)_5$ as a catalyst:

$$R-CH=CH_2+3CO+2H_2O \xrightarrow{Fe(CO)_5} R-CH_2-CH_2-CH_2OH+2CO_2$$

3.6.7.1 Mechanism

A reaction path for the carbonylation of olefins using catalytic quantities of $Co_2(CO)_8$ has been firstly proposed by Heck and Breslow [20] and is still generally accepted. A schematic illustration of the catalytic cycle, showing generation of $HCO(CO)_4$, addition of the olefin, inclusion of the olefin into the Co–H bond and attachment of CO into the Co–alkyl bond forming finally a CO–acyl group is presented in Figure 3.12. The Co–acyl bond is cleaved by the nucleophilic attack of water (or alcohols, etc.) to form an acid (or ester, etc.); hereby the catalyst is recovered.

Heck also gives a mechanistic explanation for the promotion of carbonylation by hydrogen halide in the presence of $Ni(CO)_4$ as catalyst. He assumes the development of an active species $HNi(CO)_2X$ by oxidative addition of HX to $Ni(CO)_4$. The resulting active species then serves as the active catalyst:

$$HX+Ni(CO)_4 \longrightarrow HNi(CO)_2X+2CO$$

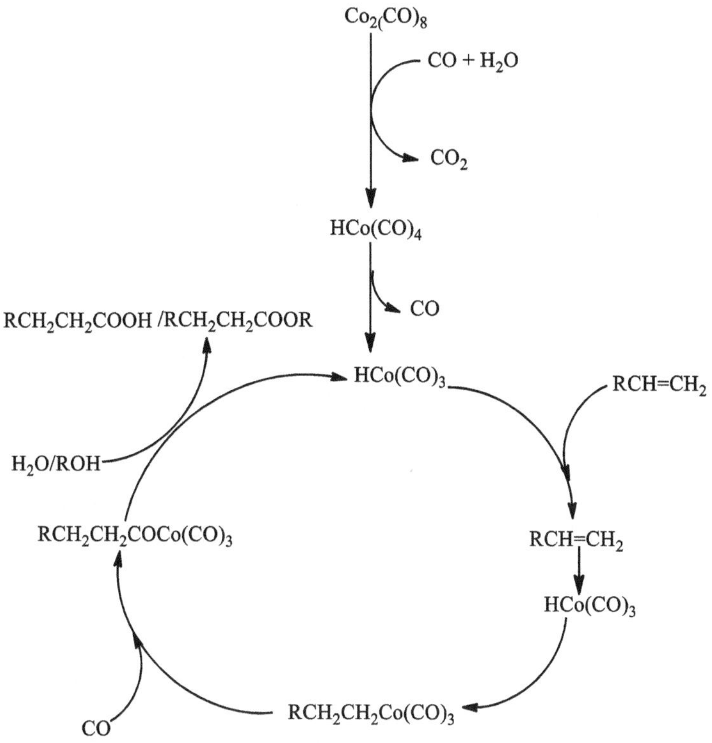

Figure 3.12: Schematic representation of the mechanism for the formation of acid/ester from olefins with CO.

3.7 Metal carbonyls: effective atomic number (EAN) rule

The number of electrons present in the atom of the metal-forming metal carbonyls plus those gained by CO molecules is called the effective atomic number (EAN) of the metals in metal carbonyls. This number in most of the metal carbonyls comes out to be equal to the atomic number of the next inert gas. This is called the EAN rule. Thus, most of the metal carbonyls obey EAN rule.

3.7.1 Mononuclear carbonyls having metallic atom with even atomic numbers

The EAN of some metal carbonyls of this category are as follows:

Metal carbonyl	No. of electrons of the metal	No. of electrons gained by COs	EAN of the next inert gas
$[Cr(CO)_6]$	24	6 × 2 = 12	24 + 12 = **36 (Kr)**
$[Fe(CO)_5]$	26	5 × 2 = 10	26 + 10 = **36 (Kr)**
$[Ni(CO)_4]$	28	4 × 2 = 08	28 + 8 = **36 (Kr)**
$[Mo(CO)_6]$	42	6 × 2 = 12	42 + 12 = **54 (Xe)**
$[Ru(CO)_5]$	44	5 × 2 = 10	44 + 10 = **54 (Xe)**
$[W(CO)_6]$	74	6 × 2 = 12	74 + 12 = **86 (Rn)**
$[Os(CO)_5]$	76	5 × 2 = 10	76 + 10 = **86 (Rn)**

Based on the EAN rule it can be explained why Ni does not form hexacarbonyl. This is because EAN in $[Ni(CO)_6]$ would be equal to 28 + 12 = 40, and this is not the atomic number of any noble gas.

3.7.2 Mononuclear carbonyls having metallic atom with odd atomic numbers

There are some metal carbonyls that contain metal atoms having odd atomic numbers. Such carbonyls do not obey EAN rule. Examples of such carbonyls are $[V(CO)_6]$, $[Mn(CO)_5]$, $[Mn(CO)_6]$ and $[Co(CO)_4]$. In order to obey EAN rule such carbonyls adopt any of the following ways:

(a) Such metal carbonyls form cationic or anionic species (called carbonylate ions) which obey EAN rule as follows:
   (i) $[V(CO)_6]$: EAN of V = 23 + 12 = 35
   $[V(CO)_6]^-$: EAN of V = (23 + 1) + 12 = **36 (Kr)**
   (ii) $[Mn(CO)_5]$: EAN of Mn = 25 + 10 = 35
   $[Mn(CO)_5]^-$: EAN of Mn = (25 + 1) + 10 = **36 (Kr)**
   (iii) $[Mn(CO)_6]$: EAN of Mn = 25 + 12 = 37
   $[Mn(CO)_6]^+$: EAN of Mn = (25–1) + 12 = **36 (Kr)**
   (iv) $[Co(CO)_4]$: EAN of Mn = 27 + 8 = 35
   $[Co(CO)_4]^-$: EAN of Mn = (27 + 1) + 8 = **36 (Kr)**

(b) Such metal carbonyls may form a single covalent bond with an atom (e.g. H and Cl) or a group (e.g. CH_3) which contributes one electron to the metal atom. Examples are:

(i) $Mn(CO)_5Cl$: EAN of Mn = 25 + 10 + 1 = **36 (Kr)**
(ii) $HMn(CO)_5$: EAN of Mn = 25 + 10 + 1 = **36 (Kr)**
(iii) $(CH_3)Mn(CO)_5$: EAN of Mn = 25 + 10 + 1 = **36 (Kr)**
(iv) $HCo(CO)_4$: EAN of Co = 27 + 8 + 1 = **36 (Kr)**

(c) Such metal carbonyls may dimerize to give dinuclear carbonyl containing metal–metal bond. The electron pair being shared between two metal atoms is counted for both the metal atoms (as in octet rule). The dinuclear carbonyls obtained by the dimerization of mononuclear carbonyls obey EAN rule as shown for $Mn_2(CO)_{10}$ add $Co_2(CO)_8$ which are dimerized form of $Mn(CO)_5$ and $Co(CO)_4$, respectively.

(i) $Mn_2(CO)_{10}$: Electrons from two Mn atoms = 25 × 2 = 50
: Electrons from 10 CO molecules = 10 × 2 = 20
: Electrons from one Mn–Mn bond = 1 × 2 = 2

EAN for two Mn atoms in $Mn_2(CO)_{10}$ = 72
∴ EAN per Mn atom = 72/2 = **36 (Kr)**

(ii) $Co_2(CO)_8$: Electrons from two Co atoms = 27 × 2 = 54
: Electrons from 8CO molecules = 8 × 2 = 16
: Electrons from one Co–Co bond = 1 × 2 = 2

EAN for two Co atoms in $Co_2(CO)_8$ = 72
∴ EAN per Co atom = 72/2 = 36 **(Kr)**

3.7.3 Polynuclear carbonyls

It has been seen that polynuclear carbonyls like $Mn_2(CO)_{10}$, $Co_2(CO)_8$, $Fe_2(CO)_9$ and $Fe_3(CO)_{12}$ follow EAN rule. EAN per metal atom in these carbonyls is calculated as follows:

(a) $Mn_2(CO)_{10}$

This compound contains one Mn–Mn bond as shown in its structure.

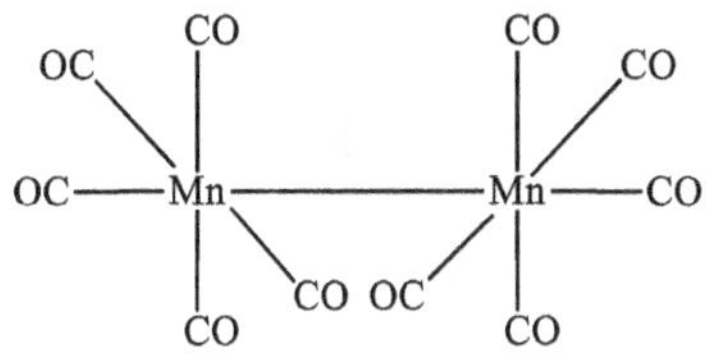

$Mn_2(CO)_{10}$

Electrons from two Mn atoms = 25 × 2 = 50
Electrons from 10 CO molecules = 10 × 2 = 20
Electrons from one Mn–Mn bond = 1 × 2 = 2

EAN for two Mn atoms in $Mn_2(CO)_{10}$ = 72
∴ EAN per Mn atom in $Mn_2(CO)_{10}$ = 72/2 = **36 (Kr)**

3.7.3.1 Another method

As each Mn atom is attached with five CO groups and with another Mn atom (one Mn–Mn bond), EAN in $Mn_2(CO)_{10}$ is given by

EAN of Mn atom = atomic number of Mn + electron donated by 5 CO + electron given by one Mn–Mn bond

= 25 + 10 + 1 = **36 (Kr)**

(b) $Co_2(CO)_8$

3.7.3.2 In solution

In solution, this carbonyl compound has a non-bridged structure shown below, in which each Co atom is coordinated with four CO moieties, and with another Co atom (one Co–Co bond):

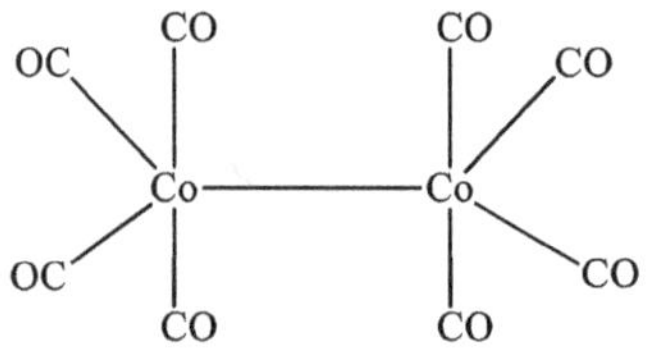

$Co_2(CO)_8$

Thus, EAN of each Co atom is given by

EAN of Co atom = atomic number of Co + [electron donated by 4 CO] + [electron given by

one Co–Co bond]

= 27 + 8 + 1 = **36 (Kr)**

3.7.3.3 In solid state

In solid state, this carbonyl compound has a bridged structure shown below, in which each Co atom is coordinated with three terminal CO moieties, two bridged CO groups and one Co atom (one Co–Co bond):

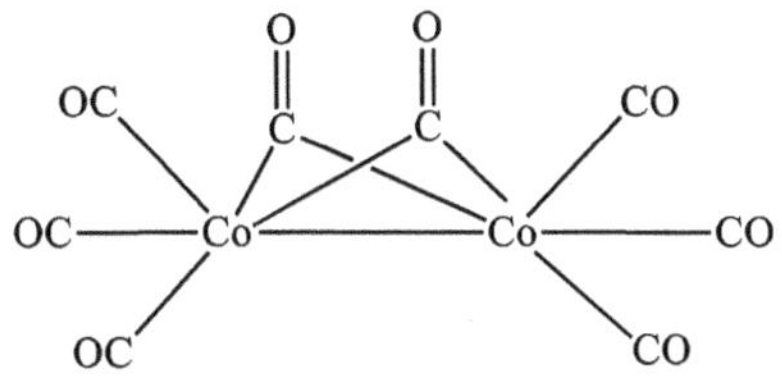

$Co_2(CO)_8$

Thus, EAN of each Co atom is given by

EAN of Co atom = atomic number of Co + [electron donated by three terminal CO] + [electron donated by two bridging CO groups] + [electron given by one Co–Co bond]

= 27 + 6 + 2 × 1 + 1 × 1 = **36 (Kr)**

(c) $Fe_2(CO)_9$

In this carbonyl compound, each Fe atom is coordinated with three terminal CO moieties, three bridging CO moieties and other Fe atom (one Fe–Fe bond). Thus, EAN of Fe atom is given by

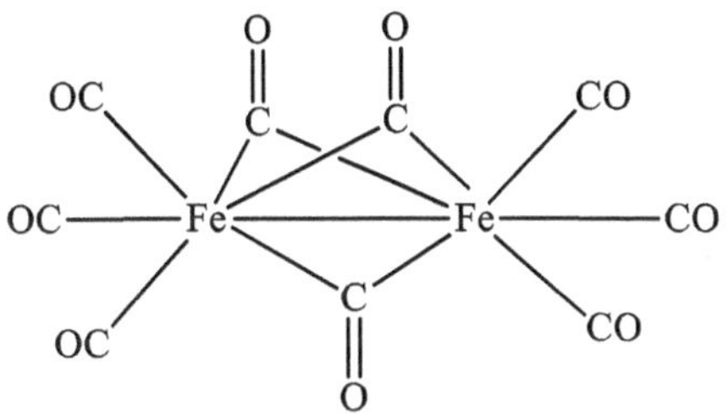

$Fe_2(CO)_9$

EAN of Fe atom = atomic number of Fe + [electron donated by three terminal CO] + [electron donated by three bridging CO groups] + [electron given by one Fe–Fe bond]

= 26 + 3 × 2 + 3 × 1 + 1 × 1 = **36 (Kr)**

(d) $Fe_3(CO)_{12}/Ru_3(CO)_{12}/Os_3(CO)_{12}$

As shown in the following structure, in these three metal carbonyls, each metal atom is coordinated with four terminal CO ligands and two other M-atoms with M–M bond:

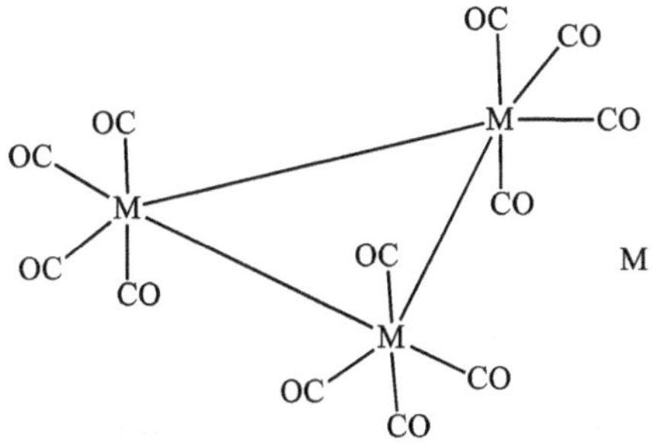

M = Fe, Ru or Os

Thus, EAN of metal atom (M) is given by

EAN of M atom = atomic number of metal + [electron donated by four terminal CO] + [electron given by two M–M bonds]

$Fe_3(CO)_{12}$:

EAN of Fe atom = 26 + 8 + 1 × 2 = **36 (Kr)**

$Ru_3(CO)_{12}$:

EAN of Ru atom = 44 + 8 + 1 × 2 = **54 (Xe)**

$Os(CO)_{12}$:

EAN of Ru atom = 76 + 8 + 1 × 2 = **86 (Rn)**

(e) $Co_4(CO)_{12}$/$Ir_4(CO)_{12}$

In these two tetrameric carbonyls, there are nine terminal CO groups, three bridging CO groups and **six M–M bonds.** Thus, EAN of four M atoms is given by

EAN of four M atom = atomic number of four M atoms + [electron donated by nine terminal CO] + [electron donated by three bridging CO] + [electron given by six M–M bonds]

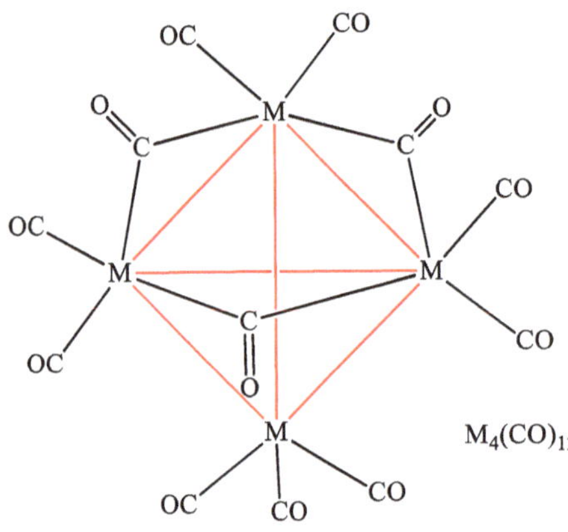

$M_4(CO)_{12}$ (M= Co or Ir)

EAN of four Co atom in **$Co_4(CO)_{12}$** = [4 × 27 + 18 + 6 + 12] = 108 + 36 = 144

∴ EAN per Co atom = 144/4 = **36 (Kr)**

EAN of four Co atom in **$Ir_4(CO)_{12}$** = [4 × 77 + 18 + 12 + 6] = 308 + 36 = 344

∴ EAN per Co atom = 344/4 = **86 (Rn)**

3.7.4 Utility of EAN rule

With the help of EAN rule, one can find out the number of metal–metal (M–M) bonds present in a given metal carbonyl. This can be described by determining the number of metal–metal bonds in $Fe_2(CO)_9$, $Fe_3(CO)_2$, $Co_4(CO)_{12}$ and [(η[5]-C_5H_5)Mo$(CO)_3$].

(a) $Fe_2(CO)_9$

Electron from two Fe atoms = 2 × 26 = 52
Electron from nine CO group = 2 × 9 = 18

EAN for two Fe atoms in $Fe_2(CO)_9$ = 52 + 18 = 70
∴ EAN for one Fe atom = 70/2 = 35

Since this EAN is less than 36(Kr) by one electron, there is **one Fe–Fe bond** in $Fe_2(CO)_9$ molecule.

(b) $Fe_3(CO)_{12}$

Electron from three Fe atoms = 3 × 26 = 78
Electron from 12 CO group = 2 × 12 = 24

EAN for three Fe atoms in $Fe_3(CO)_{12}$ = 78 + 24 = 102
∴ EAN for one Fe atoms = 102/3 = 34

Since this EAN is less than 36(Kr) by two electrons, Each Fe atom in $Fe_3(CO)_{12}$ molecule is linked with other two Fe atoms. Thus, $Fe_3(CO)_{12}$ has **three Fe–Fe bonds:**

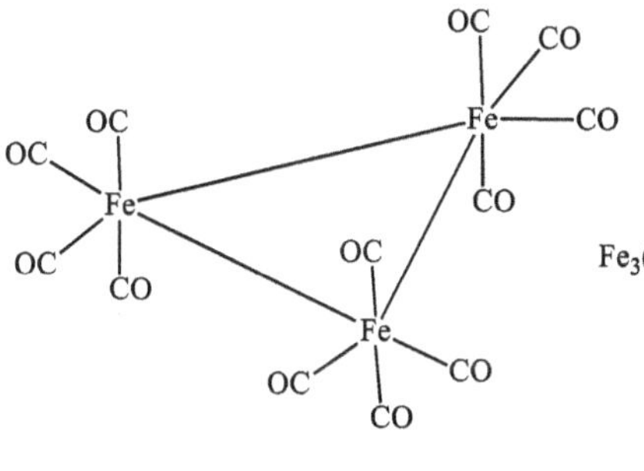

$Fe_3(CO)_{12}$

(c) $Co_4(CO)_{12}$

Electron from four Co atoms = 4 × 27 = 108
Electron from 12 CO group = 2 × 12 = 24

EAN for four Co atoms in $Co_4(CO)_{12}$ = 108 + 24 = 132
∴ EAN for one Co atom = 132/4 = 33

Since this EAN is less than 36(Kr) by three electrons, each Co atom in $Co_4(CO)_{12}$ molecule is linked with other three Co atoms. Thus, $Co_4(CO)_{12}$ has **six Co–Co bonds:**

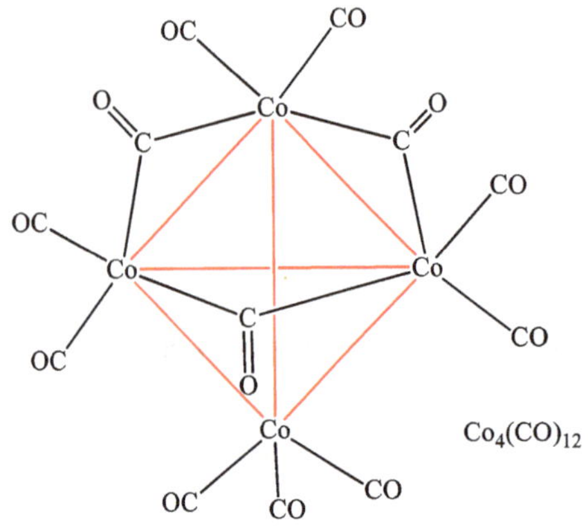

$Co_4(CO)_{12}$

(d) $[(\eta^5\text{-}C_5H_5)Mo(CO)_3]_2$

Electron from two Mo atoms = 2 × 42 = 84
Electron from six CO group = 2 × 6 = 12
Electron from two C_5H_5 group = 2 × 5 = 10

EAN for two Mo atoms in $[(\eta^5\text{-}C_5H_5)Mo(CO)_3]_2 = 106$
∴ EAN for one Mo atom = 106/2 = 53

Since this EAN is less than 54(Xe) by one electron, there is one Mo–Mo bond in the given mixed carbonyl.

3.8 Eighteen-electron rule for metal carbonyls

Since the valence electrons count in the noble gases is equal to 18 [Kr ($3d^{10}4s^24p^6$); Xe ($4d^{10}5s^25p^6$); Rn ($5d^{10}6s^26p^6$)], the noble gas formalism may be simplified to the 18-electron rule. According to this rule,

"In a stable metal carbonyl, the total (i.e. effective) number of electrons present in the valance shell of **one metal atom** is equal to 18. This number (i.e. 18) is equal to:

EN/n [EN = Effective number of electrons]

Here

EN = No. of electrons in the valence shell of metal in the given metal carbonyl = [No. of electrons in the valence shell of metal in the Free State] + [No. of electrons from CO groups/ligands + [2 × No. of M–M bonds in metal carbonyl] and n = No. of metal (M) atoms in metal carbonyl"

Examples

(a) Metal carbonyls obeying 18e rule:

The calculations of the effective number of electrons in the valence shell (V.S.) of respective one metal atom in metal carbonyls, such as $Ni(CO)_4$, $Fe(CO)_5$, $Cr(CO)_6$, $Mo(CO)_6$, $Fe_2(CO)_9$, $Co_2(CO)_8$, $Mn_2(CO)_{10}$, $Fe_3(CO)_{12}$ and $Co_4(CO)_{12}$, have shown that this number comes out to be equal to 18. Thus, these carbonyls obey 18e rule.

(i) $[Ni(CO)_4]$:Ni ($3d^8 4s^2$) = 10e
No. valence electrons in Ni = 10e
No. electrons from four CO groups = 8e

EN of electrons in the V.S. of Ni atom in $Ni(CO)_4$ = 18e

(ii) $[Fe(CO)_5]$:Fe ($3d^6 4s^2$) = 8e
No. valence electrons in Fe = 8e
No. electrons from five CO groups = 10e

EN of electrons in the V.S. of Fe atom in $Fe(CO)_5$ = 18e

(iii) $[Cr(CO)_6]$:Cr ($3d^5 4s^1$) = 6e
No. valence electrons in Cr = 6e
No. electrons from six CO groups = 12e

EN of electrons in the V.S. of Fe atom in $Cr(CO)_6$ = 18e

(iv) $[Mo(CO)_6]$:Mo ($4d^5 5s^1$) = 6e
No. valence electrons in Mo = 6e
No. electrons from six CO groups = 12e

EN of electrons in the V.S. of Mo atom in $Mo(CO)_6$ = 18e

(v) $[Fe_2CO)_9]$:Fe ($3d^6 4s^2$) = 8e
No. valence electrons in two Fe = 2 × 8e = 16e
No. electrons from nine CO groups = 9 × 2e = 18e
Electron from one Fe–Fe bond = 1 × 2e = 2e

EN of electrons in the V.S. of two Fe atoms in $Fe_2(CO)_9$ = 36e
EN of electrons in the V.S. of one Fe atom in $Fe_2(CO)_9$ = 36e/2 = 18e

(vi) $[Co_2CO)_8]$:Co ($3d^7 4s^2$) = 9e
No. valence electrons in two Co = 2 × 9e = 18e
No. electrons from eight CO groups = 8 × 2e = 16e
Electron from one Co–Co bond = 1 × 2e = 2e

EN of electrons in the V.S. of two Co atoms in $Co_2(CO)_8$ = 36e
EN of electrons in the V.S. of one Co atom in $Co_2(CO)_8$ = 36e/2 = 18e

(vii) $[Mn_2CO)_{10}]$:Mn ($3d^5 4s^2$) = 7e
No. valence electrons in two Mn = 2 × 7e = 14e
No. electrons from 10 CO groups = 10 × 2e = 20e
Electron from one Mn–Mn bond = 1 × 2e = 2e

EN of electrons in the V.S. of two Mn atoms in $Mn_2(CO)_{10}$ = 36e
EN of electrons in the V.S. of one Mn atom in $Mn_2(CO)_{10}$ = 36e/2 = 18e

(viii) $[Fe_3CO)_{12}]$:Fe ($3d^64s^2$) = 8e

No. valence electrons in three Fe = 3 × 8e = 24e

No. electrons from 12 CO groups = 12 × 2e = 24e

Electron from three Fe–Fe bond = 3 × 2e = 6e

EN of electrons in the V.S. of three Fe atoms in $Fe_3(CO)_{12}$ = 54e

EN of electrons in the V.S. of one Fe atom in $Fe_3(CO)_{12}$ = 54e/3 = 18e

(ix) $[Co_4CO)_{12}]$:Co ($3d^74s^2$) = 9e

No. valence electrons in four Co = 4 × 9e = 36e

No. electrons from 12 CO groups = 12 × 2e = 24e

Electron from six Co–Co bonds = 6 × 2e = 12e

EN of electrons in the V.S. of four Co atom in $Co_4(CO)_{12}$ = 72e

EN of electrons in the V.S. of one Co atom in $Co_4(CO)_{12}$ = 72e/4 = 18e

(b) Metal carbonyls not obeying 18e rule:

If we calculate the effective number of electrons in the valence cell of V and Co in $V(CO)_6$ and Co $(CO)_4$ molecules, respectively, we find that this number is equal to 17 as follows:

(i) $V(CO)_6$: V($3d^34s^2$) = 5e

No. valence electrons in V = 5e

No. electrons from six CO groups = 12e

EN of electrons in the V.S. of V atom in $V(CO)_6$ = 17e

(ii) $Co(CO)_4$: Co($3d^74s^2$) = 9e

No. valence electrons in Co = 9e

No. electrons from four CO groups = 8e

EN of electrons in the V.S. of Co atom in $Co(CO)_4$ = 17e

Since the effective number of electrons in the V.S. of V and Co is 17, $V(CO)_6$ and $Co(CO)_4$, both are paramagnetic.

In order to obey an 18-electron rule, $Co(CO)_4$ dimerizesto form the dimeric molecule, $Co_2(CO)_8$, in which each Co atom has 18 electrons in its V.S. as shown below. This dimeric molecule has one Co–Co bond as is evident from its structure as follows:

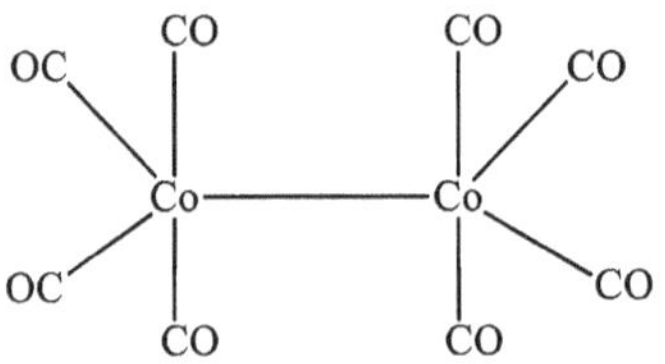

$Co_2(CO)_8$

$[Co_2CO)_8]$: Co ($3d^74s^2$) = 9e

No. valence electrons in two Co = 2 × 9e = 18e

No. electrons from eight CO groups = 8 × 2e = 16e

Electron from one Co–Co bond = 1 × 2e = 2e

EN of electrons in the V.S. of two Co atoms in $Co_2(CO)_8$ = 36e

EN of electrons in the V.S. of one Fe atom in $Co_2(CO)_8$ = 36e/2 = 18e

The structure of $Co_2(CO)_8$ shows that since the C.N. of each Co is five (maximum C.N. = Co atom is six), no extra steric strain would be created on $Co(CO)_4$ by increasing the C.N. from four in $Co(CO)_4$ to five in $Co_2(CO)_8$ during the dimerization of $Co(CO)_4$ through Co–Co bond. Hence, $Co_2(CO)_8$ would be a stable molecule.

The dimeric molecule of $V(CO)_6$, namely, $V_2(CO)_{12}$ also obeys 18e rule as shown below, if its structure is assigned as $(CO)_6$–V–V–$(CO)_6$.

[$V_2CO)_{12}$]: V ($3d^34s^2$)	= 5e
No. valence electrons in 2 V = 2 × 5e	= 10e
No. electrons from 12 CO groups = 12 × 2e	= 24e
Electron from one V–V bond = 1 × 2e	= 2e
EN of electrons in the V.S. of 2 V atoms in $V_2(CO)_{12}$	= 36e
EN of electrons in the V.S. of one V atom in $V_2(CO)_{12}$ = 36e/2	= 18e

The expected structure of $V_2(CO)_{12}$ is $(CO)_6$–V–V–$(CO)_6$ which shows that each V atom has C.N. equal to seven (maximum C.N. = Co atom is six). Due to the increase in C.N. from six [in $V(CO)_6$] to seven [in $V_2(CO)_{12}$], a lot of steric hindrance would be created on $V(CO)_6$ moiety and hence the dimer $V_2(CO)_{12}$ would be unstable. Thus, although $V_2(CO)_{12}$ obeys 18e rule, it is unstable. It is stable only in a CO matrix at extremely low temperatures.

(a) Mixed metal carbonyls obeying 18e rule:

Some mixed carbonyl complexes containing CO and organic molecules as co-ligands which obey 18-electron rule are shown below:

(i) [$Fe(CO)_3(C_4H_4)_2$]: Fe ($3d^64s^2$)	= 8e
No. valence electrons in Fe	= 8e
No. electrons from three CO groups	= 6e
No. electrons from C_4H_4 group	= 4e
EN of electrons in the V.S. of Fe atom	= 18e
(ii) [$Fe(CO)_4(PPh_3)$]: Fe ($3d^64s^2$)	= 8e
No. valence electrons in Fe	= 8e
No. electrons from four CO groups	= 8e
No. electrons from PPh_3 group	= 2e
EN of electrons in the V.S. of Fe atom	= 18e
(iii) [$Mn(CO)_5(CH_3)$]: Mn ($3d^54s^2$)	= 7e
No. valence electrons in Mn	= 7e
No. electrons from five CO groups	= 10
No. electrons from CH_3 group	= 1e
EN of electrons in the V.S. of Mn atom	= 18e
(iv) [$Mn(CO)_5(C_2H_4)$]$^+$: Mn^+ ($3d^54s^1$)	= 6e
No. valence electrons in Mn^+	= 6e
No. electrons from five CO groups	= 10e
No. electrons from C_2H_4 group	= 2e
EN of electrons in the V.S. of Mn^+ atom	= 18e

(v) $[Re(CO)_5(CH_3)]$: Re ($5d^56s^2$)	= 7e
No. valence electrons in Re	= 7e
No. electrons from five CO groups	= 10e
No. electrons from CH_3 group	= 1e
EN of electrons in the V.S. of Re atom	= 18e
(vi) $[Cr(CO)_3(\pi\text{-}C_6H_6)]$: Cr ($3d^54s^1$)	= 6e
No. valence electrons in Cr	= 6e
No. electrons from three CO groups	= 6e
No. electrons from π-C_6H_6group	= 6e
EN of electrons in the V.S. of Re atom	= 18e
(vii) $[(\eta^5\text{-}C_5H_5)Mo(CO)_3]_2$: Mo ($5d^56s^1$)	= 6e
This mixed carbonyl contains one Mo–Mo bond.	
No. valence electrons in two Mo	= 12e
No. electrons from six CO groups	= 12e
No. electrons from two (η^5-C_5H_5) group	= 10e
No. electrons from one Mo–Mo bond	= 2e
EN of electrons in the V.S. of two Mo atoms	= 36e
EN of electrons in the V.S. of one Mo atom = 36e/2	= 18e

3.8.1 Eighteen-electron rule: square planar complexes

As in square planar complexes, the C.N. of the central metal atom is four, the effective number of electrons in the valance cell of the d^8 metal atom in square planar complexes equals to: $8 + 4 \times 2 = 16$. This number shows that these complexes do not obey 18-electron rule.

Although 18-electron rule is not obeyed by square planar complexes, yet they are quite stable complexes. The stability of such complexes has been explained on the basis of crystal field theory as explained below.

In square planar complexes, splitting pattern of d-orbitals of central metal atom is shown in Figure 3.13.

The $d_{x^2-y^2}$ orbital has the highest energy and is an anti-bonding orbital. Energy gap (Δ) between d_{xy} and $d_{x^2-y^2}$ orbitals is very large. It may be seen from Figure 3.13 that d^8 electrons of the metal atom in square planar complex can be distributed as follows:

$$(d_{xz})^2 = (d_{yz})^2, (d_{z^2})^2, (d_{xy})^2 (d_{x^2-y^2})^0$$

Since the energy gap (Δ) between d_{xy} and $d_{x^2-y^2}$ orbitals is very large, the entry of ninth electron in the high-energy anti-bonding orbital is energetically unfavourable. Consequently, although square planar complexes of d^8 metal atom do not obey 18-electron rule, yet these complexes are quite stable.

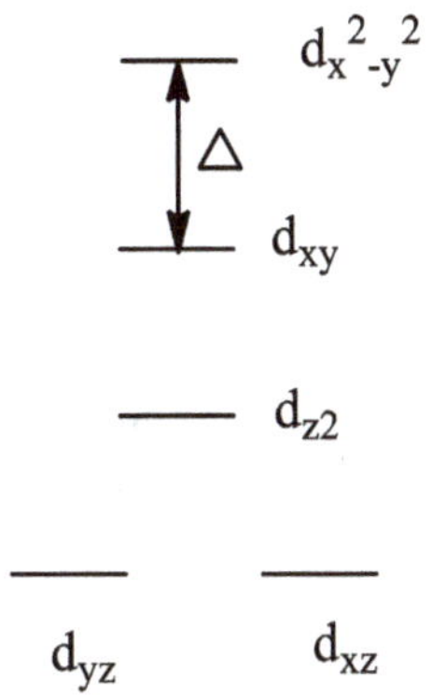

Figure 3.13: d-Orbitals splitting pattern in square planar complexes.

3.9 Types of bond present in metal carbonyls

Normally four types of bonds are found in different metallic carbonyls. Their formations are being explained here:

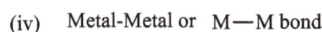

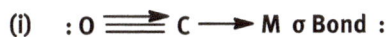

(i) $:O \overset{\rightarrow}{\equiv} C \rightarrow M$ σ Bond

(ii) $M \rightarrow CO$ π bond

(iii) 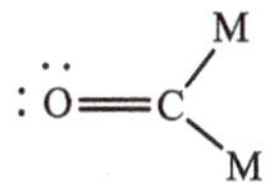

(iv) Metal-Metal or M—M bond

(i) $:O \overset{\rightarrow}{\equiv} C \rightarrow M$ σ Bond :

In this bond, CO acts as a terminal carbonyl group. This type of bond is found in all types of carbonyls. The formation of this type of bond is explained on the basis of valence bond theory as follows:

Before explaining the formation of OC→M σ-bond, one should know the structure of CO molecule on the basis of valence bond theory (Figure 3.14).

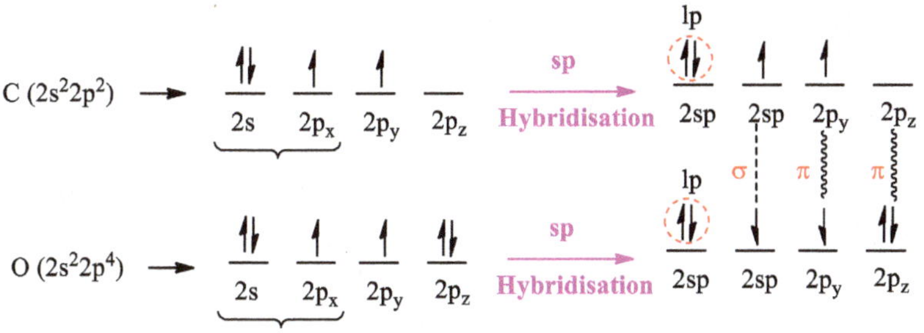

Figure 3.14: Sp hybridization in ($:O \overset{\rightarrow}{=} C:$).

It is evident from the sp hybridization shown in Figure 3.14, the lp of electrons on carbon and oxygen atoms is present in their respective sp-hybrid orbital. When a CO molecule gets linked as a terminal carbonyl group with a transition metal in a low oxidation state (0, −1 or + 1), lp of electrons present in sp-hybrid orbital of C-atom is donated to the appropriate vacant hybrid orbital of the metal atom [namely, sp^3 hybrid orbital of Ni in $Ni(CO)_4$, dsp^3 hybrid orbital of Fe in $Fe(CO)_5$, etc.] to form OC→M coordinate σ-bond. It is notable here that only the lp of electrons on C atom, and not the O atom, is donated to the vacant hybrid orbital of the metal atom. This is because of the fact that C atom is less electronegative than O atom.

(ii) M ⟶ CO π bond

The formation of this bond, that is, backdonation of electrons from metal to CO, and also OC→M σ-bond can be elucidated by molecular orbital theory (MOT), which will be discussed later on under the head bonding in metal carbonyls.

(iii) 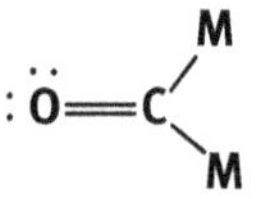

This type of bond is found in bridged polynuclear carbonyls like $Co_2(CO)_8$ in solid state, $Fe_2(CO)_9$, $Fe_3(CO)_{12}$, $Co_4(CO)_{12}$, $Ru_4(CO)_{12}$ and so on. In these carbonyls, CO molecule acts as a bridging (or ketonic) carbonyl group

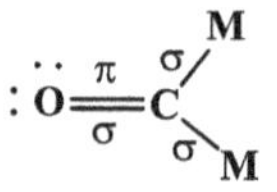

This bridging molecule system contains two C–M σ-bonds, one O–C σ-bond and one O–C π-bond. The formation of these σ- and π-bonds is shown in the following hybridization scheme (Figure 3.15).

Two half-filled sp^2 hybrid orbitals on C-atom overlap with half-filled appropriate hybrid orbital of two metal atoms and give two C–M σ-bonds. The remaining third sp^2 hybrid orbital overlaps with the appropriate 2p orbital (namely, $2p_x$) of the O-atom and forms C–O σ-bond. The O–C π-bond results in the overlap of the $2p_z$ orbitals of both C- and O-atoms.

(iv) Metal–metal or M—M bond

There are many carbonyls that contain one or more M–M bonds. Examples of carbonyl compounds containing M–M bonds are $Fe_2(CO)_9$, $Co_2(CO)_8$, $Mn_2(CO)_{10}$ and so on. Such type of bond is produced by weak overlap of the two singly filled hybrid orbitals of the two combining metal atoms. Due to weak coupling of electron spin, this bond is called fractional single bond and is represented as M–M bond. This bond is larger than the single M–M bond. The presence of M–M bond makes all the electrons paired and hence the carbonyls containing M–M bonds are diamagnetic.

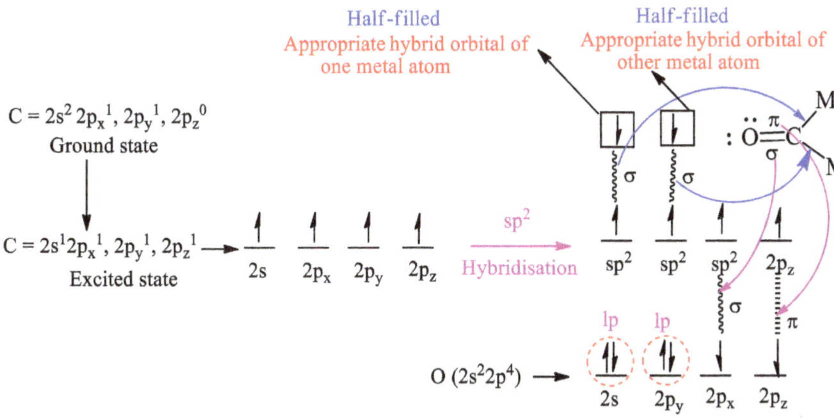

Figure 3.15: Sp^2 hybridization and formation of σ- and π-bonds in bridged polynuclear carbonyls.

3.10 Structure of metal carbonyls: valence bond (VB) approach

(i) Nickel tetracarbonyl, $Ni(CO)_4$

(a) Studies carried out on the vapour density of this compound and the freezing point of its benzene solution have shown that this metal carbonyl is of molecular formula $Ni(CO)_4$.

(b) Both the vapour-phase electron diffraction and the solid-state X-ray diffraction studies of this compound have shown that it has a tetrahedral structure (Figure 3.14) having Ni–C–O units as linear. The observed Ni–C bond length in this molecule is 1.50 Å. This bond length is shorter by 0.32 Å compared to Ni–C single bond length of 1.82 Å. The observed C–O bond length in this compound is 1.15 Å, which is longer by 0.022 Å compared to C–O bond length of 1.128 Å in CO molecules.

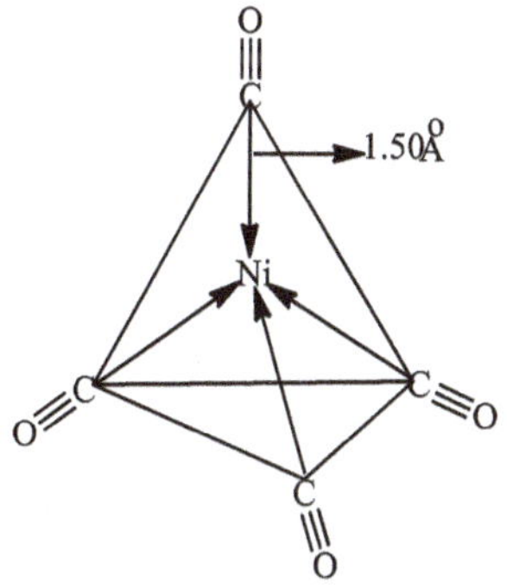

Figure 3.16: Tetrahedral structure of $Ni(CO)_4$.

(c) Tetrahedral geometry of this compound results from sp^3 hybridization (Figure 3.17) of Ni present in it as the central atom.

Here, in this hybridization scheme, the formation of O ≡ C→Ni bond is the result of the overlapping between the empty sp^3 hybrid orbital on Ni atom and doubly filled sp hybrid orbital on C-atom in CO molecule.

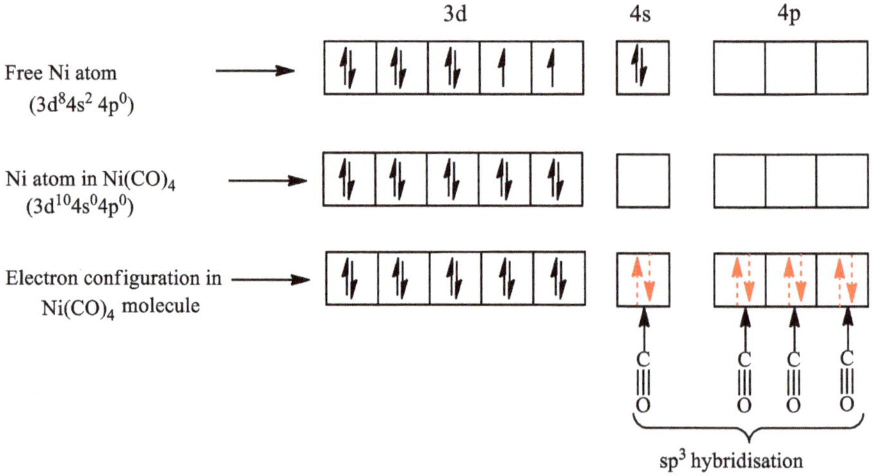

Figure 3.17: Sp^3 hybridization of Ni atom in $NiCO)_4$. Electron pair donated by CO molecule is shown by ↑↓.

(d) It is notable here that the formation of four O ≡ C→Ni bonds accumulates a large negative charge on the central Ni atom, and this is most unlikely. In such a situation, Pauling suggested that the formation of double bond between Ni and C atoms results in the backdonation of d-electron from Ni atom to CO ligands to such a magnitude that electronegativity principle is followed. Consequently, electron pair is not dispersed equally between Ni and C atoms of CO ligands but is attracted more strongly towards C atom as C is more electronegative than Ni atom (C = 2.55; Ni = 1.91). This prevents accommodation of negative charge on Ni atom.

(ii) Iron pentacarbonyl, $Fe(CO)_5$

(a) Principally, $[Fe(CO)_5]$ may have any of the two structures, namely, trigonal bipyramidal (TBP) and square pyramidal as shown in Figure 3.18. In order to decide the actual structure, the number of IR- and Raman-active vibrational bands can be calculated for each of the two geometries. Now, we record the IR and Raman spectra of this compound and tally the number of bands actually observed by calculation for both the geometries.

(b) It has been found that IR band actually observed tallies well with the number of bands calculated for TBP geometry. Thus, $[Fe(CO)_5]$ has TBP geometry, which has also been confirmed by X-ray studies. It is surprising to note that ^{13}C-NMR

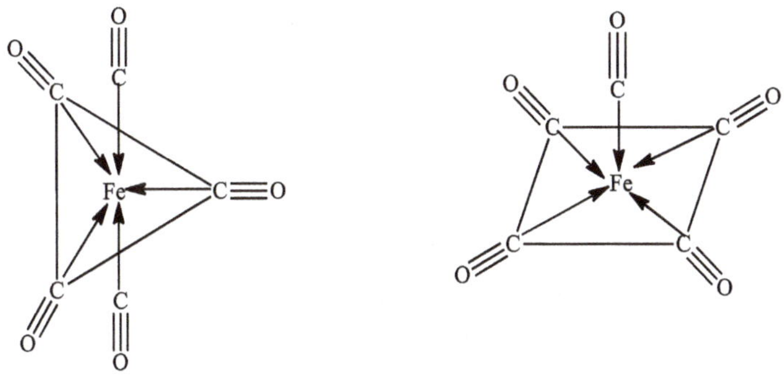

Trigonal bipyramidal structure of $Fe(CO)_5$

Square pyramidal structure of $Fe(CO)_5$

Figure 3.18: Structures of $Fe(CO)_5$.

gives only one type of signal for axial as well as equatorial CO groups. This is probably because of the fluxionality in the structure. This means that the axial and equatorial O ≡ C→Fe bonds exchange their position very fast so that an average ^{13}C resonance signal is obtained for all the five carbonyl groups attached with Fe atom. Vapour-phase electron diffraction study of $[Fe(CO)_5]$ molecule established that Fe–C axial and Fe–C equatorial bond lengths are 1.797 Å and 1.842 Å, respectively. On the other hand, X-ray study has shown that these bonds are almost of the same length.

(c) The TBP shape of $[Fe(CO)_5]$ results from dsp^3 hybridization (Figure 3.19) of Fe atom as shown below:

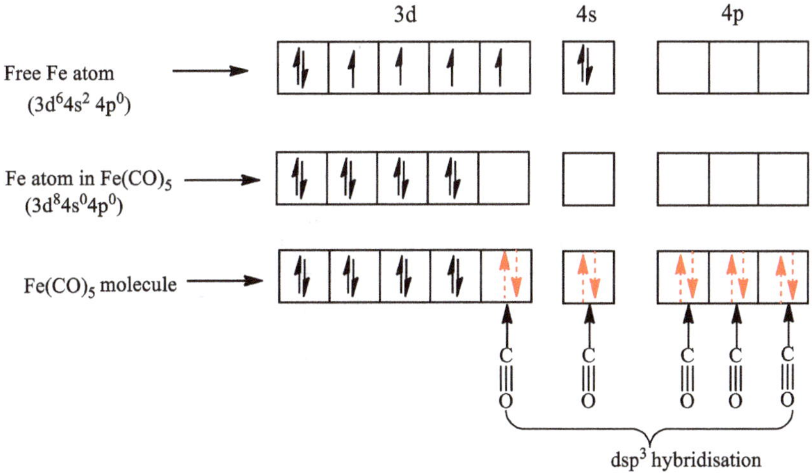

Figure 3.19: dsp^3 hybridization of Fe atom in $Fe(CO)_5$.

In this hybridization scheme, O ≡ C→Fe bond results by the overlap between the empty dsp^3 hybrid orbital on Fe atom and doubly filled sp hybrid orbital on C-atom in a CO molecule. The molecule is diamagnetic as all the electrons in d-orbitals as well in all five dsp^3 hybrid orbitals are in paired state.

(iii) Chromium hexacarbonyl, [Cr(CO)$_6$]

(a) Electron diffraction study of this molecule in the vapour state has shown that it has an octahedral shape (Figure 3.20). This study also confirms that the Cr–C and C–O bond lengths are 1.92 Å and 1.16 Å, respectively.

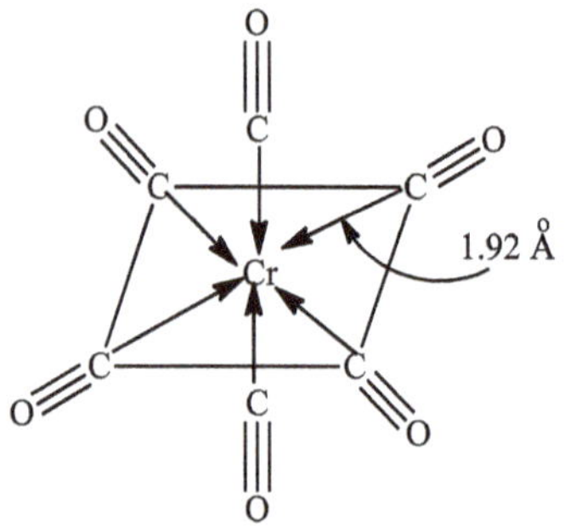

Figure 3.20: Octahedral structure of $Cr(CO)_6$ molecule.

(b) Octahedral geometry of [Cr(CO)$_6$] suggests that Cr-atom which is in zero oxidation state is d^2sp^3 hybridized as shown in Figure 3.21. Since [Cr(CO)$_6$] is diamagnetic, all the six electrons present in the V.S. of Cr atom ($3d^54s^1$) get paired in three 3d orbitals. Now, $3d_{x^2-y^2}$, $3d_{z^2}$, 4s and three 4p orbitals hybridize together to form six d^2sp^3 hybrid orbitals.

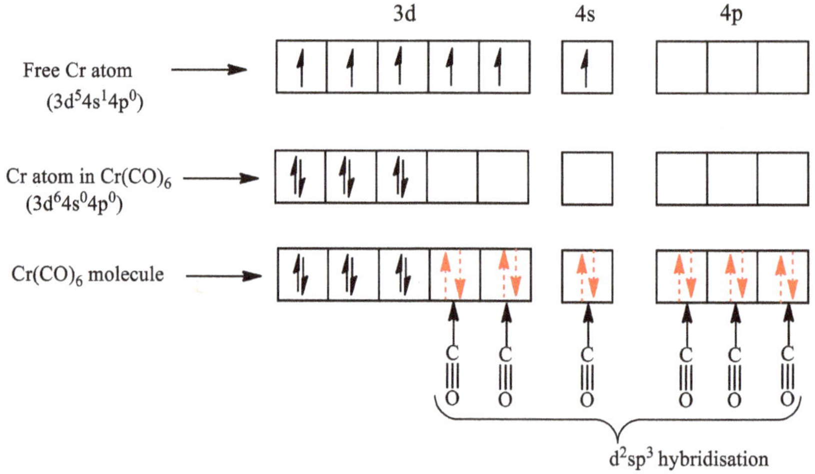

Figure 3.21: d^2sp^3 hybridization of Cr atom in $Cr(CO)_6$. Electron pair donated by CO molecule is shown by coloured electrons.

Here, O≡C→Cr bond is formed by the overlap of the empty d^2sp^3 hybrid orbital on Cr atom and doubly filled sp hybrid orbital on C atom in CO molecule.

(c) Similarly, the formation of octahedral geometry of $[Mo(CO)_6]$ and $[W(CO)_6]$ may be understood on the basis of d^2sp^3 hybridization.

(iv) Structures of $Mn_2(CO)_{10}$, $Tc_2(CO)_{10}$ and $Re_2(CO)_{10}$

All these molecules have the similar structure. As an example, let us consider the structure of $Mn_2(CO)_{10}$.

The IR spectral and X-ray diffraction studies made on $Mn_2(CO)_{10}$ have shown that each Mn atom is directly linked to other Mn atom by a σ-bond (Mn–Mn σ-bond) and to five terminal carbonyl groups (–C ≡ O) with coordinate bonds (Mn←—CO bonds) as shown in Figure 3.22. The Mn–Mn bond distance has been found to be 2.79 Å. The presence of Mn–Mn bond is also reinforced by the diamagnetic nature of the molecule.

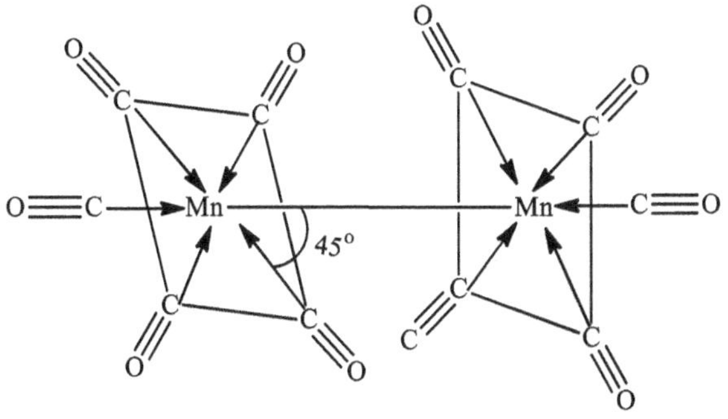

Figure 3.22: Structure of $Mn_2(CO)_{10}$: two $Mn(CO)_5$ units are staggered to each other.

It may be seen that the C.N. of each Mn atom is six and the molecule has no bridging carbonyl group in between two Mn atoms. Since C.N. of each Mn atom is six, both Mn atoms are d^2sp^3 hybridized as shown in Figure 3.23.

As seen, Mn–Mn σ-bond results by the overlap of half-filled d^2sp^3 hybrid orbital of one Mn atom with half-filled d^2sp^3 hybrid orbital of other Mn atom. The presence of no unpaired electron suggests that the molecule is diamagnetic.

(v) Structure of $Co_2(CO)_8$

(a) Non-bridged structure in solution

The IR spectral study of the solution of $Co_2(CO)_8$ has shown that this molecule has no bridging carbonyl group, that is, this molecule in solution has non-bridged structure (Figure 3.24). In this structure, each Co atom is coordinated with four terminal carbonyl ligands (Co←C ≡ O) and with other Co atom by a Co–Co σ-bond. The presence of Co–Co σ-bond has also been supported by the diamagnetic nature of this molecule. The Co–Co bond distance determined by electron diffraction study is found to be 2.52 Å.

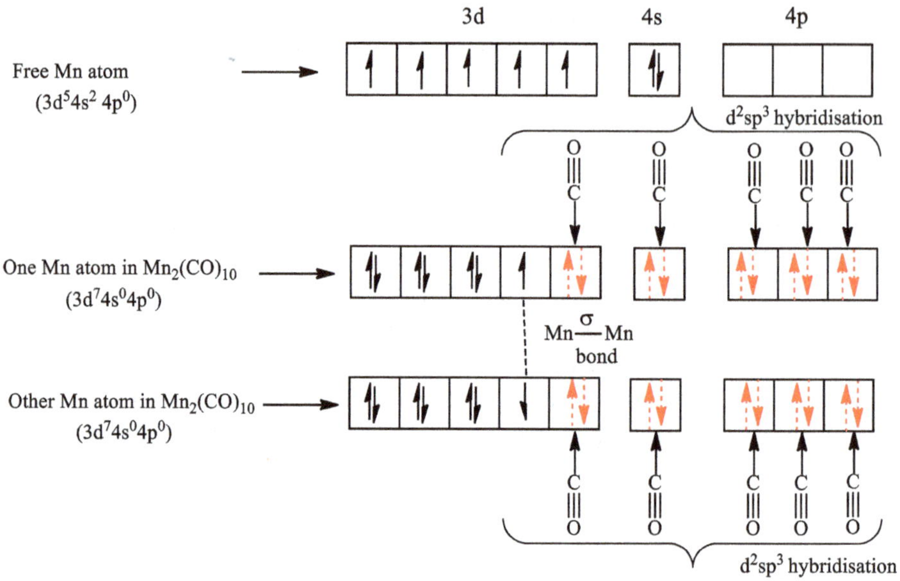

Figure 3.23: d^2sp^3 hybridization scheme for each Mn atom in $Mn_2(CO)_{10}$.

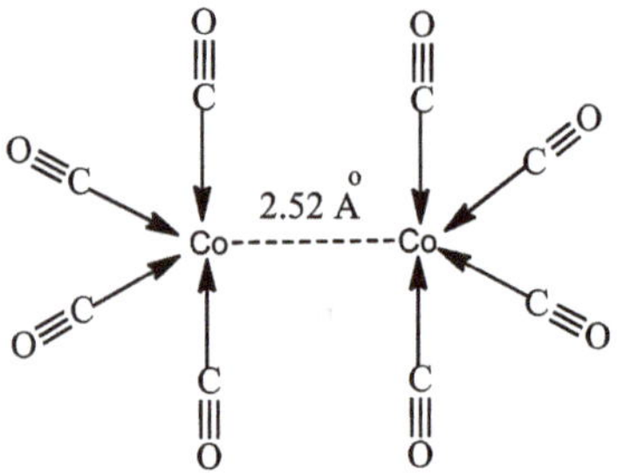

Figure 3.24: Non-bridged structure of $Co_2(CO)_8$.

Since it is evident from this structure that the C.N. of each Co atom is five, hence both Co atoms are dsp^3 hybridized as shown in the hybridizing scheme (Figure 3.25).

As seen from the above hybridization scheme, Co–Co σ-bond results by the overlap of half-filled dsp^3 hybrid orbital of one Co atom with half-filled dsp^3 hybrid orbital of other Co atom. The presence of no unpaired electron suggests that the molecule is diamagnetic.

(b) Bridged structure in solid state

(i) IR spectral study of $Co_2(CO)_8$ molecule in solid state has suggested that this compound has a bridged structure as given in Figure 3.26.

(ii) It is evident from Figure 3.26 that the bridged structure has two bridging carbonyl groups, six terminal carbonyl groups and one Co–Co bond. Each of the two Co atoms is attached with the other Co atom directly by a Co–Co

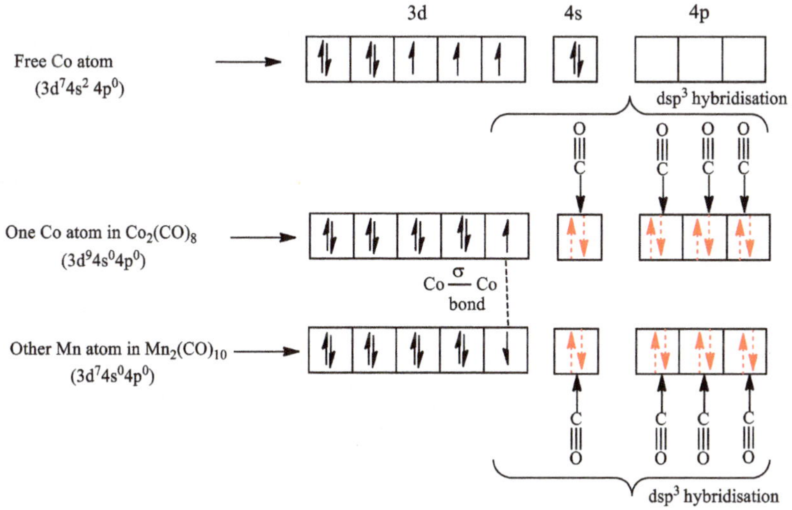

Figure 3.25: dsp^3 hybridization scheme of each Co atom in $Co_2(CO)_8$.

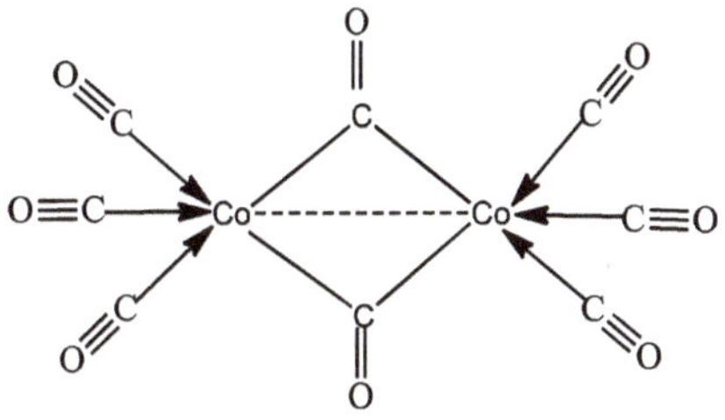

Figure 3.26: $Co_2(CO)_8$ involving bridged structure.

linkage, with two bridging carbonyl groups and three terminal CO groups. The Co–Co bond length is 2.52 Å.

(iii) In this structure, C.N. of each Co atom is six, and hence both Co atoms are d^{2}sp^3 hybridized as shown in Figure 3.27.

V.S. electronic configuration of Co atom in the free state is $3d^74s^24p^0$. When Co atom forms $Co_2(CO)_8$, one of the two electrons of 4s orbital is shifted to 3d orbital and thence the V.S. electronic configuration of Co atom becomes $3d^84s^14p^0$. Now two 3d orbitals such as $3d_{x^2-y^2}$ and $3d_{z^2}$ both singly filled, one singly filled 4s orbital and all three empty 4p orbital combine together to produce six d^{2}sp^3 hybrid orbitals. This is shown in (a) and (b) parts of the hybridization scheme (Figure 3.27). Now the formation of one Co–Co σ-bond, linking two bridging carbonyl groups with two Co atoms and the formation of three terminal CO groups with each Co atom takes place as shown in the scheme. As all the electrons in the $Co_2(CO)_8$ molecule are paired, the molecule has a diamagnetic character.

It is notable here that the Co–Co σ-bond, which is obtained through overlap of two singly occupied d^2sp^3 hybrid orbitals on two Co atoms, is bent because of the unusual overlap of two d^2sp^3 hybrid orbitals as shown in Figure 3.28.

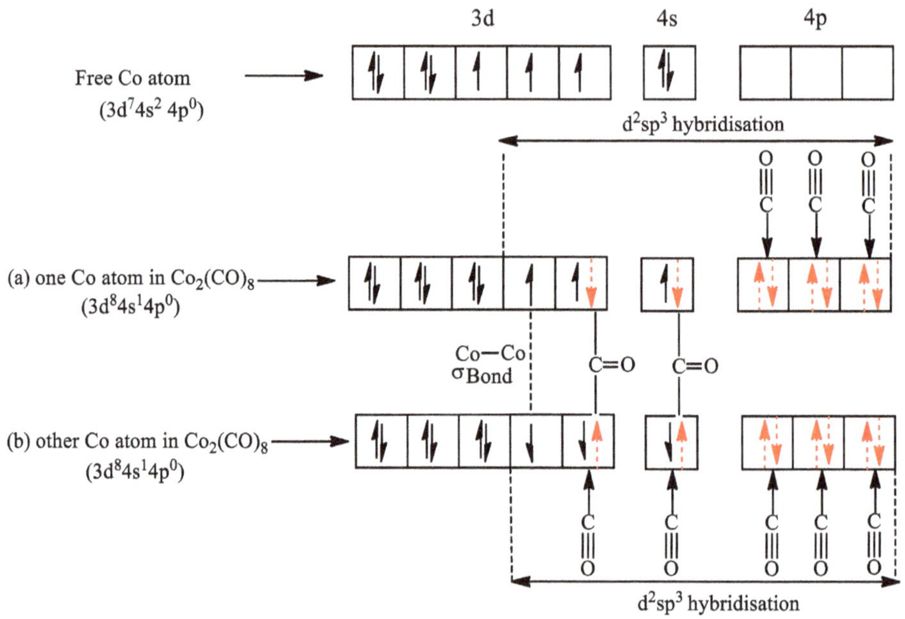

Figure 3.27: d^2sp^3 hybridization scheme of each Co atom in $Co_2(CO)_8$.

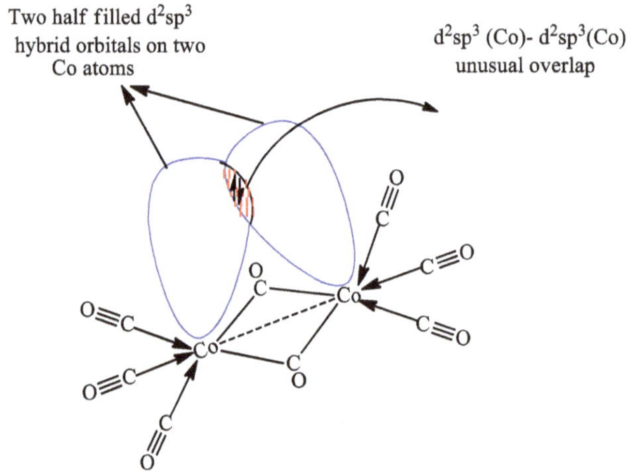

Figure 3.28: Unusual overlaps of two singly occupied d^2sp^3 hybrids on two Co atoms to form Co–Co bond in $Co_2(CO)_8$ molecule in solid state.

(iv) The two isomeric forms have little difference in their energies and exist in equilibrium with each other in a given solution:
Bridged structure $\rightleftharpoons$ Non-bridged structure
At very low temperature, the bridged structure predominates and as the temperature is raised, the non-bridged structure predominates.

(v) X-ray diffraction studies of the crystals of $Co_2(CO)_8$ has shown that the molecule in solid state has a bridged structure. On the basis of electron diffraction studies, various bond lengths found are:
Co–Co = 2.52 Å, Co–C bridging = 1.92 Å, Co–C terminal = 0.80 Å, bridging C–O = 1.21 Å, terminal C–O = 1.17 Å.

(vi) Structures of $Fe_2(CO)_9$, $Ru_2(CO)_9$ and $Os_2(CO)_9$

(a) $Fe_2(CO)_9$: IR and X-ray studies made on $Fe_2(CO)_9$ molecule have shown that each Fe atom is directly connected with other Fe atom by a Fe–Fe σ-bond, to three bridging carbonyl groups (>C=O) and three terminal Co groups (–C ≡ O) by a coordinate bond (Figure 3.29).Fe–Fe bond length = 2.46 Å. The terminal C–O bond lengths are smaller than the bridging bond lengths. The diamagnetic character of the molecule reinforced the availability of Fe–Fe bond in it.

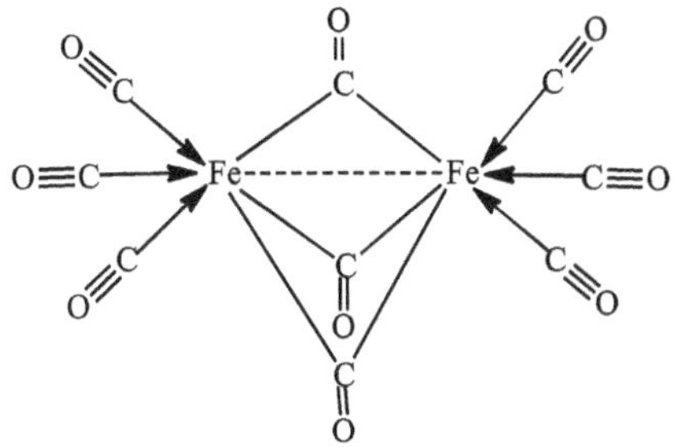

Figure 3.29: $Fe_2(CO)_9$ involving terminal and bridging carbonyl groups.

As the C.N. of each Fe atom is 7, d^3sp^3 hybridization (Figure 3.30) may be assumed to explain its formation.

(b) $Ru_2(CO)_9$ and $Os_2(CO)_9$: Unlike $Fe_2(CO)_9$,these two dinuclear carbonyls acquire structures with one bridging carbonyl group involving M–M bond as shown in Figure 3.31.
As the C.N. of each metal is 6, d^2sp^3 hybridization (Figure 3.32) may be assumed in the formation of these two dinuclear carbonyls.

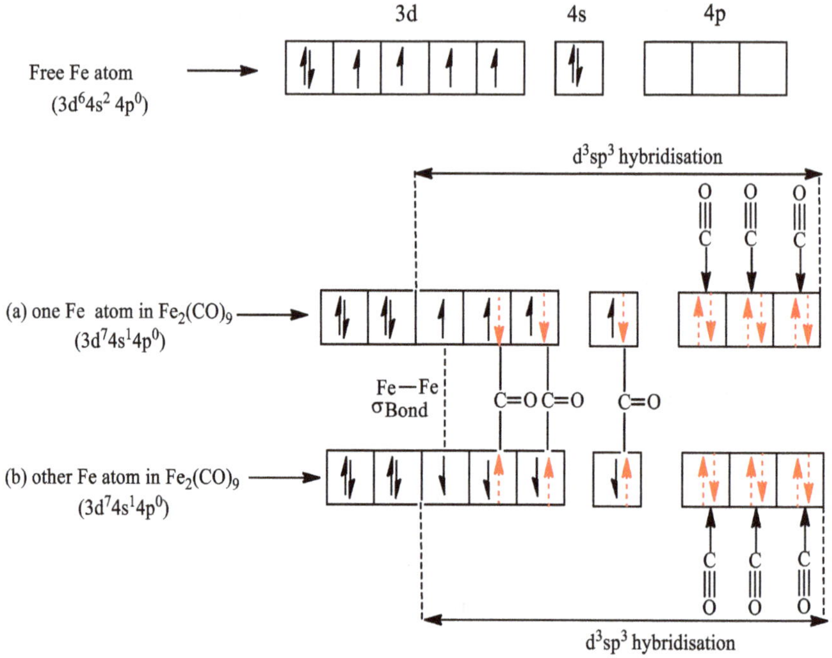

Figure 3.30: d^3sp^3 hybridization scheme of Fe in $Fe_2(CO)_9$. Electron pair donated by terminal CO group is shown by coloured electron pair (↑↓) while↑ or ↓ (coloured) represents an electron donated by C atom of bridging >C = O group.

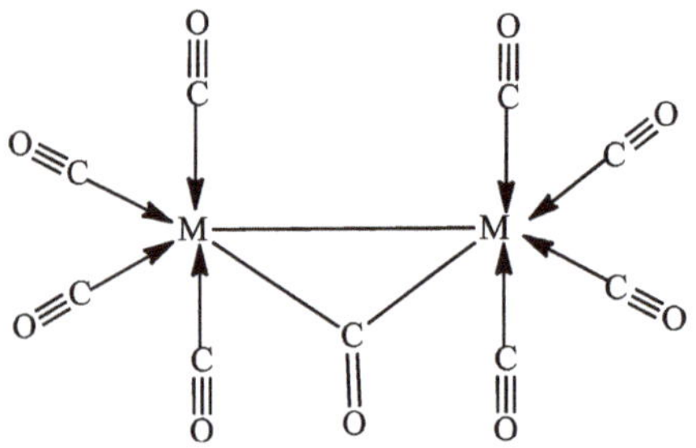

Figure 3.31: Single bridged structure of $M_2(CO)_9$ (M = Ru or Os).

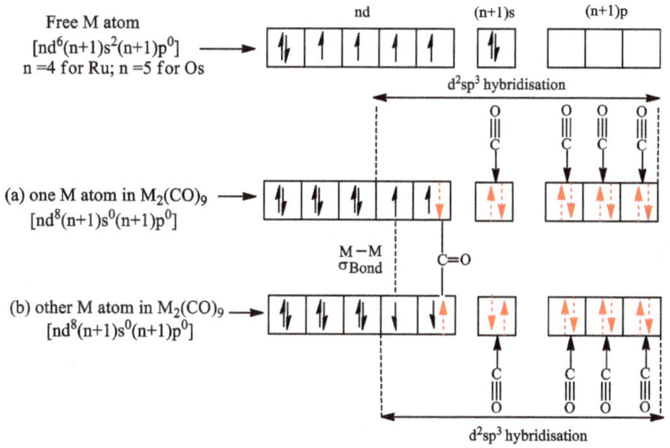

Figure 3.32: d^2sp^3 hybridization scheme of M in $M_2(CO)_9$. Electron pair donated by the terminal CO group is shown by coloured electron pair (↓↑) while ↑ or ↓ (coloured) represents an electron donated by C atom of bridging >C=O group.

(vii) Structures of $Fe_3(CO)_{12}$, $Ru_3(CO)_{12}$ and $Os_3(CO)_{12}$

(a) $Fe_3(CO)_{12}$

X-ray study made on $Fe_3(CO)_{12}$ molecule has shown the following in this molecule:

(i) Of the three Fe atoms, two Fe atoms are coordinated with three terminal CO groups, attached to two bridging CO groups and to the third Fe atom.

(ii) The remaining third Fe atom is coordinated with four-terminal CO groups and is linked to each of the two Fe atoms.

(iii) Three Fe–Fe bonds are of equal length (2.8 Å).

These features of $Fe_3(CO)_{12}$ molecule is shown in Figure 3.33.

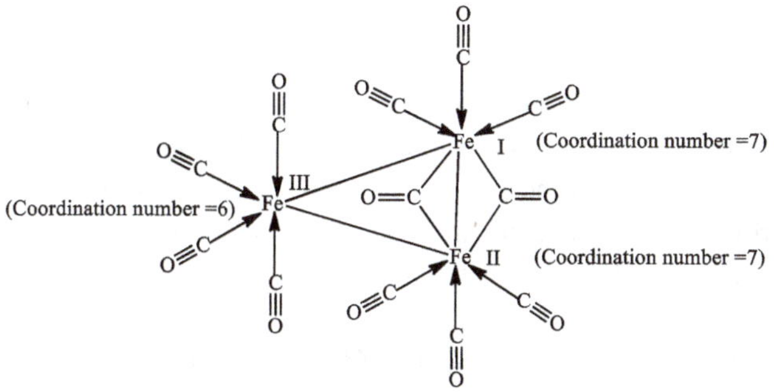

Figure 3.33: Structure of $Fe_3(CO)_{12}$.

Considering the C.N. 7, 7 and 6 of Fe^{I}, Fe^{II} and Fe^{III}, respectively, the following hybridization scheme (Figure 3.34) may be assumed in the formation of $Fe_3(CO)_{12}$ molecule.

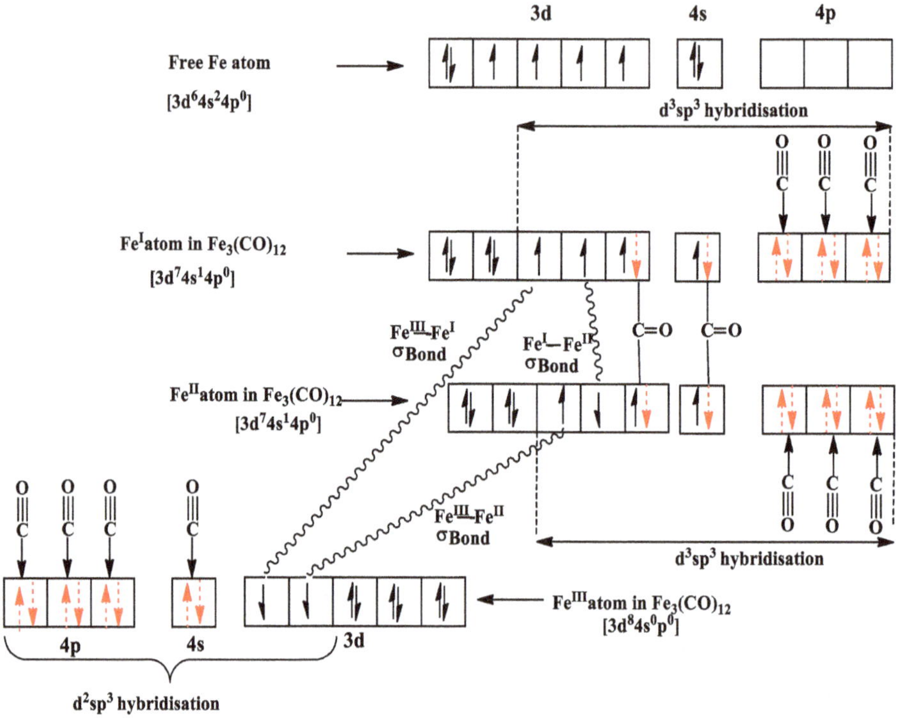

Figure 3.34: Hybridization scheme of Fe atom in $Fe_3(CO)_{12}$. Electron pair donated by terminal CO group is shown by coloured electron pair (↓↑) while ↑ or ↓ (coloured) represents an electron donated by C atom ofbridging >C=O group.

(b) $Ru_3(CO)_{12}$ and $Os_3(CO)_{12}$

X-ray diffraction studies have shown that these two trinuclear carbonyls contain no bridging CO group and each metal centre contains four terminally bonded CO group, and also bonded to two other metal atoms as shown in Figure 3.35. Thus, the C.N. of each metal centre is 6, and hence d^2sp^3 hybridization scheme shown in Figure 3.36 may be assumed in the formation of these two trinuclear carbonyls.

(viii) Structure of tetranuclear carbonyls: $Co_4(CO)_{12}$, $Rh_4(CO)_{12}$ and $Ir_4(CO)_{12}$

There are two types of structures for these molecules, each consisting of a tetrahedron of metal atoms. For $Ir_4(CO)_{12}$, the deployment of CO groups (Figure 3.37a) is such as to conserve the full symmetry of the tetrahedron. There are three terminal CO groups on each metal atom with Ir–CO bond approximately *trans* to the Ir–Ir bond.

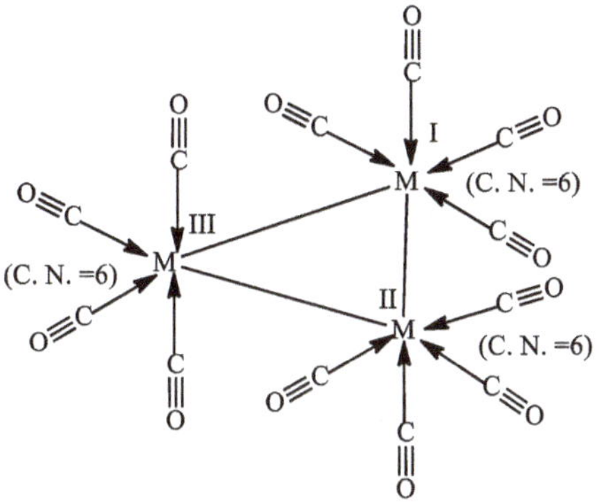

Figure 3.35: Structure of $M_3(CO)_{12}$ (M = Ru or Os).

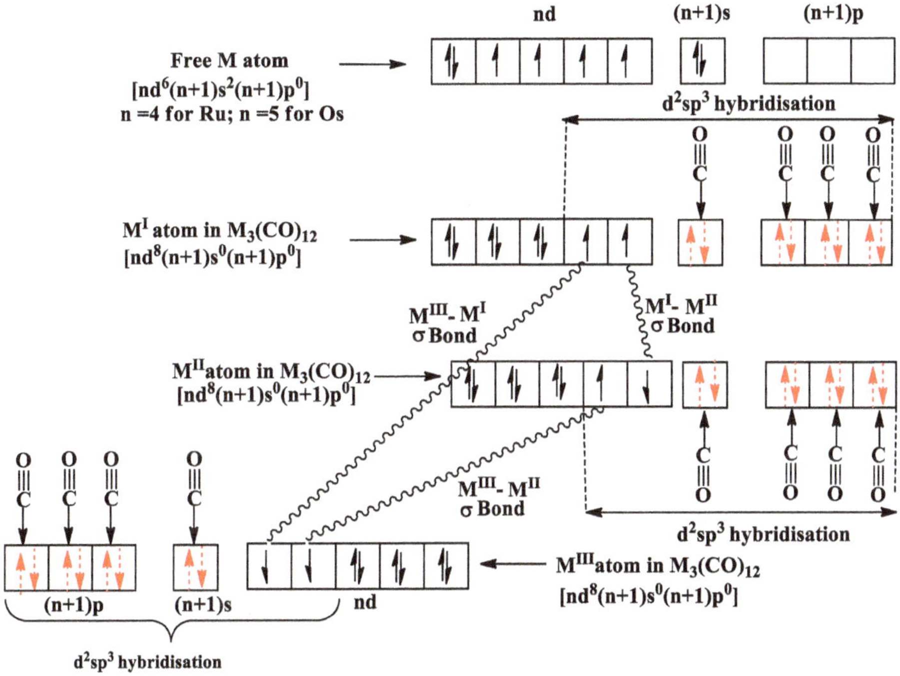

Figure 3.36: Hybridization scheme of M in $M_3(CO)_{12}$ (M = Ru or Os). Electron pair donated by terminal CO molecule is shown by (↑↓).

For the cobalt and rhodium compounds, the structure has lower symmetry. One metal atom has three terminal CO groups as in $Ir_4(CO)_{12}$, and the remaining nine CO groups occupy both symmetrical bridging positions and terminal positions around the triangle formed by other three metals (Figure 3.37b).

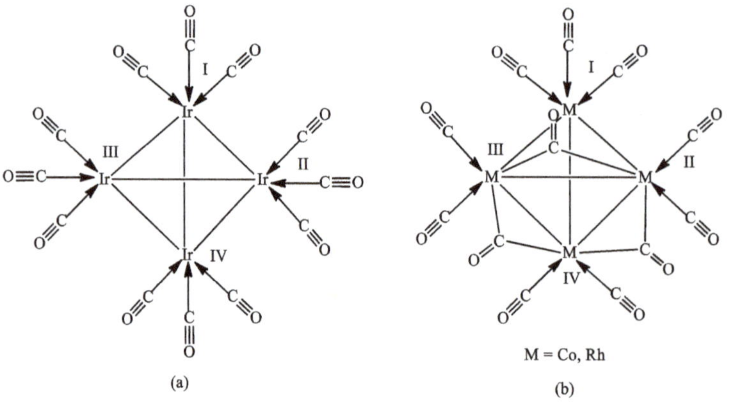

Figure 3.37: Structure of tetrameric carbonyl complexes, $M_4(CO)_{12}$ (where M = Co, Rh or Ir).

Hybridization schemes

(a) Tetrameric $Ir_4(CO)_{12}$

As the C.N. of each Ir is 6, d^2sp^3 hybridization (Figure 3.38) may be assumed at each Ir centre for the formation of $Ir_4(CO)_{12}$. As all the electrons in the $Ir_4(CO)_{12}$ molecule are paired, the molecule has a diamagnetic character.

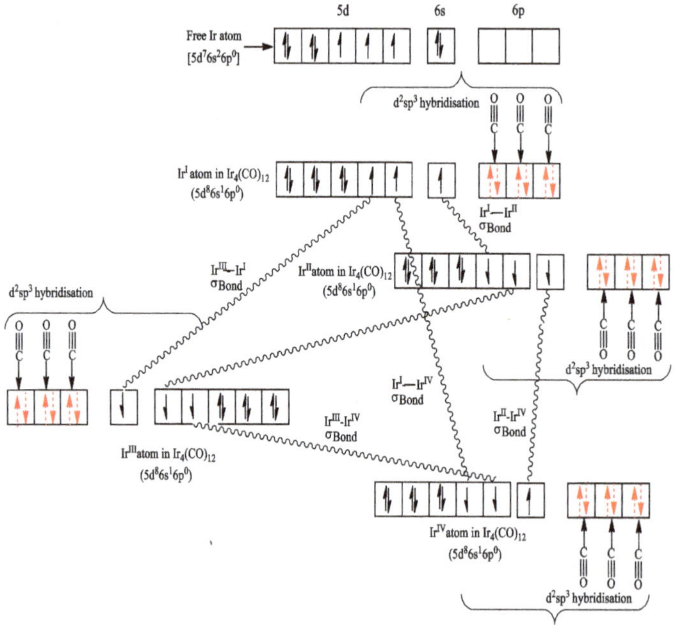

Figure 3.38: Hybridization scheme of Ir atom in $Ir_4(CO)_{12}$. Electron pair donated by terminal CO molecule is shown by (↑↓).

(b) Tetrameric $M_4(CO)_{12}$ (where M = Co, Rh)

The formation of tetrameric $M_4(CO)_{12}$ (where M = Co, Rh) may be understood by assuming d^2sp^3 hybridization at M^I involving C.N. 6, and d^3sp^3 hybridization at M^{II}, M^{III} and M^{IV} considering their hepta-coordination (Figure 3.39).

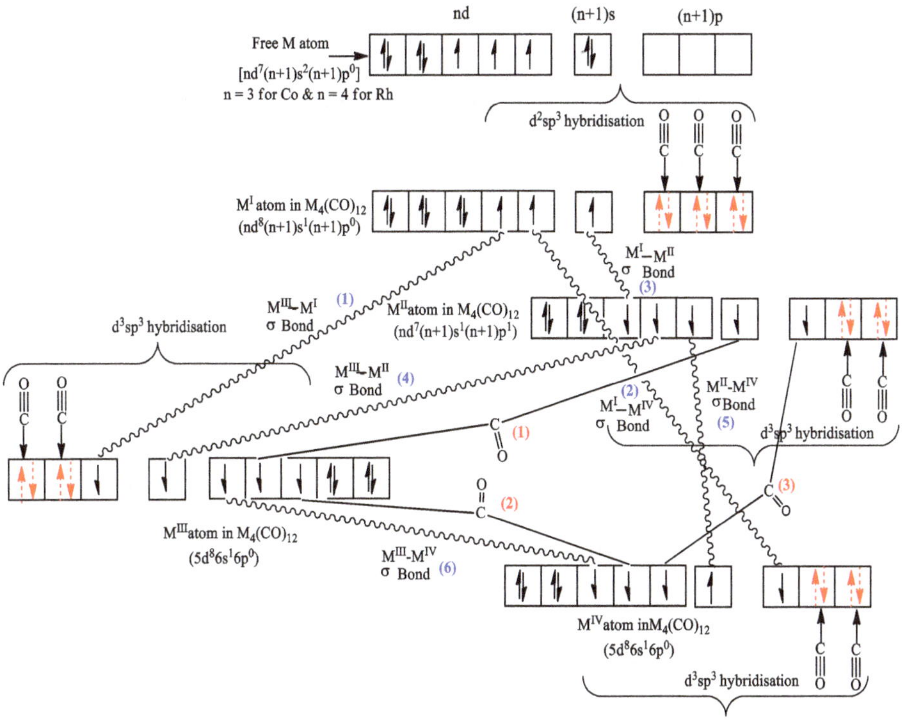

Figure 3.39: Hybridization scheme of M in $M_4(CO)_{12}$ (M = Co or Rh).

3.11 Bonding in metal carbonyls: nature of M–C and C–O bonds in metal carbonyls

Most of the physical studies on metal carbonyls for their structural elucidation conclude that the metal–carbon bonds (M–CO) in metal carbonyls are of multiple natures. For instance, Ni–C bond length in $Ni(CO)_4$ is found to be 1.50 Å which is 0.32 Å shorter compared to Ni–C single bond length of 1.82 Å. Moreover, the C–O bond length in metal carbonyls [1.15 Å in $Ni(CO)_4$] is longer than C≡O bond length in CO molecule (1.28 Å). Based on the bond length data, it was concluded that metal carbonyls have OC⟶M σ-bond as well as M⟶CO π-bonds, that is, metal carbonyls have double bonding which can be represented as

$$OC \underset{\pi}{\overset{\sigma}{\rightleftharpoons}} M$$

So, the two facts, that is, multiple (double bond) natures of M–CO bond and longer C–O bond length compared to CO molecule, must be explained by any bonding theory.

3.11.1 Formation of OC⟶M σ-bond: valence bond theory (VBT) approach

It is obvious that in the Lewis structure of CO molecule shown below, the loan pair of

$$:C \equiv O:$$

electron on carbon atom in CO molecule resides in sp-hybrid orbitals. All carbonyls contain terminal CO groups that are attached to the metal atom by

$$:O \equiv C: \longrightarrow M \ \sigma\text{-bond.}$$

This OC→M σ-bond is formed by the overlap between sp-hybrid orbital on C atom containing lp of electrons and the appropriate empty hybrid orbitals of the metal in metal carbonyls. For example:

(i) In $Cr(CO)_6$ molecule, formation of each of the OC⟶Cr σ-bond takes place by the overlap of completely filled sp-hybrid orbital on C atom and empty d^2sp^3 hybrid orbital of Cr atom.

(ii) In $Fe(CO)_5$ molecule, formation of each of the OC⟶Co σ-bond occurs by the overlap of completely filled sp-hybrid orbital on C atom and empty dsp^3 hybrid orbital of Fe atom.

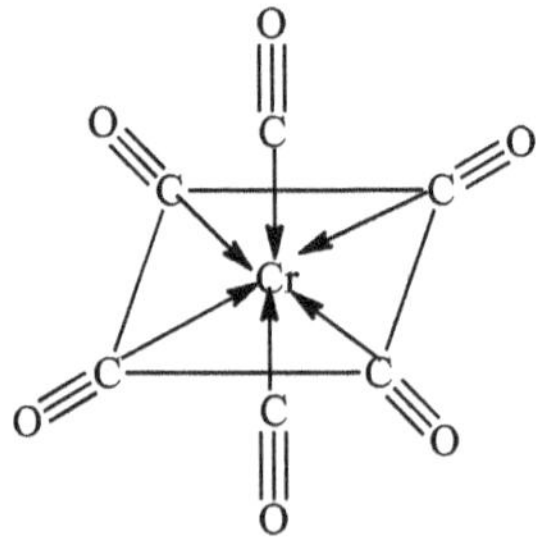

Octahedral structure of $Cr(CO)_6$ molecule

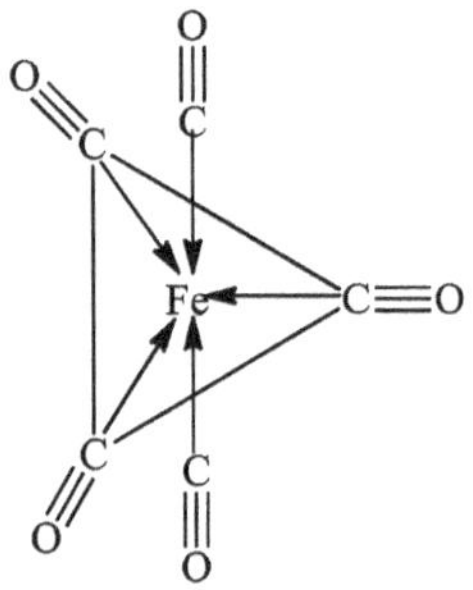

Trigonal bipyramidal structure of $Fe(CO)_5$

3.11.2 Formation of OC⟶M σ-bond and M⟶CO π bond: molecular orbital theory (MOT) approach

Before explaining the formation of σ- and π-bonds in metal carbonyls on the basis of MOT, let us have a look over M.O. diagram of CO (6 + 8 = 14e system) as shown in Figure 3.26.

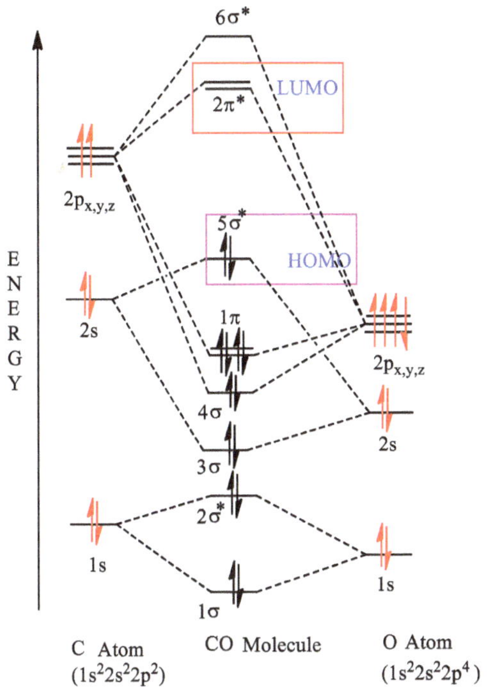

Figure 3.40: M. O. diagram of carbon monoxide (taken from Ref. [21]).

This energy-level diagram contains different molecular orbitals of CO molecule in increasing order of their energies. X-axis has been considered as the molecular axis in the formation of this molecular orbital diagram.

(i) Formation of OC→M σ-bond

The highest occupied molecular orbital (HOMO), namely, 5σ* present in the M.O. diagram of CO is anti-bonding orbital localized on C atom and thus has significant carbon (C) 2s character. This molecular orbital $(5\sigma^*)^2$ has an lp of electrons and hence overlaps with the vacant suitably directed σ-type orbital on transition metal M to form OC$(5\sigma^*)^2$→M σ-bond (Figure 3.41a). As the 5σ* orbital has substantial C 2s character, CO makes bond to a metal as a σ-donor through the C atom and not the O atom.

(ii) Formation of M→CO π-bond

The lowest occupied molecular orbitals (LUMOs), namely, 2π* present in the M.O. diagram of CO are two degenerate vacant anti-bonding π-molecular orbitals, namely, π^*_{2py} and π^*_{2pz} (consider *x*-axis as the molecular axis of CO). These π-molecular orbitals have proper symmetry and hence one of these orbitals overlaps with the suitable non-bonding filled d-orbital (d_{xz} or d_{yz}) of the metal to accept the electrons from the filled d-orbitals. This overlap results in the formation of M(d)→(CO)(π^*_{2py} orπ^*_{2pz}) π-bond (Figure 3.41b). This is called π-backdonation and the bonding is called π-backbonding.

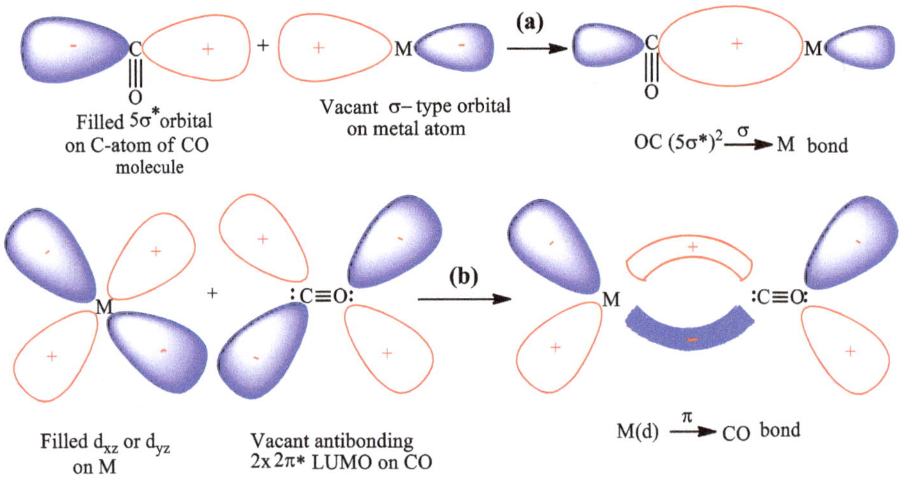

Figure 3.41: Formation of σ- and π-bonds in metal carbonyls: (a) formation of OC $(5\sigma^*)^2$Mσ bond and (b) formation of M(d)πCO (π*) bond.

Since the non-bonding occupied d-orbitals of the metal (d_{xz} and d_{yz}) make π-bond, these orbitals are called dπ-orbitals and the π-bond is represented as M(dπ)→(CO)(π*) π-bond. The formation of M→(CO) π-bond stabilizes or strengthens OC→M σ-bond. Since electron density from the metal is accepted by CO ligand in its π* molecular orbital, CO is called π-accept or (π-acid) ligand.

The transfer of electron density from CO to metal in the formation of OC→M σ-bond decreases electron density in CO. This enhances π-acceptor power of CO. At the same time, the transfer of electron density from metal to CO in the formation of M→(CO)π bond increases electron density on CO. This enhances σ-donor power of CO. Thus, σ- and π-bond formations reinforce each other. The term **synergic** has been used to describe such reinforcement.

(iii) Effect of π-backbonding on M–C and C–O bond orders

During the formation of π-bond, electronic charge is transferred from the filled non-bonding d-orbital of the metal to $\pi^\star(\pi^\star_{2py}$ or$\pi^\star_{2pz})$ molecular orbital of the CO. The transfer of electronic charge increases M–C bond order in metal carbonyl and decreases C–O bond order in CO. Thus, the increase in metal d-electron transfer to $\pi^\star$ orbital of CO will increase M–C bond order in metal carbonyls and will decrease the C–O bond order in CO. This is well evident from the observed bond order data in iso-electronic and iso-structural metal carbonyl species, such as $[Ni(CO)_4]^0$, $[Co(CO)_4]^-$ and $[Fe(CO)_4]^{2-}$.

Explanation

Since the –ve charge in $[Ni(CO)_4]^0$, $[Co(CO)_4]^-$ and $[Fe(CO)_4]^{2-}$ increases as $[Ni(CO)_4]^0 < [Co(CO)_4]^- < [Fe(CO)_4]^{2-}$, the ease of electron transfer from the metal d-orbital to$\pi^\star$-orbital of CO also increases in the same direction. The enhancement in the facility of electron transfer also increases M–C bond order as

$$\underset{1.33}{Ni-C} < \underset{1.89}{Co-C} < \underset{2.16}{Fe-C}$$

and C–O bond order decreases as

$$\underset{2.64}{[Ni(CO)_4]^0} < \underset{2.14}{[Co(CO)_4]^-} < \underset{1.85}{[Fe(CO)_4]^{2-}}$$

3.11.3 Bonding in metal carbonyl versus π-complexes of unsaturated organic ligands

The difference in bonding in metal carbonyls and that in π-complexes of unsaturated organic ligands like alkene, alkynes and cyclopentadienes is that in π-complexes of organic ligands both donation and back-acceptance of electrons by the unsaturated ligands involves the use of π-orbitals of the unsaturated ligand. Contrary to this, in metal carbonyls donation of electron pair takes place from $5\sigma^*$ molecular orbital of CO to the appropriate metal σ-type of orbitals and backdonation takes place from the filled metal d-orbital to $\pi^\star$ molecular orbital of CO.

3.11.4 Metal carbonyls and IR spectra

The IR spectral data of metal carbonyls have been used in several cases. Some of them are as follows:

(i) Determination of geometry

If theoretically a metal carbonyl can have more than one geometry, we can decide which geometry is most stable. For this, we calculate the number of IR-active together with Raman-active vibrational bands for each of the geometry. Thereafter, IR and Raman spectra are recorded and finally tally the number of bands observed with the number of bands theoretically predicted.

The geometry of pentacoordinated carbonyl compound, [$Fe(CO)_5$], had been an issue of long deliberation owing to its two possible geometries: (a) TBP having D_{3h} point group and (b) square pyramidal belonging to C_{4v} point symmetry as shown in Figure 3.42. While the dipole moment of magnitude 0.64 Debye of this compound favours the square pyramidal structure, the electronic diffraction study displays TBP as the correct structure.

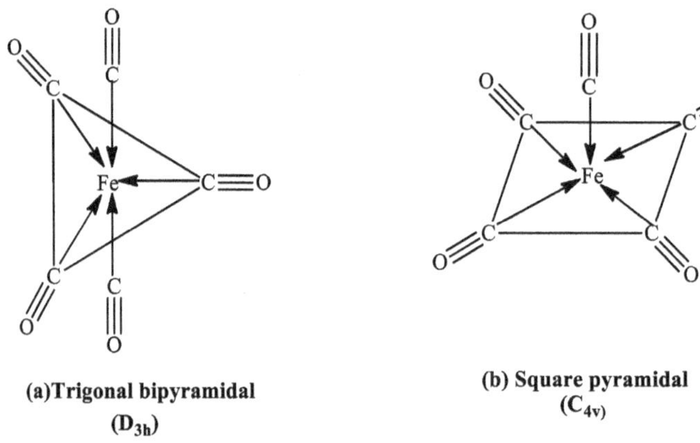

Figure 3.42: Two possible structures of [$Fe(CO)_5$].

As per the prediction of group theory considering TBP structure of [$Fe(CO)_5$] (D_{3h}), there should be two IR-active (A_2'', E') and three Raman-active ($2A_1'$, E') fundamental modes of vibrations for both ν(CO) and ν(Fe–C). On the other hand, for square pyramidal structure (C_{4v}), it should display three IR-active ($2A_1$, E) and four Raman-active ($2A_1$, B_1, E) CO stretching vibrations (see Table 3.1).

Experimentally observed two IR- and three Raman-active bands (see Table 3.1) are consistent with TBP (D_{3h}) structure of [$Fe(CO)_5$]. The totally symmetric mode (A_1') at 2,114 cm^{-1} is found to be absent in IR spectrum but appeared as a strong band in the Raman spectrum (Figure 3.43).

Table 3.1: The predicted and observed infrared- and Raman-active modes for TBP and SP geometry in $Fe(CO)_5$.

Vibrational spectrum	Trigonal bipyramidal (D_{3h})	Square pyramidal (C_{4v})	Experimental ν(CO) bands (cm^{-1})
IR	A_2'', E'	$2A_1$, E	2,088, 1,994
Raman	$2A_1'$, E'	$2A_1$, B, E	2,114, 2,031, 1,984

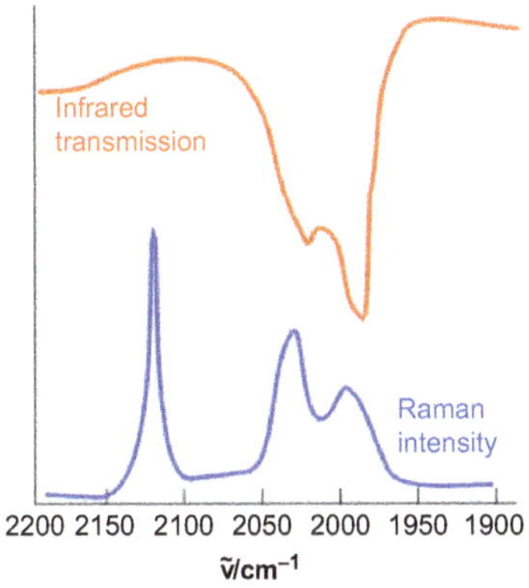

Figure 3.43: The observed infrared and Raman spectra of $[Fe(CO)_5]$ (liquid) (only in the CO stretching region).

(ii) Bond order of coordinated terminal CO group

It has been observed that IR absorption frequency due to stretching vibration of coordinated CO is directly proportional to its bond order. In other words, IR absorption band due to the stretching vibration of coordinated CO with a higher bond order would occur at a higher frequency and at lower IR frequency for lower bond order.

As per IR data, the $\nu(CO)$ for free CO occurs at 2,250 cm^{-1} while that for coordinated CO lies in the range 2,220–1,700 cm^{-1}. This is due to lowering of bond order of coordinated CO because of the backdonation of electrons form metal $d\pi$ orbital to π^* orbitals of ligated CO.

Let us consider the observed IR data in some metal carbonyls:

(i)	Metal carbonyl	$[V(CO)_6]^-$	$[Cr(CO)_6]^0$	$[Mn(CO)_6]^+$
	$\nu(CO)$ (cm^{-1})	1,860	1,980	2,090

Since the presence of positive charge on $[Mn(CO)_6]^+$ resists the flow of metal$d\pi$ electrons into the π^* orbitals of CO, the bond order of CO is little reduced compared to $[Cr(CO)_6]^0$ and $[V(CO)_6]^-$ in which the charge on the carbonyl is0 and –1, respectively. On the other hand, the presence of –ve charge on $[V(CO)_6]^-$ facilitates flow of metal $d\pi$ electrons into the π^* orbitals of CO, and so the bond order of CO is much reduced compared to $[Cr(CO)_6]^0$ and $[Mn(CO)_6]^+$. This is why the trend in $\nu(CO)$ values in these three carbonyls is as given above.

(ii)	Metal carbonyl	$[Ni(CO)_4]^0$	$[Co(CO)_4]^-$	$[Fe(CO)_4]^{2-}$
	$\nu(CO)$ (cm^{-1})	2,040	1,920	1,790

Owing to the increase in negative charge on the metal carbonyl, the π-backdonation of electrons from the metal d-orbitals to the π^* orbitals of CO weakens the CO bond and hence reduces $\nu(CO)$. More is the –ve charge, more is the reduction in $\nu(CO)$.

(iii) Differentiation of terminal and bridging CO groups

Since OC–M bond in a terminal carbonyl group (O ≡ C→M) is stronger than the bridging carbonyl group, the IR absorption for a terminal CO group would occur at a higher frequency than for a bridging CO group.

The CO groups may be linked to the metal centre in the form of terminal as well as bridging ligands in polynuclear metal carbonyl compounds. The IR spectra of polynuclear carbonyls are found to be very supportive in distinguishing two types of carbonyl ligands, whether terminal or bridging. While the terminal CO ligands in neutral molecules have ν(CO) in the range 2,100–1,900 cm^{-1}, the two bridged COs between two metal atoms have stretching frequency fall in the region 1,900–1,700 cm^{-1}. In case of the three COs bridged between two metal atoms, ν(CO) is observed below 1,700 cm^{-1}. The successive decrease in energy with the increase in the number of bridging CO indicates successive decrease in C–O bond order.

The following representative examples illustrate how the IR spectroscopy is cooperative in ascertaining bridging and terminal carbonyl groups in metal carbonyls.

(i) The golden yellow-coloured dinuclear carbonyl compound $Mn_2(CO)_{10}$ may have two potential structures as shown in Figure 3.44.

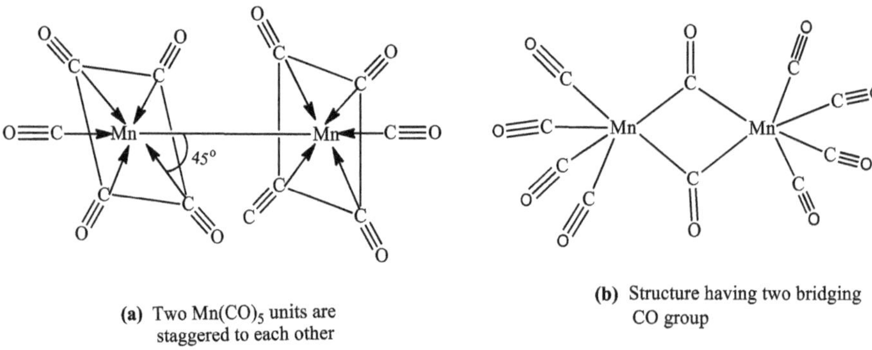

45°
(a) Two Mn(CO)5 units are staggered to each other
(b) Structure having two bridging CO group

Figure 3.44: Two probable structures of $Mn_2(CO)_{10}$.

The IR spectrum of this carbonyl compound displays no band consistent with the bridging CO group and shows only bands corresponding to CO stretching vibration in the range 2,100–2,000 cm^{-1} for terminal COs. This rules out the possibility of bridging structure (b) of $Mn_2(CO)_{10}$.

(ii) IR spectral data of $Fe_2(CO)_9$ and $Fe_3(CO)_{12}$ indicate that both the compounds contain the terminal and the bridging of CO ligands. Instead, the IR spectrum of $Ru_3(CO)_{12}$ or $Os_3(CO)_{12}$ exhibits only one band type consistent with the terminal CO groups. Based on the IR spectral data, their structures are shown in Figure 3.45.

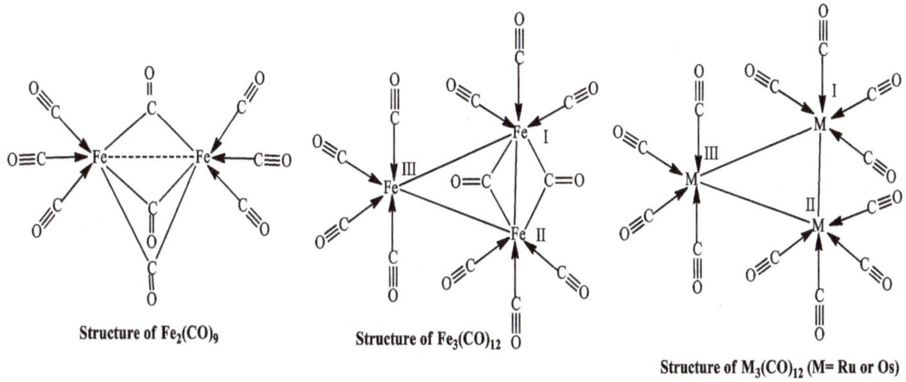

Structure of $Fe_2(CO)_9$
Structure of $Fe_3(CO)_{12}$
Structure of $M_3(CO)_{12}$ (M= Ru or Os)

Figure 3.45: Chemical structures of dimeric $Fe_2(CO)_9$, $Fe_3(CO)_{12}$ and trimeric carbonyls $M_3(CO)_{12}$ (M = Ru or Os).

The observed CO stretching frequencies ν(CO)s are also suggestive of the electron donor abilities of the metal and other ligands in metal carbonyls containing terminal COs. The greater is the electron density delivered by the metal and other ligands, the greater would be the backdonation, and therefore, the lower will be the CO bond order. Consequently, the observed ν(CO) is reduced.

(iv) Distinction in substituted and non-substituted metal carbonyls

Let us take the example of $[Mo(CO)_6]$ and $[Mo(dieta)(CO)_3]$ [where dieta = diethylenetriamine $NH(CH_2CH_2NH_2)_2$]. Since diethylenetriamine is a tridentate ligand, three CO groups can be replaced from $[Mo(CO)_6]$ by one dieta molecule to get $[Mo(dieta)(CO)_3]$.

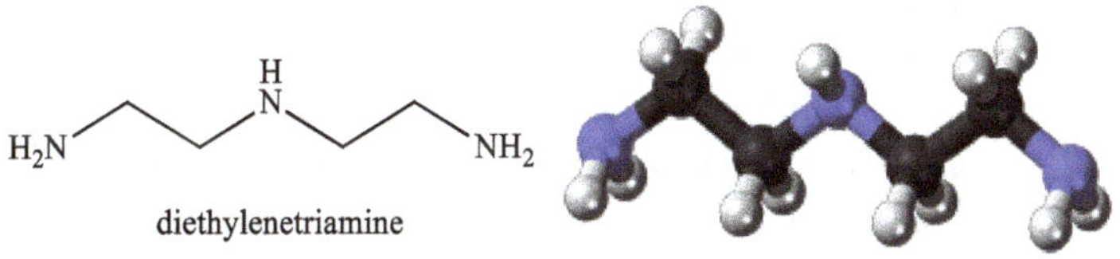

IR spectra of $[Mo(CO)_6]$ shows ν(CO) at 2,004 cm^{-1} while $[Mo(dieta)(CO)_3]$ exhibits ν(CO) at 1,760 cm^{-1}. Thus, on the basis of IR spectrum substituted and non-substituted metal carbonyls can be distinguished. This lowering of ν(CO) in $[Mo(dieta)(CO)_3]$ can be understood as follows.

Since the ligand dieta is not capable of back-accepting electrons from the Modπ-orbital because of the absence of π^* orbitals, the remaining CO groups present in $[Mo(dieta)(CO)_3]$ share dπ-electrons from the Mo to prevent accumulation of negative charge of the Mo atom. Consequently, ν(CO) in $[Mo(dieta)(CO)_3]$ is lowered and appears at 1,760 cm^{-1}.

3.12 Metal carbonyl compounds: a new class of metallopharmaceuticals

3.12.1 Introduction

CO is a colourless, odourless and tasteless diatomic molecule in gas phase. This gaseous molecule has long been pondered as a toxic by-product of environmental and industrial processes. Its toxic effect is well acknowledged [22] and exists in its strong affinity for haemoglobin, which is nearly 245 times more than that of O_2. Moreover, paltry attached CO at the haeme binding sites prevents the release of O_2 from the remaining haeme groups and assists in movement of the O_2 dissociation curve to the left. These actions of CO reduce the O_2-carrying capacity and delivery potential leading to tissue hypoxia. Higher levels of CO also bind to other enzyme/protein molecules, such as cytochromes P450, cytochrome c and myoglobin. This further intensifies the damaging actions of CO.

Like NO (NO^+ being isoelectronic with CO), it has been well established that CO also acts as a messenger molecule within the mammalian system tracked from HO-based metabolic pathway, and has been ranked as second gasotransmitter after NO. Indeed, in normal and unaffected way, CO is generated in human being at a rate of 3–6 cm^3 per day. This rate of CO production is increased significantly by certain inflammatory states and pathological conditions accompanying with red blood cell haemolysis (destruction of red blood cells). Over the last two decades, the importance of biological effects of CO has greatly increased in mammals. It is now well recognized in the medical literature that CO has a key role as a signalling molecule in mammals. The amazing discovery of CO as a signalling molecule in mammalian physiology has recently heightened attention in this toxic gas among researchers of biochemical and pharmaceutical community. Principally, from catabolism of haeme via the enzyme HO, CO is produced endogenously. The CO so produced takes part in numerous anti-inflammatory, anti-proliferative and vasoregulatory pathways. Low amounts of CO have demonstrated beneficial effects in animal models in preventing organ graft rejection and safeguarding the heart during reperfusion after cardiopulmonary (relating to the heart and lungs) bypass surgery. Moreover, the beneficial effects of CO have drawn attention of the pharmaceutical industry for its use as a cytoprotective agent.

Safety-related issues for the use of this toxic gas have provoked research in the area of syntheses of CORMs. So far, several metal carbonyls have been used as CORMs in supporting prolonged and safe delivery of low doses of CO to cellular targets. As many carbonyl complexes release CO upon light illumination, coordination chemists and biologists have recently began to explore the possibility of "controlled CO delivery" through the use of light. During the past two decades, a number of photoactive CORMs or "photoCORMs" have been synthesized that release CO upon illumination with UV or visible light. The efficacy of these photoCORMs in CO delivery has also been confirmed. Advanced design principles for isolation of photoCORMs have started to appear in recent findings. Search of the literature reveals the advent of a new exciting area of drug development in such efforts. The potential of CORMs and photoCORMs as CO donating pharmaceuticals along with a brief overview of the physiological roles of CO will be presented in the coming section.

3.12.2 Sources of CO in the human body

As per the metabolic pathways concerned with the CO biosynthesis, almost 14% of 500 μmol per day is obtained from lipid peroxidation and from photo oxidation plus self-activation of cytochrome p-450. Bacteria and xenobiotics also contribute the same minor percentage [23–25]. Major contribution (almost 86%) is generated by the erythrocyte breakdown, wherein the HO catalyses this oxidation. Like NOS, HO also exists in two isoforms, namely, HO-1 and HO-2. These are also called as inducible and constitutive, respectively. Both the isoforms show the same rate-limiting step while catabolizing haeme, the difference lies with the regulation, amino acid sequence and distribution in the tissues. Another HO has been recently identified and named as HO-3. This form of HO was detected in the several organs of rats. Till date, no haeme degradation study has been reported for this newly detected HO member [26].

The metabolic pathway of HO-catalysed haeme oxidation involves several important stages as has been illustrated in Figure 3.46a, b. In addition to CO, other intermediary products like α-meso-hydroxyhaeme, verdohaeme, biliverdin (converts to bilirubin as excretory product conjugated by glucuronic acid shown in Figure 3.47) are also involved [27, 28]. The main source of the yellow colour of urine is bilirubin. Overproduction of bilirubin leads to jaundice, a syndrome of hyper-bilirubinaemia.

HO's activity is frequently observed in bruises (skin discolouration caused by injury). When bruise formation occurs during injury, haemoglobin is liberated from lysed red blood cells. This results in a dark red/purple area at the place of injury in the skin, which is linked with deoxygenated haemoglobin. Thereafter, the haeme released from haemoglobin is oxidized by HO of the vascular tissue to biliverdin along with the release of CO and Fe^{2+} ions. This has several apparent results. In the extravascular space, oxygen is consumed to give blue oxygen-free haemoglobin, and a green colour, indicating the presence of biliverdin, looks in the underlying tissues. The reduction of biliverdin to bilirubin yields a yellow shade, which then slowly fades (Figure 3.48). Some of the CO that is produced binds to the iron centre of haemoglobin to produce the bright-red COHb, which is very obvious in Figure 3.48d.

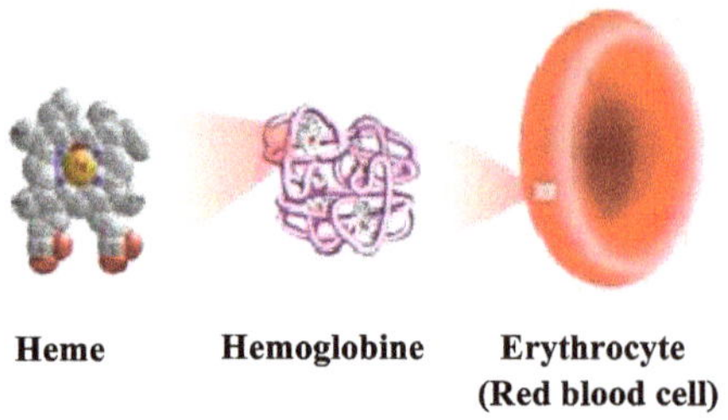

(a)

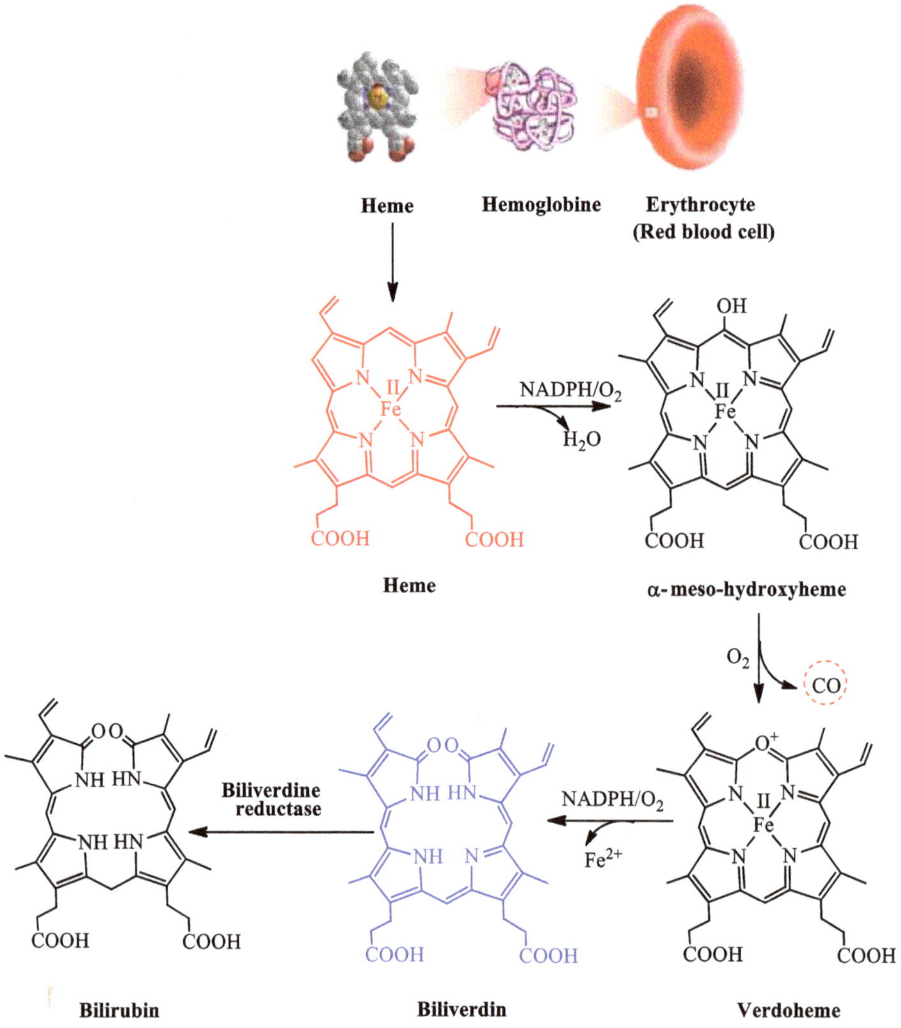

(b)

Figure 3.46: Oxidation of haeme by haeme oxygenase (HO) forming CO as a by-product.

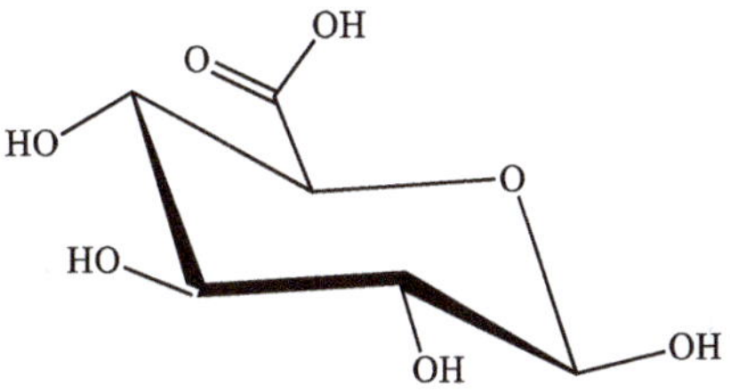

Figure 3.47: Chemical structure of glucuronic acid.

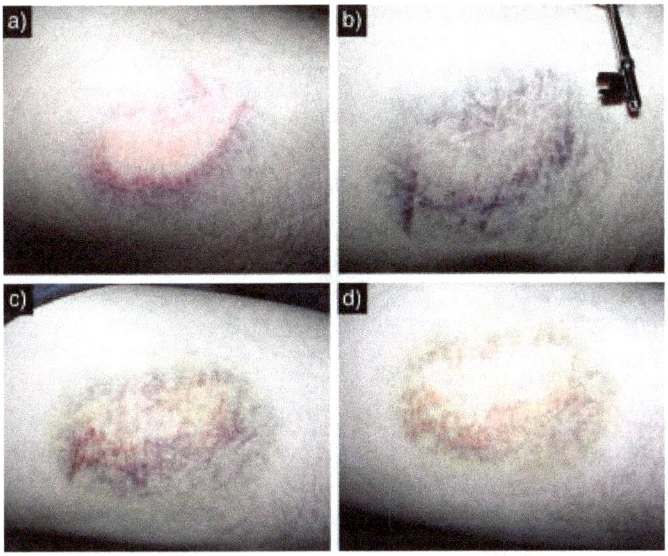

Figure 3.48: The variations in the look of a bruise formation in thigh of a hockey player taken at different times after injury (a) after 12 h, showing the red and purple colour of haeme. (b) After 2 days, showing green tinges owing to the formation of biliverdin. (c) After 5 days showing yellow colouration because of the formation of bilirubin. (d) After 10 days showing some residual bilirubin. The bright red colour in (d) is owing to CoHb (taken from Ref. [25]).

The bioaction of HO-1 under stressful situation gets enhanced and the CO production gets increased than the optimal value [29]. Therefore, such an elevation in the concentration can be used as a sign convention medically to read the associated behaviour. The similar correlation has been found in several diseases wherein a patient is expected to suffer from stress and strain conditions. For instance, in bronchiectasis, asthma, cystic fibrosis, hyperglycaemia and other diseases CO level appears higher than the normal [29]. Hence, the detection level of CO because of inducible HO-1 can help in diagnosis of pathophysiological state.

3.12.3 Generated CO in mammals: target sites

(i) Soluble guanylyl cyclase (sGC)

Well demonstrated by Ignarro et al. [30], NO activates soluble guanylyl cyclase (sGC) by binding to a haeme molecule and replacing a coordinated histidine. The activated sGC so obtained catalyses the transformation of guanosine triphosphate (GTP) to cyclic guanosine monophosphate, cGMP, as per Figure 3.49.

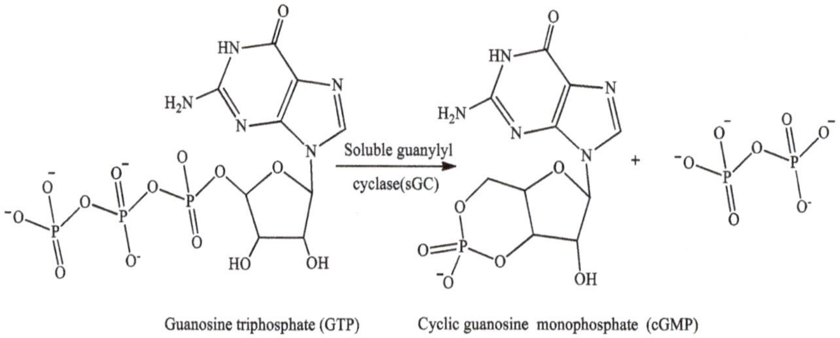

Figure 3.49: sGC catalysing the conversion of GTP to cGMP.

The cGMP now goes into a series of pathways. Some of them are associated with medical conditions given in Figure 3.50are shown in red.

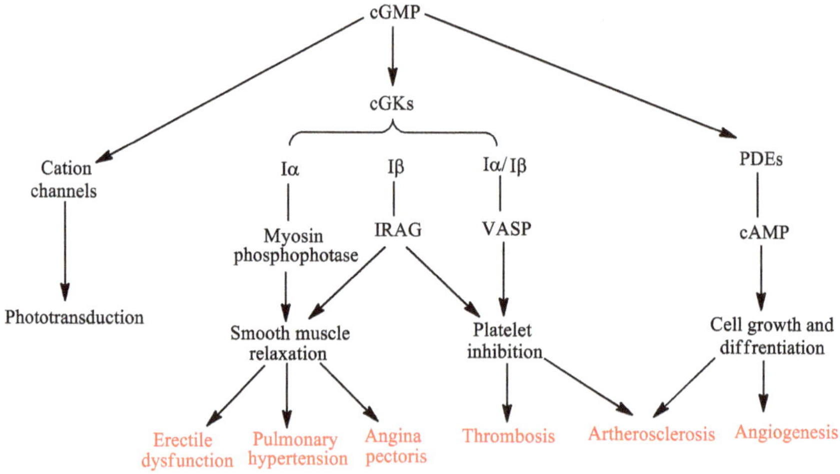

Figure 3.50: Cyclic guanosine monophosphate (cGMP) routes improving medical conditions shown as red. cGKs, cGMP-dependent protein kinases, Iα and Iβ; IRAG, inositol 1,4,5-triphosphate (IP_3) receptor-associated cGKIβ substrate; VASP, vasodilator-stimulated phosphoprotein; PDEs, phosphodiesterases; cAMP, cyclic adenosine monophosphate (adapted from Ref. [31]).

Analogous to NO, CO triggers guanylyl cyclase to yield cGMP. This improves medical conditions given in Figure 3.50. But it is approximately 1/80th as effective as NO. However, there is a family of synthetic compounds, of which YC-1 is an example that when combined with CO are as effective as NO in activating guanylyl cyclase. This supports the possibility that there is a natural intracellular signalling molecule similar toYC-1 (Figure 3.51).

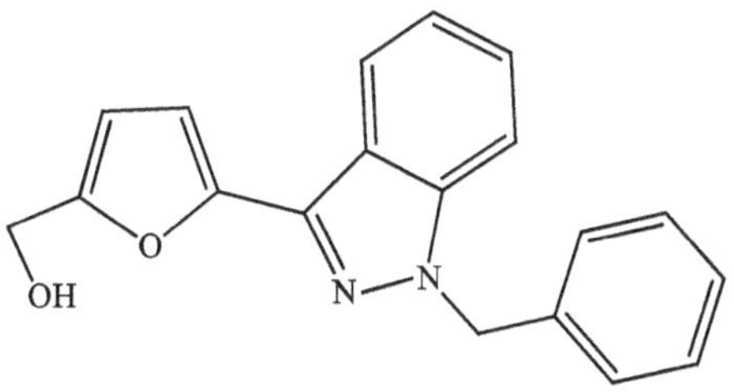

Figure 3.51: Structure of YC-1 [1-benzyl-3-(5′-hydroxymethyl-2-furyl)indazole].

Being a heterodimer, sGC enzyme comprises α- and β-subunits having formula weight kDa of 73–88 and 70, respectively. The latter subunit has a link with the prosthetic group of haeme-b (an iron–protoporphyrin IX) that allows enzyme–substrate activity. Meanwhile, NO plays the role of a primary substrate for this action. By binding to NO, the sGC bioaction gets enhanced by 200 times. Hence, the pathway followed thus stimulates sGC causing vasodilatory effect by NO has been elaborated to a considerable extent. This mechanistic approach is generally targeted in the design and development of drugs meant for vasodilation. Similarly, very low sGC activity has been reported for CO-mediated (induced) action of this form.

Since the structural composition of *haeme b* indicates that the β-domain represents a coordination sphere of five species, and sGC stimulation by NO involves binding of the molecule with *haeme b* forming a complex of six-coordinated sites having very low stability (more reactive), having proximal attachment of His105. This reactive complex attains a stage to let His105 uncoordinate and form a five-coordinated complex retaining haeme-NO in its composition (Figure 3.52). Thus, a biologically communicate persuasion reaches to bring the necessary geometric changes within the active site of sGC for the respective enzymatic action [32].

There is difference in opinion as to whether the NO is on the same or the opposite side to His-105 in the activated form. It may be possible that the NO binds to the haeme in guanylyl cyclase to give what is in effect of Fe^{III} and NO^-. Even though there is no crystal structure, the structure of the five-coordinate NO adduct of cytochrome c′ indicates a bent Fe–N–O bond with angles of 124° and 132° for the two conformers with a half-occupancy for each. This is consistent [33] with coordination of NO^- in the NO adduct of cytochrome c′.

Since, the haeme–NO complexes in both sGC and cytochrome c′ are very similar [34], it is expected that the bond Fe–N–O is also bent in the five-coordinate NO adduct of guanylyl cyclase. This is further substantiated by the appearance of a NO stretching frequency [ν(NO)] [35] at 1,677 cm^{-1}.

Contrary to the above, after coordination of CO to guanylyl cyclase, the iron becomesa six coordinate and that the metal centre almost remains as Fe^{II}. When YC-1 interacts with the α-domain, the His-105 is disconnected to produce a five-coordinate iron in the haeme-carbonyl and YC-1 [36] (Figure 3.52).

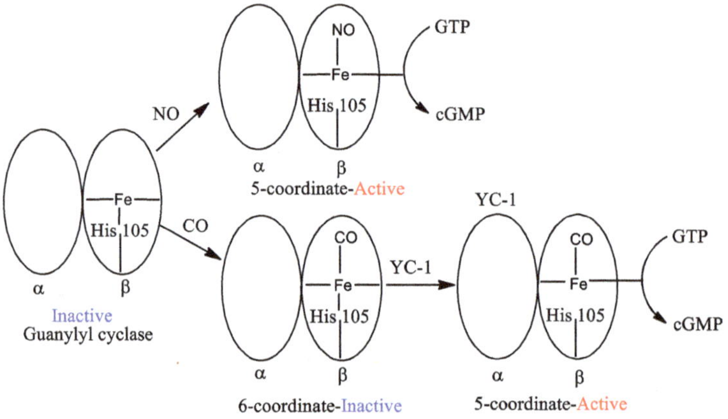

Figure 3.52: Diagrammatic illustration of the activation of guanylyl cyclase by NO and CO/YC-1 (taken from Ref. [36]).

CO-induced sGC stimulation has been described mechanistically in a view supporting different conformational changes dissimilar to NO-dependent pathway. This dissimilarity happens in the form of forming six-coordination sphere and retention of His-105 during the vasodilatory effect, and such an enzymatic action pathway appears four times more efficient as compared to the suggested NO-based reaction course [24]. The inexpression of five-coordinate system by the pathway following CO dependence has been supposed to be because of the effect of increased radius of the ion, and hence an electronegativity gets decreased. Therefore, sGC action-based CO-induced vasorelaxation follows a different mechanistic pathway as compared to NO.

(ii) Non-cGMP pathways

Amid other possible targets of CO that result in vasorelaxation, the big conductance of potassium channels (BK_{Ca}) is of much consideration. These channels are found in nearly every tissue comprising vascular smooth muscle cells (SMCs). These are enthused by a variety of modulators. Contributory factors that attenuate to BK_{Ca} are protein kinases, haeme and endogenously produced NO and CO. Some regulatory processes related to physiology, namely, neuronal excitability, muscle contractility and regulation of vascular tone are results of BK_{Ca} activity. BK_{Ca} channels are made up of two components: (i) a pore-forming α-subunit and (ii) an accessory β-subunit. Each component facilitates the flux of ions that produce an action potential [37]. Motivating effects of endogenous NO and CO gases on BK_{Ca} channels that result in a vasodilatory effect are based on the interactions of these two gases on specific subunits of BK_{Ca} channels.

The function of CO in modulating vasodilatory accomplishment in rat-tail artery, which possesses all three possible targets of CO (sGC, cytochrome P450 and K-channels) was studied by Wang and co-workers. On administration of the non-specific K^+ channel inhibitor tetraethylammonium ion, they found complete suppression of CO-induced vasorelaxation. This suggests the involvement of K^+ channels in this physiological response [38]. As potassium channels are among the most common type of ion-selective channels taking part in cell function and regulation of blood flow, it is vital to know the specific type of K^+ channel responsible for the observed CO-dependent response. Employing both large and small conductance channel inhibitors (charybdotoxin and apamin), Wang et al. demonstrated that the large conductance of Ca^{2+}-activated potassium channels (BK_{Ca}) is the primary target of CO-induced vasodilation [39].

Jaggar et al. [40] suggested that CO is capable to directly modify BK_{Ca} via binding to a channel-bound haeme area confined in the α-subunit. This channel-bound haeme works as a BK_{Ca} inhibitor by sensing CO and prevents pore closing. Such interaction increases the opening probability of these channels, and thus allowing for smooth muscle relaxation and increased blood flow. Interestingly, such CO-mediated controlling of BK_{Ca} channels and hence of vasodilation appears to be negligibly affected by endogenously produced NO and by reactive oxygen species (ROS), such as H_2O_2 and peroxynitrite [41]. These findings confirm the integrity of CO as a potent vasoregulatory molecule which acts on a signalling pathway dissimilar from that of NO.

3.13 CO signalling in anti-inflammatory responses

NO and CO are comparatively inert compared to the free radical species. Notably, the substrates of endogenously produced CO as a signalling molecule are restricted to metalloproteins holding haeme groups. Mostly, CO-mediated results that comprise vasorelaxation and anti-inflammatory responses are due to the interaction of CO with haeme-containing proteins involved in important physiological responses. An overview of some of the advantageous results associated with CO application, as well as the protein targets associated with each pathway is given in Figure 3.53.

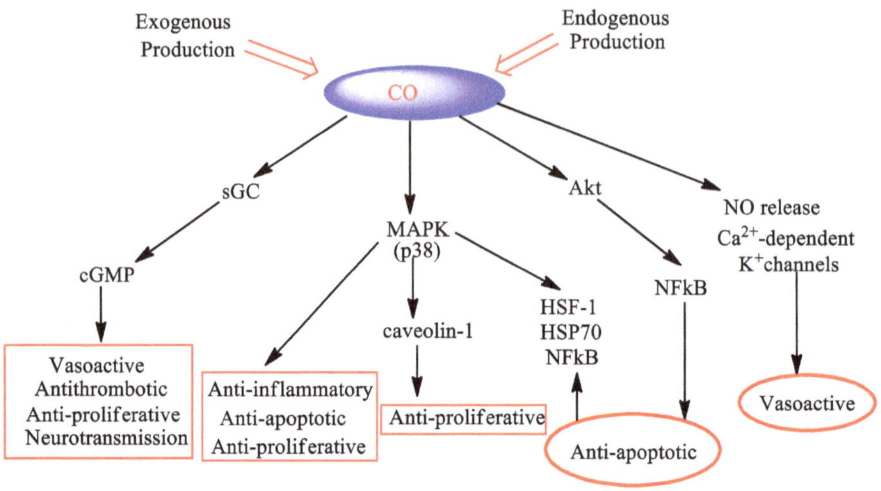

Figure 3.53: CO signal transduction pathways leading to beneficial results (adapted from Ref. [24]).

For instance, it is well known that interaction of CO with BK_{Ca}-bound haeme proteins demonstrates how CO achieves the vasoregulatory effect through protein inhibition. The terminal enzyme of the mitochondrial electron transport chain, namely, cytochrome c oxidase, is another crucial haeme protein that is inhibited by CO. Non-toxic concentrations of exogenously used CO have been found to activate a slight increase in mitochondrial ROS. The ROS in turn activates the vascular adaptations of the system to reduced oxygen conditions [42]. Such responses to hypoxic environments comprise up-regulation of transcription factors like *hypoxia-inducible factor1-alpha* (*HIF-1α*) and nuclear factor kappa-light-chain enhancer of activated B cells (NF-κB). Both serve to promote survival by producing anti-inflammatory cytokines and chemokines. When CO has been administered prior to events of stress, it is capable of initiating this type of redox (ROS) signalling mechanism, thereby functioning as a potent pre-conditioning agent. Vieira and co-workers [43], in a model of brain ischemia, have demonstrated that a CO pre-conditioning step averts neuronal apoptosis via increased ROS production.

Notably, the most interesting signalling pathway modified by CO is that of mitogen-activated protein kinases (MAPKs). In reply to stress and external stimuli, this protein group initiates signalling cascades. Basically, MAPKs include (i) extracellular signal-regulated kinases 1/2 (ERK1/2), (ii) c-Jun (N)-terminal kinases 1/2/3 (JNK1/2/3) and (iii) p38 (α, β, γ and δ) isoforms. While the MAPK family is divided into many subgroups depending on their participation in physiological and pathological processes, all MAPKs have a conserved serine–threonine domain. This on phosphorylation activates cell signalling [44]. CO is observed to precisely affect α and β isoforms of p38 MAPK.

In a hepatic (liver) inflammation model, Brugger and co-workers [45] demonstrated that the antioxidant results of CO are dependent on p38 MAPK stimulation. The antioxidant properties of CO were continued in spite of HO deactivation. This suggests that exogenously applied CO has the same anti-inflammatory properties as endogenously produced CO in this model. Likewise, the anti-proliferative properties of exogenous CO have been detected in a model of chronic pancreatitis/pancreatic inflammation and were found to proceed via p38 MAPK activation [46].

In spite of the cumulative substantiation on the role of MAPK in CO-assisted protection, the molecular basis for this interaction is still uncertain exclusively because MAPKs are not identified having haeme as prosthetic groups. However, it is fairly apparent that the beneficial effects of CO lie in its capacity to participate in a number of signalling pathways. Consequently, this unusual signalling molecule could potentially affect numerous physiological processes not restricted to vasorelaxation.

3.14 Therapeutic scope of CO

The CO gas is identified for its prominent role as a molecular messenger in the physiological process for the nervous system. It is also well known for some important therapeutic usages [47].

CO has the prospective for anti-inflammatory, anti-proliferative, anti-atherogenic, anti-allodynia, anti-nociceptive, anti-hyperalgesia and anti-apoptotic effects. It is crucial for vasodilatory phenomena reducing intraocular pressure (IOP), immunosuppressive administrated medications. It also has the capability to develop the pathological cellular process (Figure 3.38). CO also has many advantages for different biological organs such as the heart, kidney, liver, lungs, pancreatic islet and the small intestine in organ transplantation, protection and preservation. It is helpful to reduce the intensity of the ischemia/reperfusion injury (IRI), and lessen the myocardial infarction and allograft (tissue graft between genetically different organisms) rejection. CO is known to stimulate the cytoprotective activity, and it is also involved in anti-microbial and anti-hypertensive activities It has a modulated utility for haeme-dependent proteins like mitochondrial cytochromes and NADPH.

3.14.1 Ways of CO delivery in human body as a therapeutic agent

Principally, CO molecules can be used as a therapeutic agent inside the human body in two ways, and these are direct and indirect CO inclusion. The direct insertion has not been perfect, owing to its increase in the COHb level above 10% and lack of tissue selectivity (Figure 3.54). Moreover, it provides a direct interaction of the CO and lungs only while detainment of CO is also observed in this method. These limitations do not permit CO to approach other

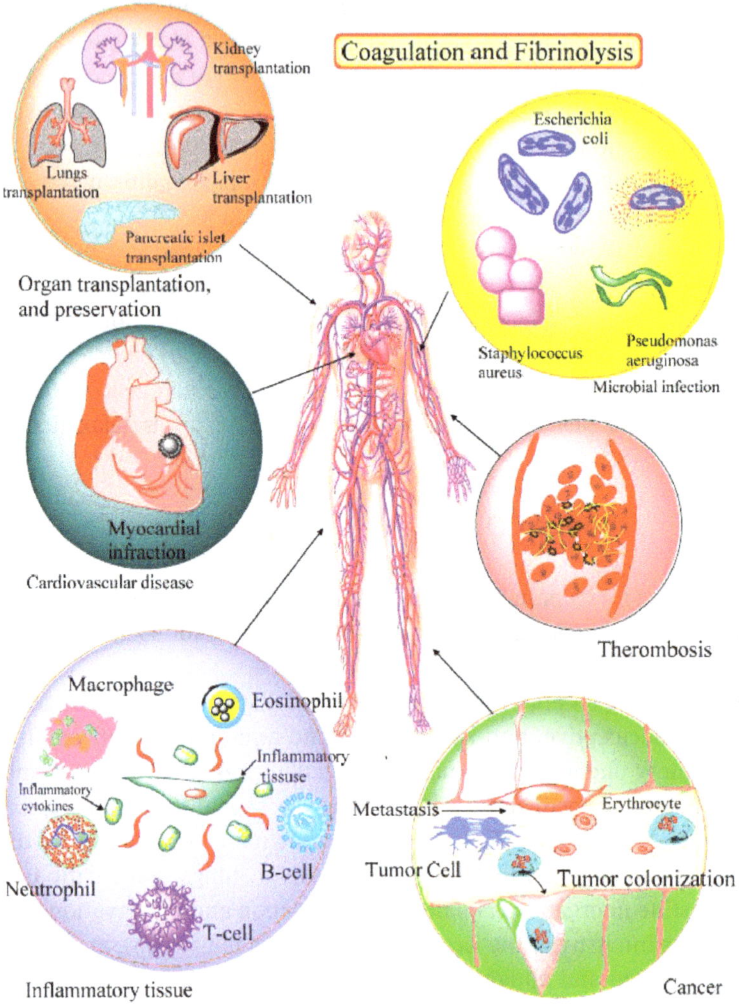

Figure 3.54: Utility of CO-releasing materials in coagulation and fibrinolysis (adapted from Ref. [47]).

biological organisms for therapy. In order to overcome these problems, researchers have developed an alternate strategy called "exogenous endeavour" wherein CORMs serve as therapeutic agents [31].

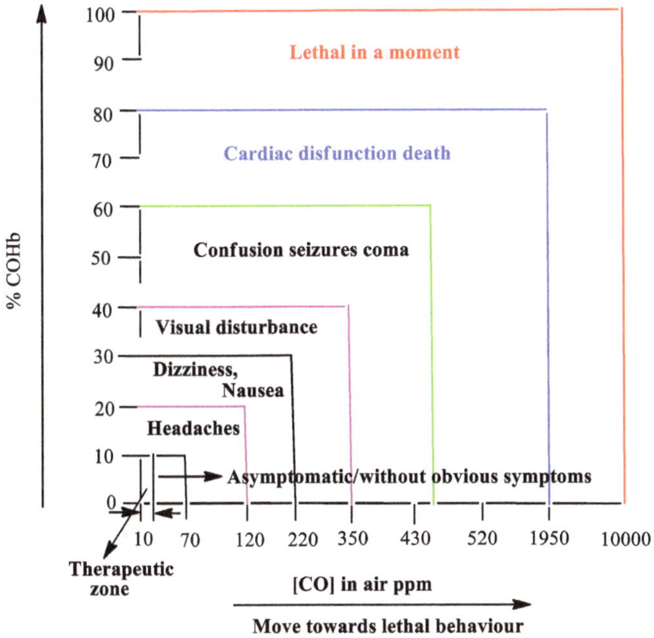

Figure 3.55: The direct CO inhalation and increase in % COHb after the therapeutic zone (~10%) during the main stream blood circulation (based on data available in Refs. [48, 49]).

3.14.2 Requirement of exogenous endeavour, why?

The CO-releasing fragment is basically an "exogenous endeavour" that has started the paths for therapeutic medications (Figure 3.56). The exogenous CO donor makes space for searching the affected sites, reaches at the diseased tissue site and struggles with the selected tissues of damaged organs or/and diseased cells. If required, the CO-releasing rate can be controlled and modified according to specs. In order to break down CORMs and CO-releasing materials (CORMats) into CO and metal residue, numerous activators are being used for controlling the CO liberation rate. This has already been practiced in photoCORMs and photoCORMats through ultraviolet (UV), visible and near-IR light with on/off switching facility [50, 51]. The photon energy is also utilized to extract CO from its parent organometallic ligand.

The key benefit of CO's exogenous interactions with the mammalian organism is that it reduces the CO moiety to be directly induced into blood streams for maintaining the COHb under permissible serum levels of up to 10%. Without an exogenous CO administration, it is rather challenging to get fruitful results.

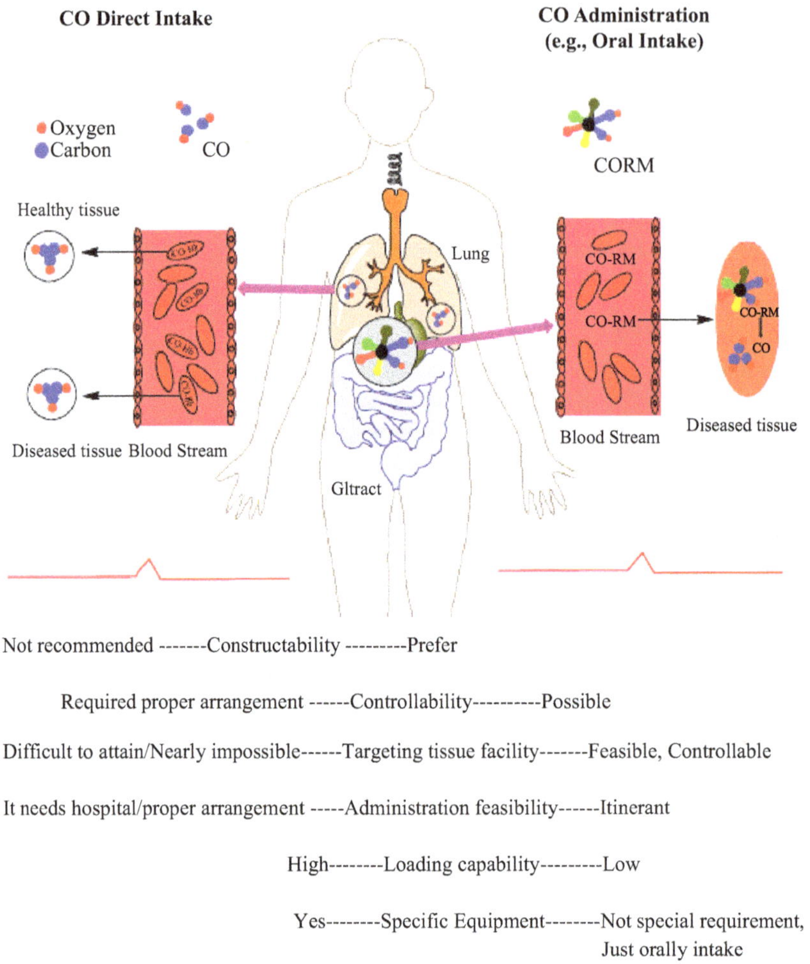

Figure 3.56: The direct and indirect inhalation of CO showing their different biological observation inside a human body and their feasibility.

3.15 Therapeutic applications of CO and CO resealing molecules/ materials

CO despite labelled as "silent killer" for being responsible in forming COHb resulting in CO poisoning emerged as a surprise with its role as a signalling molecule in mammals. This surprising discovery in mammalian physiology has recently elevated interest in this toxic gas among researchers in biochemical and pharmaceutical community. As a result, the design and development of CORMs has interestingly became a huge field in gasotransmitter research. With this aim, the following section mainly involves the description on physiological and therapeutic applied aspects of CO with special emphasis on the related CORMs.

3.15.1 Role of exogenously applied CO gas

3.15.1.1 Reduction in POI complications

Despite its harmful nature, direct inhalation of CO in a controlled way has been found to have some therapeutic benefits also. A clinical trial of controlled CO quantity was conducted on healthy volunteers for temporary paralysis of intestines, which is known as postoperative ileus (POI). Ordinarily, every patient was engaged in this POI after surgery of the abdomen. This clinical study has shown that serious POI complications could be significantly reduced if ~ 250 ppm CO is inhaled before and after the colon surgery [52].

3.15.1.2 Graft survival in organ transplantation

Most remarkable of the CO-assisted results is its capacity to stimulate graft survival as observed in animal models of organ transplantation [53, 54]. Much of the injury related with organ transplant procedures originate from IRI, a form of tissue damage resulting from a combination of oxygen deficiency (ischemia reperfusion injury) during organ preservation and subsequent restoration of blood flow (reperfusion) upon transplantation.

As a strategy, use of CO gas of the concentration of 200–400 ppm at different stages of organ transplantation has been shown to lessen host inflammatory response, thereby reducing the extent of IRI. When donor animals were delivered a dose of CO gas prior to organ procurement, the said organs were found to better resist hypoxic conditions as evidenced by the minimal ischemia–reperfusion-associated damage [55, 56]. It is notable that pre-treatment of CO appears to minimize a variety of pathological indicators comprising T-lymphocytes, tumour necrosis factor, interleukin (IL)-1β and macrophage inflammatory proteins [57–59]. Similarly, administration of CO gas to the recipients after transplantation has been found to boost graft tolerance and uphold survival.

Soares et al. [60] verified that long-term survival of mouse-to-rat cardiac transplants was guaranteed when recipients were exposed to CO gas at different time points after transplantation. In this study, it was found that CO prevented myocardial infarction in the presence of tin protoporphyrin IX (SnPPIX), an HO inhibitor. Therefore, this study supports the view that exogenously delivered CO can act as a graft-promoting agent in place of endogenously produced CO. A number of animal models of cardiac diseases also tell that CO dosage initiates mitochondrial biogenesis due to the anticoagulative properties of CO. Moreover, by reducing platelet aggregation due to IRI in transplantation procedures, CO fosters recovery of vascular endothelial cells (ECs) [61]. In general, the ability of CO to augment graft survival is ascribed to its anti-inflammatory and anti-apoptotic property. Its role in apoptosis, a process of programmed cell death (PCD), indicates even greater significance of CO in signalling mechanisms. Regrettably, application of CO gas in a hospital setting raises technical and safety-related problems. These issues have stimulated the search for exogenous CORMs to deliver CO to biological targets in recent years

3.15.1.3 Blood pressure regulation

Remarkably, studies in the past few decades have established that CO is an important regulator of vasomotor (affecting diameter of blood vessels) tone. Exogenously administered CO relaxes isolated vessels from numerous tissues and animal species [62]. Likewise, administration of CO dilates resistance vessels in several organs, comprising liver, heart, kidney and lung.

Like NO, CO relaxes numerous vascular tissues by activating soluble guanylate cyclase (sGC) in vascular smooth muscle leading to the production of cGMP. However, unlike NO, which forms a penta-coordinate complex with the haeme moiety of the enzyme, CO forms a hexa-coordinate complex and this likely contributes to the lower potency of CO for sGC activation and vasodilation [63]. Although cGMP plays a major role in CO-induced dilation in large vessels, such as the aorta (main artery leaving heart), CO helps relaxation in resistance vessels by stimulating calcium-activated potassium channels in vascular SMC. This kind of action of CO was first identified in rat tail arteries, where CO was shown to directly interact with histidine residues in the channel to increase their open probability [64]. This has subsequently been extended to include renal (kidneys) interlobular arteries and porcine (pig) cerebral arterioles.

Too much production of CO may be damaging. For example: (i) overproduction of CO has been involved in the vascular collapse during septic shock, (ii) endotoxaemia results in the widespread generation of HO-1 within the SMC and EC of large vessels and arterioles [65]. It has been observed that the administration of the HO inhibitor, zinc protoporphyrin-IX annuls endotoxin-persuaded hypotension. This suggests that CO may help in systemic blood pressure decrease. Interestingly, while HO-1 null mice are better able to maintain their systemic blood pressure compared to wild-type animals, they exhibit increased oxidative stress, end organ damage and mortality during endotoxaemia [66].

Coherent with the above findings, the exogenous administration of CO has been demonstrated to mitigate/lessen the worsening of the lung, kidney and liver function during septic shock [67]. Therefore, while CO may participate in the hypotensive response to sepsis it may play an important role in preserving organ function.

3.15.2 Carbon monoxide–releasing molecules (CORMs) and photoactive CORMs or photoCORMs

As a ligand, coordination chemistry of CO has been extensively explored with regard to organometallic chemistry and catalysis. The innovation of the beneficial roles of CO has now added additional interest in metal carbonyl complexes of transition metals as CORMs for targeted CO delivery. As many transition metal carbonyl complexes exhibit CO release upon light illumination, the very desirable property of control on CO delivery is also possible with these CO donors. In the past few years, a number of photoCORMs have been synthesized that release CO upon illumination with UV or visible light.

3.15.2.1 Carbon monoxide–releasing molecules (CORMs)

3.15.2.1.1 Effect of CORMs on bacteria

Due to well defined bactericidal implications of several reported CORMs by having the potentiality to cross the cell surface (cell membrane/wall) and interfere with the main machinery of the bacterial cell. Many bacterial strains including *E. coli, H. pylori, S. aureus* and *P. aeruginosa* have been studied to determine their sensitivity against some selected CORMs, and the main role of CO in such bioactions has been fascinatingly documented. For instance, Bang et al. [68] studied the very significant antibacterial effect of CORM-2, that is, $Ru_2Cl_4(CO)_6$ (Figure 3.41),against multidrug resistant *E. coli* related to uropathogenesis,and satisfactory results were concluded using biofilms of the target material for selective surface actions.

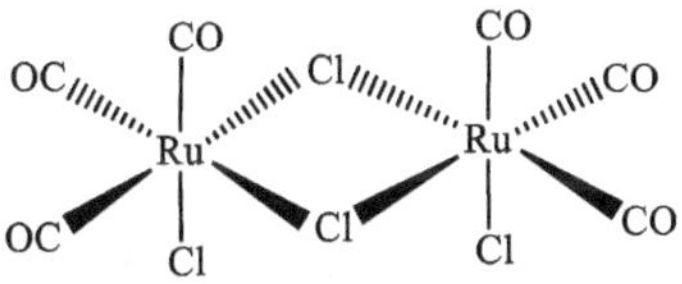

Figure 3.57: Chemical structure of CORM-2.

With a different and more meaningful anti-infectious potentiality of several CORMs (Figure 3.58) selected by Desmard et al. [69] against *P. aeruginosa* to find the dependence of respective activity on the structural composition remarked that CORM-2 and CORM-3 showed CO-releasing phenomenon after invading the biological cell and started the behaviour after 1 min within the cell. However, boron-based CORM, CORM-A1, was shown to represent a slow releaser. In addition to CORMs given in Figure 3.57, another CO-releaser namely, CORM-371 (Figure 3.58) containingMn as a central metal was also found to be a slow CO releaser during the same study. All the forms of CORMs inhibited the growth of *P. aeruginosa*, but the Ru-based compounds were shown to have the best anti-*Pseudomonas* action.

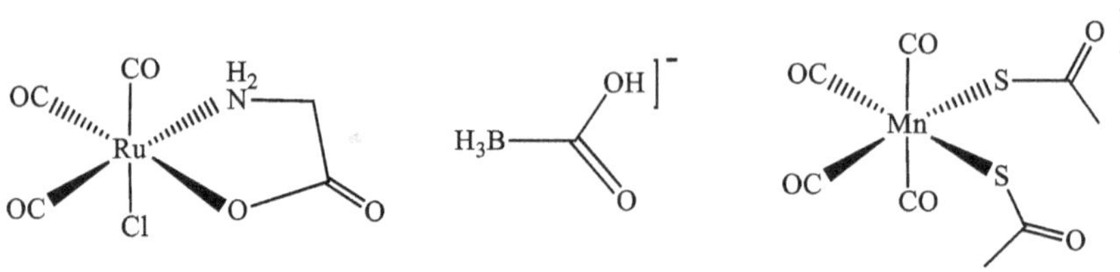

Figure 3.58: Chemical structures of CORM-3, CORM-A1 and CORM-371.

3.15.2.1.2 CORMs and diabetic neuropathy

The surprising biological relevance of CORMs motivated several research groups to find their applicability in the neuropathy problems of diabetic (type I) people. More importantly, it may be noted that for such evaluations, cannabinoid receptor 1 (CB1R) and 2 (CB2R) (agonists) are generally targeted to reduce pain. Castany and co-workers in the similar study used CORM-2 and CB2R (JWH-015) in a combinatory fashion, and for CORM-2 separately for STZ rat models (Figure 3.59). The main effect of CORM-2 in reducing mechanical allodynia and thermal allodynia was found in notable expression. The combinatory treatment of CORM-2 with JWH-015 was found to have enhancing stimulation upon the levels of HO-1 and CB2R, and induced decreasing trend in the NOS1 levels. Therefore, the related biological parameters of neuropathy in special reference to diabetic conditions could be treated using CORM-2 [70].

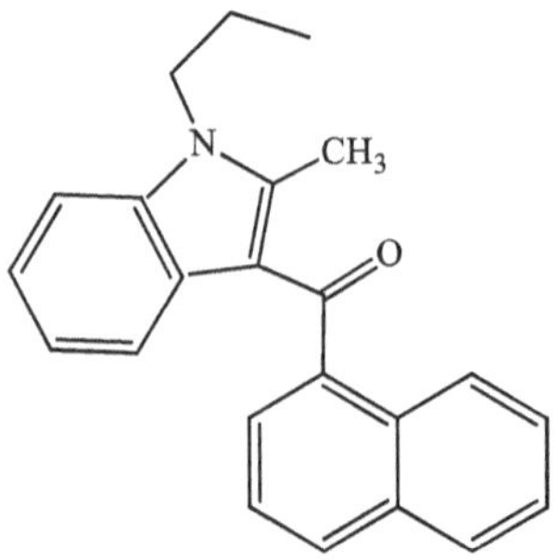

Figure 3.59: Structure of JWH-015.

Hervera et al. in 2013 noted the beneficial use of CORM-2 as a treatment option for nociceptive condition in the murine rat models associated with sciatic nerve injury [71]. From the results collected from the study it is clear that the medication using CORM causes reduction in pain caused by injured sciatic nerve showing hypersensitivity in the form of mechanical and thermal sense. Similarly, the applied interest reported by Nikolic et al. [72] using CORM-A1 shows averting diabetic results obtained via the beta cell regenerative effect. Remarkably, CORM-A1 also showed good response in modulating the immune system by dropping the level of T-helper cells which is very helpful in diabetes treatment. Therefore, the diabetic treatment and the mitigation of associated severity could be controlled medically by using CORMs.

3.15.2.1.3 Anti-inflammatory response of CORMs

It is a known fact that pathological state is accompanied by symptomatic expression in the form of inflammation developed by the affected tissue, an organ or a whole system. In association with other medicinal relevances noted for CORMs, the well-remarked anti-inflammatory response recorded against colitis (colon inflammation) by Steiger and co-workers [73], using CORM-2 for 2 days, suggest the efficient CO-releasing role in anti-inflammatory medication. Moreover, Nagao et al. [74] focused their study on application of nanodimensional CO donor, CO-HbV (CO-haemoglobin vesicles), upon the same colitis inflammation found in rat models and observed the considerable reduction in the harmful effects developed therein.

Inflammatory response is intervened by several factors including the main pivotal role of dendritic cells also, as dendritic cells serve as a protective tool during viral invasion of a host, and also the exposure of dendritic cells to bacterial lipopolysaccharide (LPS) bring forth inflammatory response through binding with toll-like receptor 4 (TLR4) or via the complex "myeloid differentiation factor-2" (MD2) located on the cell surface. In finding the link of this inflammatory approach with the usage of CO, the work reported by Riquelme and co-workers [75] indicates that CO prevents the LPS-based inflammatory effect, and hence acts as an efficient anti-inflammatory drug. Moreover, the reduction in surface expression of TLR4/MD2 was also recorded, but no effect was noted in control cells. The LPS-received animal models in the form of mice regained normalcy on CO exposure.

A general perception accompanying with inflammatory diseases is excessive blood clotting. Maruyama et al. [76] investigated the CORM-2 application for a blood coagulatory role by passing the compound through a tissue factor and plasminogen activator inhibitor type 1 expressiveness, and determined the respective signalling pathways (MAPK and NF-κB), which all stimulate thrombosis (formation of blood clot). Taking the findings concluded from the study into consideration, it was concluded that CORM-2 intake decreases blood clotting induced by inflammation.

Another inflammatory disorder is uveitis, which is an inflammation of uvea of the eye. Inflammation under such context is blindness. Fagone and co-workers [77] have examined the role of the CORM for the treatment of inflammation of uvea of the eye. Accordingly, a CO liberating molecule, CORM-A1, is effective for retinal activity in the eye. CORM-A1 induced the expression showed lower levels of interferon gamma and IL-17A (inflammatory cytokine) and increased amounts of IL-10 (anti-inflammatory cytokine) in rat models in comparison with the non-CORM-treated uveitis-induced mice.

3.15.2.1.4 CORMs and sepsis: a correlation with systemic inflammation

One of the serious medical conditions characterized by dysregulated systemic inflammatory responses followed by immunosuppression is known as sepsis. This has now become a major challenge for scientists and clinicians. In order to study the pathophysiology of sepsis, different animal models have been developed. Of these models, polymicrobial sepsis induced by caecal ligation and puncture (CLP) is the most extensively used model. This is because of its close resemblance with the progression and characteristics of human sepsis. Lee and co-workers [78] demonstrated that when mice were impaired with CLP leading to systemic inflammation, CO was seen to enhance survival through a series of processes involving autophagy (degrading damaged intracellular contents) and another way called as phagocytosis (bacteria engulfing by cells). It was noticed that CO exposure induces autophagy proteins, Atg7, beclin 1 and LC3B. Also, the study reveals that CO causes the number of autophagosomes to increase.

The acute kidney injury (AKI) is a general problem persuaded by sepsis. A harmful effect of AKI is seen to be reversed by CO medication. Generally, blood urea nitrogen and serum (liquid part of blood) creatinine are taken as indicators in the onset of AKI. The concentration of these two indicators is found to be increased after CLP-based sepsis. Surprisingly, when CORM-2 treatment resulted in the reduction of concentrations of these two, Ismailova and co-workers reported that sepsis with AKI bring about a larger death rate, killing half of the selected animal models (rats) in 3 days. However, CORM-2-treated rats showed higher longevity by decreasing markers for AKI. Hence, this study concludes that CORM-2 can be used to neutralize AKI-induced toxicity [79].

A study over application of CORM-3 in myocardial dysfunction conducted by Zhang et al. [80] demonstrates to have an implicative effect in lessening this sepsis problem. The same application of this CO-liberating drug provides satisfactory therapeutic benefit for lung sepsis. Notably, the mechanistic approach myocardial applied aspect is indicative of the role of CO in inhibiting the NLRP3 activation-fibrobast inflammasome in cardiac tissue. Hence, for prevention of myocardial apoptosis and protection for lung injury, CORM-3 administered intraperitoneally restores the normal functioning.

3.15.2.1.5 Obesity affected by CORMs

Inflammatory response is not only because of human body invasion by foreign bodies or injury but prolonged state of obesity can also cause inflammation. In context with the health problems associated with such a condition in special reference to inflammation, Hosick and co-workers [81] in 2016 reported chronic treatmental investigation using CORM-A1 to find the respective medical use for obesity and associated hyperglycaemia and diabetes. The study targets four sets of mice based on medication condition: (i) first group was retained with a dietary control confined to high-fat intake; (ii) second group was supposed to receive a fat diet along with a saline injection (intraperitoneal); (iii) third set comprised a set fed with a diet having fatty composition along with CORM-A1; and (iv) the fourth one was just like the third one but with inactivated form of the CORM (iCORM). The results showed that CORM-A1-treated mice followed a considerable lack of gaining weight despite high-fat feeding for 18 weeks comparing with the other medication sets. CORM-A1 also resulted in 45% mass gain in case of lean rats. The overall study suggests an optimal effect upon the fat metabolic pathways and also a well-defined anti-inflammatory response was concluded. In another study reported by Zheng et al. [82] to find positive correlation of CORMs with obesity, similar results of decreasing obesity by the application of CORMs were noted.

3.15.2.1.6 CORMs and lung functioning

Presently, the advantageous outcome of CO on lung complications is in much dispute, although in vivo studies over this subject line persuaded by acid nasal infusion [83], aeroallergens [84], hyperoxia [85] (excessive oxygen supply, opposite of hypoxia) and mechanical stretch [86] specify effectiveness of inhaled CO. However, several other studies [87] proved any non-defensive effect of CO in the context of lung health. Abid et al. [88] have recently demonstrated the beneficial effects of CORM, CORM-3, on reversal pulmonary hypertension with the fact that pH change occurs as a result of certain diseases.

3.15.2.1.7 Role of CORMs in vascularization and cancer

CORM-2 has been found to stop angiogenesis [89]. In other words, this means that CORM-2 halts the initialization generally brought up by VEGF (vascular endothelial growth factor). VEGF is supposed to enhance actin stress in ECs resulting in spindle-like networking. The CORM-2 inhibits this spindle formation. In a comprehensive way, this can be expressed by referring CORM-2 as a preventive candidate that does not allow VEGF-dependent cellular proliferation to happen like retinoblastoma protein-tumour suppressor creates cell cycle restrictions. In another study [90] using CORM-401 (Figure 3.60), it was noted that this CORM shows a promotional effect on vasorelaxation of pre-contracted aortic rings (3-fold active than CORM-A1). Moreover, the bioactivity of this fashion got enhanced in H_2O_2 presence. The proangiogenic potentiality of CORM-401 was found to be genetically expressive in the form of VEGF and also in IL-8. In another report of CORM-2 applied for remarking connection with pancreatic fibrosis, Schwer et al. [91] emphasized the use in the prevention of pancreatic stellate cells responsible for protein synthesis linked with fibrosis [91]. These findings also suggest the anticancer role of the selected CORM by cell aggregation.

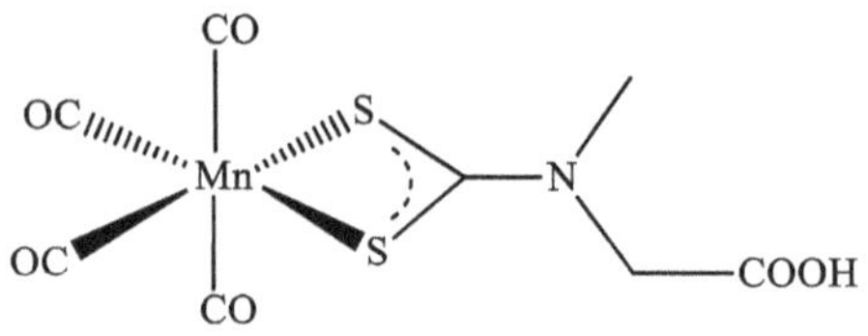

Figure 3.60: Structure of CORM-401.

3.15.2.1.8 Cardiovascular effects of CORMs

Since the vascularization and other blood componental regulation role of CORMs as discussed above puts forth an opinion to find the relevance of these drugs in heart-associated problems, several investigations [92] are evident of the fact to declare CORMs in special reference to CORM-3 and CORM-2 as CO releasers of having cardioprotective properties, improving functioning and structural stability of the heart supported by the experimental studies led upon rat models. Recently, Tsai and co-workers [93] successfully established this relation CORM-2 with the smooth muscle of aortic zone by finding persuasion related to angiotensin II reduction. The mechanistic details reveal here the inhibitory role of the CORM for metalloproteinases-9 expression along with ROS/IL-6 production. Similarly, Soni and co-workers [94] have demonstrated the use of the same CORM in heart functioning in reference to doxorubicin (DXR) as a reference drug (anticancer) having side effects of cardiotoxicity. The study reveals the use of CORM-2 to eliminate the cardiotoxic effects of DXR and hence proves the well-remarked cardiovascular role of CORM-2 in a very beneficial way. Also, it may be noted that cardiac functioning since being a electrophysiological process, and the role of CORMs in terms of action potential and sinus rhythm studied by Abramochkin and co-workers [95] indicates that CORMs behave in chronotropic drug style in a positive way rendering the effect to enhance heart rate by intervening as an electrochemically active component.

3.15.2.1.9 Procoagulant and anticoagulant effects of CORMs

Due to the well-defined relation of CORMs with blood regulatory effects, the investigation to find the effect of CO releasing on blood clotting or coagulation was determined by Nielsen and co-workers [96] employing the plasma contact with CORM-2. The findings recorded by the group indicates the enhanced thrombi growth. Both the activated and inactivated forms of the target CORM was experimented, and it was observed that the iCORM is half active as compared to CORM-2 and the blood coagulation rate gets increased by CORM exposure because of thrombin–fibrinogen-enhanced interactions. Similar work was carried out by Nielsen and co-workers [97] and Machovec and co-workers [98] to validate the earlier observations. Some of the observations found so were contrary to the results reported earlier [96], indicating no substantial effect of thrombus formation as a result of CORM-2. Though the already reported enhanced rate of coagulation upon CORM-2 treatment was shown in the validation report as well, the outcomes reported for CORM-2 indicate a well-remarked blood coagulatory role. Later in an indirect way to find relation of CORM-2 with blood coagulation in relevance with the same assumptions, Nielsen and Bazzell [99] observed snake venom weakening effect displayed by CORM-2.

3.15.2.1.10 Visionary role of CORMs (eye health)

Triggered by the death of the retinal ganglion cells (RGCs), glaucoma is a continuing optic neuropathy. It is the leading cause of irretrievable blindness globally. The pathway through which this progressive RGC death occurs is not fully understood. Nevertheless, it is clear that numerous factors are responsible for the common effect of ganglion cell death. Clinically, it is well established that the major risk factor for glaucoma is the higher IOP [100]. It is notable that in open-angle glaucoma, higher IOP occurs from an imbalance between production and outflow of aqueous humour (AqH). Although scanty data are available in literature on the beneficial role of CORMs in ophthalmic diseases, yet the works reported so far in this direction are presented below:

(i) Recently, it has been observed that CO represents a crucial molecular recognition for the healthy status of ocular system supported by the tests reported by Bucolo and Drago [101] using CO at a very small concentration.
(ii) Stagni and co-workers [102] while administrating CORM-3 observed considerable effect on decreasing IOP in rabbit animal models. CORM-3 is structurally represented in Figure 3.61. The CO-mediated experimental record was observed for 30 min following 24 h exposure to CORM-3. This suggests that the CO has an affirmed relation with IOP-lowering phenomenon.

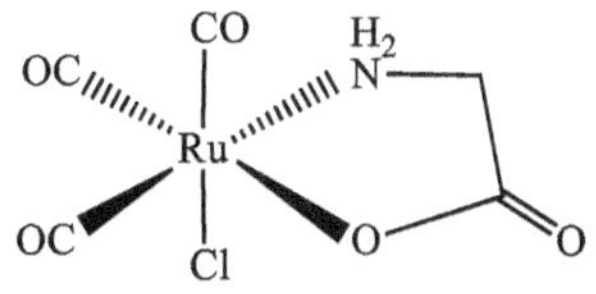

Figure 3.61: Structure of CORM-3.

The effects of CORMs on IOP are a result of increase in activation of sGC. The activation of sGC by CO-dependent donors (e.g. CORM-3) increases the rate of AqH flow through the trabecular meshwork (TM) and Schlemm's canal (Figure 3.62). In fact, the change in outflow facility occurs concurrently with sGC-persuaded decreases in TM cell volume. sGC consists of an α-subunit and a smaller haeme-containing β-subunit, and both constitute the active enzyme. The binding of CO to sGC results in the formation of cGMP from guanosine 5′-triphosphate (GTP). Increased cGMP activates protein kinase G (PKG) and causes the subsequent phosphorylation of target proteins. It is believed that CO, acting through the sGC, cGMP and PKG pathways, decreases the TM cell volume.

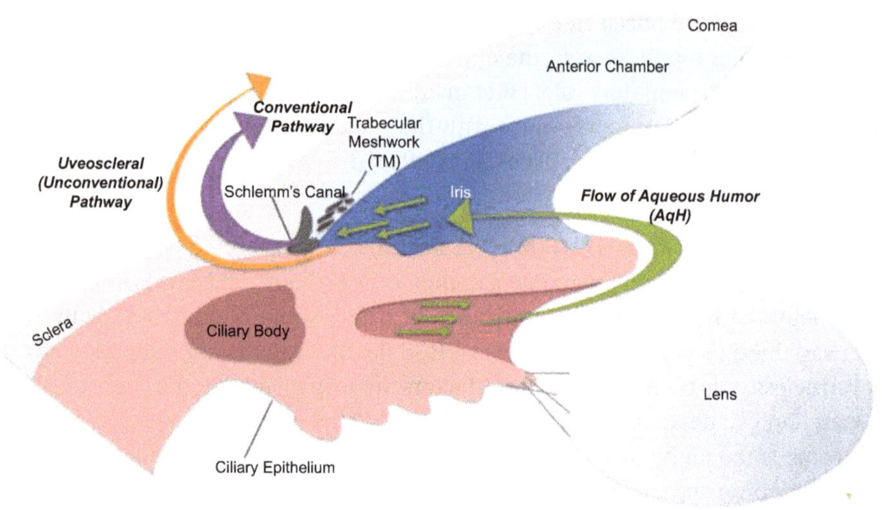

Figure 3.62: Diagrammatic representation of aqueous humour (AqH) equality: (a) AqH at the ciliary body in the eye flows (**green arrows**) through two routes independently that control AqH dynamics: (i) via the trabecular meshwork (TM) and Schlemm's canal (**purple arrow**) (conventional route) and (ii) via the uveoscleral tract (**orange arrow**) (non-conventional route). (b) The balance of (AqH) production at the ciliary body and elimination in the anterior chamber establishes intraocular pressure (IOP) in the eye.

The TM is supposed to be a smooth muscle-like tissue with contracting properties, and it is enriched with BK_{Ca} channels. In the TM cells, the probability that the BK_{Ca} channels will open is increased by membrane stretching, depolarization and intracellular Ca^{2+} levels [103]. Contraction and relaxation of the cells are believed to regulate AqH outflow. This mechanism could play an important role in mediating the CO-persuaded decrease in IOP.

(iii) Uveitis, a sight-threatening inflammatory disease of the eye, is the third leading cause of blindness in developed countries. Usually, it occurs at the age between 20 and 60 years, and males and females are usually affected in the same way. The yearly rate of occurrence is expected to be between 17 and 52 per 100,000 persons, with a frequency of 38–714 per 100,000 persons [104]. Thus, the burden of this eye disease is very significant. Uveitis may be classified as anterior, intermediate, posterior or panuveitis based on the anatomical influence of the eye.

The conventional pharmacological treatment of uveitis includes corticosteroids and immunosuppressive agents, which are faulty by their side effects and only efficient in half of the patients with repeated disease. New therapeutic strategies are thus strongly required.

From outside, administered CO may represent an effective treatment for conditions characterized by a dysregulated inflammatory response. CORMs are a novel class of compounds having capacity of carrying and liberating controlled quantities of CO. Among CORMs, CORM-A1 (sodium boranocarbonate) (Figure 3.63) represents the first example of water-soluble CO releaser (Figure 3.64).

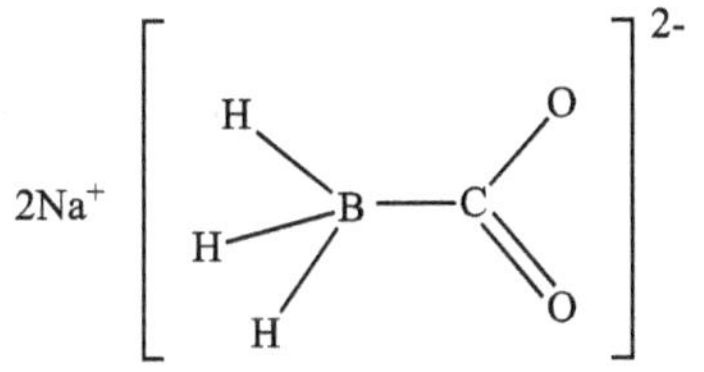

Figure 3.63: Chemical structure of CORM-A1.

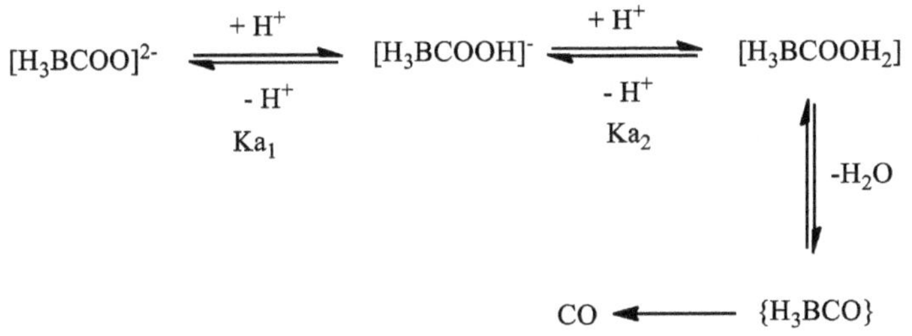

Figure 3.64: CO-release pathways from CORM-A1.

A study conducted by Nicoletti and co-workers [77] demonstrates that CORM-A1 under a late prophylactic regime is able to considerably improve the natural course of experimental autoimmune uveoretinitis, a rodent model of immunoinflammatory posterior uveitis. This study strongly supports the development of CORM-A1 as a prospective new drug for treatment of patients with non-infectious posterior uveitis.

3.15.2.2 PhotoCORMs

The term "photoCORM" was coined for the first time by Ford et al. They observed the delivery of small gasotransmitters (NO and CO) from transition metal complexes, especially a tungsten(0) complex $Na_3[W(CO)_5(TPPTS)]$ (Figure 3.65) displaying CO release and solubility in water [105]. Motterlini and co-workers [106] reported the first biological use of light-persuaded CO release from chemical compounds, namely, the pure metal carbonyl complexes, iron pentacarbonyl and dimanganese decacarbonyl (Figure 3.65).

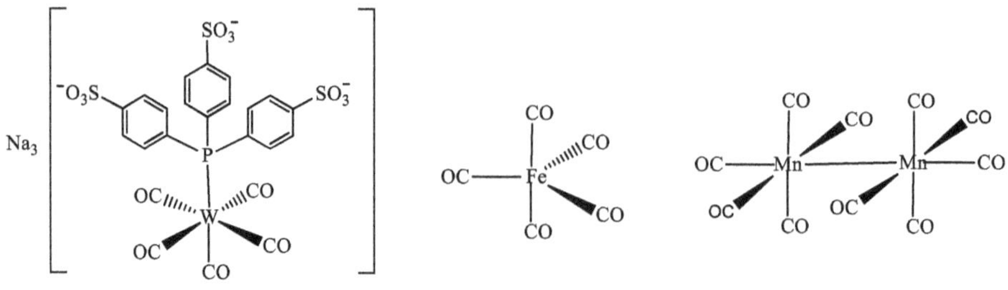

Sodium tris(sulphonatophenyl)-phosphinepentacarbonyltungsten(0)

Iron pentacarbonyl

Dimanganese decacarbonyl

Figure 3.65: Some photoactive CO-releasing metal carbonyls (photoCORMs).

3.15.2.2.1 Therapeutic applications of photoCORMs

(i) The first compound used as CORM was commercially available homoleptic metal carbonyl, namely, $[Mn_2(CO)_{10}]$ (dimanganese decacarbonyl) owing to its well-known photoactivity. In a revolutionary study, Motterlini et al. established that irradiation of $[Fe(CO)_5]$ and $[Mn_2(CO)_{10}]$ with a cold light source leads to CO release and such compounds persuade vasorelaxation in pre-contracted rat aortic rings much comparable to endogenously produced CO [106]. Loss of CO from these compounds proceeds through dissociative processes that produce short-lived intermediates. These short-lived species further react with solvents and oxygen to yield a variety of products. For instance with $[Fe(CO)_5]$, quantum yield measurements indicate that the parent carbonyl compound forms both the mono- and disubstituted carbonyl species $[Fe(CO)_4L]$ and $[Fe(CO)_3L_2]$ (L = other ligands present in the solution) upon illumination with UV radiation [107]. On the other hand, $[Mn_2(CO)_{10}]$ displays CO loss producing the transient species $Mn_2(CO)_9$, as opposed to the typical metal–metal bond cleavage observed during continuous photolysis of dinuclear metal carbonyls [108].

Additional carbonyl compound derived from bidentate 2-aminoethanethiol or cysteamine, namely, $[Fe(SCH_2CH_2NH_2)_2.(CO)_2]$ (Figure 3.66), known as CORM-S1, displays CO photorelease when exposed to visible light [109] of wavelength λ = 400 nm generating two CO molecules per CORM-S1. This photoCORM has been used to motivate voltage and Ca^{2+}-activated potassium channels in a light-

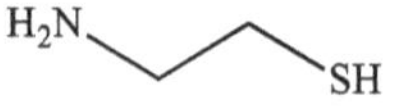

2-aminoethanethiol (or cysteamine)

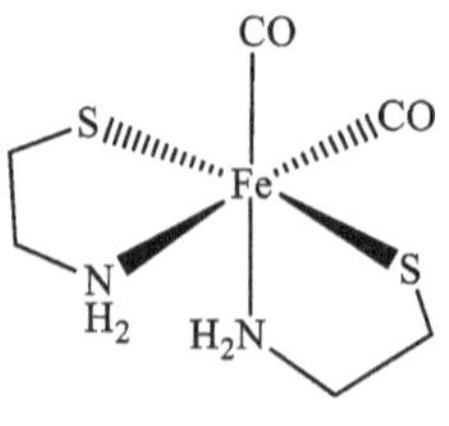

Figure 3.66: Chemical structure of 2-aminoethanethiol and $[Fe(SCH_2CH_2NH_2)_2(CO)_2]$.

dependent manner. Undoubtedly, carbonyl compounds of this type could find use as sustainable photoCORMs for physiological applications.

(ii) Niesel et al. [110] testified light-persuaded CO release from *fac*-$[Mn(tpm)(CO)_3]PF_6$ [tpm = tris(pyrazolyl)methane] (Figure 3.67). This manganese carbonyl compound liberates CO when exposed to the light of wavelength 365 nm. This compound has validated its potential as a photodynamic therapy agent. The CO release from this compound has been used for cytotoxic purposes (HT29 colon cancer cells). The light-induced CO release at 365 nm only led to considerable activity as compared to non-light-induced forms. Later, a nanodimensional approach led upon *fac*-$[Mn(tpm)(CO)_3]PF_6$ by Dördelmann et al. [111] in 2011 reported surprising results in the context of the earlier observations confining the drug-likeness more pronouncing because of light activation (Figure 3.67).

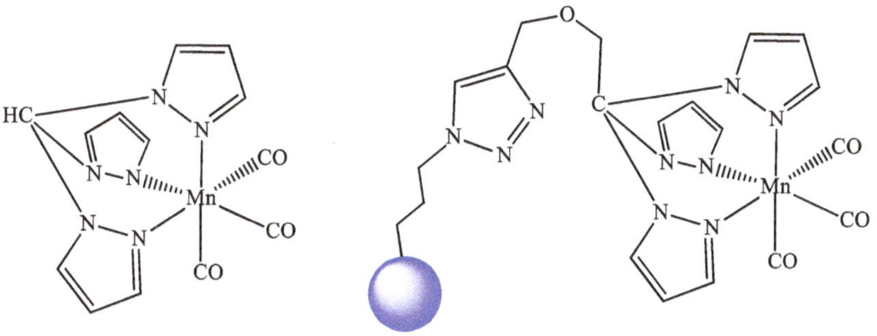

$[Mn(tmp)(CO)_3]PF_6$ SiO_2NP-$[Mn(tmp)(CO)_3]PF_6$

Figure 3.67: Structural representation of fac-$[Mn(tpm)_3(CO)_3]PF_6$ and its SiO_2-based nanoform.

(iii) In the motivation towards enhancing photoactivity among CORMs, Mascharak et al. [112] suggested the increase of conjugation in the ligational structures while targeting $[LMn(CO)_3]^+$ complexes (L refers to a tripodal ligand). A set of polypyridine and/or quinoline derivatives were used to develop benzenoid system with an enhanced conjugation (Figure 3.68). The photolability of CO in manganese tricarbonyls was observed to follow the same in fac-$[Mn(dpa)(CO)_3]$ ClO_4, fac-$[Mn(tpa)(CO)_3]ClO_4$ and fac-$[Mn(pqa)$-$CO)_3]ClO_4$, regarding lower energy requirement to ease the CO-releasing phenomenon. The compound fac-$[Mn(pqa)$ $(CO)_3]ClO_4$ shown to absorb in 300–400 nm energy range under low power consumption ~ 100 mW towards visible light to release the CO. Being water-soluble CORM and easy CORM the complex has been used as photoCORM for vasorelaxation studies in mice animal models. Satisfactory results were gained.

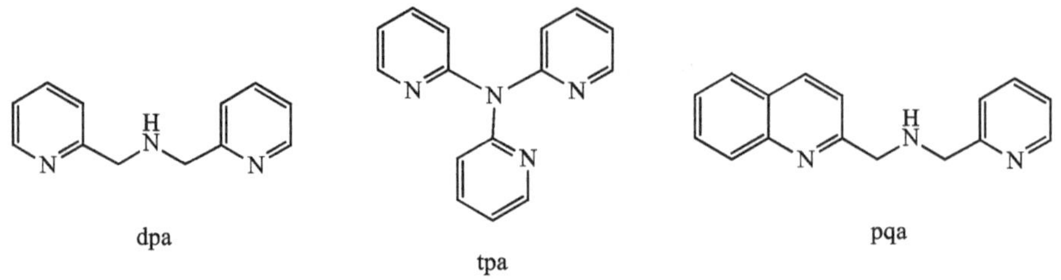

Figure 3.68: Structures of polypyridine ligands used in the synthesis of fac-Mn(I).

Tricarbonyl complexes as CORMs under light irradiation

(iv) Based on the search of efficacious CORM constructed on the conjugation concept, Mascharak and co-workers [113] extended this applied aspect by employing conjugated Schiff bases to serve as a ligand system behaving as electron-rich anchoring molecules towards Mn(I)CO complexes to display visible range-driven CO release. While studying the two compounds of this fashion, namely, *fac*-[MnBr(pmtpm)$(CO)_3$] (where pmtpm = 2-pyridyl-*N*-(2′-methylthiophenyl)-methyleneimine)) and [MnBr(qmtpm)$(CO)_3$] (where qmtpm = 2-quinoline-*N*-(2′-methyl-thiophenyl)methyleneimine) (Figure 3.69), it was found that the respective MLCT (metal to ligand charge transfer transition) excitation patterns follow to absorb at 500 and 535 nm, respectively. Theoretical computational results involving the molecular orbital composition analysis around the coordination sphere suggest that the nature of such transition(s) pertaining to Mn–CO bonding orbital(s) is predominantly ligand-based molecular orbital (LUMO in both cases), and hence contributes towards photobands for photoinduced CO release.

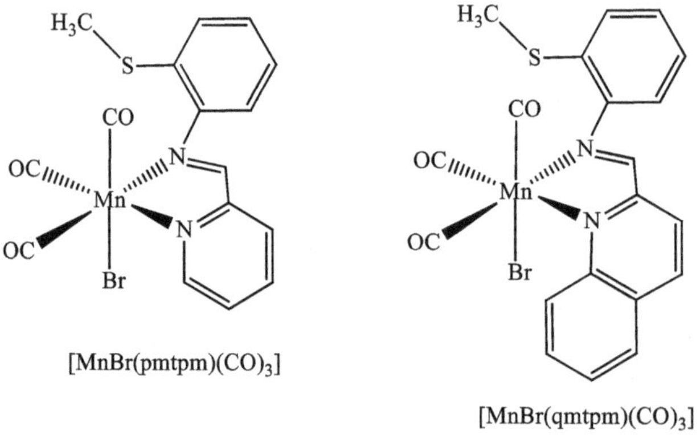

Figure 3.69: PhotoCORMs derived from designed Schiff base ligands.

The legitimacy of the above hypothesis has been further established [114] in the two Ru(II) carbonyl complexes [RuCl(qmtpm)(CO)(PPh_3)]BF_4 and [RuCl(qmtpm)$(CO)_2$]ClO_4 [qmtpm = 2-quinoline-*N*-(2′-methylthiophenyl)methyleneimine]. Even though the low affinity of the thioether group for Mn(I) centre caused in coordination behaviour of the N,N,S-ligand: pmtpm [2-pyridyl-N-(2′-methylthio-phenyl)methyleneimine] and qmtpm in the Mn(I) carbonyls as bidentate [113], qmtpm coordinates in a tridentate fashion in these two Ru(II) carbonyl complexes (Figure 3.70). Both the ruthenium complexes in ethanol rapidly release CO under light illumination. Remarkably, $[RuCl(qmtpm)(CO)(PPh_3)]BF_4$ releases one equivalent of CO upon exposure to visible light. DFT and TDDFT results reveal that the 460 nm band of this photoCORM also corresponds to MLCT transitions from two molecular orbitals of predominantly Ru–CO bonding character to the LUMO comprised of the π^* orbitals of the quinoline, the imine functionality and the thiophenyl moiety of the qmtpm ligand structure. Additionally, the DFT data reveal that the addition of the second CO ligand to the Ru(II) centre in $[RuCl(qmtpm)(CO)_2]ClO_4$ shifts the MLCT band to 380 nm (blue shift) from 460 nm in case of the monocarbonyl $[RuCl(qmtpm)(CO)(PPh_3)]BF_4\}$ due to lowering of the HOMO levels. Altogether, these findings demonstrate that (a) photoCORMs with significant CO-donating capacities could be synthesized with the use of designed ligands that allow MLCT transitions in the visible (or near IR) region and (b) fewer CO ligands must be included in photoCORMs for better sensitivity to visible light.

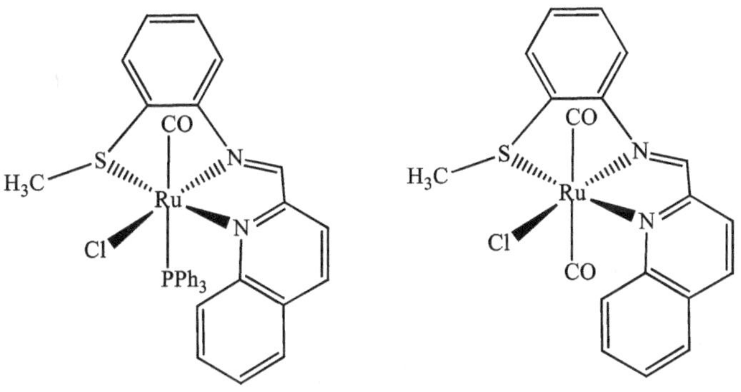

$[RuCl(qmtpm)(CO)(PPh_3)]BF_4$ $[RuCl(qmtpm)(CO)_2]ClO_4$

Figure 3.70: Ru-based photoCORMs involving Schiff base ligand.

3.16 CO as a signalling molecule in plants: a vegetation echo of CO

3.16.1 Introduction

CO is recognized as a second member of gasotransmitter family that renders essential bioprocesses discovered after NO. As has been discussed in the NO relation with plants, the gaseous molecule in plants is playing so many vital roles and is an indirect indicator of biotic and abiotic stresses in plants. Similarly, the associated role of CO has been discovered surprisingly in plants performing major physiological actions in plants. Like in NO case, the environmental stresses together with other growth-specific conditions have been found to have profound implications on the production of CO in plants. Till now, the use of CO is facilitated externally by virtue of fumigant forms of CO, water solutions or in the other synthesized CORMs including haematin and haemin (Figure 3.71). In genetic approach, CO has been found to act as a crucial plant growth regulator and hence the plant development. The CO role begins to appear expressive starting from the germination stage and is equally important till the plant death/growing old. Therefore, as mentioned earlier, these results suggest the significance of CO in plant world in a similar fashion as has been described in NO role in plant kingdom [115]. So, it is logical to state that like "NO is good news for plants" can also be titled "CO is good news for plants".

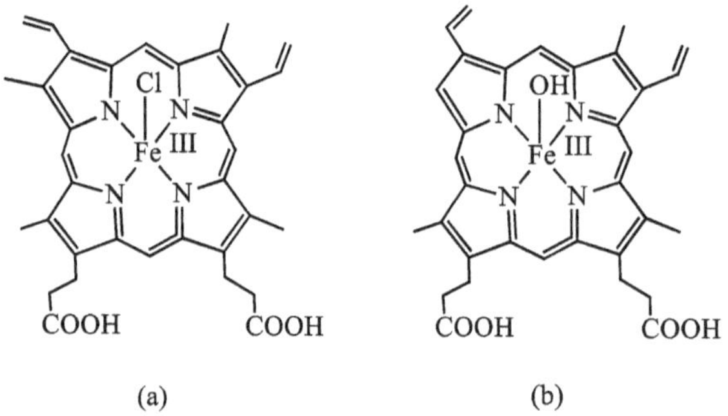

Figure 3.71: Structure of (a) haemin and (b) haematin.

3.16.2 Synthesis of CO in plants

The appearance of CO biosynthesis in plants was reported for the first time in 1959 by Wilks [116]. Thereafter, the photoproduction of CO in living plants was also acknowledged in 1999 by Schade and co-workers [117]. Moreover, in 2002, Muramoto and co-workers [118] observed a plastid (specialized plant cell) HO (AtHO1) recombinant protein, which was able to catalyse the formation of CO in plants from haeme molecules in vitro.

Studies conducted by Xuan and co-workers [119] in 2008 on cucumber provided pharmacological, physiological and molecular evidence that auxin (a plant hormone) rapidly triggers HO activity and this produces CO. The CO so produced then triggers the signal transduction events that lead to the auxin responses of adventitious root (AR) formation in cucumber.

Researches carried out in the recent past have shown that the genes for HOs have been recognized in a range of plant species. It includes a small family of four members in all. This can be classified into two sub-families: HY1, HO1, HO3 and HO4, all belonging to the HO1 sub-family, while HO2 is the only number of the HO2 sub-family [120]. All members of the HO1 sub-family (HY1, HO3 and HO4) can transform haeme proteins to biliverdin-IXα with a simultaneous release of CO and Fe^{2+}, while the sub-family HO_2 does not exhibit HO activity. The biliverdin-IXα (BV-IXα) can be reduced to bilirubin-IXα (BR-IXα) by an enzyme biliverdin reductase.

As discussed above, HOs have been known as the main route of CO production in plants from haeme proteins, and the opposite results have been proposed by Zilli et al. [121] in soybean. These authors proposed that HOs are not the main source of CO generation in soybean plants. Instead, lipid peroxidation and ureide metabolism could also be assumed as potential sources of CO. Another possible route of CO release in plants is the breaking of haeme methylene bridges in the presence of H_2O_2 or ascorbic acid [122]. Altogether, synthesis of CO is a complex physiological process and more non-enzymatic biosynthetic routes of CO need to be elucidated. The complete routes of CO generation in plants are shown in Figure 3.72.

3.16.3 Role of CO in plant growth and development

Remarkably, a high concentration of external CO is toxic to a variety of plants. Nevertheless, CO at an appropriate concentration involves in many important physiological processes as an active signalling mediator. CO has been studied by several research groups to explain the roles of this signalling molecule in plant growth and development. Moreover, increasing evidence in plants for CO has revealed that it is used in a number of intercellular and intracellular biological functioning. For instance, CO is found to delay gibberellins (GA) (plant hormone)-activated PCD in wheat aleurone cells byup-regulating ascorbate peroxidase (APX) and catalase (CAT) expression, and decreasing H_2O_2 overproduction [123]. So far, the studies concerning the roles of CO in plants are generally focussing on seed germination, root development and stomata (plant pore) closure.

3.16.3.1 Seed germination

Particularly, seed germination is a highly specific phase in plant life. This is the first vital step for seedling formation. Moreover, it is a critical step in a plant's life cycle and is controlled by a wide range of endogenous and environmental factors. It has been demonstrated by numerous researches that CO contributes a beneficial effect in several plants for promoting seed germination in a dose-dependent manner. The application of low concentrations of exogenous CO (0.1% or 1%) stimulated seed germination of foxtail (*Setaria faberi*, a type of grass) under favourable temperature

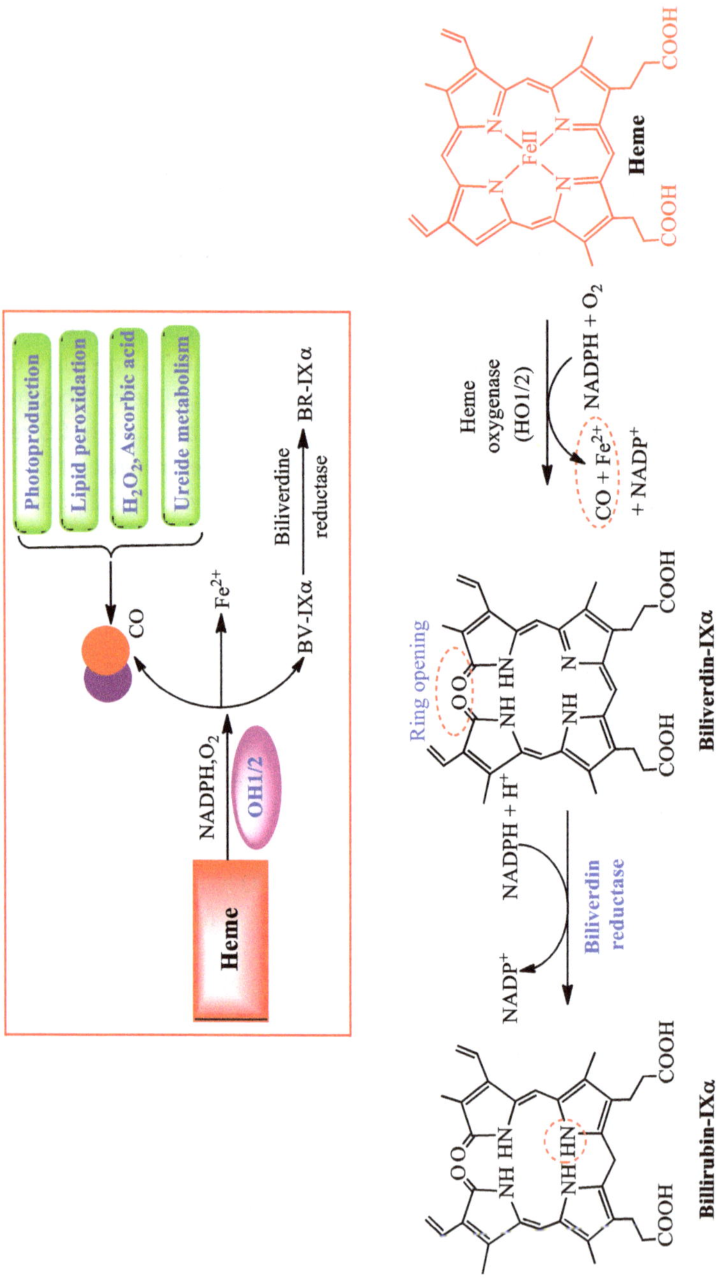

Figure 3.72: The complete routes of CO production in plants.

and moisture conditions, while germination decreased with the addition of 75% CO due to the inhibition of mitochondrial respiration [124].

In another study, it is found that both CO donor haeme and aqueous solution of CO dose dependently accelerate the physiological process of seed germination in *Oryza sativa* (a type of grass) [125]. This is achieved by activation of amylase activity and by increasing the formation of energy resources. Similarly, CO serves as a positive regulator and is also involved in the process of seed germination in wheat [126] and *Brassica nigra* (black mustard) [127].

3.16.3.2 Root development

The root systems are found to play vital roles in plant nourishment and acquisition of water. CO has shown positive results in regulating plant root development. For instance, the stimulating effects of auxin [indole-3 acetic acid (IAA)] or NO on root elongation were duplicated by application of aqueous solution of CO with different saturations in wheat seedlings [128]. Remarkably, exogenous application of CO in tomato stimulated root hair density and elongation with increase of 3.38- and 2.48-folds, respectively, as compared to the control. As per genetic analyses, CO is competent to affect the root hair formation by up-regulation of *LeExt1* gene expression [129].

Studies made so far on CO-persuaded root development typically focussed on lateral root (LR) and AR. Studies made on these two types of root developments are illustrated below:

(i) LR is derived from the pericycle of parent root. Herein, mature cells are stimulated to dedifferentiate and proliferate to form an LR primordium (early tissue growth). This finally leads to the development of LR. LR plays an essential role in the development of plant root system accountable for water-use efficiency and the extraction of nutrients from soils. Interestingly, CO has been found to encourage the formation of LR. In rapeseed seedlings, the total length and number of LR increased considerably in a dose-dependent manner using the haematin (a CO donor) or aqueous solution of CO. On the other hand, the positive effects were fully inverted by the addition of a CO scavenger, haemoglobin (Hb) or the CO-specific synthetic inhibitor zinc protoporphyrin-IX (ZnPPIX) [130]. This has been observed that treatment with exogenous CO up-regulates HO-1 (*LeHO-1*) expression and the amount of LeHO-1 proteins, and then stimulates the formation of tomato LR [130]. The above results indicate that exogenous CO in someway is involved in the formation process of LR in plants.

(ii) AR development is an essential step for vegetative propagation. The exploration of CO-induced AR formation began in 2006 when Xu et al. [131] testified that CO displays positive effects on AR formation in mung bean seedling. Thereafter, more attention has been given to highlight AR formation persuaded by CO. In 2008, Xuan et al. [119] observed that CO stimulated AR number and length dose dependently in IAA-depleted cucumber seedlings. It has also been demonstrated that the initiation of AR formation by methane-rich water (MRW)* was blocked by ZnPPIX, and further reversed by 20% CO [132] aqueous solution**.

3.16.3.3 Stomatal closure

Stomatal movement controls the plant water status, and it can be activated by numerous environment or hormonal factors. Among these, the stress hormone abscisic acid (ABA) (Figure 3.54) is a key player in regulating stomatal movement under drought and humidity stress [133]. When ABA treatment was found to increase CO content and HO activity in *Vicia faba* leaves, then different research groups began to examine the association between CO and stomatal closure.

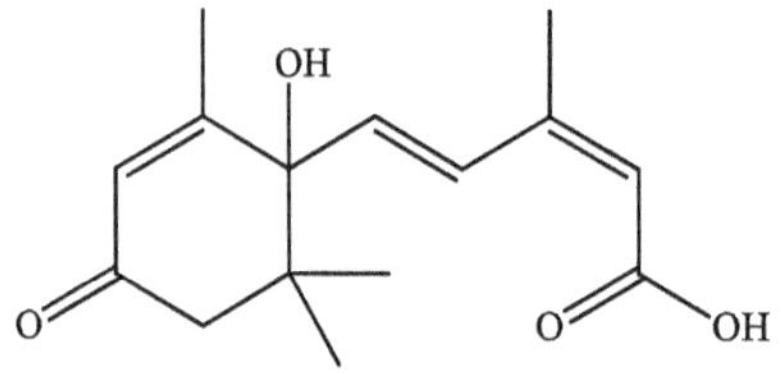

Figure 3.73: Structure of abscisic acid.

Interestingly, further investigation revealed that use of haematin and CO aqueous-solution** externally, not only resulted in the enhancement of CO release but also persuaded stomatal closure in dose- and time-dependent manners [134]. Notably, the results of CO in stomatal movement are comparable to NO and H_2O_2 [135, 136].

*Preparation of methane-rich water

MRW is prepared from CH_4 gas (released from the pure gas cylinder) bubbled into 500 mL distilled water at a rate of 160 mL min^{-1} for at least 30 min. This is 100% saturated stock solution of MRW, which can be diluted to the required concentrations (1%, 10%, 50% and 80% MRW [v/v]).

**Preparation of CO aqueous solution

CO gas is prepared by heating formic acid (HCOOH) with concentrated H_2SO_4 according to the following reaction:

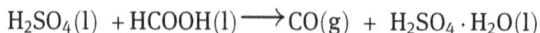

$$H_2SO_4(l) + HCOOH(l) \longrightarrow CO(g) + H_2SO_4 \cdot H_2O(l)$$

Additionally, an aqueous solution of CO is further obtained by bubbling CO gas gently through a glass tube into 300 mL of fresh CO_2-free BES-KCl buffer (pH 6.15) in an open bottle for at least 30 min. This much time is long enough to ensure that the solution is saturated with CO gas. Then the saturated stock solution (100% saturation) is immediately diluted with fresh CO_2-free BES-KCl buffer to the concentration as per experiment requirement (0.01%, 10%, 20% saturation, etc.). Structure of BES is shown in Figure 3.57 for convenience.

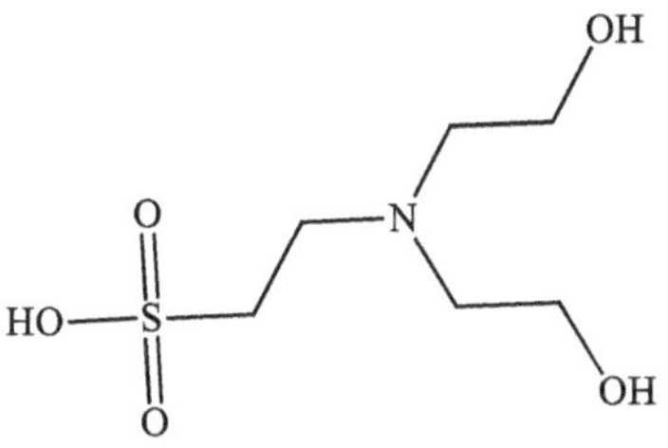

Figure 3.74: Bis(2-hydroxyethyl)-2-aminoethanesulphonic acid (BES).

3.16.4 Role of CO in abiotic stresses

Unfortunately, abiotic stresses are foremost limitations to plant growth, survival, yield and distribution. Owing to the disturbance of cellular redox homeostasis, abiotic stresses give rise to the oxidative stress and overproduction of ROS. It has been known that ABA is a key regulator involved in plant developmental processes and responses to biotic and abiotic stresses [137]. Identical to ABA, CO is also required for the mitigation of abiotic stress-persuaded oxidative stress.

3.16.4.1 Salt stress

Salt stress has become a permanent threat to crop yields in some way or other. It often causes many unfortunate consequences in plants, namely, growth inhibition, ionic phytotoxicity and ROS overproduction. It has been demonstrated by different research groups that in wheat even low levels of CO improves the inhibition in seed germination and the damage of seedling leaves produced by salt stress through enhancing antioxidant enzyme activities including superoxide dismutase (SOD), APX, CAT and guaiacol peroxidase [138, 139]. Similar result was established in *Oryza sativa* (Asian rice). Remarkably, CO enhances the activities of CAT and SOD and upregulates the expression of *CAT* and *Cu/Zn-SOD* genes. These result in improving salt-induced oxidative damage and finally decreasing the inhibition [125] in seed germination.

Zhang et al. observed that in *Cassia obtusifolia*, haematin or CO-saturated aqueous solution increases the concentration of cytosolic osmotic substances (total soluble sugars, free proline and soluble protein) and antioxidant enzyme activities and lightens the damage of photosynthetic system under salt stress. As a result, this treatment improves the inhibition of seed germination and seedling growth derived from salinity stress [140].

3.16.4.2 Drought stress

Drought is also a key constraint to food production. This is because it confines growth and development of plants. Also drought or osmotic stress persuades the accumulation of ROS [141] in plants. It has been noticed that the pre-treatment of ≤0.1 μM haematin (a CO donor) could delay wheat leaf chlorophyll loss mediated by further treatment of H_2O_2 and paraquat (an organic herbicide) (Figure 3.75) in a dose- and even time-dependent [142] manner.

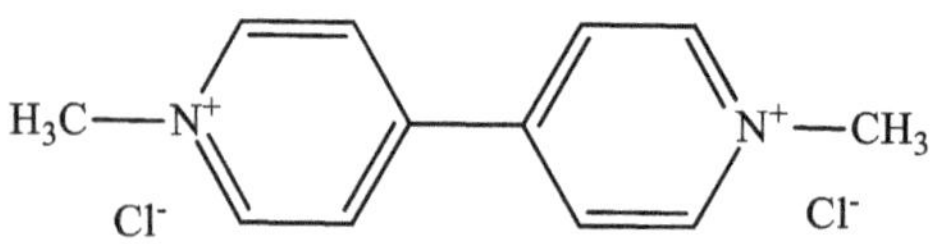

Figure 3.75: Structure of *N*, *N'*-dimethyl-4,4'bipyridinium dichloride (paraquat).

The investigation carried out by Liu and co-workers concludes that the endogenous HO/CO signalling system is essential for the lessening of osmotic stress-persuaded inhibition of wheat seed germination and lipid peroxidation. This might have a possible interaction with NO [126].

3.16.4.3 UV radiation stress

The depletion of ozone layer in stratosphere results in more UV-B (280–320 nm) radiation reaching the Earth's surface. In fact, the UV-B exposure to plants increases the magnitude of ROS and oxygen-derived free radicals. This leads to cellular damage and apoptosis in plants. Indeed, UV-B radiation triggered an increase in the expression of HO-1 and its transcript levels in a dose-dependent manner. This HO-1 expression was observed as a cell protection mechanism against UV-B radiation-persuaded oxidative damage [143]. As the CO production was closely related with HO-mediated haeme catabolism, it is obvious to assume that most probably CO exerted potential functions in modulating the defence response of plants to UV-B stress. Study conducted by Kong et al. concludes that CO can regulate iron homeostasis in iron-starved *Arabidopsis* and exogenous CO can prevent iron deficiency-induced chlorosis (condition in which leaves produce insufficient chlorophyll) and improves chlorophyll accumulation [144].

3.16.4.4 Heavy metal stress

Particularly, heavy metal toxicity such as mercury (Hg), cadmium (Cd), iron (Fe) and copper (Cu) arising from contaminations is known to affect the biosphere. And this induces health hazards for animals and plants. Heavy metal–persuaded oxidative stress in plants could be reduced in the presence of NO, small reactive gaseous molecules. Similar to NO, the dynamic role of CO in dismissing [145, 146] heavy metal stress in plants has been confirmed.

Han and co-workers observed that addition of haematin and CO to $HgCl_2$-treated alfalfa root reduces lipid peroxidation and increased root elongation via activating antioxidant enzymes including glutathione reductase, monodehydroascorbate reductase and SOD activities as well as decreasing lipoxygenase activity [145]. Notably, CO is also found to augment the tolerance of algae to Hg exposure, and it was closely related to the lower accumulation of Hg and free radical species [147]. As per the observation of Meng et al. [148], the harmful effect by Hg stress could be partially inverted by the intake of CO in Indian mustard plants via suppressing the production of O_2^- and H_2O_2 and increasing the accumulation of proline (Figure 3.76).

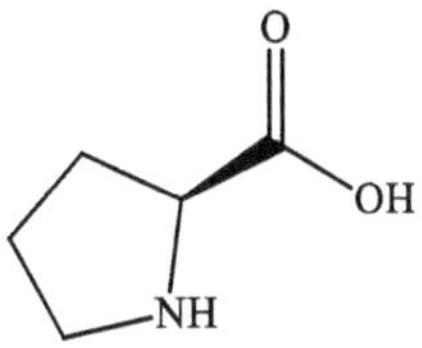

Figure 3.76: Chemical structure of proline.

Cd-persuaded oxidative damage was found to improve by the pre-treatment of CO via modifying glutathione metabolism in alfalfa (*Medicago sativa*). This accelerated the exchange of oxidized glutathione (GSSG) to reduced glutathione (GSH) to restore GSH: GSSG ratio and further decreased the oxidative damage [149]. Furthermore, Zheng and co-workers [146] reported that Cu-persuaded oxidative damage in algae (*Chlamydomonas reinhardtii*) was lessened by CO, principally via the improvement of CAT activity. Moreover, Kong et al. observed that the up-regulation of expression of genes related to Fe acquisition, namely, AtIRT1, AtFRO2, AtFIT1 and AtFER1 by CO is responsible for preventing the Fe-deficient-persuaded chlorosis and improving chlorophyll accumulation in *Arabidopsis* (small flowering plant) [144].

3.17 Cross-talk between CO and other signalling molecules

As a signalling molecule, CO is known for its adaptable properties and this control variety of physiological process in animals and plants. Numerous studies have shown that CO signal transduction is extremely complex. This usually does not operate as the linear pathways but extensive cross-talk occurs between various signal transductions. Therefore, this section presents a detailed view of the interaction between CO signalling molecule and other such signalling molecules in plants.

3.17.1 Cross-talk between CO and NO

CO signalling transportation pathways always do not work alone, but it is somewhat closely associated with NO. These two endogenously generated gasses share several common downstream signalling pathways and have some similar properties. As an illustration, similar to NO, CO binds to the iron atom of haeme proteins of sCG to activate the enzyme and increase intracellular second messenger cGMP production, thus exerting many of their biological functions including regulating vascular tone, inhibiting platelet aggregation and decreasing blood pressure [150]. But the existence of this phenomenon in plants does not have sufficient proof for its confirmation.

Song and co-workers [135] observed that CO is capable to duplicate the result of NO to some extent in dose-dependent manner persuading stomatal closure and K-to-Na ratios [151]. Accumulating proofs in animals support that there exists a close association in the expression of HO and NOS accountable for producing CO and NO. This suggests a possible interaction between the CO- and NO-generating systems. It has become gradually clear that CO can improve the activity of NO synthase (NOS) in plants. Song and co-workers [135] in 2008 proposed that there might exist HO-1 and NOS-like enzyme activity in *Vicia faba* guard cells. Moreover, CO was found to involve in darkness-persuaded NO synthesis via the NOS-like enzyme.

In 2008, Xie et al. [151] testified that NaCl-treated wheat seedling roots stemmed in a moderate enhancement of endogenous NO concentration, while a very strong increase of NO seemed on addition of 50% CO-saturated aqueous solution. On the other hand, CO could directly bind to and inactivate NOS, reducing the enzyme activity possibly due to the competition of CO with NO for binding to its targets such as sCG [152].

Studies carried out by Jeandroz et al. [153] have shown that NOS enzymes only exist in a few algal species but seem to be not conserved in land plants. This suggests that the generation of NO may depend mainly on nitrate assimilation in land plants. Therefore, NO synthesis is a complex process in plants and the interaction between the CO- and NO-generating systems via NOS enzymes also needs additional authentication. Moreover, a functional interaction of NO and CO has been confirmed in regulating plant growth and development. As an illustration, CO improves osmotic-persuaded wheat seed germination inhibition and lipid peroxidation that requires participation of NO [126].

Santa-Cruz and co-workers [154] suggested that NO was involved in the HO signalling pathway which might directly increase UV-B-persuaded HO-1 transcription in soybean plants. In the meantime, NO might act as a downstream signalling molecule in haemin-persuaded cucumber [155] adventitious rooting process.

3.17.2 Cross-talk between CO and phytohormone

Guo et al. [129] have shown that CO may somewhat involve in IAA-persuaded tomato LR development via altering biosynthesis in some way. In the meantime, Xuan et al. [119] established that there exists an association with the series IAA→HO/CO→AR. IAA (indole-3 acetic acid) first activated the HO/CO signalling system, and then triggers the signal transduction events, thus heading to AR formation in cucumber. Cao et al. [134] reported that CO is found to be involved in ABA-persuaded stomatal closure wherein NO and cGMP may function as downstream intermediates in CO signalling transduction network. As per recent report of Xie and co-workers [156], HO/CO may be a part or signalling system of H_2S-induced cytoprotective role against gibberellic acid (GA)-persuaded PCD. From the above studies, it appears that phytohormone induces numerous different developmental responses in plants, wherein CO takes part in some way or other. The chemical structure of GA is shown in Figure 3.77 for convenience of readers.

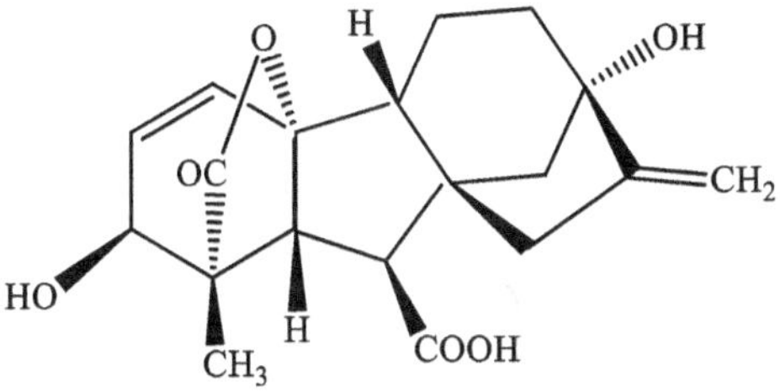

Figure 3.77: Chemical structure of gibberellic acid.

3.17.3 Cross-talk between CO and other small signalling molecules

Resembling NO, H_2O_2 and other small signalling molecules also play a crucial role in CO-intervened physiological responses. As the first illustration, She and Song for the first time reported [136] that CO-persuaded stomatal closure was possibly arbitrated by H_2O_2 signalling pathways in *Vicia faba*. Interestingly, up-regulation of *HO* expression protected aleurone (plant protein) layers against GA-induced PCD in wheat involves a modulation of H_2O_2 metabolism as reported by Wu et al. [123]. Accumulating evidence supports that HO/CO signalling system-arbitrated cucumber AR formation interacts closely with H_2S, H_2 and CH_4. For instance, HO-1 as a downstream component was involved in H_2S-persuaded AR cucumber formation through the alternation in expression of *DNAJ-1* and *CDPK1/5* genes [157]. In the similar way, hydrogen-rich water-persuaded AR formation was HO-1/CO dependent by up-regulating target genes related to auxin signalling and AR formation including *CsDNAJ-1, CsCDPK1/5, CsCDC6* and *CsAUX22B/*. Cui et al. [132] recently reported that MRW may act as a stimulator of AR that has partly arbitrated by HO-1/CO and Ca^{2+} pathways.

As discussed above, the overall signalling pathways involving CO and other signalling molecules in plant growth and development is pictorially shown in Figure 3.78.

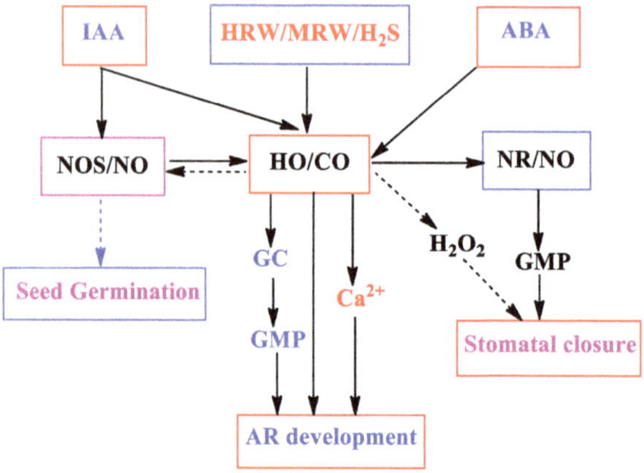

Figure 3.78: Pictorial view of signalling routes of CO and other signalling molecules in plant growth and development. The pathways shown by dashed arrows are not fully clear.

3.18 Concluding remarks

CO being the second gasotransmitter after NO has thus been found to be highly significant in the living world, although the metabolic pathways related with HO is the main origin which opened persuasion among scientific community to design and develop molecular models capable of releasing CO. The role of this molecule though found on need basis in mammalian tissue cannot be underestimated, and the synthetic strategies are employed carefully to avail the mimicking models of such systems. CO signalling transportation pathways do not work in an independent fashion, but possess close association with NO. These two endogenously generated gasses share several common downstream signalling pathways and have some similar properties. As an illustration, similar to NO, CO binds to the iron atom of haeme proteins of sCG to activate the enzyme and increase intracellular second messenger cGMP production, thus exerting many of their biological functions including regulating vascular tone, inhibiting platelet aggregation and decreasing blood pressure. But the existence of this phenomenon in plants does not have sufficient proof for its confirmation.

Although the carriage of CO could be organic as well as inorganic form, but the systems based on transition metal centres have proven quite beneficial in addressing several problems related to industry and medicine. Meanwhile, Ru-, Mn- and B-based CORMs have shown highly significant use in research. Photolability like in NORMs have been attentively dealt and several isoforms of light-activated molecules have shown promising results. The surprising results that CORMs have shown in industrial reactions, mammalian connective tissue regulation, sense organ maintenance and plant regulatory role signify the vast applicability of such scaffolds. The plant-based investigations in relevance to CO has suggested good implications to declare the molecule as a significant CO echo from plants. By the help of direct or indirect explanatory investigations, the molecular recognition of CO in living world is thus a very significant. To stress here that natural components are always symmetric in terms of every applied aspect, but similar gaseous molecule if produced by unnatural way is adding a pollutant to natural resources and harmful to living world. In the context of its vital role in living things, the generation of CO at the cost of signalling cascade proves that the natural phenomenon is always perfect and symmetric. Therefore, to develop molecular machineries of CO and subjecting them to release CO under various conditions represent highly economical, significant and ever-beneficial area of research coming under a broad branch of chemical science referred to as organometallic chemistry.

References

[1] T. Sjostrand, Endogenous formation of carbon monoxide in man, Nature, 164 (1949) 580–581.
[2] R. Tenhunen, H. S. Marver and R. Schmid, The enzymatic conversion of heme to bilirubin by microsomal heme oxygenase, Proc. Natl. Acad. Sci., U. S. A., 61 (1968) 748–755.
[3] T. Katayama, Y. Ikeda, M. Handa, T. Tamatani, S. Sakamoto, M. Ito, Y. Ishimura and M. Suematsu, Immuno-neutralization of glycoprotein Ibalpha attenuates endotoxin-induced interactions of platelets and leukocytes with rat venular endothelium *in vivo*, *Circ. Res.*, 86 (2000) 1031–1037.
[4] S. Errico, R. Shohreh, E. Barone, A. Pusateri, N. Mores and C. Mancuso, Heme oxygenase-derived carbon monoxide modulates gonadotropin-releasing hormone release in immortalized hypothalamic neurons, *Neurosci. Lett.*, 471 (2010) 175–178.
[5] P. Schützenberger, Sur un Nouveau Composé de Platine, Compt. Rendus, 66 (1868) 666–668.
[6] A. F. Hill, simple carbonyls of ruthenium. new avenues from the Hieber base reaction, *Angew. Chem. Int. Edit.*, 39 (2000) 130–133.
[7] J. P. Collman, S. R. Winter and D. R. Clark, *J. Am. Chem. Soc.*, 94 (1972) 1788 W. Siegel and J. P. Collman, *J. Am. Chem. Soc.*, 94 (1972) 2516.
[8] M. Wrighter, Photochemistry of metal carbonyls, *Chem. Rev.*, 74 (1974) 401–430.
[9] W. Reppe, *Justus Liebig's Ann. Chem.*, 582 (1953) 1.
[10] H. Hohenschutz, N. Von Kupetow and W. Himmele, *Hydrocarbon Process*, 45 (1966) 141.
[11] J. Hjortkajaer and V. W. Jensen, *Ind. Eng. Chem., Prod. Res. Dev.*, 15 (1976) 46.
[12] J. Gauthier-Lafaye and R. Perron, *Methanol and Carbonylation (English Trans.)*, Editions Technip, Paris, France and Rhone-Poulenc Recherches, Courbevoie, France, (1987), 149.
[13] N. Rizkalla, German patent 2.610.036 (1976); H. Kuckertz, German patent 2.450.965 (1976).
[14] Chem. Week 126 (1980) 40.
[15] D. Forster, Mechanistic pathways in the catalytic carbonylation of methanol by rhodium and iridium complexes, *Adv. Organomet. Chem.*, 17 (1979) 255–267.
[16] R. Kummer, H. W. Schneider, F. J. Weiss and O. Lemon, German Patent, 2 837 815 (1980), Baden Aniline and Soda Factory (BASF).
[17] Belgian Patent 770 615 (1972) to Baden Aniline and Soda Factory (BASF).

[18] J. Tsuji, I. Kiji and S. Hosaka, *Tetrahedron Lett.*, 5 (1964) 605–608 K. Bittler, N. V. Kupetow, D. Neubauer and H. Reis, *Angew. Chem.*, 80 (1968) 352–359.

[19] I. Wender and P. Pino eds., *Organic Synthesis via Metal Carbonyls*, John Wiley and Sons, New York, II, (1977).

[20] R. F. Heck and D. S. Breslow, *J. Am. Chem. Soc.*, 83 (1963) 2023–2027.

[21] C. C. Romão, W. A. Blättler, J. D. Seixas and G. J. L. Bernardes, *Chem. Soc. Rev.*, 41 (2012) 3571–3583.

[22] R. P. Smith, Toxic responses of the blood, In: *Casarett and Doull's Toxicology, The Basic Science of Poisons*, 3rd edited by C. D. Klaassen, M. O. Amdur and J. Doull, MacMillan Publishing Company, New York, (1986), 233–244.

[23] C. L. Hartsfield, Cross talk between carbon monoxide and nitric oxide, *Antioxid. Redox Signal*, 4 (2002) 301–307.

[24] M. A. Gonzales and P. K. Mascharak, Photoactive metal carbonyl complexes as potential agents for targeted CO delivery, *J. Inorg. Biochem.*, 133 (2014) 127–135.

[25] T. R. Johnson, B. E. Mann, J. E. Clark, R. Foresti, C. J. Green and R. Motterlini, Metal carbonyls: a new class of pharmaceuticals?, *Angew. Chem. Int. Ed.*, 42 (2003) 3722–3729.

[26] M. D. Maine, The heme oxygenase system: a regulator of second messenger gases, *Annu. Rev. Pharmacol. Toxicol.*, 37 (1997) 517–554.

[27] R. Tenhunen, H. S. Marver and R. Schmid, Microsomal heme oxygenase: characterization of the enzyme, *J. Biol. Chem.*, 244 (1969) 6388–6394.

[28] R. Schmid and A. F. McDonagh, *The Porphyrins*, VI, Ed. D. Dolphin, Academic Press, New York, (1979), 257–293.

[29] R. Motterlini, A. Gonzales, R. Foresti, J. E. Clark, C. J. Green and R. M. Winslow, Heme oxygenase-1–derived carbon monoxide contributes to the suppression of acute hypertensive responses in vivo, *Circ. Res.*, 83 (1998) 568–577.

[30] L. J. Ignarro, G. Cirino, A. Alessandro and C. Napoli, Nitric oxide as a signalling molecule in the vascular system: an overview, *J. Cardiovasc. Pharmacol.*, 34 (1999) 879–886.

[31] B. E. Mann and R. Motterlini, CO and NO in medicine, *Chem. Commun.*, 2007 4197–4208.

[32] L. J. Ignarro, *Nitric Oxide: Biology and Pathobiology*, Academic Press, San Diego, 2000.

[33] D. M. Lawson, C. E. M. Stevenson, C. R. Andrew and R. E. Eady, Unprecedented proximal binding of nitric oxide to heme: implications for guanylate cyclase, *EMBO J.*, 19 (2000) 5661–5671.

[34] J. R. Stone and M. A. Marletta, Soluble guanylate cyclase from bovine lung: activation with nitric oxide and carbon monoxide and spectral characterization of the ferrous and ferric states, *Biochemistry*, 33 (1994) 5636–5640.

[35] G. Deinum, J. R. Stone, G. T. Babcock and M. A. Marletta, Binding of nitric oxide and carbon monoxide to soluble guanylate cyclase as observed with resonance Raman spectroscopy, *Biochemistry*, 35 (1996) 1540–1547.

[36] E. Martin, K. Czarnecki, V. Jayaraman, F. Murad and J. Kincaid, Resonance raman and infrared spectroscopic studies of high-output forms of human soluble guanylyl cyclase, *J. Am. Chem. Soc.*, 127 (2005) 4625–4631.

[37] S. Ghatta, D. Nimmagadda, X. Xu and S. T. O'Rourke, Large-conductance, calcium-activated potassium channels: structural and functional implications, *Pharmacol. Therapeut.*, 110 (2006) 103–116.

[38] L. Wu, Z. Wang and R. Wang, Tetraethylammonium-evoked oscillatory contractions of rat tail artery: A K-K model, *Can. J. Physiol. Pharmacol.*, 78 (2000) 696–707.

[39] L. Wu, K. Cao, Y. Lu and R. Wang, Different mechanisms underlying the stimulation of K(Ca) channels by nitric oxide and carbon monoxide, J. *Clin. Invest.*, 110 (2002) 691–700.

[40] J. H. Jaggar, A. Li, H. Parfenova, J. Liu, E. S. Umstot, A. M. Dopico and C. W. Leffler, Heme is a carbon monoxide receptor for large-conductance Ca^{2+}-activated channels, *Circ. Res.*, 97 (2005) 805–812.

[41] D. L. Dong, Y. Zhang, D. H. Lin, J. Chen, S. Patschan, M. S. Goligorsky, A. Nasjletti, B. F. Yang and W. H. Wang, Carbon monoxide stimulates the Ca^{2+}-activated big conductance K channels in cultured human endothelial cells, *Hypertension*, 50 (2007) 643–651.

[42] C. A. Piantadosi, Carbon monoxide, reactive oxygen signalling, and oxidative stress, *Free Radical Bio. Med.*, 45 (2008) 562–569.

[43] H. L. A. Vieira, C. S. F. Queiroga and P. M. Alves, Pre-conditioning induced by carbon monoxide provides neuronal protection against apoptosis., *J. Neurochem.*, 107 (2008) 375–384.

[44] M. Cargnello and P. P. Roux, Activation and function of the MAPKs and their substrates, the MAPK-activated protein kinases, *Micobiol. Mol. Biol. Rev.*, 75 (2011) 50–83.

[45] J. Brugger, M. A. Schick, R. W. Brock, A. Baumann, R. M. Muellenbach, N. Roewer and C. Wunder, Carbon monoxide has antioxidative properties in the liver involving p38 MAP kinase pathway in a murine model of systemic inflammation, *Microcirculation*, 17 (2010) 504–513.

[46] C. I. Schwer, M. Mutschler, P. Stoll, U. Goebel, M. Humar, A. Hoetzel and R. Schmidt, Carbon monoxide releasing molecule-2 inhibits pancreatic stellate cell proliferation by activating p38 mitogen-activated protein kinase/heme oxygenase-1 signalling, Mol, *Pharmacol*, 77 (2010) 660–669.

[47] M. Faizan, N. Muhammad, K. U. K. Niazi, Y. Hu, Y. Wang, Y. Wu, H. Sun, R. Liu, W. Dong, W. Zhang and Z. Gao, CO-releasing materials: An emphasis on therapeutic implications, as release and subsequent cytotoxicity are the part of therapy, *Materials*, 12 (2019) 1643–1684.

[48] L. E. Otterbein, *Respir. Care*, 54 (2009) 925–932.

[49] S. T. Omaye, *Toxicology*, 180 (2002) 139–150.

[50] S. Yang, M. Chen, L. Zhou, G. Zhang, Z. Gao and W. Zhang, Photo-activated CO-releasing molecules (PhotoCORMs) of robust sawhorse scaffolds [μ^2-OOCR^1, η^1-NH_2CHR^2 (C=O] OCH_3, $Ru(I)_2CO_4$, *Dalton Trans.*, 45 (2016) 3727–3733.

[51] A. E. Pierri, P. J. Huang, J. V. Garcia, J. G. Stanfill, M. Chui, G. Wu, N. Zheng and P. C. Ford, A photoCORM nanocarrier for CO release using NIR light, *Chem. Commun.*, 51 (2015) 2072–2075.

[52] ClinicalTrials.gov identifier: NCT01050712, September 2013.

[53] A. Nakao, H. Toyokawa, M. Abe, T. Kiyomoto, K. Nakahira, A. M. K. Choi, M. A. Nalesnik, A. W. Thomson and N. Murase, Protective effect of carbon monoxide in transplantation, *Transplantation*, 81 (2006) 220–230.

[54] T. Kaizu, A. Nakao, A. Tsung, H. Toyokawa, R. Sahai, D. A. Geller and N. Murase, Carbon monoxide inhalation ameliorates cold ischemia/reperfusion injury after rat liver transplantation, *Surgery*, 138 (2005) 229–235.

[55] P. N. A. Martins, A. Reutzel-Selke, A. Jurisch, C. Denecke, K. Attrot, A. Pascher, K. Kotsch, J. Pratschke, P. Neuhaus, H. D. Volk and S. G. Tullius, Induction of carbon monoxide in donor animals prior to organ procurement reduces graft immunogenicity and inhibits chronic allograft dysfunction, *Transplantation*, 82 (2006) 938–944.

[56] H. Wang, S. S. Lee, W. Gao, E. Czismadia, J. McDaid, R. Öllinger, M. P. Soares, K. Yamashita and F. H. Bach, Donor treatment with carbon monoxide can yield islet allograft survival and tolerance, *Diabetes*, 54 (2005) 1400–1406.

[57] R. Song, R. S. Mahidhara, Z. Zhou, R. A. Hoffman, D. W. Seol, R. A. Flavell, T. R. Billiar, L. E. Otterbein and A. M. K. Choi, Carbon monoxide inhibits T-lymphocyte proliferation via caspase-dependent pathway, *J. Immunol.*, 172 (2004) 1220–1226.

[58] D. Morse, S. E. Pischke, Z. Zhou, R. J. Davis, R. A. Flavell, T. Loop, S. L. Otterbein, L. E. Otterbein and A. M. K. Choi, Suppression of inflammatory cytokine production by carbon monoxide involves the JNK pathway and AP-1, *J. Biol. Chem.*, 278 (2003) 36993–36998.

[59] L. E. Otterbein, F. H. Bach, J. Alam, M. Soares, H. T. Lu, M. Wysk, R. J. Davis, R. A. Flavell and A. M. K. Choi, Carbon monoxide has anti-inflammatory effects involving the mitogen-activated protein kinase pathway, *Nat. Med.*, 6 (2000) 422–428.

[60] K. Sato, J. Balla, L. E. Otterbein, R. N. Smith, S. Brouard, Y. Lin, E. Csizmadia, J. Sevigny, S. C. Robson, G. Vercellotti, A. M. K. Choi, F. H. Bach and M. P. Soares, Oxygenase-1 suppresses the rejection of carbon monoxide generated by heme mouse-to-rat cardiac transplants, *J. Immunol.*, 166 (2001) 4185–4194.

[61] K. S. Ozaki, S. Kimura and N. Murase, Use of carbon monoxide in minimizing ischemia reperfusion injury in transplantation, *Transplant. Rev.*, 26 (2012) 125–139.

[62] J. F. Ndisang, H. E. N. Tabien and R. Wang, Carbon monoxide and hypertension, *J Hypertens.*, 22 (2004) 1057–1074.

[63] V. G. Kharitonov, V. S. Sharma, R. B. Pilz, D. Magde and D. Koesling, Basis of guanylate cyclase activation by carbon monoxide, *Proc. Natl. Acad. Sci., USA.*, 92 (1995) 2568–2571.

[64] R. Wang and L. Wu, The chemical modification of K_{Ca} channels by carbon monoxide in vascular smooth muscle cells, *J. Biol. Chem.*, 272 (1997) 32804–32809.

[65] S. F. Yet, A. Pellacani, C. Patterson, L. Tan, S. C. Folta, L. Foster, W. S. Lee, C. M. Hsieh and M. A. Perrella, Induction of heme oxygenase-1 expression in vascular smooth muscle cells: A link to endotoxic shock, *J. Biol. Chem.*, 272 (1997) 4295–4301.

[66] P. Wiesel, A. P. Patel, N. DiFonzo, P. B. Marria, C. U. Sim, A. Pellacani, K. Maemura, B. W. LeBlanc, K. Marino, C. M. Doerschuk, S. F. Yet, M. U. Lee and M. A. Perrella, Endotoxin-induced mortality is related to increased oxidative stress and end-organ dysfunction, not refractory hypertension in heme oxygenase-1-deficient mice, *Circulation*, 102 (2000) 3015–3022.

[67] S. Mazzola, M. Forni, M. Albertini, M. L. Bacci, A. Zannoni, F. Gentilini, M. Lavitrano, F. H. Bach and M. G. Clement, Carbon monoxide pretreatment prevents regulatory derangement and ameliorates hyperacute endotoxic shock in pigs, *FASEB J.*, 19 (2005) 2045–2047.

[68] C. S. Bang, R. Kruse, K. Johansson and K. Persson, Carbon monoxide releasing molecule-2 (CORM-2) inhibits growth of multidrug-resistant uropathogenic *Escherichia coli* in biofilm and following host cell colonization, *BMC Microbiol.*, 16 (2016) 64–73.

[69] M. Desmard, R. Foresti, D. Morin, M. Dagoussat, A. Berdeaux, E. Denamur, S. Crook, B. Mann, D. Scapens and P. Montravers, Differential antibacterial activity against *Pseudomonas aeruginosa* by carbon monoxide-releasing molecules, *Antioxid. Redox Signal.*, 16 (2012) 153–163.

[70] S. Castany, M. Carcolé, S. Leánez and O. Pol, The role of carbon monoxide on the anti-nociceptive effects and expression of cannabinoid 2 receptors during painful diabetic neuropathy in mice, *Psychopharmacology*, 233 (2016) 2209–2219.

[71] A. Hervera, S. Leánez, R. Motterlini and O. Pol, Treatment with carbon monoxide-releasing molecules and an HO-1 inducer enhances the effects and expression of µ-opioid receptors during neuropathic pain, *Anesthesiology*, 118 (2013) 1180–1197.

[72] I. Nikolic, T. Saksida, M. Vujicic, I. Stojanovic and S. Stosic-Grujicic, Anti-diabetic actions of carbon monoxide releasing molecule (CORM)-A1: immunomodulation and regeneration of islet beta cells, *Immunol. Lett.*, 165 (2015) 39–46.

[73] C. Steiger, K. Uchiyama, T. Takagi, K. Mizushima, Y. Higashimura, M. Gutmann, C. Hermann, S. Botov, H. G. Schmalz, Y. Naito and L. Meinel, Prevention of colitis by controlled oral drug delivery of carbon monoxide, *J. Control. Release*, 239 (2016) 128–136.

[74] S. Nagao, K. Taguchi, Y. Miyazaki, et al., Evaluation of a new type of nano-sized carbon monoxide donor on treating mice with experimentally induced colitis, *J. Control. Release*, 234 (2016) 49–58.

[75] S. A. Riquelme, S. M. Bueno and A. M. Kalergis, Carbon monoxide down-modulates toll-like receptor 4/MD2 expression on innate immune cells and reduces endotoxic shock susceptibility, *Immunology*, 144 (2015) 321–332.

[76] K. Maruyama, E. Morishita, T. Yuno, A. Sekiya, H. Asakura, S. Ohtake and A. Yachie, Carbon monoxide (CO)-releasing molecule-derived CO regulates tissue factor and plasminogen activator inhibitor type 1 in human endothelial cells, *Trombosis Res.*, 130 (2012) e188–e193.

[77] P. Fagone, K. Mangano, S. Mammana, E. Cavalli, R. D. Marco, M. L. Barcellona, L. Salvatorelli, G. Magro and F. Nicoletti, Carbon monoxide-releasing molecule-A1 (CORM-A1) improves clinical signs of experimental autoimmune uveoretinitis (EAU) in rats, *Clin. Immunol.*, 157 (2015) 198–204.

[78] S. Lee, S. J. Lee, A. A. Coronata, L. E. Fredenburgh, S. W. Chung, M. A. Perrella, K. Nakahira, S. W. Ryter and A. M. Choi, CO confers protection in sepsis by augmenting beclin 1-dependent autophagy and phagocytosis, *Antioxid. Redox Signalling*, 20 (2014) 432–442.

[79] A. Ismailova, D. Kuter, D. S. Bohle and I. S. Butler, Bioinorganic Chemistry and Applications, Volume 2018, Article ID 8547364, doi: org/10.1155/2018/8547364, 10.

[80] W. Zhang, A. Tao, T. Lan, G. Cepinskas, R. Kao, C. M. Martin and T. Rui, Carbon monoxide releasing molecule-3 improves myocardial function in mice with sepsis by inhibiting NLRP3 inflammasome activation in cardiac fibroblasts, *Basic Res. Cardiol.*, 112 (2017) 16–25.

[81] P. A. Hosick, A. A. AlAmodi, M. W. Hankins and D. E. Stec, Chronic treatment with a carbon monoxide releasing molecule reverses dietary induced obesity in mice, *Adipocyte*, 5 (2016) 1–10.

[82] M. Zheng, Q. Zhang, Y. Joe, S. Kim, M. J. Uddin, H. Rhew, T. Kim, S. W. Ryter and H. T. Chung, Carbon monoxide releasing molecules reverse leptin resistance induced by endoplasmic reticulum stress, *Am. J. Physiol-Endocrinol. Metab.*, 304 (2013) E780–E788.

[83] J. A. Nemzek, C. Fry and O. Abatan, Low-dose carbon monoxide treatment attenuates early pulmonary neutrophil recruitment after acid aspiration, *Ame. J. Physiol. Lung Cell Mol. Physiol.*, 294 (2008) L644–L653.

[84] J. T. Chapman, L. E. Otterbein, J. A. Elias and A. M. K. Choi, Carbon monoxide attenuates aeroallergen-induced inflammation in mice, *Ame. J. Physiol. Lung Cell Mol. Physiol.*, 281 (2001) L209–L216.

[85] L. E. Otterbein, S. L. Otterbein and E. Ifedigbo, MKK3 mitogen-activated protein kinase pathway mediates carbon monoxide-induced protection against oxidant-induced lung injury, *Am. J. Pathol.*, 163 (2003) 2555–2563.

[86] T. Dolinay, M. Szilasi, M. Liu and A. M. K. Choi, Inhaled carbon monoxide confers antiinflammatory effects against ventilator-induced lung injury, *Am. J. Respir. Crit. Care Med.*, 170 (2004) 613–620.

[87] L. A. Mitchell, M. M. Channell, C. M. Royer, S. W. Ryter, A. M. K. Choi and J. D. McDonald, Evaluation of inhaled carbon monoxide as an anti-inflammatory therapy in a nonhuman primate model of lung inflammation, *Am. J. Physiol. Lung Cell Mol. Physiol.*, 299 (2010) 891–897.

[88] S. Abid, A. Houssaini and N. Mouraret, p21-dependent protective effects of a carbon monoxide-releasing molecule-3 in pulmonary hypertension, *Arterioscler. Thromb. Vasc. Biol.*, 34 (2014) 304–312.

[89] S. Ahmad, P. W. Hewett, T. Fujisawa, S. Sissaoui, M. Cai1, G. Gueron, B. Al-Ani, M. Cudmore, S. F. Ahmed, M. K. K. Wong, B. Wegiel, L. E. Otterbein, L. Vítek, W. Ramma, K. Wang and A. Ahmed, Carbon monoxide inhibits sprouting angiogenesis and vascular endothelial growth factor receptor-2 phosphorylation, *Thromb. Haemost.*, 113 (2015) 329–337.

[90] S. Fayad-Kobeissi, J. Ratovonantenaina, H. Dabiré, J. L. Wilson, A. M. Rodriguez, A. Berdeaux, J.-L. Dubois-Randé, B. E. Mann, R. Motterlini and R. Foresti, Vascular and angiogenic activities of CORM-401, an oxidant-sensitive CO-releasing molecule, *Biochem. Pharmacol.*, 102 (2016) 64–77.

[91] C. I. Schwer, P. Stoll, S. Rospert, E. Fitzke, N. Schallner, H. Bürkle, R. Schmidt and M. Humar, Carbon monoxide releasing molecule-2 CORM-2 represses global protein synthesis by inhibition of eukaryotic elongation factor eEF2, *Int. J. Biochem. Cell Biol.*, 45 (2013) 201–212.

[92] H. Segersvärd, P. Lakkisto, M. Hänninen, H. Forsten, J. Siren, K. Immonen, R. Kosonen, M. Sarparanta, M. Laine and I. Tikkanen, Carbon monoxide releasing molecule improves structural and functional cardiac recovery after myocardial injury, *Eur. J. Pharmacol.*, 818 (2018) 57–66.

[93] M. H. Tsai, C. W. Lee, L. F. Hsu, et al., CO-releasing molecules CORM2 attenuates angiotensin II-induced human aortic smooth muscle cell migration through inhibition of ROS/IL-6 generation and matrix metalloproteinases-9 expression, *Redox Biol.*, 12 (2017) 377–388.

[94] H. Soni, G. Pandya, P. Patel, A. Acharya, M. Jain and A. A. Mehta, Beneficial effects of carbon monoxide releasing molecule-2 (CORM-2) on acute doxorubicin cardiotoxicity in mice: Role of oxidative stress and apoptosis, *Toxicol. Appl. Pharmacol.*, 253 (2011) 70–80.

[95] D. V. Abramochkin, O. P. Konovalova, A. Kamkin and G. F. Sitdikova, Carbon monoxide modulates electrical activity of murine myocardium via cGMP-dependent mechanisms, *J. Physiol. Biochem.*, 71 (2015) 107–119.

[96] V. G. Nielsen, J. K. Kirklin and J. F. George, Carbon monoxide releasing molecule-2 increases the velocity of thrombus growth and strength in human plasma, *Blood Coagulation and Fibrinolysis*, 20 (2009) 377–380.

[97] V. G. Nielsen, N. Chawla, D. Mangla, et al., Carbon monoxide-releasing molecule-2 enhances coagulation in rabbit plasma and decreases bleeding time in clopidogrel/aspirin-treated rabbits, *Blood Coagulation and Fibrinolysis*, 22 (2011) 756–759.

[98] K. A. Machovec, D. S. Ushakumari, I. J. Welsby and V. G. Nielsen, The pro-coagulant properties of purified fibrinogen concentrate are enhanced by carbon monoxide releasing molecule-2, *Thromb. Res.*, 129 (2012) 793–796.

[99] V. G. Nielsen and C. M. Bazzell, Carbon monoxide attenuates the effects of snake venoms containing metalloproteinases with fibrinogenase or thrombin-like activity on plasmatic coagulation, *Med. Chem. Commun.*, 7 (2016) 1973–1979.

[100] A. Sommer, J. M. Tielsch, J. Katz, H. A. Quigley, J. D. Gottsch, J. Javitt and K. Singh, Relationship between intraocular pressure and primary open angle glaucoma among white and black Americans: the Baltimore eye survey, *Arch. Ophthalmol.*, 109 (1991) 1090–1095.

[101] C. Bucolo and F. Drago, Carbon monoxide and the eye: Implications for glaucoma therapy, *Pharmacol. Ther.*, 130 (2011) 191–201.

[102] E. Stagni, M. G. Privitera, C. Bucolo, G. M. Leggio, R. Motterlini and F. Drago, A water-soluble carbon monoxide-releasing molecule (CORM-3) lowers intraocular pressure in rabbits, *Br. J. Ophthalmol.*, 93 (2009) 254–257.

[103] X. Gasull, E. Ferrer, A. Llobet, A. Castellano, J. M. Nicolás, J. Palés and A. Gual, Cell membrane stretch modulates the high-conductance Ca^{2+}-activated K^+ channel in bovine trabecular meshwork cells, *Invest. Ophthalmol. Vis. Sci.*, 44 (2003) 706–714.

[104] D. Wakefield and J. H. Chang, Epidemiology of uveitis, *Int. Ophthalmol. Clin.*, 45 (2005) 1–13.

[105] R. D. Rimmer, R. Richter and P. C. Ford, A photochemical precursor for carbon monoxide release in aerated aqueous media, *Inorg. Chem.*, 49 (2009) 1180–1185.

[106] R. Motterlini, J. E. Clark, R. Foresti, P. Sarathchandra, B. E. Mann and C. J. Green, Carbon monoxide-releasing molecules characterization of biochemical and vascular activities, *Circ. Res.*, 90 (2002) e17–e24.

[107] S. K. Nayak, G. J. Farrell and T. J. Burkey, Photosubstitution of two iron pentacarbonyl CO's in solution via a single-photon process: dependence on dispersed ligands and role of triplet Intermediates, *Inorg. Chem.*, 33 (1994) 2236–2242.
[108] R. S. Herrick and T. L. Brown, Flash photolytic investigation of photoinduced carbon monoxide dissociation from dinuclear manganese carbonyl compounds, *Inorg. Chem.*, 23 (1984) 4550–4553.
[109] R. Kretschmer, G. Gessner, H. Görls, S. H. Heinemann and M. Westerhausen, Dicarbonyl-bis (cysteamine)iron(II): a light induced carbon monoxide releasing molecule based on iron (CORM-S1), *J. Inorg. Biochem.*, 105 (2011) 6–9.
[110] J. Niesel, A. Pinto, H. W. Peindy N'Dongo, K. Merz, I. Ott, R. Gust and U. Schatzschneider, Photoinduced CO release, cellular uptake and cytotoxicity of a tris(pyrazolyl)methane (tpm) manganese tricarbonyl complex, *Chem. Commun.*, 2008 1798–1800.
[111] G. Dördelmann, H. Pfeiffer, A. Birkner and U. Schatzschneider, Silicium dioxide nanoparticles as carriers for photoactivatable CO-releasing molecules (PhotoCORMs), *Inorg. Chem.*, 50 (2011) 4362–4367.
[112] M. A. Gonzalez, M. A. Yim, S. Cheng, A. Moyes, A. J. Hobbs and P. K. Mascharak, Manganese carbonyls bearing tripodal polypyridine ligands as photoactive carbon monoxide-releasing molecules, *Inorg. Chem.*, 51 (2012) 601–608.
[113] M. A. Gonzalez, S. J. Carrington, N. L. Fry, J. L. Martinez and P. K. Mascharak, Syntheses, structures, and properties of new manganese carbonyls as photoactive CO-releasing molecules: Design strategies that lead to CO photolability in the visible region, *Inorg.Chem*, 51 (2012) 11930–11940.
[114] M. A. Gonzalez, S. J. Carrington, I. Chakraborty, M. M. Olmstead and P. K. Mascharak, Photoactivity of mono- and di-carbonyl Complexes of Ru(II) bearing an N,N,S-donor ligand: Role of ancillary ligands on the capacity of CO photorelease, *Inorg. Chem.*, 52 (2013) 11320–11331.
[115] Q. Jin, W. Cui, Y. Xie and W. Shen, Carbon monoxide: A ubiquitous gaseous signalling molecule in plants, In: *Gasotrnasmitters in plants: The rise of a new paradigm in cell signalling*, edited by L. Lorenzo and G.-M. Carlos, Springer, (2016).
[116] S. S. Wilks, Carbon monoxide in green plants, *Science*, 129 (1959) 964–966.
[117] G. W. Schade, R.-M. W. Hofmann and P. J. Crutzen, CO emissions from degrading plant matter, *Tellus: Chem. Phys. Meteorol.*, 51B (1999) 889–908.
[118] T. Muramoto, N. Tsurui, M. J. Terry, A. Yokota and T. Kohchi, Expression and biochemical properties of a ferredoxin-dependent heme oxygenase required for phytochrome chromophore Synthesis, *Plant Physiol.*, 30 (2002) 1958–1966.
[119] W. Xuan, F. Y. Zhu, S. Xu, B. K. Huang, T. F. Ling and J. Y. Qi, The heme oxygenase/carbon monoxide system is involved in the auxin- induced cucumber adventitious rooting process, *Plant Physiol.*, 148 (2008) 881–893.
[120] G. S. Shekhawat and K. Verma, Heme oxygenase (HO): An overlooked enzyme of plant metabolism and defence, *J. Exp. Bot.*, 61 (2010) 2255–2270.
[121] C. G. Zilli, D. M. Santa-Cruz and K. B. Balestrasse, Heme oxygenase-independent endogenous production of carbon monoxide by soybean plants subjected to salt stress, *Environ. Exp. Bot.*, 102 (2014) 11–16.
[122] J. Dulak and A. Józkowicz, Carbon monoxide: a "new" gaseous modulator of gene expression, *Acta Biochim. Pol.*, 50 (2003) 31–48.
[123] M. Z. Wu, J. J. Huang, S. Xu, T. F. Ling, Y. J. Xie and W. B. Shen, Haem oxygenase delays programmed cell death in wheat aleurone layers by modulation of hydrogen peroxide metabolism, *J. Exp. Bot.*, 62 (2010) 235–248.

[124] J. Dekker and M. Hargrove, Weedy adaptation in Setaria spp. V. Effects of gaseous environment on giant foxtail (*Setaria faberii*) (Poaceae) seed germination, *Am. J. Bot.*, 89 (2002) 410–416.
[125] K. L. Liu, S. Xu, W. Xuan, T. F. Ling, Z. Y. Cao and B. K. Huang, Carbon monoxide counteracts the inhibition of seed germination and alleviates oxidative damage caused by salt stress in *Oryza sativa*, *Plant Sci.*, 172 (2007) 544–555.
[126] Y. H. Liu, S. Xu, T. F. Ling, L. L. Xu and W. B. Shen, Heme oxygenase/carbon monoxide system participates in regulating wheat seed germination under osmotic stress involving the nitric oxide pathway, *J. Plant Physiol.*, 167 (2010) 1371–1379.
[127] R. Amooaghaie, F. Tabatabaei and A. M. Ahadi, Role of hematin and sodium nitroprusside in regulating *Brassic anigra seed* germination under nanosilver and silver nitrate stresses, *Ecotox. Environ. Safe*, 113 (2015) 259–270.
[128] W. Xuan, L. Q. Huang, M. Li, B. K. Huang, S. Xu, H. Liu, Y. Gao and W. Shen, Induction of growth elongation in wheat root segments by heme molecules: a regulatory role of carbon monoxide in plants?, *Plant Growth Regul.*, 52 (2007) 41–51.
[129] K. Guo, K. Xia and Z. M. Yang, Regulation of tomato lateral root development by carbon monoxide and involvement in auxin and nitric oxide, *J. Exp.Bot.*, 59 (2008) 3443–3452.
[130] Z. Y. Cao, W. Xuan, Z. Y. Liu, X. N. Li, N. Zhao and P. Xu, Carbon monoxide promotes lateral root formation in rapeseed, *J. Integr. Plant Biol.*, 49 (2007) 1070–1079.
[131] J. Xu, W. Xuan, B. K. Huang, Y. H. Zhou, T. F. Ling and S. Xu, Carbon monoxide-induced adventitious rooting of hypocotyl cuttings from mung bean seedling, *Chinese Sci. Bull.*, 51 (2006) 668–674.
[132] W. Cui, F. Qi, Y. Zhang, H. Cao, J. Zhang, R. Wang and W. Shen, Methane-rich water induces cucumber adventitious rooting through heme oxygenase1/carbon monoxide and Ca^{2+} pathways, *Plant Cell Rep.*, 34 (2015) 435–445.
[133] A. Grondin, O. Rodrigues, L. Verdoucq, S. Merlot, N. Leonhardt and C. Maurel, Aquaporins contribute to ABA-triggered stomatal closure through OST1-mediated phosphorylation, *Plant Cell*, 27 (2015) 1945–1954.
[134] Z. Y. Cao, B. K. Huang, Q. Y. Wang, W. Xuan, T. F. Ling and B. Zhang, Involvement of carbon monoxide produced by heme oxygenase in ABA-induced stomatal closure in *Vicia faba* and its proposed signal transduction pathway, *Chinese Sci. Bull.*, 52 (2007) 2365–2373.
[135] X. G. Song, X. P. She and B. Zhang, Carbon monoxide-induced stomatal closure in *Vicia faba* is dependent on nitric oxide synthesis, *Physiol. Plant.*, 132 (2008) 514–525.
[136] X. P. She and X. G. Song, Carbon monoxide-induced stomatal closure involves generation of hydrogen peroxide in *Vicia faba* guard cells, *J. Integr. Plant Biol.*, 50 (2008) 1539–1548.
[137] A. S. Raghavendra, V. K. Gonugunta, A. Christmann and E. Grill, ABA perception and signalling, *Trends Plant Sci.*, 15 (2010) 395–401.
[138] B. K. Huang, S. Xu, W. Xuan, M. Li, Z. Y. Cao and K. L. Liu, Carbon monoxide alleviates salt-induced oxidative damage in wheat seedling leaves, *J. Integr. Plant Biol.*, 48 (2006) 249–254.
[139] S. Xu, Z. Sa, Z. S. Cao, W. Xuan, B. K. Huang and T. F. Ling, Carbon monoxide alleviates wheat seed germination inhibition and counteracts lipid peroxidation mediated by salinity, *J. Integr. Plant Biol.*, 48 (2006) 1168–1176.
[140] C. P. Zhang, Y. C. Li, F. G. Yuan, S. J. Hu and P. He, Effects of hematin and carbon monoxide on the salinity stress responses of *cassia obtusifolia* L. seeds and seedlings, *Plant Soil.*, 359 (2012) 85–105.
[141] R. Mittler, Oxidative stress, antioxidants and stress tolerance, *Trends Plant Sci.*, 7 (2002) 405–410.

[142] Z. S. Sa, L. Q. Huang, G. L. Wu, J. P. Ding, X. Y. Chen and T. Yu, Carbon monoxide: A novel antioxidant against oxidative stress in wheat seedling leaves, *J. Integr. Plant Biol.*, 49 (2007) 638–645.

[143] G. G. Yannarelli, G. O. Noriega, A. Batlle and M. L. Tomaro, Heme oxygenase up-regulation in ultraviolet-B irradiated soybean plants involves reactive oxygen species, *Planta*, 224 (2006) 1154–1162.

[144] W. W. Kong, L. P. Zhang, K. Guo, Z. P. Liu and Z. M. Yang, Carbon monoxide improves adaptation of *Arabidopsis* to iron deficiency, *Plant Biotechnol.* J., 8 (2010) 88–99.

[145] Y. Han, W. Xuan, T. Yu, W. B. Fang, T. L. Lou and Y. Gao, Exogenous hematin alleviates mercury-induced oxidative damage in the roots of *Medicago sativa*, *J. Integr. Plant Biol.*, 49 (2007) 1703–1713.

[146] Q. Zheng, Q. Meng, Y. Y. Wei and Z. M. Yang, Alleviation of copper- induced oxidative damage in *chlamydomonas reinhardtii* by carbon monoxide, *Arch. Environ. Con. Tox.*, 61 (2011) 220–227.

[147] Y. Y. Wei, Q. Zheng, Z. P. Liu and Z. M. Yang, Regulation of tolerance of Chlamydomonas reinhardtii to heavy metal toxicity by hemeoxygenase-1 and carbon monoxide, *Plant Cell Physiol.*, 52 (2011) 1665–1675.

[148] D. K. Meng, J. Chen and Z. M. Yang, Enhancement of tolerance of Indian mustard (*Brassica juncea*) to mercury by carbon monoxide, *J. Hazard. Mater.*, 186 (2011) 1823–1829.

[149] Y. Han, J. Zhang, X. Chen, Z. Gao, W. Xuan, S. Xu, X. Ding and W. Shen, Carbon monoxide alleviates cadmium-induced oxidative damage by modulating glutathione metabolism in the roots of *medicago sativa*, *New Phytol.*, 177 (2008) 155–166.

[150] S. H. Snyder, S. R. Jaffrey and R. Zakhary, Nitric oxide and carbon monoxide: Parallel roles as neural messengers, *Brain Res. Rev.*, 26 (1998) 167–175.

[151] Y. J. Xie, T. F. Ling, Y. Han, K. L. Liu, Q. S. Zheng and L. Q. Huang, Carbon monoxide enhances salt tolerance by nitric oxide-mediated maintenance of ion homeostasis and up-regulation of antioxidant defence in wheat seedling roots, *Plant Cell Environ.*, 31 (2008) 1864–1881.

[152] Y. Ding, W. K. McCoubrey Jr and M. D. Maines, Interaction of heme oxygenase-2 with nitric oxide donors. Is the oxygenase an intracellular 'sink' for NO?, *Eur. J. Biochem.*, 264 (1999) 854–861.

[153] S. Jeandroz, D. Wipf, D. J. Stuehr, L. Lamattina, M. Melkonian, Z. Tian, Y. Zhu, E. J. Carpenter, G. K.-S. Wong and D. Wendehenne, Occurrence, structure, and evolution of nitric oxide synthase-like proteins in the plant kingdom, *Sci. Signal*, 9 (2016) re2 1–9.

[154] D. M. Santa-Cruz, N. A. Pacienza, A. H. Polizio, K. B. Balestrasse, M. Tomaro and G. G. Yannarelli, Nitric oxide synthase-like dependent NO production enhances heme oxygenase up-regulation in ultraviolet-B-irradiated soybean plants, *Phytochemistry*, 71 (2010) 1700–1707.

[155] W. Xuan, S. Xu, M. Y. Li, B. Han, B. Zhang, J. Zhang, Y. Lin, J. Huang, W. Shen and J. Cui, Nitric oxide is involved in hemin-induced cucumber adventitious rooting process, *J. Plant Physiol.*, 169 (2012) 1032–1039.

[156] Y. J. Xie, C. Zhang, D. Lai, Y. Sun, M. K. Samma, J. Zhang and W. Shen, Hydrogen sulphide delays GA-triggered programmed cell death in wheat aleurone layers by the modulation of glutathione homeostasis and heme oxygenase-1 expression, *J. Plant Physiol.*, 171 (2014) 53–62.

[157] Y. T. Lin, M. Y. Li, W. T. Cui, W. Lu and W. B. Shen, Haem oxygenase-1 is involved in hydrogen sulfide-induced cucumber adventitious root formation, *J. Plant Growth Regul.*, 31 (2012) 519–528.

[158] Y. T. Lin, W. Zhang, F. Qi, W. T. Cui, Y. J. Xie and W. B. Shen, Hydrogen-rich water regulates cucumber adventitious root development in a heme oxygenase-1/carbon monoxide-dependent manner, *J. Plant Physiol.*, 171 (2014) 1–8.

Exercises

Multiple-choice questions/fill in the blanks

1. The toxicity of CO was first time revealed in 1857 by
   (a) Claude Bernard
   (b) Joseph Priestley
   (c) Henry Cavendish
   (d) Sir Humphrey Davy

2. Reduction in protein's ability to shuttle and transfer oxygen into tissues causing hypoxia is due to interaction of iron haemoglobin with
   (a) NO (b) CO (c) H_2S (d) All of these

3. The endogenous production of CO in man from the oxidative metabolism of haeme was first time reported in 1949 by
   (a) R. Schmid
   (b) R. Motterlini
   (c) T. Sjostrand
   (d) None of these

4. IOP is related to
   (a) Heart (b) Lungs (c) Eye (d) None of these

5. Sort out a crystalline metal carbonyl from the following:
   (a) $Ni(CO)_4$ (b) $Fe(CO)_5$ (c) $Ru(CO)_5$ (d) $Mn_2(CO)_{10}$

6. Which one of the following is a golden yellow metal carbonyl?
   (a) $V(CO)_6$ (b) $Te_2(CO)_{10}$ (c) $Mn_2(CO)_{10}$ (d) $Re_2(CO)_{10}$

7. The high stability of *trans*-hydridocarbonyl(triphenylphosphine)rhodium(I), *trans*-$[RhH(CO)(PPh_3)_3]$ is because it follows an electron rule called
   (a) 16-electron rule
   (b) 17-electron rule
   (c) 18-electron rule
   (d) None of these.

8. Hydroformylation of alkene by rhodium catalyst, $[RhH(CO)(PPh_3)_3]$, gives straight-chain aldehyde through
   (a) Markovnikov addition of alkene to the catalyst
   (b) anti-Markovnikov addition to the catalyst
   (c) Oxidative addition of H_2 to the catalyst
   (d) None of these

9. Sort out a metal carbonyl having EAN 54 from the following:
   (a) $[Ni(CO)_4]$ (b) $[Fe(CO)_5]$ (c) $[Cr(CO)_6]$ (d) $[Mo(CO)_6]$

10. Which one of the following metal carbonyl does not follow EAN rule?
    (a) $[W(CO)_6]$ (b) $[Ru(CO)_5]$ (c) $[Mo(CO)_6]$ (d) $[V(CO)_6]$

11. Sort out a metal carbonyl from the following which does not have EAN equal to the atomic number of the next inert gas.
    (a) $[Co(CO)_4]^-$ (b) $[Mn(CO)_6]^+$ (c) $[Mn(CO)_5]^-$ (d) $[Mn(CO)_6]$

12. In the structure of $Mn_2(CO)_{10}$, the two $Mn(CO)_5$ units are
    (a) Eclipsed to each other
    (b) Staggered to each other
    (c) Gauche to each other
    (d) None of these

13. IR spectral study of $Co_2(CO)_8$ molecule in solid state has suggested that this compound has structure, namely
    (a) Non-bridged structure
    (b) Bridged structure
    (c) Equilibrium of both bridged and non-bridged structures
    (d) All of these

14. In metal carbonyls, such as $[Ni(CO)_4]^0$, $[Co(CO)_4]^-$ and $[Fe(CO)_4]^{2-}$, the correct trend in M–C bond order is
    (a) Ni–C > Co–C > Fe–C (b) Ni–C < Co–C < Fe–C
    (c) Ni–C ~ Co–C < Fe–C (d) Ni–C < Co–C > Fe–C

15. In metal carbonyls, namely, $[Ni(CO)_4]^0$, $[Co(CO)_4]^-$ and $[Fe(CO)_4]^{2-}$, the correct trend in C–O bond order is
    (a) $[Ni(CO)_4]^0 \sim [Co(CO)_4]^- < [Fe(CO)_4]^{2-}$
    (b) $[Ni(CO)_4]^0 < [Co(CO)_4]^- < [Fe(CO)_4]^{2-}$
    (c) $[Ni(CO)_4]^0 < [Co(CO)_4]^- \sim [Fe(CO)_4]^{2-}$
    (d) $[Ni(CO)_4]^0 > [Co(CO)_4]^- > [Fe(CO)_4]^{2-}$

16. Among the metal carbonyls, $[V(CO)_6]^-$, $[Cr(CO)_6]^0$ and $[Mn(CO)_6]^+$, the ν(CO) is maximum in
    (a) $[V(CO)_6]^-$
    (b) $[Mn(CO)_6]^+$
    (c) $[Cr(CO)_6]^0$
    (d) None of these

17. The principal source of the yellow colour of urine is
    (a) Biliverdin
    (b) Bilirubin
    (c) Glucuronic acid
    (d) None of these

18. The suitable filled metal d-orbital to back donate electrons to the anti-bonding π-molecular orbitals, namely, $\pi^{\star}_{2py}$ or $\pi^{\star}_{2pz}$ of CO to form M(d) →(CO) π bond in metal carbonyls is
    (a) d_{z^2} (b) $d_{x^2-y^2}$ (c) d_{xz} or d_{yz} (d) d_{xy}

19. The endogenous production of CO in mammals is due to oxidation of haeme catalysed by a family of enzyme called
    (a) Haeme oxygenases
    (b) Haeme reductases
    (c) Biliverdin reductases
    (d) None of these

20. The biosynthesis of CO in plants was first time reported in 1959 by
    (a) G. W. Schade (b) S. S. Wilks (c) M. J. Terry (d) P. J. Crutzen

21. Haemin is a compound which is a donor of
    (a) NO (b) H_2S (c) CO (d) None of these

22. Overproduction of bilirubin leads to ________, a clinical syndrome of hyperbilirubinaemia.

Short-answer-type questions

1. What are metal carbonyls? Explain with suitable examples.
2. Present a brief note on the reactivity of metal carbonyls with reference to the following:
   (a) Reactions with NO
   (b) Formation of anionic carbonyl complexes
   (c) Formation of cationic carbonyl complexes
   (d) Insertion reactions in metal carbonyls
   (e) Nucleophilic attack
3. Describe the mechanism of hydrogenation reaction of alkenes taking a suitable carbonyl compound.
4. What is hydroformylation reaction? Why is it is called oxo reaction? Give an example of this reaction.
5. Draw the catalytic cycle for hydroformylation of alkenes by ruthenium catalyst, $[Ru(CO)_3(PPh_3)_2]$.
6. Draw the catalytic cycle for the formation of acid/ester from olefins with CO in the presence of $Co_2(CO)_8$ as a catalyst.
7. Just draw the catalytic cycle for synthesis of acetic acid by BP Cativa process using iridium carbonyl as a catalyst. What are the benefits of using iridium as a catalyst compared to rhodium?
8. What is EAN rule in metal carbonyls? Work out the EAN of $[Cr(CO)_6]$ and $[Ni(CO)_4]$.
9. Work out the EAN of $Co_4(CO)_{12}$ and $Ir_4(CO)_{12}$
10. By calculating the EAN of each Co atom in $Co_4(CO)_{12}$, justify that this compound contains six Co–Co bonds.
11. How is the calculation of EAN per metal atom in a polynuclear carbonyl helpful to know the number of metal–metal (M–M) bonds present in that metal carbonyl? Explain this taking at least two examples of polynuclear carbonyls.
12. What is 18-electron rule for metal carbonyls? How is the total (i.e. effective) number of electrons present in the valance shell of one metal atom in a polynuclear carbonyl calculated? Give four examples of metal carbonyl obeying18-electron rule with justification.
13. Give some examples of mixed metal carbonyls obeying 18e rule with justification.
14. Although 18-electron rule is not obeyed by square planar complexes, yet they are quite stable complexes. Give justification behind the stability of such complexes.
15. Draw the two possible structures, TBP and square pyramidal, of $Fe(CO)_5$, and present evidence in favour of the TBP structure. The ^{13}C-NMR gives only one type of signal for axial as well as equatorial CO groups. Give the possible reason for this.
16. Draw the structures of $Mn_2(CO)_{10}$, $Tc_2(CO)_{10}$ and $Re_2(CO)_{10}$. Are they diamagnetic?

17. Draw the structures of $Co_4(CO)_{12}$, $Rh_4(CO)_{12}$ and $Ir_4(CO)_{12}$ and comment on their magnetic behaviour.
18. Describe the effect of π-back bonding on M–C (M = metal) and C–O bond orders in metal carbonyls with suitable examples.
19. Draw the M.O. diagram of CO molecule and point out the HOMO and LUMO in that which participate in bonding in metal carbonyls.
20. Differentiate the bonding in metal carbonyls and that in π-complexes of unsaturated organic ligands like alkene, alkynes or cyclopentadienes.
21. Of the two possible structures, (i) TBP (D_{3h}) and (ii) square pyramidal (C_{4v}), how the vibrational spectroscopy supports in favour of TBP (D_{3h}).
22. Draw the formation of biliverdin and CO from the oxidation of haeme by HO.
23. Present a brief account of therapeutic potential of CO.
24. Give the names of structures of some CORMs.
25. Highlight the role of CORMs in helpful effects in various ocular conditions.
26. Draw the structures along with names of some photoCORMs.
27. Show the overall pathways of generation of CO in plants diagrammatically.
28. Draw the structures of haemin and haematin. Do they donate CO?

Long-answer-type questions

1. Discuss in detail the different reactivity of metal carbonyls.
2. Present a detailed view of the catalytic aspects of metal carbonyls.
3. What is hydroformylation reaction? Highlight the importance of this reaction. Present the detailed view ofcatalysts used for hydroformylation reaction.
4. Describe the large-scale synthesis of CH_3COOH by carbonylation of methanol using metal carbonyl as catalyst.
5. Present the detailed view of manufacturing of acetic acid by BP Cativa process using iridium carbonyl as catalyst.
6. Describe the manufacturing of acetic anhydride by carbonylation of methyl acetate using metal carbonyls as catalyst. Also highlight the uses of acetic anhydride in different sectors.
7. Present an explanatory note on carbonylation of olefins and acetylenes to carboxylic acids or esters or alcohols using metal carbonyls as catalyst.
8. What is 18-electron rule in metal carbonyls? Give examples of metal carbonyls obeying and not obeying18-electron rule with justification.
9. How many different types of bonds may be present in metal carbonyls? Explain their formations on the basis of suitable theories.
10. Draw and discuss the structure of $Mn_2(CO)_{10}$ on the basis of hybridization. Is it diamagnetic or paramagnetic?

11. Explain the formation of bridging and non-bridging structures of $Co_2(CO)_8$ on the basis of hybridization. What is the reason behind the bent Co–Co σ-bond in its bridged structure?
12. Point out the difference in structures of $Fe_2(CO)_9$, $Ru_2(CO)_9$ and $Os_2(CO)_9$. Draw the hybridization scheme for each of the dimeric carbonyl compound.
13. Draw and discuss the structures of $Fe_3(CO)_{12}$, $Ru_3(CO)_{12}$ and $Os_3(CO)_{12}$ using suitable hybridization schemes.
14. Differentiate the structures of tetranuclear carbonyls: $Co_4(CO)_{12}$, $Rh_4(CO)_{12}$ and $Ir_4(CO)_{12}$ on the basis of hybridization.
15. Present the detailed view of OC→M σ-bond and M→CO π-bond in metal carbonyls.
16. Describe the utility of IR spectra of metal carbonyls with special reference to:
    (i) Determination of geometry of metal carbonyls
    (ii) Bond order of coordinated terminal CO group
    (iii) Differentiation of terminal and bridging CO groups
    (iv) Distinction in substituted and non-substituted metal carbonyls
17. Describe the different sources of CO in human body. How the action of HO is observed in bruises (skin decolouration caused by injury)?
18. What are the possible targeting sites of CO in mammals to ameliorate medical conditions and CO-mediated vasorelaxation?
19. Present an overview of CO signalling in anti-inflammatory responses.
20. What are the ways of CO delivery in human body for therapeutic use?
21. Describe the therapeutic applications of CO and CO resealing molecules or materials.
22. What are photoCORMs? Describe thetherapeutic applications of PhotoCORMs.
23. Describe the biosynthesis of CO in plants.
24. Describe the role of CO in plant growth and development.
25. Present the detailed view of CO in abiotic stresses.
26. What do you understand by cross-talk between CO and other signalling molecules? Explain.

Chapter IV
Advantageous role of gaseous signalling molecule, H_2S: hydrogen sulphide and their respective donors, in ophthalmic diseases and physiological implications in plants

4.1 Introduction

Three gaseous molecules of leading effect in biology are NO, CO and H_2S. Irrespective of their toxicity, these three gaseous molecules have prominent effects on mammalian physiology and key involvements in therapeutics. At very low concentrations, they participate in key signalling and regulatory functions in human biology, and so are now termed as "gasotransmitters".

Just after the discovery of the role of nitric oxide (NO) as a crucial endogenous signalling mediator in 1987, research on this molecule has rapidly prospered and expanded in many directions. With the knowledge that NO and its second messenger cyclic guanosine monophosphate (cGMP) are distributed in most tissues with a varied range of biological effects, it has become evident that the NO pathway is also important for multiple functions in the eye. Glaucoma is a progressive optic neuropathy caused by degeneration of the retinal ganglion cells (RGCs) and is the leading cause of irreparable blindness worldwide. Numerous risk factors for the disease have been detected, but elevated intraocular pressure (IOP) remains the primary risk factor accountable to treatment. There is increasing evidence that NO-releasing molecules is a direct regulator of IOP and that dysfunction of the NO–guanylate cyclase (GC) pathway is associated with glaucoma incidence. NO has shown promise as a novel therapeutic with targeted effects which (i) lower IOP, (ii) increase ocular blood flow (OBF) and (iii) confer neuroprotection. The various effects of NO in the eye appear to be facilitated through the activation of the GC-3′–5′-cGMP pathway and its effect on downstream targets, such as protein kinases and Ca^{2+} channels.

In the late 1990s, the scientific community saw a very unusual phenomenon, the conversion of NO from harmful gases into an important chemical messenger. In the same time period, carbon monoxide (CO), another gas usually associated with environmental pollution, air poisoning and suicidal behaviour, was also undergoing a similar change in image, although not as closely followed as NO. It had been known for several decades that the human body generated CO upon the decomposition of haemoglobin, which was determined by the discovery that heme oxygenase is the enzymatic source of CO. However, its role as an endogenous neurotransmitter was suggested in the early 1990s. Thereafter, the biological activity of CO has been extensively studied through both the direct intake of CO and in the form of alleged

https://doi.org/10.1515/9783110727302-004

"carbon monoxide releasing molecules (CORMs)". The importance of CO molecule to the ocular system has only recently been recognized. Low concentrations of CO can provide favourable effects in various ocular conditions, as an illustration, in glaucoma. Recently, it has recently been established that a water-soluble CO-releasing molecule (CORM) lowers IOP in glaucomatous rabbits. Also CORM is a prospective new drug for treatment of patients with non-infectious posterior uveitis. It is a sight-threatening inflammatory eye disease, which represents the third leading cause of blindness in developed countries.

Hydrogen sulphide (H_2S) is primarily known for its unpleasant smell of rotten eggs. It is very noxious for both animals and human beings. Continued exposure of approximately 5 ppm H_2S causes warning signs. This includes headaches and loss of sleep. Remarkably, 500 ppm H_2S can be life threatening, and concentrations higher than 1,000 ppm cause almost instantaneous death. It has been observed that its toxicity is linked with the inhibitory effect on cytochrome c oxidase. Notably, cytochrome c oxidase is one of the significant enzymes of the mitochondrial respiratory chain. Importantly, H_2S is another internally produced gas generated by way of enzymatic reactions in mammals. This participates in maintenance of mammalian homeostasis. In mammals, L-cysteine (an amino acid) is responsible for generation reaction of H_2S. This reaction is catalysed by four main enzymes, namely, (i) cystathionine β-synthase (CBS), (ii) cystathionine γ-lyase (CSE), (iii) 3-mercaptopyruvate sulphur transferase (3-MST) and (iv) cysteine aminotransferase (CAT). These enzymes are available inside the mammalian eyeballs at different locations.

It is seen that the abnormal levels of H_2S are linked with multifarious diseases. These are neurodegenerative diseases, myocardial injury and ophthalmic diseases. In case of excessive exposure of H_2S, cellular toxicity takes place. This increases the risk of various diseases. Remarkably, under physiological status, H_2S plays a vital role in maintaining cellular physiology and preventing damages to tissues. The abnormal distribution of CBS, CSE, 3-MST and CAT and the accumulation of substrates and intermediates can change the concentration of generated H_2S.This causes abnormal structures or functions in the eyes. Detailed studies have shown that intake of H_2S donors could control IOP. Moreover, H_2S donors protect retinal cells, inhibit oxidative stress and improve inflammation by modifying the function of intra or extracellular proteins in ocular tissues. Thus, a number of slow-releasing H_2S donors are found to be effective drugs for handling various diseases.

Studies have shown that H_2S is a key player in the regulation of normal plant physiological processes. These include seed germination, root morphogenesis, photosynthesis and flower senescence. At physiological concentrations, H_2S has also found as an important messenger in plant defence signalling against several abiotic stresses.

Therefore, a detailed view with regard to the progress and our understanding of H_2S synthesis and signalling functions in plants along with the role of H_2S donors in ocular diseases would be presented in the upcoming section.

4.2 Introductory view of gasotransmitters: endogenous signalling molecules

Based on differences in their properties, signalling molecules are generally categorized as neurotransmitter and gasotransmitter. The size, shape and function of different types of signalling molecules can vary significantly.

In general, gasotransmitters refer to the distinctive class of molecules like NO, CO and H_2S, responsible for communication amongst body cells for a particular biological action. Even though these molecules exist in solvated form while in biological medium, the respective differences in size, action, shape and bio-membrane interactions stems their multitude biological roles reported so far. The signal transduction pathway among such carriers may range from short to long distances to transmit the required information [1]. The properties and functional diversity found in these bioessential signalling molecules, therefore gave rise to coin a new term in reference to their biological relevance as "gasotransmitters".

Several parameters may be found differentiating neurotransmitters from gasotransmitters. From the cellular biology it is clear that neurotransmitters stored in vesicles get released by the intervention of a suitable stimulus (Figure 4.1, A top). These responses are receptor-specific in nature and depend on the molecular signalling

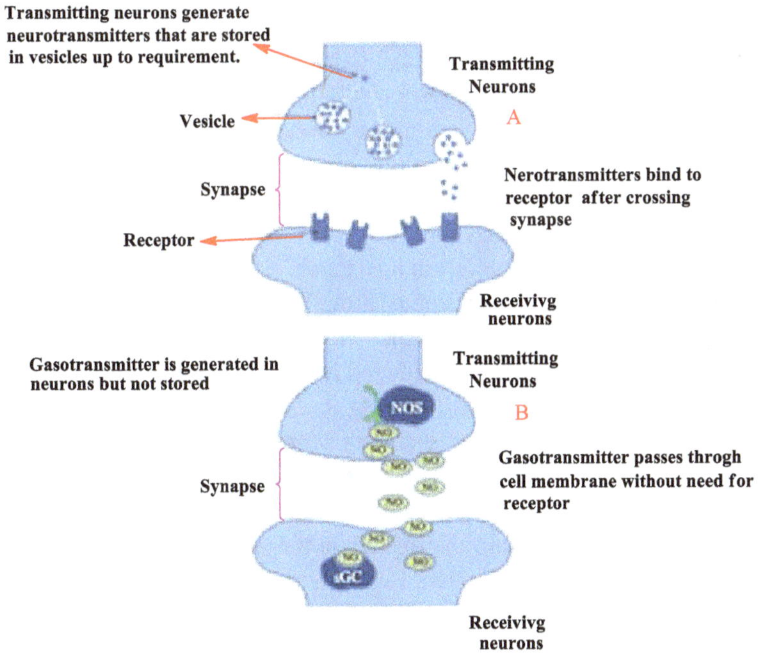

Figure 4.1: Diagrammatic representation of mechanistic action of neurotransmitter (A, top) and gasotransmitter (B, bottom).

to bring forth a physiological move. Hence, synaptic vesicles behave as a reservoir of information required at the time of safety or normal physiological functioning. On the other hand, gasotransmitters are endogenously availed small molecules of signalling potentiality ("gaso" refers to their gaseous nature under normal conditions) [1, 2]. Gasotransmitters have the main characteristic feature of diffusing through cellular membrane without the aid of any receptor (Figure 4.1, B bottom). No reservoir is required (like vesicles in neurotransmitters), but are rapidly produced in response to a stimulus [3] when needed. Moreover, here in gasotransmitters cell exocytosis fashion followed in neurotransmitters is not pronounced at all. Therefore, a separate term "gasotransmitter" coined by Wang [4] in 2002 is suitable to distinguish them from neurotransmitters. The vasorelaxant and gasotransmitter labelling of NO enhanced scientific vigour to an extraordinary fashion and activated the seek for other molecules [5, 6] of this class. Gases other than these are also under interrogation to add further possible members to this group.

4.3 Biosynthesis of H_2S

4.3.1 In ocular tissues

H_2S is considered as the recently recognized third member among gasotransmitters. This gas is also produced endogenously through enzymatic reactions (Figure 4.2) and is responsible for maintaining physiological balance. Recently, non-enzymatic pathways have also been reported to be responsible for biosynthesis of hydrogen sulphide. The optimal concentration of H_2S under normal conditions is said to be present in micromoles (μM) [7]. L-Cysteine is the important bio-essential source of this gas involving the role of CBS, cytosolic/pyridoxal-5′-phosphate-dependent enzymes and CSE [8, 9] or the tandem enzymes, CAT and 3-MST, mainly confining the activity in mitochondria. Absolute H_2S is produced from CSE and CBS, whereas 3-mercaptopyruvate transfers sulphur to cysteine through the catalytic application of 3-MST resulting in the formation of persulphide [10, 11]. Thereafter, persulphide acts as a H_2S releaser endogenously causing the reduction of disulphides, for example, reduction of dihydrolipoic acid or thioredoxin under physiological terms [12]. Strictly, therefore it is thioredoxin reductase acting sequence wise in combination with glutathione (GSH) and sulphur oxidase in this phenomenon. The significance of H_2S (oxidized form) in respiratory chain is well documented acting as electron donor in Q, III and IV steps of the chain that results in the generation of energy currency (ATP) and cellular oxygen consumption. Hence, biosynthesis and H_2S-mediated sensing of oxygen represent essential role of H_2S biochemistry.

Research groups of Persa and Pong [14, 15] in 2006 reported that the H_2S-generating enzymes, CBS and CSE, are present in ocular tissues, particularly in the **retina**. As per

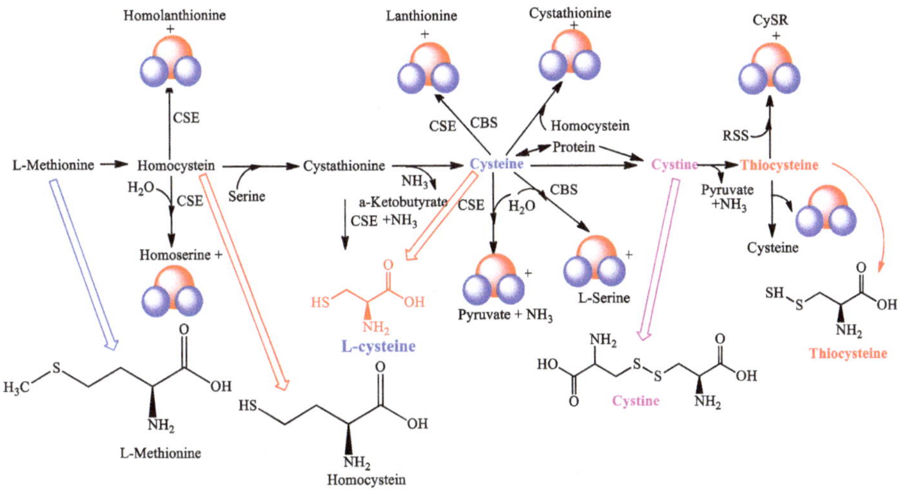

(i) Production of H_2S in tissues via cystosolic enzymes CBS and CSE

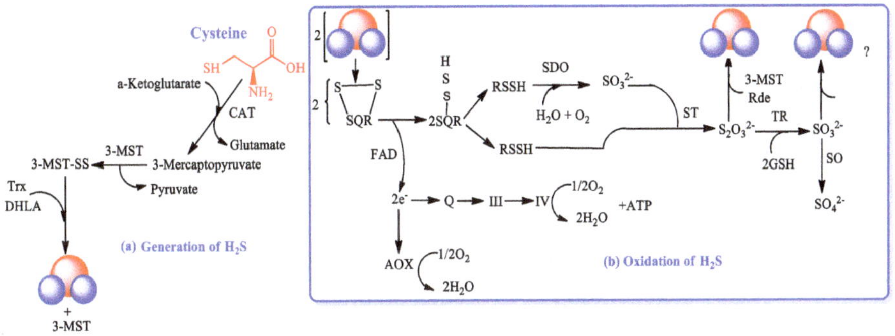

(ii) (a) Generation of H_2S via tandem enzymes, CAT and (3-MST) and (b) H_2S oxidation in the mitochondria

Figure 4.2: Biosynthesis of H_2S and its oxidation (adapted from Ref. [13]).

report of Shibuya et al. [16], D-cysteine also generates H_2S in presence of catalysts as shown in Figure 4.3.

Remarkably, the generation of H_2S in cattle eye's tissues is explored. This includes tissues of cornea, iris, ciliary, aqueous humour (AqH) muscle, lens, choroid of the eye and retina, excluding vitreous humour. Kulkarni et al. [17] observed that the maximum generation of H_2S occurs in cornea and retina of cattle. CBS is highly distributed in the cornea, conjunctiva and iris. Instead, in retina and optic nerve, very small amount of CBS has been observed. Reasonably smaller amount CBS is detected in lens. But it is absent in the vitreous humour. Persa and co-workers observed that throughout the lifecycle, the distribution of CBS is always high in anterior segments. However, it has a usual trend of age

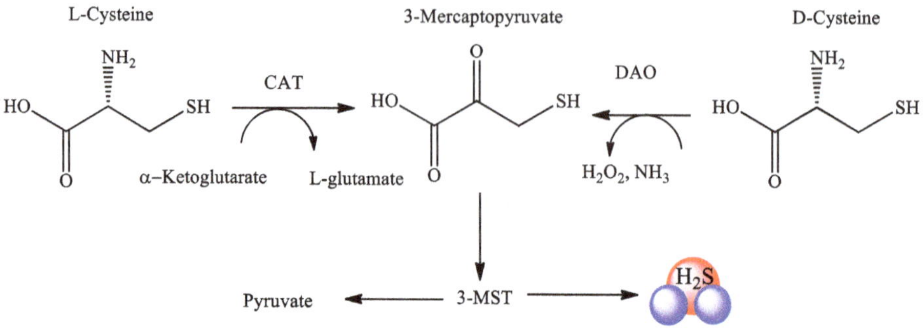

Figure 4.3: Pathways of H_2S generation from D-cysteine and L-cysteine.

dependent increase in retina [14]. Moreover, CSE has been detected in retina of amphibians and mammals. Its activity can be found [15] there.

Mikami et al. [18] observed that the 3-MST/CAT path is the foremost route to generate H_2Sin mammalian retina. This is due to the presence of these two tandem enzymes both in the retinal neurons. Moreover, they increase at low concentrations of Ca^{2+} in brightness. It is observed that the deficiency of H_2S or its substrates is much related with ectopia lentis, myopic, cataract, optic atrophy and retinal detachment [19, 20] (Figure 4.4).

4.3.2 In plants

Both enzymatic and non-enzymatic routes are responsible for generation of H_2S in plants. However, the generation of hydrogen sulphide via non-enzymatic route occurs in a very small amount. In *Arabidopsis thaliana,* the generation of hydrogen sulphide via enzymatic routes is shown in Figure 4.5.

Sulphur incorporates in plants via two reaction systems: (i) the reduction of sulphate and (ii) the synthesis of cysteine. The reduction of sulphate is initiated first through the absorption of SO_4^{2-} into the plant from the soil. This process is completed by specific sulphate transporters [22] proteins. The ATP sulphurylase first activates the absorbed sulphate (SO_4^{2-}) in the plant. This results in adenosine-5′-phosphosulphate (APS). The APS so obtained is reduced in a two-step reaction by APS reductase and sulphite reductase (SiR). In this two-step reduction process, the activated sulphate (SO_4^{2-}) is first reduced to sulphite (SO_3^{2-}), and thereafter the resulting sulphite (SO_3^{2-}) is reduced to sulphide (S_2^-; H_2S). The sulphide so obtained incorporates into O-acetylserine (OAS) to form cysteine. This process is catalysed by *O*-acetylserine thiol lyase (OAS-TL). In fact, cysteine serves as a source of sulphur for all organic molecules containing reduced sulphur. For example, GSH, proteins, cofactors and vitamins [23, 24]. These reactions take place in chloroplasts/plastids. Remarkably, OAS-TL catalyses the reverse reaction. Here, cysteine is converted to H_2S, and OAS is formed as a secondary product. Moreover, H_2S is produced via the interaction of

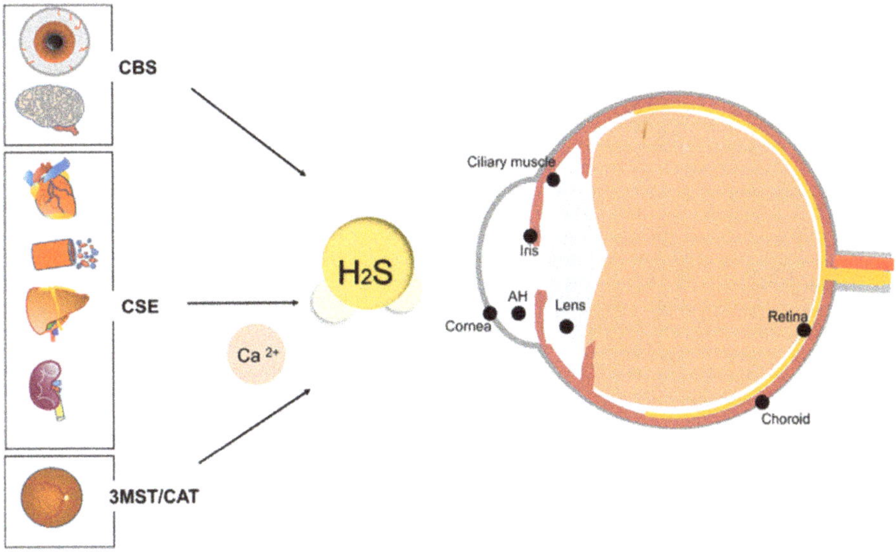

Figure 4.4: Production of H_2S principally controlled by four enzymes, CBS, CSE, 3-MST and CAT (adapted from Ref. [21]).

D-cysteine with D-cysteine desuphfhydrase (DCDES), and L-cysteine with L-cysteine desulphhydrase 1 (DES1). In these two reactions, ammonia and pyruvate are formed as by-products. Furthermore, H_2S is also generated as a primary product in a reaction of cyanide and cysteine in presence of cyanoalanine synthase (CAS) as a catalyst. In this reaction, cyanoalanine is formed as a secondary product. The overall generation of H_2S in plants is shown in Figure 4.5.

Experimentally, it is seen that lower levels of H_2S exhibit an encouraging effect in germination, size and fresh weight in several plants. Contrary to this, higher levels of H_2S display phytotoxic effects. Both effects are shown in the bottom of Figure 4.5.

4.4 Biological chemistry of H_2S

The biological synthesis of H_2S, as in the case of NO and CO, must be followed by the consumption or target phenomenon. This gas has no colour, is flammable and smells like rotten eggs. Its acid strength is weak (p*K*a value of 6.98 at 25 °C and 6.76 at 37 °C). The dissociation of this gas in water may be represented as shown in Figure 4.6. The values [25] of Ka_1 and Ka_2 are 1.3×10^{-7} M and 1×10^{-19}, respectively.

The undissociated H_2S is volatile in nature, while the dissociated form HS^- is not volatile. In physiological medium, these dissociation patterns are pH dependant. At the 7.4 pH, one-third of H_2S remains undissociated. Physiological pH does not support the substantial presence of S^{2-} (because high pH is required for it). As

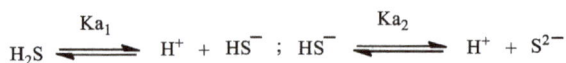

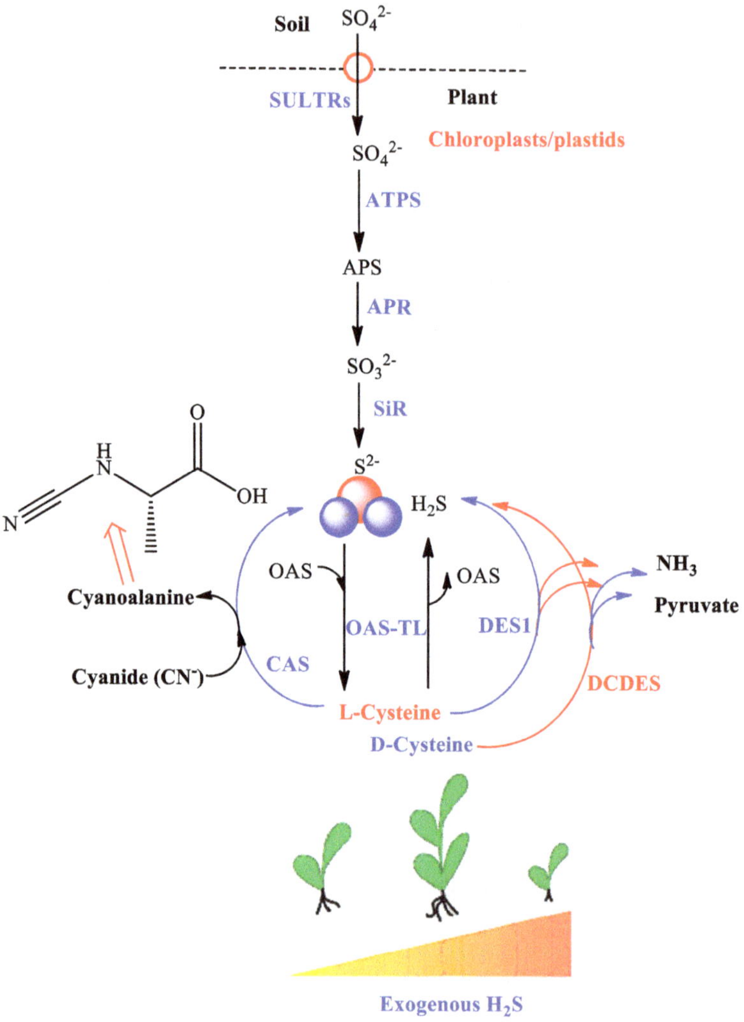

Figure 4.5: Diagrammatic representation of generation of H_2S in higher plants.

$$H_2S \underset{}{\overset{Ka_1}{\rightleftharpoons}} H^+ + HS^- \; ; \; HS^- \underset{}{\overset{Ka_2}{\rightleftharpoons}} H^+ + S^{2-}$$

Figure 4.6: Dissociation reactions of hydrogen sulphide in aqueous solution.

per another scientific observation, H_2S mainly persists in HS^- form (82%) because of having strength (pKa_1: 6.76; pKa_2: 19.6) [26] as weak acid. Therefore, both H_2S and SH^- are contributory under biological activity of hydrogen sulphide, in spite of the fact that SH^-, represents more nucleophilic potential than cysteine (Cys) or reduced form of GSH, that swiftly coordinates with bio-metallic centres or interacts with other compounds [27].

In mitochondrial chain reactions, H_2S displays sequential oxidation trend. Initially, it gets oxidized to thiosulphate, followed by conversion to sulphite and subsequently to sulphate. The first oxidation stage is non-enzymatic in nature, while the rest steps are carried out enzymatically using a biocatalyst, thiosulphate cyanide sulphur transferase. In spite of the observation showing SO_4^{2-} as a key final product of the metabolic pathways followed by H_2S, urinary thiosulphate represents a non-specific indicator to sense the quantity of H_2S production [28] within a body. Overall, the target of H_2S is highly influenced by several factors especially the rapid oxidation disfavouring the long-distance transport. Therefore, the development of efficient storage system endogenously applicable is suggested. For instance, in NO association, H_2S gets stored as nitrosothiols (RS–NO) implying the dual gas-combinatory fashion. This opens several areas of interest to seek answer for the unexplored queries regarding the isolated and combined gasotransmitters research. Meanwhile, the solvated fashion of H_2S reveals different solubility trend based on the nature of the solvent. Due to lipophilic feature of this gas, H_2S easily crosses cell membrane. The tested solubility experiments have shown it to be five times more soluble in lipophilic solvent as compared to water. The data furnished from its solubility behaviour in various solvents other than water have been keenly recorded [29, 30].

Harmful effects of hydrogen sulphide at less than 100 ppm levels have come out [31] in the mode of sore throat, irritation in eyes, dizziness and so on. The exposure at greater than 1,000 ppm affects CNS, respiratory chain and may even cause death [32].

4.5 Implications of H_2S in ophthalmic diseases

4.5.1 Introduction

Similar to the two well-known gasotransmitters, NO and CO, H_2S has also been recognized as the third gasotransmitter. It plays significant roles in normal physiological conditions as well as in the process/progress of several diseases [33].

Notably, as the third gasotransmitter, H_2S controls organ growth and also upkeeps of equilibrium in tissues. Generated uncommon concentrations of H_2S are associated with diverse human diseases. These include neurodegenerative diseases, myocardial injury and ophthalmic diseases. Excessive exposure of H_2S can be responsible for cellular toxicity and orchestrate pathological process. Moreover, this increases the risk of numerous diseases. Interestingly, H_2S plays a crucial role in the maintenance of cellular physiology in physiological condition. Additionally, it participates in the regulation of tissue damages. It is well discussed earlier that in mammals, the generation of H_2S is catalysed by four enzymes: CBS, CSE, 3-MST and

CAT. Importantly, inside the mammalian eyeballs, these enzymes are identified at different sites. Unexpectedly, abnormal presence of these enzymes and the assemblage of substrates and intermediates can modify the levels of H_2S by several orders of scale. This happening is responsible for unusual structures/functions of the eyes. Detailed studies have shown that ingestion of H_2S donors' controls IOP, protects retinal cells, prevents oxidative stress and lessens inflammation. Such reliefs are achieved by modifying the function of intra or extracellular proteins in ocular tissues. Thus, several slow-releasing H_2S donors have been found as effective drugs for medicating multiple diseases. Therefore, the biological functioning of H_2S metabolism and its role in ophthalmic diseases are presented in the coming section of this chapter.

4.5.2 Ocular drug delivery

Two anatomical regions are there in human eye. The first one is called the anterior segment. This involves cornea, conjunctiva, sclera and anterior uvea. The second one is known as posterior segment, and this includes retina, vitreous and choroid. These are different parts of eye with diverse structures. Unique challenges, therefore, exist for beneficial drug delivery to each of these sites. The most widely accepted route of ocular drug delivery is the topical administration of the drug in the form of eye drops into the conjunctival cul-de-sac. Regardless of its obvious ease of access, the eye is well secure from external materials along with therapeutic substances. This is owing to a number of mechanisms operative there efficiently. These mechanisms include closing and reopening of eyes (blinking), induced production of tears, tear turnover and nasolacrimal drainage. All these activities help in the quick removal of substances from the eye's surface. Moreover, cornea also protects the eye. In fact, this forms the physical–biological barrier [34, 35] [Figure 4.7]. Consequently, these protections reduce the drug levels at the intended sites.

4.5.3 Ocular bioavailability

Routinely, the human eye can hold ~30 microlitre (µL) of eye drops. But, after a single closing and reopening of the eye, this volume is reduced to 7–10 µL via nasolacrimal (nose and tear glands) drainage. In this process, ophthalmic drug is methodically absorbed across the nasal mucosa or the gastrointestinal tract [36]. Thereafter, topically applied drug undergoes a systemic loss substantially. This is owing to the absorption of drug from conjunctiva into the local flow. In the meantime, tear turnover is stimulated by the pH and the tonicity of the drug solution. This removes drug solutions from the conjunctival cul-de-sac in a very little time. Another factor of low absorption of ophthalmic drug solution is the poor permeable nature of the cornea. The cornea consists of five dissimilar layers. Three of these five layers, namely, (i) epithelium, (ii)

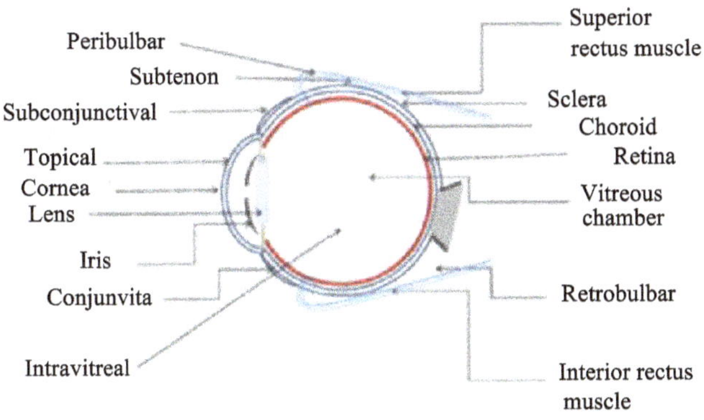

(a)

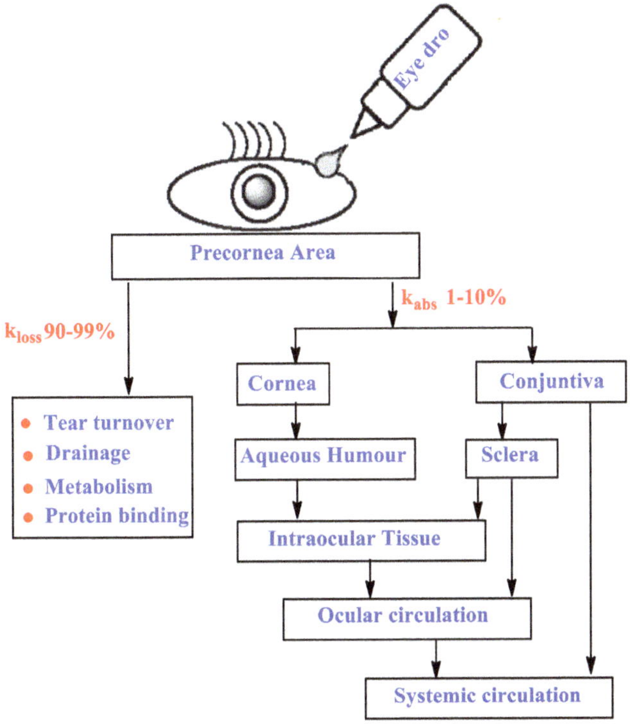

(b)

Figure 4.7: (a) Schematic presentation of the ocular structure and (b) the scheme showing topical drug movement into the eye.

stroma and (iii) endothelium, are the main obstacles for the absorption of ophthalmic solution. The hurdles behind the absorption of ophthalmic solution by these three corneal layers are explained as follows:

Being lipophilic in nature, corneal epithelium possesses five to seven layers of cells. These are linked by tight junctions. This represents the rate-limiting barrier for the transcorneal diffusion of most of the hydrophilic drugs. The stroma is primarily made up of hydrated collagen. This exercises a diffusional obstruction to highly lipophilic drugs. The endothelium is not a significant barrier to trans corneal diffusion. Its permeability depends on the molecular weight and not on the nature of the compound making ocular drug. Usually, 1–10% of the introduced dose is absorbed ocularly. Finally, less than 1% of it reaches to the AqH [37].

4.5.4 H_2S and glaucoma

4.5.4.1 Intraocular pressure: its reduction

It is well established that in glaucoma patients, IOP is the main cause of optic neuropathy in human eye. This damages the neurons of retina and heads [38] of optic nerve. The steady IOP is based on the equilibrium of AqH generation in the ciliary body and AqH outflow in the chamber angle, particularly in trabecular meshwork [39].

Neufeld et al. [40] observed that the administration of cyclic adenosine monophosphate (cAMP) in anterior chamber increases AqH outflow. This helps in maintenance of IOP in rabbit eye. It has been found by Robinson and co-workers that H_2S-liberating compounds, which are, L-cysteine and sodium hydrosulphide (NaHS) have an effect on adenylyl cyclase and ATP-sensitive potassium channels (K_{ATP}) in eyes. This increases cAMP concentrations in porcine ocular anterior segments. Consequently, the increased cAMP levels intervenes the outflow of AqH [41].

Studies conducted by Módis et al. [42] in or on tissue from an organism in an external environment have demonstrated that H_2S contributes in the prevention of phosphodiesterase (PDE). Moreover, it partakes in enrichment of intra mitochondrial cAMP concentrations. The PDE prevention activity assisted by H_2S is responsible for the gradual increase of cAMP and cGMP. Investigations carried out by Bucolo and Drago [35] indicate that elevated intraocular cGMP level reduces the cell volume of trabecular meshwork. This promotes outflow of AqH, and thereby lowering the IOP. The well-known H_2S donors, such as GYY4137 and ACS 67 [Figure 4.8], are well-investigated by two research groups [43, 44] for stabilizing IOP. This is because of the fact that intake of these two H_2S donors upregulate the intraocular GSH expression with increase of cGMP levels.

As per report of Monjok and co-workers [45], H_2S donors effect on anterior uvea, and thereby relaxing iris smooth muscles. This results in IOP lowering. Instead, norepinephrine (a chemical) (Figure 4.9) is released by intraocular degenerating sympathetic nerve terminals. This causes a decrease in AqH outflow facility with a long

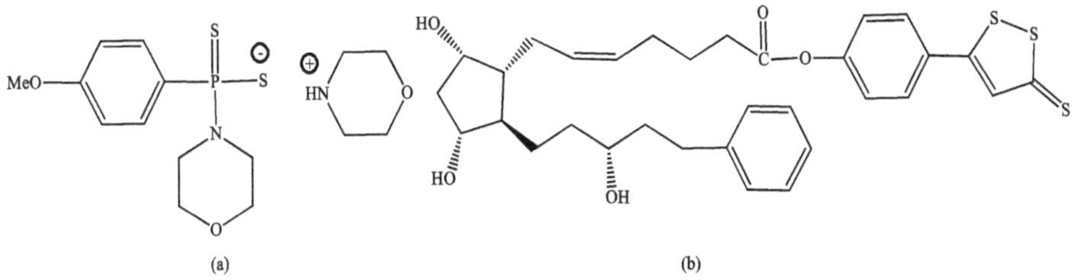

Figure 4.8: Chemical structures of (a) GYY4137 and (b) ACS 67.

term sequential increase in IOP. In another study conducted by Zhan et al., it was observed that norepinephrine may cause an acute increase in outflow facility of AqH. This helps in lowering IOP [46]. It has been observed that level of norepinephrine in AqH increases during night [47] in rabbits. This is related to an increase rather than a decrease in IOP. Interestingly, Yoshitomi et al. [47] observed that the release of the chemical norepinephrine, is reduced by the H_2S from sympathetic nerves. This causes stabilization in the IOP.

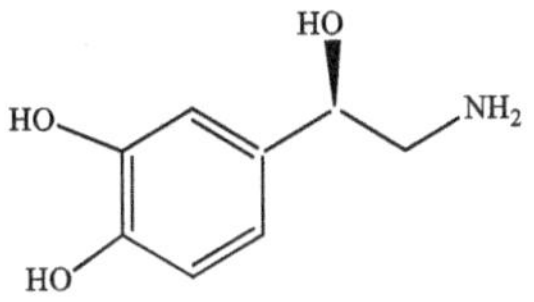

Figure 4.9: Chemical structure of norepinephrine.

4.5.4.2 Effect of H_2S donors on ocular blood supply

It is notable that the damage of the optic nerve in glaucoma may occur due to ischemia (lack of blood supply) with or without unusual IOP. In vivo studies conducted by Flammer [48] discloses that insufficient blood supply can cause optic nerve head deterioration along with ganglion cell death. This suggests that metabolic processes are basically affected by unusual OBF in order to adjust to visual functional needs.

Salomone and co-workers [49] testified that taking 1 mM concentration of GYY4137 significantly nurture phenylephrine-induced tone in the ophthalmic arteries of rabbits. However, quite a few evidences support that novel H_2S donors, AP67 and AP72 (Figure 4.10) effectively dilate pre-contracted posterior ciliary arteries (PCAs) [50]. These two events are vital to OBF. Kulkarni-Chitnis and co-workers [51] reported that even low concentrations of GYY4137 (100 nM to 100 μM) may cause relaxations in PCAs in the presence of phenylephrine induced tone. This happens because of inside production of both prostanoids and H_2S. Moreover, protonated analogues of GYY4137, such as AP67 and AP72, effectively dilate phenylephrine-induced PCAs depending on the concentration of the two H_2S donors AP67 and AP72 in question. These effects are mainly based on the action taken by H_2S on

K_{ATP} channels. The above studies establish the role of H_2S in modifying the OBF in glaucoma.

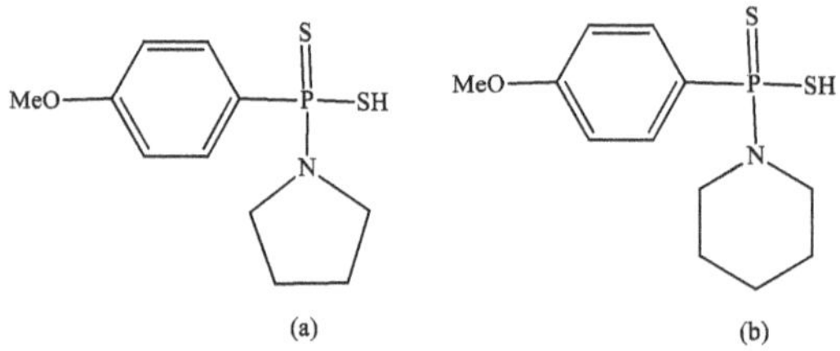

Figure 4.10: Chemical structure of H_2S donors: (a) AP67 and (b) AP72.

4.5.4.3 Protection on neurons: effect of H_2S donor

The foremost characteristics of glaucoma comprise continuing *death of RGCs* along with damage of optic nerve [52]. These two events are generally persuaded by some factors. These include the loss of neurotrophic factors, intracellular and extracellular toxicity of glutamate (anion of glutamic acid; Figure 4.11) and neuro-inflammation [53, 54]. White et al. [55] reported that H_2S serves as a neurotransmitter. Moreover, it is capable of preventing cell death and breakdown of neurons in the nervous system.

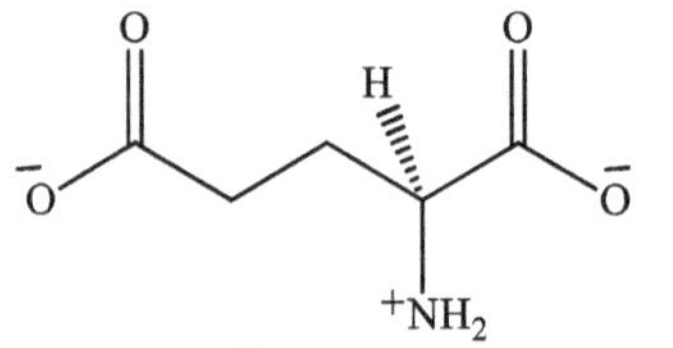

Glutamate

Figure 4.11: Chemical structure of glutamate.

Osborne et al. [56] have shown that the two H_2S donors, which are, ACS1 and ACS14 [Figure 4.12] are found to make the amount/concentration of the intracellular GSH better. Moreover, they encourage the protection of neurons via opening K_{ATP} channels.

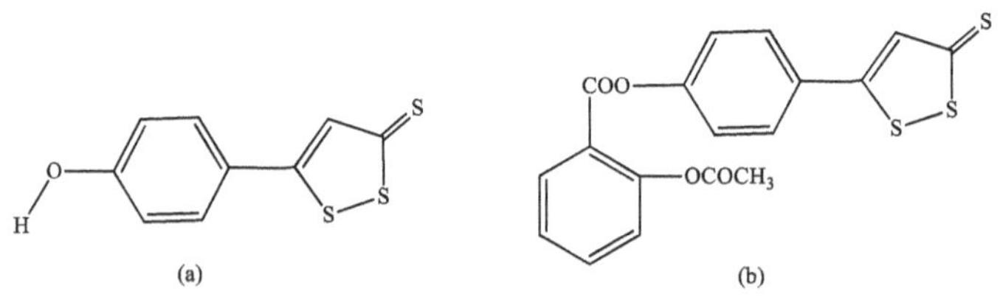

Figure 4.12: Chemical structure of H_2S donors: (a) ACS1 and (b) ACS14.

Compared to the glutamate-treated RGCs, intake of H_2S increases the phosphorylation of Akt. Additionally, it stimulates cell sustainability in response to oxidative stress [57]. Huang and co-workers demonstrated [58] that H_2S is able to lessen RGC apoptosis in a prolonged ocular hypertensive rat. This becomes possible through balancing mitochondrial function, suppressing glial activation and downregulating the autophagy process. It is observed by Huang et al. [59] that the intake of NaHS in the form of intracameral injection in rats having glaucoma prevents the loss of RGCs. This occurs via improving the amounts of H_2S in retina. A continuous release of H_2S from GYY4137 in combination with the in situ gel forming poly(lactic-co-glycolic acid)-based system, for up to 72 h, has shown a great potential in medicating glaucoma [60]. The various roles of H_2S in the treatment of glaucoma are pictorially shown in Figure 4.13.

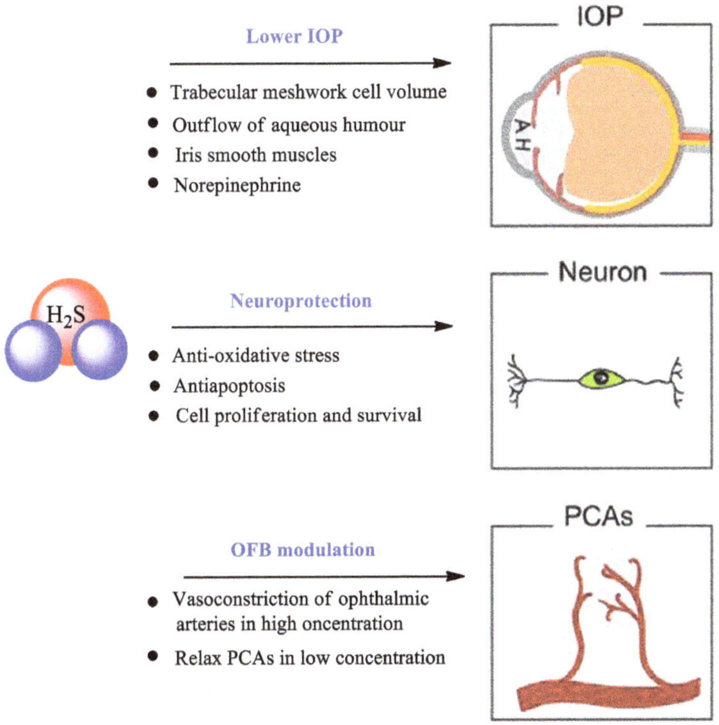

Figure 4.13: Roles of H_2S in the treatment of glaucoma.

4.5.5 H_2S and diabetic retinopathy

The cause of diabetic retinopathy (DR) involves several different factors. A variety of pathways related to hyperglycaemia are involved in the commencement and advancement of this condition. All the retinal cells are affected in the diabetic environment.

In view of such disease and tissue intricacy, it is improbable that any single route is exclusively liable for retinal pathophysiology [61]. Different beneficial roles of H_2S in DR are summarized below:

4.5.5.1 Effects of advanced glycation end products in diabetic retinopathy: reduction using H_2S donors

Extraordinarily high level of sugar in the blood causes the condensation reaction between glucose and the amino end of protein in a non-enzymatic way. This results in the collection of AGEs' large molecules, usually called macromolecules. Notably, this is closely related with the incident of DR [62, 63].

AGEs can be connected with intracellular proteins. This connection disturbs their usual functioning. AGEs also obstruct in normal metabolic pathways, which are, ATP production. They eradicate the inner blood–retinal barrier (BRB) in eye. This results in oxidative stress and inflammation [64, 65].

Liu et al. [66] reported that H_2S stimulates galactose metabolism to reduce the generation AGEs in neuronal cells. Moreover, it also plays a vital role in preventing oxidative stress. Usui and co-workers [67] demonstrated that H_2S reduces ROS production and lipid peroxidation. Moreover, H_2S enhances the appearance of superoxide dismutase (SOD) and glutathione peroxidase (GPX). These are the two inside generated antioxidant enzymes. Additionally, H_2S converses the excessive glucose induced increase in the appearance of aldehyde oxidase 1. Notably, it decreases the glutathione synthetase level. Finally, this opposes the oxidative stress induced by AGEs in cells.

4.5.5.2 Oxidative stress and inflammation: inhibitory role of H_2S donors

It is well known that H_2S is responsible for the origin/cause of numerous diseases owing to its toxicity. Nevertheless, H_2S is identified for anti-inflammatory properties in vivo/in vitro.

M. Brownlee [68] reported that in diabetic patients, the electron transfer system of the cellular mitochondrial respiratory chain is highly perturbed by hyperglycaemia. This results in easy generation of oxygen free radical (O^-) and superoxideradical [O_2^-]. The O^-/O_2^- so obtained transforms NO into more toxic species peroxynitrite ($ONOO^-$). This irretrievably binds to cytochrome c and weakens the mitochondrial functions. Geng and Kimura et al. [69] reported that when the concentration of intracellular reactive oxygen species (ROS) increases, it instantly participates in lipid peroxidation enormously. This can be suppressed by H_2S in DR animal models.

Several research groups [70, 71] reported anti-inflammatory property of H_2S. In fact, it participates in scavenging of the pro-inflammatory oxidants, which are, $ONOO^-$, HOCl, superoxide (O_2^-) and H_2O_2. Gemici and co-workers [72] reported that H_2S-releasing drugs move the pro-inflammatory response to anti-inflammatory one. Such drugs decrease the concentrations of TNF-α, IL-8 and IFN-γ but increase the concentrations of cyclooxygenase (COX)-2 and eicosanoids. Whiteman et al. [73] reported that the GYY4137, a H_2S-releasing drug, prevents lipopolysaccharide-persuaded production of inflammatory agent by macrophages. Moreover, it upregulates the release of anti-inflammatory cytokine, IL-10. According to Li et al. [74], H_2S controls on inflammatory cytokine production. This can be attributed to the suppressive function of H_2S on NF-κB activation.

4.5.5.3 Protective effect of H_2S donors on retinal neurons

Lieth and co-workers [75] reported that in the progress of the pathological process in DR, neuron damage usually occurs. Thereafter, this accumulates into visible retinal vascular lesions. Osborne et al. [76] reported that after ischemia–reperfusion or exposure to oxidative stress, a H_2S donor, namely, ACS67 (Figure 4.14) is found to be effective in preventing RGC apoptosis and reactive gliosis in Muller cells. According to Si and co-workers [77] intake of H_2S donors recovers the appearance of brain-derived neurotrophic factor and retinal synaptic vesicle protein in streptozotocin (STZ)-persuaded diabetic rats. This suggests that H_2S possibly block neuronal degeneration of retina in diabetic patients. The neuroprotective effect of H_2S in retina is also related to its control on the intracellular GSH content.

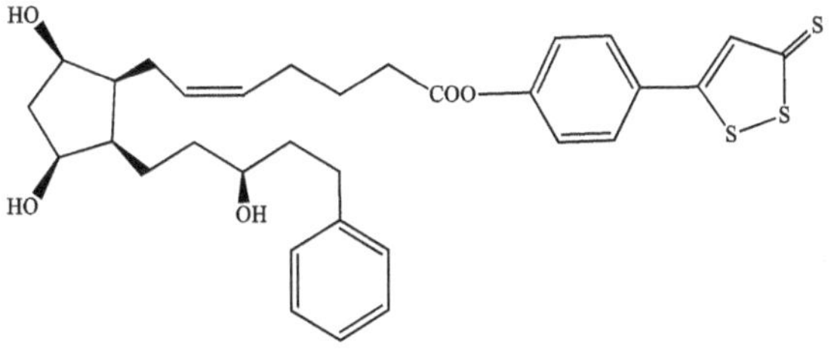

Figure 4.14: Chemical structure of H_2S donors ACS67 (S-Latanoprost).

4.5.5.4 Multiple effects of H_2S donors on retinal blood vessels

Qaum et al. [78] reported that in the course of DR, the failure in normal functioning of BRB is a main cause of retinal vascular injuries. In the development of DR, both retinal ischemia and hypoxia persuade the appearance of hypoxia inducible factor (HIF-1α).Thereafter, activation of successive vascular endothelial growth factor (VEGF) signalling occurs. Remarkably, the signalling pathway, HIF-1α-VEGF-VEGFR2, is liable for diabetes-induced failure in usual functioning of BRB and also in excessive angiogenesis. In vivo experiments conducted by Si et al. [77] shows that in the retina of STZ-persuaded diabetic rats, there occurs reduction in BRB permeability and decrease in acellular capillaries. In this situation, intake of NaHS treatment results in the reduction in VEGF content of vitreous and gene expression of VEGFR2 and HIF-1α. Moreover, this increases the expression of occluding.

Oshitari et al. [79] observed that H_2S intake in the form of H_2S donor prevents too much deposition of laminin and collagen IVα3. These maintain the vascular consistency in the retinas of diabetic rats. Papapetropoulos and co-workers [80] reported that in intraocular tissues VEGF stimulates endothelial cells to generate and thereafter release H_2S. It has been observed that at the beginning stage of diabetes, H_2S protects against oxidative or nitrosative stress in the retina as well as in vitreous humour. Moreover, it is likely that H_2S plays a protective role on BRB in hyperglycaemic circumstance. Szabó et al. [81] in 2011 demonstrated that intake of exogenous NAHS (a H_2S donor) may enhance the effect of VEGF on vascular endothelial cells, as well as in the angiogenesis process. Examinations of H_2S concentration in the vitreous and plasma of proliferative diabetic retinopathy (PDR) patients have shown a very high appearance. This suggests the potential consequences of H_2S in the pathogenesis of PDR [82].

4.5.6 H_2S and retinal degeneration

Retinal degenerative diseases generally involve (i) age-related macular degeneration (AMD), (ii) retinitis pigmentosa (RP) and (iii) glaucomatous retinal neuropathy. These are the foremost grounds relating to disease for anomalous structure and function of neurons in the retina at all stages. This causes an irreparable loss to vision clarity [83].

Busskamp et al. [84] reported that continuing deterioration of the photoreceptor cells in the retina is the cause of the severe visual injury of RP and AMD. Two research groups [85] cited below reported that the apoptosis in photoreceptor cell is the fundamental base of human retinal deterioration and visible radiation persuaded retinal deterioration models.

A study conducted by Mikami et al. [86] reveals that intake of a H_2S donor, NaHS, inhibited the visible radiation-persuaded photoreceptor deterioration. This also decreases photoreceptor cell apoptosis, and DNA destruction in the outer nuclear layer (ONL) of retina in a mice retinal deteriorating model triggered by extreme visible radiation exposure. At the same time, NaHS hindered high K^+-evoked Ca^{2+} influx in mice retinal ONL and outer plexiform layer. This study suggests that restoration of Ca^{2+} homeostasis is possibly concerned with the defensive outcome of H_2S on the photoreceptor cell apoptosis. The aforesaid results indicate that hydrogen sulphide possibly act as a neuroprotector in order to prevent from retinal deterioration. Instead, unusual function of retinal pigment epithelium (RPE) cells is a key disease process of RP and AMD [87].

Njie-Mbye and co-workers [88] demonstrated that exogenous H_2S donor NaHS and endogenous H_2S increase the cAMP concentrations in rat RPE cells dose-dependently. However, it was observed by two research groups [89, 90] that the increase in intracellular cAMP concentration results in the downregulation of phagocytic activity of RPE cells. This will result in the development of RP and AMD. The aforesaid results put forward that H_2S possibly intensify the process of RP and AMD. The above studies conclude that additional researches are required to confirm regarding the protective or detrimental role of H_2S on RP and AMD.

Glaucoma is not generally considered as a retinal disease. But glaucomatous retinal neuropathy [91] is the leading cause of vision impairment in glaucoma. It was observed by Huang et al. [59] in 2017 that in a prolonged ocular hypertension rat model, H_2S concentrations and general tone of CBS, CSE and 3MST (internal enzymes responsible for generation of H_2S) in retinal tissues are significantly reduced along with the damage of RGCs. However, medication with a H_2S donor (NaHS) improves the survival of RGCs with no effect on the IOP. Interestingly, Perrino et al. [43] observed that H_2S-releasing agent ACS67 reduces the IOP in carbomer-persuaded glaucoma in rabbits. Moreover, it increases the quantity of GSH and cGMP in AqH. The aforesaid results give a better insight for the cause and development of glaucomatous retinal neuropathy and provide a possible therapeutic goal for glaucoma.

4.6 Physiological functions of H_2S in plants

4.6.1 Introduction

Hydrogen sulphide (H_2S) is a colourless, flammable and water-soluble gas. It smells like that of the decomposed eggs. Bearing a tag of an air pollutant and a life threatening gas, it has been labelled as the third gasotransmitter in animal and plant cells. It is equally important as NO, CO and hydrogen peroxide (H_2O_2) [92]. These small molecules are high permeable in nature. This permits them to cross biological membranes and to act as signalling molecules. H_2S dissociates easily under physiological conditions. It is, therefore, assumed that hydrogen sulphide pools include [93] H_2S, HS^- and S_2^-. However, the active form of hydrogen sulphide in cells has not been fully clarified. H_2S is involved in many physiological and pathological processes in animals. These include cell proliferation, apoptosis, inflammatory processes, hypoxia protection, neuromodulation and cardioprotection. These have been reported earlier by three research groups [94–96].

The effects of H_2S in plants were first noticed during the 1960s. Around that time, H_2S has been found to show effect on the plant growth and to affect disease resistance [97, 98] in plants. More recently, controlling properties of H_2S have become known in plants along with protective effects. The protective effects of H_2S against oxidative and metal stresses, drought and heat tolerance and osmotic and saline stresses have been reported by Aroca and co-workers [99]. Moreover, H_2S is found to be associated in controlling important physiological processes in plants. These are stomatal closure, modulation of photosynthesis and autophagy (self-degradation of cellular components) regulation. Interestingly, inside generation of

H_2S in plants was also identified. Moreover, owing to the similarity in the physiological effects of H_2S and NO, a possible cross-talk of this gaseous molecule with NO in plants has been proposed. The upcoming section is, therefore, targeted to highlight on the latest understanding of hydrogen sulphide as a signalling molecule, its role in plant physiology and mechanism of action.

4.6.2 Improvement in seed germination and plant growth

The enhancement in rates of seed germination owing to exogenous H_2S treatments was confirmed by Zhang et al. [100] in 2008. In fact, H_2S or hydrosulphide (HS^-) (rather than other sulphur-containing components) resulting from the external H_2S donor, NaHS, helps in promoting germination of seeds. Zhang and co-workers [101] observed that NaHS, particularly, stimulates the activity of endosperm β-amylase and preserves lower levels of malondialdehyde (MDA) and hydrogen peroxide (H_2O_2) in sprouting seeds. Moreover, consumption of NaHS to seedling cuttings of sweet potato motivates the adventitious roots [102] in numbers as well as lengths.

Carter et al. [103] reported that the dialkyldithiophosphates capable of slowly releasing H_2S upsurge the weight of corn plant up to 39% after 4.5 weeks treatment of 1–200 mg of dialkyldithiophosphate. In fact, the ultimate breakdown products of dibutyldithiophosphate (Figure 4.15) in the soil were shown to be phosphoric acid and butanol. These two chemicals are found in the environment. Thus, the biocompatibility of the breakdown products, are important advances in the field of H_2S in agriculture.

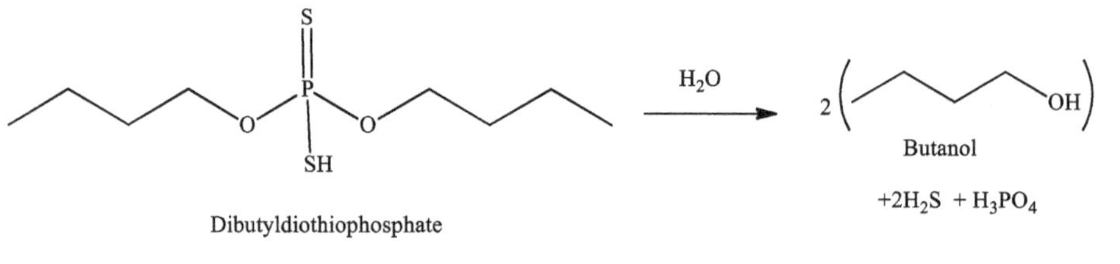

Figure 4.15: Degradation products of dibutyldithiophosphate.

Chen et al. [104] reported that NaHS (a H_2S donor) controls the expression of genes involved in photosynthesis and thiol redox alteration in *Spinacia oleracea* seedlings to regulate its photosynthesis. It is expected that an increase in the stomatal density [105] also contributes to the improvement of photosynthesis. Zhang et al. [106] reported that the osmotic persuaded decrease in the chlorophyll concentration improves by spraying the NaHS solution.

It is found by Zhang et al. [107] that H_2S delays flower opening and senescence in cut flowers and branches. They also observed that these two effects occur in a dose-dependent manner. Dooley and co-workers [108] reported that H_2S strongly affects plant metabolism at most stages of life and causes statistically significant increases in biomass, including higher fruit yields. Thus, the increased fruit/crop yields reported here has the potential to affect the world's agricultural output.

4.6.3 H_2S in fruit ripening and post-harvest damage to fresh produce

So far, limited research data are available on endogenous H_2S metabolism in fruits and vegetables [109]. Muñoz-Vargas et al. [110] in 2018 observed that inside H_2S content in non-climacteric sweet pepper (*Capsicum annuum* L.) fruits (Figure 4.16) increases during the transition from green immature to red ripe.

Figure 4.16: Green and red sweet pepper.

On the other hand, the number of studies concentrating on the economic impact of biotechno-logical applications of H_2S on fruit ripening and post-harvest storage (for preventing the loss of fresh produce caused by fungi, bacteria, viruses and low temperatures used to store fruits and vegetables) has augmented much more over the last one decade. In fact, all these factors are usually associated with oxidative stress. Studies carried out by Ziogas and co-workers revealed that the exogenous application of H_2S have a beneficial effect on the shelf life of a varied range of fruits, vegetables and flowers [111]. Remarkably, the exogenous application of H_2S to fruits and vegetables facilitates their quality be maintained. Studies conducted by Aghdam et al. conclude that a common effect observed after exogenous treatment with H_2S is an increase in antioxidant systems. These prevent ROS overproduction and consequently controlling oxidative damage [112, 113].

4.6.4 Abiotic stresses in plants

Plants are exposed to a wide range of environmental stresses. This reduces and limits the output of agricultural crops. Usually two types of environmental stresses are come across to plants which can be characterized as (i) abiotic stress and (ii) biotic stress. The abiotic stress causes the loss of major crop plants worldwide. This includes radiation, salinity, floods, drought, extremes in temperature, heavy metals (HMs) and so on. Resembling to other second messengers, which are Ca^{2+}, H_2O_2, abscisic acid and NO, the speedy generation of H_2S internally can be activated by numerous stresses (Figure 4.17) in several species of plant. In fact, it is a general reaction of plants to various abiotic stresses, and is closely linked with the achievement of cross-adaptation in plants. The upcoming section will focus of some abiotic stresses and H_2S signalling in them.

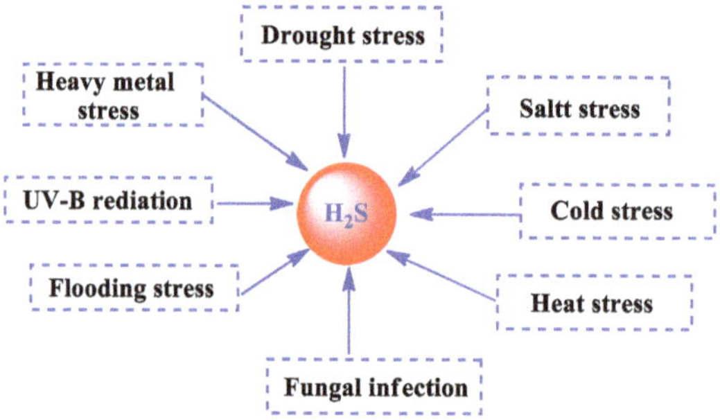

Figure 4.17: Numerous environmental stresses inducing H_2S production internally in plants.

4.6.4.1 Heavy metal stress: activation of H_2S signalling

Generally, interaction of plants with various HM stresses rapidly produces H_2S. Of the HMs, Cd is liable for extremely unpleasant stress owing to its toxicity and stability [114]. In rice seedlings, 0.5 mM Cd stress resulted in an increment of H_2S content from ~5 μmol g^{-1} fresh weight (FW) to ~6 μmol g^{-1} FW. Contrary to this, addition of 0.1 mM NaHS (an H_2S donor molecule) resulted an even further increase in the level of H_2S (~ 8 μmol g^{-1} FW) as compared with Cd treatment alone. Exposure to 0.2 mM hypotaurine (H_2S scavenger) (Figure 4.18) with NaHS decreased H_2S level compared with NaHS alone. This clearly indicates that this elevated level of H_2S is correlated with the enhanced Cd tolerance [115].

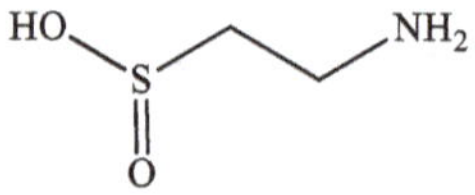

Figure 4.18: Chemical structure of hypotaurine.

Zhang et al. [116] observed that the Cd stress in Chinese cabbage stimulates the emission of endogenous H_2S emission. Moreover, the relative expression of genes, DCD1) and DES1, OAS-TL, accountable for H_2S synthesis, were found to be increased when treated with Cd with a series of concentrations 0, 5, 10 and 20 mM, receptively, for 24 h.

4.6.4.2 Salt stress: activation of H_2S signalling

Usually, salt stress results to an osmotic stress response, and activates speedy production of H_2S. Lai and co-workers demonstrated that with the increase in the concentration of NaCl commencing from 50 to 300 mM in alfalfa seedlings gradually causes the induction of total DES1 activity. Consequently, the induced DES1 activity [117] increases the internal production of H_2S from 30 to 70 nmol g^{-1} fresh weight. Christou and co-workers [118] testified when strawberry seedlings is exposed to salinity of 100 mM NaCl and non-ionic osmotic stress of 10% PEG-6000, a significant enhancement of H_2S concentration occurs between 48 and 50 nmol g^{-1} FW in leaves. Moreover, 0.1 mM NaHS pre-treated plants when exposed to both stresses for 7 days, results a collection of considerably higher amounts of H_2S (55 nmol g^{-1} FW) in their leaves compared to only NaCl-stressed plants.

4.6.4.3 Drought stress: activation of H_2S signalling

Drought is the most severe abiotic stress affecting production of agricultural crops globally. Shen et al. [119] reported that the treatment of *Arabidopsis* seedlings with polyethylene glycol of average molecular weight 8000 (PEG, 8000) simulates drought stress. This causes an increase in the rate of generation of internal H_2S by 0.8 nmol mg^{-1} protein min^{-1}. Zhang and co-workers [120] demonstrated that exposure of osmotic stress using PEG-6000 for 2 days, the internal generation of H_2S in wheat seeds rapidly increases from 1.5 to 3.5 μmol g^{-1} dry weight (DW).

4.6.4.4 Low-temperature stress: activation of H_2S signalling

Unusual lowering in temperature is also a leading environmental stress. This restricts plant growth, development and distribution. In 2013, Fu et al. [121] testified that in grape (*Vitis vinifera* L.) seedlings, chilling stress at 4 °C induces the expression of L/DCD genes. Therefore, this increases the activities of these two genes. The increased activity of genes successively improves internal H_2S accumulation from 7 to 15 μmol g^{-1} FW. In the same way, Shi et al. [122] reported that a cold stress of 4 °C in Bermuda grass seedlings persuades the collection of inside H_2S level of 14 nmol g^{-1} FW.

Ma and co-workers [123] investigated the adaptive strategies of alpine plants to the extremely cold conditions at high altitudes. They use a comparative proteomics method (large scale study of proteins), and studied the dynamic patterns of protein expression in *Lamiophlomis rotata* plants grown at three different altitudes of 4,350, 4,800 and 5,200 m. They found that the concentrations and activities of protein enzymes, OAS-TL, CAS and L/DCD participated in H_2S biosynthesis markedly increase at the altitudes of 4,800 and 5,200 m. Moreover, the collection of H_2S increases to 12, 22 and 24 nmol g^{-1} FW, respectively. This study confirms that during environmental stress at higher altitudes, H_2S plays an important role in adaptation of *Lamiophlomis rotata*.

4.6.4.5 High-temperature stress: activation of H_2S signalling

It is observed that in several plants, high temperature can also persuade inside generation of H_2S, similar to so many stresses mentioned above. Chen et al. [124] reported that exposure of three-week-old tobacco seedlings with high temperature at 35 °C increases the activity of L-cysteine desulphhydrase (LCD). This successively persuades the generation of 8 nmol g^{-1} FW of H_2S internally in tobacco seedlings. Moreover, the H_2S production persists at higher level even after 3 days of high-temperature exposure. Remarkably, the generation of H_2S at high temperature motivates the gathering of jasmonic acid (Figure 4.19a), and thereafter promoting nicotine (Figure 4.19b) synthesis. These data suggest that biosynthesis of H_2S and nicotine is linked to tobacco plants under high-temperature stress. In strawberry seedlings, Christou et al. reported that the heat stress results in a marked variation of H_2S content. This is indicated by a significant increase in H_2S content after 1, 4, 8 h of exposure to 42 °C compared with control plants. When 0.1 mM NaHS-pre-treated plants are exposed to heat stress, a significant increase in H_2S content was observed [125] after 1 h exposure to heat stress. Thereafter, a gradual lowering in H_2S content occurs to control level.

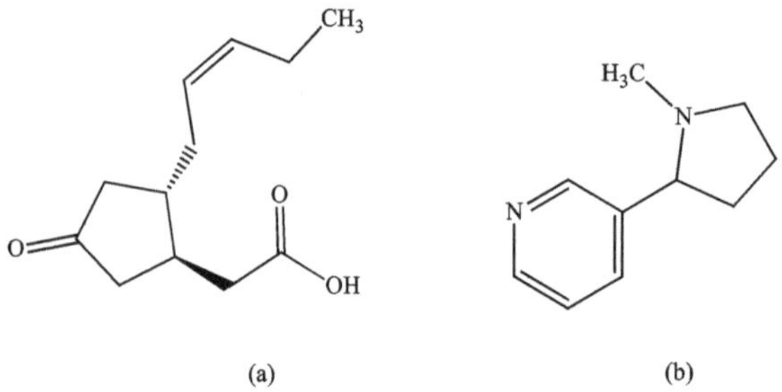

Figure 4.19: Chemical structure of (a) jasmonic acid and (b) nicotine.

4.6.4.6 UV-B radiation stress: activation of H_2S signalling

Exposure of UV-B radiation on barley seedlings leaves motivates generation of H_2S, It was observed by a study conducted by Li and co-workers [126]. They found that production of H_2S reaches a peak of ~230 nmol g^{-1} FW after 12 h of exposure. This sequentially promotes the accumulation of flavonoids and anthocyanins (two UV-absorbing compounds). Notably, H_2S started to drop gradually. However, it is substantially higher than that of the control (~125 nmol g^{-1} FW) at the exposure of 48 h. As for as the activity of LCD is concerned, alike trend was noticed. It was validated by the application of DL-propargylglycine, an inhibitor of LCD. This resulted in complete prevention of the H_2S production and the growth of UV-absorbing compounds persuaded by UV-B radiation.

4.6.4.7 Hypoxia and fungal infection: activation of H_2S signalling

Notably, flooding frequently causes hypoxia in plant roots. This considerably restrains crop production enormously. Cheng and co-workers [127] demonstrated that in pea seedlings, hypoxia activates biosynthetic system of H_2S, which are, LCD, DCD, OAS-TL and cysteine synthase. These successively persuades the growth of inside H_2S generation from ~0.9 (control) to 5.1 µmol g^{-1} FW (hypoxia for 24 h). This suggests that H_2S is possibly a signalling to hypoxia. In fact, this activates the tolerance of the pea seedlings to hypoxic stress. In addition, this proposition was substantiated by external application of a H_2S donor, NaHS.

Fungal infection is considered as a common biotic stress in plant species. Bloem et al. reported that in oilseed rape seedlings, fungal infection with *Sclerotinia sclerotiorum* results a stronger increase in H_2S release, reaching a peak of 3.25 nmol g^{-1} DW min^{-1} after two days infection. This indicates that the generation of H_2S seems to be a part of the response to the fungal infection [128].

4.6.5 Crop plants and biotic stress

Various kinds of biotic stresses in plant species are instigated by different living organisms like fungi, virus, bacteria, nematodes and insects. These biotic stressing agents are responsible for numerous types of diseases, infections and damages to crop plants. In the prevailing circumstances, the crop productivity is ultimately affected. However, different mechanisms have been developed through research to defeat biotic stresses. By carefully studying the genetic mechanism of the agents instigating these stresses, the biotic stresses in plants can be overcome. Genetically modified plants have proven to be the great effort against biotic stresses in plants by developing resistant varieties of crop plants.

4.6.6 Exogenously applied NaHS: activation of H_2S signalling

Different routes of H_2S signalling have been discussed in the preceding sections Apart from these, H_2S signalling in plant cells under normal growth conditions can be activated by the use of NaHS from outside. Another way of H_2S signalling in plant cells is to up-regulate the expression of genes (LCD/DCD) involved in H_2S biosynthesis. Christou and co-workers found [118] that the treatment of roots in strawberry seedlings with 0.1 mM NaHS results irrelevantly raised H_2S level (35 nmol g^{-1} FW) in leaves compared with control plants (25 nmol g^{-1} FW). Likewise, when wheat seeds is treated with NaHS, the inside H_2S level of 4.5 µmol g^{-1} DW was found to be slightly higher [101] than that of control having H_2S concentration 1.7 µmol g^{-1} DW. These results clearly signify that H_2S is easy to pass through plant cells and can be transported to other tissues or organs owing to its highly lipophilic property. This sequentially exercises its physiological role in plants [129].

Jin and co-workers [129] demonstrated that the expression of LCD/DCD in *Arabidopsis* seedlings display inside H_2S content of 6 nmol mg^{-1} protein min^{-1} and 14 nmol mg^{-1} protein min^{-1} in the normal and drought stress conditions, respectively. These two H_2S contents are found to be higher compared to the control (3 nmol mg^{-1} protein min^{-1}). The pattern of LCD/DCD expression in normal case was analogous to the drought related genes persuaded by dehydration. Moreover, outside use of H_2S (80 μM) in the form of H_2S donor was also observed to motivate further the expression of drought linked genes.

4.6.7 Cross-adaptation: persuaded by H_2S

Remarkably, cross-adaptation is the way wherein plants exposed to one stress can persuade the resistance to other stresses. Interestingly, this is applicable to abiotic and biotic stresses also. We have seen in the discussion given above that a wide array of environmental stressors was capable to activate H_2S signalling in plants. On the other hand, pre-treatment of plants with outside use of H_2S in the form of H_2S donor provided added strength to further stress coverage. The upcoming part, therefore, presents the scope of H_2S for cross-adaptation to various abiotic and abiotic stresses.

4.6.7.1 Metal and metalloid tolerance: persuaded by H_2S

HMs belong to a group of metal that have density more than 6 g cm^{-3}. This includes Cr, Cu, Zn, Hg, Pb, As and so on. Owing of their toxicity and stability, these HMs have become the leading abiotic stress in plant species. Moreover, these HMs have threatened human health via way of the food chains. Such stress gives rise the formation of oxidative stress, that is, the excessive build-up of ROS. This leads to lipid peroxidation, protein oxidation, enzyme inactivation and DNA damage [130]. Contrary to this, higher plants have developed an advanced antioxidant defence system to scavenge [131, 132] too much ROS and maintain its equilibrium in plants.

(i) Among metalloids, arsenic is an extremely toxic. It is a key pollutant of the soil. It has been observed that arsenic treatment in peas seedlings upsurges the collection of ROS. This accumulated ROS successively damages to lipids, proteins and biomembranes. For the time being, higher cysteine level was observed in arsenic-stress seedlings compared to all other treatments including As-free, NaHS and As + NaHS. The effects caused by ROS were diminished by the intake of NaHS. Experimental studies conducted by Singh et al. have shown that arsenic treatment hampers the activity of the enzymes participated in the ascorbic acid–GSH cycle. On the other hand, treatment of NaHS improves [133] the enzyme activities.

(ii) Zhang and co-workers [134] reported that intake of NaHS from outside in Cr stress, improves the sprouting rate of wheat seeds depending on dose. Moreover, intake of NaHS in Cr stress improves the activities of enzymes amylase, esterase. Additionally, the activities of antioxidant enzymes SOD, catalase (CAT), ascorbate peroxidase (APX) and GPX also improve by administration of NaHS. Instead, it reduces the activity of lipoxygenase and overproduction of MDA as well as H_2O_2 induced by Cr. Additionally, NaHS maintains higher internal H_2S level in wheat.

(iii) Zhang and co-workers [135] reported that NaHS in a dose-dependent manner improves the inhibitory effect of Cu stress in wheat. Additionally, H_2S or HS^- derived from NaHS plays the potential role in promoting seed germination under Cu stress. Additional experiments conducted by the same authors demonstrated that NaHS increases the activities of enzymes amylase and esterase. NaHS reduces the disruption in integrity of plasma membrane persuaded by Cu. Moreover, it maintains lower MDA and H_2O_2 levels in sprouting seed.

(iv) Particularly in acidic soil, aluminium is an unwanted component for plants. It badly affects the plant growth, enlargement and endurance. Chen et al. reported that in barley saplings, aluminium strain stops the lengthening of roots. On the other hand, pre-treatment of NaHS temperately saves the prevention of root enlargement persuaded by aluminium. This saving was closely interrelated with the decrease of Al deposit in saplings. Moreover, intake of NaHS considerably lessened citrate secretion and oxidative stress persuaded by aluminium via stimulating the antioxidant [136] assembly. Zhang and co-workers [137] also reported that NaHS lessens aluminium toxicity in germinating wheat seedlings.

(v) Zinc is a crucial component for plant species. However, its noxious activities can be seen when it is deposited excessively in plants. Liu and co-workers observed that in saplings of *Solanum nigrum* L., H_2S improves the stoppage of development by extra Zn, particularly, in roots and an upsurge in free content of cytosolic Zn^{2+} in roots. This activity in presence of H_2S was correlated well with the decrease of Zn uptake and equilibrium related genes expression [138]. Additionally, H_2S further improves the expression of the metallothioneins to chelate excessive Zn and lessens Zn-oxidative stress by regulating the genes expression of antioxidant enzymes.

4.6.7.2 Salt tolerance: persuaded by H_2S

Remarkably, direct and indirect injuries to plants are persuaded by salt strain. These include ion toxicity, osmotic stress, nutrient imbalance and oxidative stress [139]. Several defensive methodologies are worked out by plants for fighting with salt injury. These include osmotic modification by synthesizing osmolytes, which are, proline, glycine betaine, trehalose and total soluble sugar. Other defensive practices are ion and nutrient balance by regulating transporter, and augmentation of antioxidant capacity by stimulating the activity of antioxidant enzymes, such as SOD, CAT, APX, GPX and GSH reductase. Moreover, the internal synthesis of antioxidants, which are, ascorbic acid and GSH is also as a part of motivating the activity of antioxidant enzymes.

Bao and co-workers [140] reported that in salt-sensitive wheat cultivar, seed pre-treating taking varied concentrations of NaHS (0.01, 0.05, 0.09, 0.13 mM) for 12 h significantly improves the stoppage of seed sprouting and seedling development persuaded by 100 mM NaCl depending upon the concentration. This is shown by rate of sprouting and development in seedlings of wheat.

Wang et al. [141] observed that in alfalfa seed, NaHS pre-treatment differentially activate total and isoenzymatic activities as well as corresponding transcripts of antioxidant enzymes (SOD, CAT, peroxidase (POD) and APX) under 100 mM NaCl stress. This results in the improvement of oxidative harm persuaded by NaCl. Additionally, NaCl stress prevents seed sprouting and seedling progression. However, pre-treatment with NaHS considerably decreases the preventive effect and increases potassium (K) to sodium (Na) ration in the root parts.

4.6.7.3 Drought tolerance: persuaded by H_2S

Ordinarily, shortage of water due to drought induces osmotic and oxidative stresses. These badly affect plant growing, development and production. Plants maintain water balance and ROS equilibrium by osmotic adjustment and antioxidant system. These issues are reported by different research groups [132, 142]. Zhang and co-workers [143] reported that the wheat sprouting rate slowly decreases with the increasing concentrations of PEG-6000 (mimicking osmotic stress). On the other hand, NaHS treatment promotes seed sprouting of wheat under osmotic stress in a dose-dependent manner. This is further reflected by LCD/DCD pattern of *Arabidopsis thaliana* seedling sharing similarity with linked genes confined to drought conditions. The H_2S/NaHS treatment behaves as an inducing for seedlings displaying a higher survival rate in addition to reducing stomatal aperture size [144].

4.6.7.4 Cold tolerance: persuaded by H_2S

Like other factors, the cooling atmosphere also induces behavioural change addressed in the form of biochemical modifications. Based on whether the temperature >0 °C or <0 °C, the stress in the former range is termed as low-temperature stress and latter is the freezing stress. This fashion of stress is usually expressive in the form of oxidative and osmotic stresses. In other words this refers to the physiological modifications in plants due to cold adjustable through osmotic adjustment in combination with activation of antioxidant system [122]. In connection with the implications of sulphur in the form of NaHS Shi et al. recorded enhanced tolerance to cold-induced stress (freezing stress) by taking Bermuda grass as subject model. These resisting features were mechanistically elaborated as lowering the leakage of electrolytes, and hence results in increasing the survival rate of the plant subject to the freezing conditions. Also, the NaHS treatment was shown to develop the prevention from ROS burst with the additional effect of saving cells from getting damaged in the freezing temperature. These anti-oxidating modifications do induce feasibility among enzymes like GPX, CAT and GR. Meanwhile, GSH pool also intervenes in a non-enzymatic fashion to avail the changing redox state with respect to the stress.

In another example reported by Fu and co-workers [121] the similar form of sulphide in grape seedlings (for chilling stress) have been shown to express highly active SOD behaviour in connection with genetic expressions which are VvICE1 and VvCBF3, causing reduction in the concentration of superoxide radicals. On the other hand, treatment of hypotaurine (HT, an H_2S scavenger) (Figure 4.18) in grape seedlings displayed opposite effect under the chilling stress.

4.6.7.5 Heat tolerance: persuaded by H_2S

Besides global warming, high temperature is a well-known abiotic stress in plants worldwide. The mechanism lying behind the heat based injury in plants and the respective heat tolerance have gained considerable scientific temper [145]. In context with gaining the scientific insights Christou and co-workers [146] reported NaHS and heat treatment of Strawberry roots and accumulated the evidences in favour of the positive role of NaHS in activating a network of physiological pathways as a defence against heat shock (42 °C). The induced heat tolerance was found to express the role in a cooperative way involving the combinatory role of chlorophyll, ASA, GSH and several other antioxidant cytosolic biocatalysts. Additionally, Christou et al. [125] conducted NaHS experiment over the similar subject plant models and supports the findings previously made.

4.6.7.6 Flooding tolerance and pathogen resistance: persuaded by H_2S

From the accumulated evidences available in literature it is clear that floods results in hypoxia and anoxia of root system of a plant. Being able to adjust in environmental changes, generally plants have been shown to possess capability of improving hypoxia tolerance through mitigating oxidative damage [147]. As per the view updated by Cheng and co-workers [148], hypoxia results in the death of root tips while studying pea seedlings. However, facilitating exogenous H_2S can prevent plant from hypoxia by reducing ROS damage and by inhibiting the production of ethylene. In case of restricting H_2S biosynthesis in plants, triggers hypoxia. Therefore, sulphide reduces the hypoxia risk in flood prone plants.

In all the above described stress factors that can be possibly controlled through exogenous H_2S administration, the different forms of stresses in plants, have been shown to adopt an optimal concentration of NaHS ranging from 0.05 to 1.5 mM. It may also be noted that, higher concentration, that is, >1.5 mM [149] results in adverse effects on plant survival and growth, and even tolerance to stressing factors may get diminished. Hence, optimal concentration of NaHS must be retained to avoid any suffering in plant species.

4.7 Conclusions

H_2S is thus the youngest member of well-established gasotransmitters. Like NO and CO, this molecule plays signalling role in mammalian physiology. Moreover, the role that the molecular scaffolds of H_2S or ^-SH releasers play in the industrial, medicinal along with environmental fields is conclusive to declare H_2S chemistry as a major field of gasotransmitter modelling. The profound role in living world from microbial to animal world signifies the essence of this molecule in biology. Among several molecular recognition in plant kingdom, H_2S is also a prominent warrior. Like in NO- and CO-based metallic carriage systems, here too metal complexes represent less energy requiring H_2S donors.

References

[1] A. K. Mustafa, M. M. Gadalla and S. H. Snyder, Signaling by gasotransmitters, *Sci. Signal*, 2 (2009) 1–8.

[2] L. Li, P. Roseand and P. K. Moore, Hydrogen sulfide and cell signaling, *Annu. Rev. Pharmacol. Toxicol.*, 51 (2011) 169–187.

[3] R. Wang, 2004, *Signal Transduction and the Gasotransmitters: NO, CO, and H_2S in Biology and Medicine*, Humana, Totowa, (2004).

[4] R. Wang, 2002. Two's company, three's a crowd: Can H_2S be the third endogenous gaseous transmitter?, *FASEB J.*, 16 (2002) 1792–1798.

[5] R. F. Furchgott and J. V. Zawadzki, The obligatory role of endothelial cells in the relaxation of arterial smooth muscle by acetylcholine, *Nature*, 288 (1980) 373–376.

[6] A. Hermann, G. F. Sitdikova and T. M. Weiger Eds., *Gasotransmitters: Physiology and Pathophysiology*, Springer-Verlag, (2012).

[7] W. Zhao, J. Zhang, Y. Lu and R. Wang, The vasorelaxant effect of H_2S as a novel endogenous gaseous K_{ATP} channel opener, EMBO J., 20 (2001) 6008–6016.

[8] L. Li, P. Rose and P. K. Moore, Hydrogen sulfide and cell signaling, *Ann. Rev. Pharmacol. Toxicol.*, 51 (2011) 169–187.

[9] R. Wang, Physiological implications of hydrogen sulfide: a whiff exploration that blossomed, *Physiol. Rev.*, 92 (2012) 791–896.

[10] M. Ishigami, K. Hiraki, K. Umemura, Y. Ogasawara, K. Ishii and H. Kimura, A source of hydrogen sulfide and a mechanism of its release in the brain, *Antioxid. Redox Signaling*, 11 (2009) 205–214.

[11] N. Shibuya, M. Tanaka, M. Yoshida, Y. Ogasawara, T. Togawa, K. Ishii and H. Kimura, 3-Mercaptopyruvate sulfurtransferase produces hydrogen sulfide and bound sulfane sulfur in the brain, *Antioxid. Redox Signal.*, 11 (2009) 703–714.

[12] Y. Mikami, N. Shibuya, Y. Kimura, N. Nagahara, Y. Ogasawara and H. Kimura, Thioredoxin and dihydrolipoic acid are required for 3-mercaptopyruvate sulfurtransferaseto produce hydrogen sulphide, *Biochem. J.*, 439 (2011) 479–485.

[13] K. R. Olsona, J. A. Donald, R. A. Dombkowski and S. F. Perry, *Respir. Physiol. Neurobiol.*, 184 (2012) 117–129.

[14] C. Persa, K. Osmotherly, C. W. Chen, S. Moon and M. F. Louabcd, The distribution of cystathionine β-synthase in the eye: Implication of the presence of a transsulfuration pathway for oxidative stress defense, *Exp. Eye Res.*, 83 (2006) 817–823.

[15] W. W. Pong, R. Stouracova, N. Frank, J. P. Kraus and W. D. Eldred, Comparative localization of cystathionine β-synthase and cystathionine γ-lyase in retina: differences between amphibians and mammals, *J. Comp. Neurol.*, 505 (2010) 158–165.

[16] N. Shibuya, S. Koike, M. Tanaka, M. Ishigami-Yuasa, Y. Kimura, Y. Ogasawara, K. Fukui, N. Nagahara and H. Kimura, A novel pathway for the production of hydrogen sulfide from D-cysteine in mammalian cells, *Nat. Commun*, 4 (2013) 1–7.

[17] M. Kulkarni, Y. F. Njie-Mbye, I. Okpobiri, M. Zhao, C. A. Opere and S. E. Ohia, Endogenous production of hydrogen sulfide in isolated bovine eye, *Neurochem. Res.*, 36 (2011) 1540–1545.

[18] Y. Mikami, N. Shibuya, Y. Kimura, N. Nagahara, M. Yamada and H. Kimura, Hydrogen sulfide protects the retina from light-induced degeneration by the modulation of Ca^{2+}influx, *J. Biol. Chem.*, 286 (2011) 39379–39386.

[19] J. D. Picker and H. L. Levy, *Homocystinuria caused by Cystathionine Beta-Synthase Deficiency*, University of Washington, Seattle, (2010).

[20] M. Yu, G. Sturgill-Short, P. Ganapathy, A. Tawfik, N. S. Peachey and S. B. Smith, Age-related changes in visual function in cystathionine-beta- synthase mutant mice, a model of hyperhomocysteinemia, *Exp. Eye Res.*, 96 (2012) 124–131.

[21] Y. Han, Q. Shang, J. Yao and Y. Ji, *Cell Death Dis.*, 10 (2019) 293–304.

[22] P. Buchner, H. Takahashi and M. J. Hawkesford, Plant sulphate transporters: Co-ordination of uptake, intracellular and long-distance transport, *J. Exp. Bot.*, 55 (2004) 1765–1773.

[23] H. Takahashi, S. Kopriva, M. Giordano, K. Saito and R. Hell, Sulfur assimilation in photosynthetic organisms: Molecular functions and regulations of transporters and assimilatory enzymes, *Annu. Rev. Plant Biol.*, 62 (2011) 157–184.

[24] H. Birke, L. J. De Kok, M. Wirtz and R. Hell, The Role of compartment-specific cysteine synthesis for sulfur homeostasis during H_2S exposure in Arabidopsis, *Plant Cell Physiol.*, 56 (2015) 358–367.

[25] J. M. Mir and R. C. Maurya, Physiological and pathophysiological implications of hydrogen sulfide: A persuasion to change the fate of the dangerous molecule, *J. Chinese Adv. Mater. Soc.*, (2018); doi: 10.1080/22243682.2018.1493951.

[26] C. Bayse, *ACS Symposium Series, American Chemical Society*, Washington DC, (2013).

[27] M. N. Hughes, N. M. Centelles and K. P. Moore, Making and working with hydrogen sulfide: the chemistry and generation of hydrogen sulfide *in vitro* and its measurement *in vivo*: a review, *Free Radic. Biol. Med.*, 47 (2009) 1346–1353.

[28] M. C. Belardinelli, A. Chabli, B. Chadefaux-Vekemans and P. Kamoun, Urinary sulphur compounds in Down syndrome, *Clin. Chem.*, 47 (2001) 1500–1501.

[29] K. Fischer, J. Chen, M. Petri and J. Gmehling, Solubility of H_2S and CO_2 in N-octyl-2-pyrrolidone and of H_2S in methanol and benzene, *AIChE J*, 48 (2002) 887–893.

[30] E. A. Guenther, K. S. Johnson and K. H. Coale, Direct ultraviolet spectrophotometric determination of total sulfide and iodide in natural waters, Anal. Chem., 73 (2001) 3481–3487.

[31] R. O. Beauchamp Jr., J. S. Bus, J. A. Popp, C. J. Boreiko and D. A. Andjelkovich, A critical review of the literature on hydrogen sulfide toxicity, *Crit. Rev. Toxicol.*, 13 (1964) 25–97. R. J. Reiffenstein, W. C. Hulbert, S. H. Roth. Toxicology of hydrogen sulphide, *Annu. Rev. Pharmacol. Toxicol*, 32 (1992) 109–134.

[32] S. Kage, S. Kashimura, H. Ikeda, K. Kudo and N. Ikeda, Fatal and nonfatal poisoning by hydrogen sulfide at an industrial waste site, *J. Forensic Sci*, 47 (2002) 652–655.

[33] C. Szabo, Hydrogen sulphide and its therapeutic potential, Nat. Rev. Drug Discov., 6 (2007) 917–935.

[34] R. Gaudana, H. K. Ananthula, A. Parenky and A. K. Mitra, Ocular drug delivery, *AAPS J.*, 12 (2010) 348–360.

[35] C. Bucolo and F. Drago, Carbon monoxide and the eye: Implications for glaucoma therapy, Pharmacol. Ther., 130 (2011) 191–201.

[36] V. H. Lee and J. R. Robinson, Topical ocular drug delivery: Recent developments and future challenges, *J. Ocul. Pharmacol.*, 2 (1986) 67–108.

[37] S. Macha, A. K. Mitra and P. M. Hughes, Overview of ocular drug delivery, in: *Ophthalmic drug delivery systems*, A. K. Mitra Ed., 1–12, New York, Marcel Dekker, Inc, (2003).

[38] F. Maggio, Glaucomas, *Top. Companion Anim. Med.*, 30 (2015) 86–96.

[39] A. V. Mantravadi and N. Vadhar, Glaucoma, *Prim. Care Clin. Off. Pract.*, 42 (2015) 437–449.

[40] H. Neufeld, D. K. Dueker, T. Vegge and M. L. Sears, Adenosine 3, 5-monophosphate increases the outflow of aqueous humour from the rabbit eye, *Invest. Ophthalmol.*, 14 (1975) 40–42.

[41] J. Robinson, E. Okoro, C. Ezuedu, L. Bush, C. A. Opere, S. E. Ohia and Y. F. Njie-Mbye, Effects of hydrogen sulfide-releasing compounds on aqueous humour outflow facility in porcine ocular anterior segments, ex vivo, *J. Ocul. Pharmacol.*, 33 (2017) 91–97.

[42] K. Módis, P. Panopoulos, G. Olah, C. Coletta, A. Papapetropoulos and C. Szabo, Role of phosphodiesterase inhibition and modulation of mitochondrial cAMP levels in the bioenergetic effect of hydrogen sulfide in isolated mitochondria, *Nitric oxide*, 31 (2013) S28–S28.

[43] E. Perrino, C. Uliva, C. Lanzi, P. D. Soldato, E. Masini and A. Sparatore, New prostaglandin derivative for glaucoma treatment, *Bioorg. Med. Chem. Lett.*, 19 (2009) 1639–1642.

[44] A. Salvi, P. Bankhele, J. Jamil, M. Kulkarni Chitnis, Y. F. Njie-Mbye, S. E. Ohia and C. A. Opere, Effect of hydrogen sulfide donors on intraocular pressure in rabbits, *J. Ocular Pharmacol. Therapeut.*, 32 (2016) 371–375.

[45] E. M. Monjok, K. H. Kulkarni, G. Kouamou, M. McKoy, C. A. Opere, O. N. Bongmba, Y. F. Njie and S. E. Ohia, Inhibitory action of hydrogen sulfide on muscarinic receptor-induced contraction of isolated porcine irides, *Exp. Eye Res.*, 87 (2008) 612–616.

[46] G. L. Zhan, P. Y. Lee, D. C. Ball, C. J. Mayberger, M. E. Tafoya, C. B. Camras and C. B. Toris, Time dependent effects of sympathetic denervation on aqueous humour dynamics and choroidal blood flow in rabbits, *Curr. Eye Res.*, 25 (2002) 99–105.

[47] T. Yoshitomi, B. Horio and D. S. Gregory, Changes in aqueous norepinephrine and cyclic adenosine monophosphate during the circadian cycle in rabbits, *Invest. Ophthalmol. Vis. Sci.*, 32 (1991) 1609–1613.

[48] J. Flammer, The impact of ocular blood flow in glaucoma, *Progress. Retin. Eye Res.*, 21 (2002) 359–393.

[49] S. Salomone, R. Foresti, A. Villari, G. Giurdanella, F. Drago and C. Bucolo, Regulation of vascular tone in rabbit ophthalmic artery: cross talk of endogenous and exogenous gas mediators, *Biochem. Pharmacol.*, 92 (2014) 661–668.

[50] M. Kulkarni-Chitnis, Y. F. Njie-Mbye, L. Mitchell, J. Robinson, M. Whiteman, M. E. Wood, C. A. Opere and S. E. Ohia, Inhibitory action of novel hydrogen sulfide donors on bovine isolated posterior ciliary arteries, *Exp. Eye Res.*, 134 (2015) 73–79.

[51] M. Kulkarni-Chitnis, Y. F. Njie-Mbye, C. A. Opere, M. E. Wood, M. Whiteman and S. E. Ohia, Pharmacological actions of the slow release hydrogen sulfide donor GYY4137 on phenylephrine-induced tone in isolated bovine ciliary artery, *Exp. Eye Res.*, 116 (2013) 350–354.

[52] A. Denoyer, C. Roubeix, A. Sapienza, A. R. L. Goazigo, S. M. Parsadaniantz and C. Baudouin, Retinal and trabecular degeneration in glaucoma: New insights into pathogenesis and treatment, *J. Fr. Ophtalmol.*, 38 (2015) 347–356.

[53] A. Goyal, A. Srivastava, R. Sihota and J. Kaur, Evaluation of oxidative stress markers in aqueous humor of primary open angle glaucoma and primary angle closure glaucoma patients, *Curr. Eye Res.*, 39 (2014) 823–829.

[54] N. N. Osborne and S. D. Olmo-Aguado, Maintenance of retinal ganglion cell mitochondrial functions as a neuroprotective strategy in glaucoma, *Curr. Opin. Pharmacol.*, 13 (2013) 16–22.

[55] B. J. O. White, P. A. Smith and W. R. Dunn, Hydrogen sulphide–mediated vasodilatation involves the release of neurotransmitters from sensory nerves in pressurized mesenteric small arteries isolated from rats, *Br. J. Pharmacol.*168, (2013), 785–793.

[56] N. N. Osborne, J. Dan, A. Shah, A. Majid, P. D. Soldata and A. Sparatore, Glutamate oxidative injury to RGC-5 cells in culture is necrostatin sensitive and blunted by a hydrogen sulfide (H_2S)-releasing derivative of aspirin (ACS14), *Neurochem. Int.*, 60 (2012) 365–378.

[57] A. S. A. Majid, A. M. S. A. Majid, Z. Q. Yin and D. Ji, Slow regulated release of H_2S inhibits oxidative stress induced cell death by influencing certain key signaling molecules, *Neurochem. Res.*, 38 (2013) 1375–1393.

[58] S. Huang, P. Huang, Z. Lin, X. Liu, X. Xu, L. Guo, X. Shen, C. Li and Y. Zhong, Hydrogen sulfide supplement attenuates the apoptosis of retinal ganglion cells in experimental glaucoma, *Exp. Eye Res.*, 168 (2018) 33–48.

[59] S. Huang, P. Huang, X. Liu, Z. Lin, J. Wang, S. Xu, L. Guo, C. K.-S. Leung and Y. Zhong, Relevant variations and neuroprotecive effect of hydrogen sulfide in a rat glaucoma model, *Neuroscience*, 341 (2017) 27–41.

[60] A. Patil, S. Singh, C. Opere and A. Dash, Sustained-release delivery system of a slow hydrogen sulfide donor, GYY 4137, for potential application in glaucoma, *AAPS PharmSciTech.*, 18 (2017) 2291–2302.

[61] A. W. Stitt, AGEs and Diabetic Retinopathy, *Invest. Ophthalmol. Vis. Sci.*, 51 (2010) 4867–4874.

[62] L. Yang, Q. Wu, Y. Li, X. Fan, Y. Hao, H. Sun, Y. Cui and L. Han, Association of the receptor for advanced glycation end products gene polymorphisms and circulating RAGE levels with diabetic retinopathy in the Chinese population, *J. Diabetes Res.*, 2013 (2013), 1–8.

[63] P. S. Padayatti, C. Jiang, M. A. Glomb, K. Uchida and R. H. Nagaraj, High concentrations of glucose induce synthesis of argpyrimidine in retinal endothelial cells, *Curr. Eye Res.*, 23 (2001) 106–115.

[64] S. A. Kandarakis, C. Piperi, F. Topouzis and A. G. Papavassiliou, Emerging role of advanced glycation-end products (AGEs) in the pathobiology of eye diseases, *Progress Retin. Eye Res.*, 42 (2014) 85–102.

[65] C. Hirata, K. Nakano, N. Nakamura, Y. Kitagawa, H. Shigeta, G. Hasegawa, M. Ogata, T. Ikeda, H. Sawa, K. Nakamura, K. Ienaga, H. Obayashi and M. Kondo, Advanced glycation end products induce expression of vascular endothelial growth factor by retinal müller cells, *Biochem. Biophys. Res. Commun.*, 236 (1997) 712–715.

[66] -Y.-Y. Liu, B. V. Nagpure, P. T. H. Wong and J.-S. Bian, Hydrogen sulfide protects SH-SY5Y neuronal cells against d-galactose induced cell injury by suppression of advanced glycation end products formation and oxidative stress, *Neurochem. Int.*, 62 (2013) 603–609.

[67] S. Usui, B. C. Oveson, T. Iwase, L. Lu, S. Y. Lee, Y.-J. Jo, Z. Wu, E.-Y. Choi, R. J. Samulski and P. A. Campochiaro, Overexpression of SOD in retina: need for increase in H_2O_2-detoxifying enzyme in same cellular compartment, *Free Radical Biol. Med.*, 51 (2011) 1347–1354.

[68] M. Brownlee, The pathobiology of diabetic complications: A unifying mechanism, *Diabetes*, 54 (2005) 1615–1625.

[69] B. Geng, L. Chang, C. Pan, Y. Qi, J. Zhao, Y. Pang, J. Du and C. Tang, Endogenous hydrogen sulfide regulation of myocardial injury induced by isoproterenol, *Biochem. Biophys. Res. Commun.*, 318 (2004) 756–763. H. Kimura, The physiological role of hydrogen sulfide and beyond, *Nitric Oxide*, 41 (2014) 4–10.

[70] M. Whiteman, N. S. Cheung, Y.-Z. Zhu, S. H. Chu, J. L. Siau, B. S. Wong, J. S. Armstrong and P. K. Moore, Hydrogen sulphide: a novel inhibitor of hypochlorous acid-mediated oxidative damage in the brain?, *Biochem. Biophys. Res. Commun.*, 326 (2005) 794–798.

[71] S. Muzaffar, N. Shukla, M. Bond, A. C. Newby, G. D. Angelini, A. Sparatore, P. D. Soldato and J. Y. Jeremy, Exogenous hydrogen sulfide inhibits superoxide formation, NOX-1 expression and Rac1 activity in human vascular smooth muscle cells, *J. Vasc. Res.*, 45 (2008) 521–528.

[72] B. Gemici, W. Elsheikh, K. B. Feitosa, S. K. P. Costa, M. N. Muscara and J. L. Wallace, H_2S-releasing drugs: Anti-inflammatory, cytoprotective and chemopreventative potential, *Nitric Oxide*, 46 (2015) 25–31.

[73] M. Whiteman, L. Li, P. Rose, C.-H. Tan, D. B. Parkinson and P. K. Moore, The effect of hydrogen sulfide donors on lipopolysaccharide-induced formation of inflammatory mediators in macrophages, *Antioxid. Redox Signal.*, 12 (2010) 1147–1154.

[74] L. Li, G. Rossoni, A. Sparatore, L. C. Lee, P. D. Soldato and P. K. Moore, Anti-inflammatory and gastrointestinal effects of a novel diclofenac derivative, *Free Radic. Biol. Med.*, 42 (2007) 706–719.

[75] E. Lieth, T. W. Gardner, A. J. Barber and D. A. Antonetti, Retinal neurodegeneration: Early pathology in diabetes, *Clin. Exp. Ophthalmol*, 28 (2000) 3–8.

[76] N. N. Osborne, D. Ji, A. S. A. Majid, R. J. Fawcett, A. Sparatore and P. D. Soldato, ACS67, a Hydrogen sulfide–releasing derivative of latanoprost Acid, attenuates retinal ischemia and oxidative Stress to RGC-5 cells in culture, *Invest. Opthalmology Vis. Sci.*, 51 (2010) 284–294.

[77] Y.-F. Si, J. Wang, J. Guan, L. Zhou, Y. Sheng and J. Zhao, Treatment with hydrogen sulfide alleviates streptozotocin induced diabetic retinopathy in rats, *Br. J. Pharmacol.*, 169 (2013) 619–631.

[78] T. Qaum, Q. Xu, A. M. Joussen, M. W. Clemens, W. Qin, K. Miyamoto, H. Hassessian, S. J. Wiegand, J. Rudge, G. D. Yancopoulos and A. P. Adamis, VEGF-initiated blood-retinal barrier breakdown in early diabetes, *Invest. Ophthalmol. Vis. Sci.*, 42 (2001) 2408–2413.

[79] T. Oshitari, P. Polewski, M. Chadda, A.-F. Li, T. Sato and S. Roy, Effect of combined antisense oligonucleotides against high-glucose- and diabetes-induced overexpression of extracellular matrix components and increased vascular permeability, *Diabetes*, 55 (2006) 86–92.

[80] A. Papapetropoulos, A. Pyriochou, Z. Altaany, G. Yang, A. Marazioti, Z. Zhou, M. G. Jeschke, L. K. Branski, D. N. Herndon, R. Wang and C. Szabó, Hydrogen sulfide is an endogenous stimulator of Angiogenesis, *Proc. Natl Acad. Sci., USA*, 106 (2007) 21972–21977.

[81] C. Szabó and A. Papapetropoulos, Hydrogen sulphide and angiogenesis: Mechanisms and applications, *Br. J. Pharmacol.*, 164 (2011) 853–865.

[82] R. Ran, L. Du, X. Zhang, X. Chen, Y. Fang, Y. Li and H. Tian, Elevated hydrogen sulfide levels in vitreous body and plasma in patients with proliferative diabetic retinopathy, *Retina*, 34 (2014), 2003–2009.

[83] S. Cottet and D. F. Schorderet, Mechanisms of apoptosis in retinitis Pigmentosa, *Curr. Mol. Med.*, 9 (2009) 375–383.

[84] V. Busskamp, J. Duebel, D. Balya, M. Fradot, T. J. Viney, S. Siegert, A. C. Groner, E. Cabuy, V. Forster, M. Seeliger, M. Biel, P. Humphries, M. Paques, S. M. Said, D. Trono, K. Deisseroth, J. A. Sahel, S. Picaud and B. Roska, Genetic reactivation of cone photoreceptors restores visual responses in retinitis pigmentosa, *Science*, 329 (2010) 413–417.

[85] A. Wenzel, C. Grimm, M. Samardzija and C. E. Reme, Molecular mechanisms of light-induced photoreceptor apoptosis and neuroprotection for retinal degeneration, *Prog. Retin. Eye Res.*, 24 (2005) 275–306.

[86] Y. Mikami, N. Shibuya, Y. Kimura, N. Nagahara, M. Yamada and H. Kimura, Hydrogen sulfide protects the retina from light-induced degeneration by the modulation of Ca^{2+} influx, *J. Biol. Chem.*, 286 (2011) 39379–39386.

[87] C. L. Da, F. K. Chen, A. Ahmado, J. Greenwood and P. Coffey, RPE transplantation and its role in retinal disease, *Prog. Retin. Eye Res.*, 26 (2007) 598–635. X. Liu, Y. Zhang, Y. He, J. Zhao and G. Su, Progress in histopathologic and pathogenetic research in a retinitis pigmentosa model, *Histol. Histopathol*, 30 (2015) 771–779.

[88] Y. F. Njie-Mbye, O. Y. N. Bongmba, C. C. Onyema, A. Chitnis, M. Kulkarni, C. A. Opere, A. M. LeDay and S. E. Ohia, Effect of hydrogen sulfide on cyclic AMP production in isolated bovine and porcine neural retinae, *Neurochem. Res.*, 35 (2010) 487–494.

[89] M. O. Hall, T. A. Abrams and T. W. Mittag, The phagocytosis of rod outer segments is inhibited by drugs linked to cyclic adenosine monophosphate production, *Invest. Ophthalmol. Vis. Sci.*, 34 (1993) 2392–2401.

[90] S. Kuriyama, M. O. Hall, T. A. Abrams and T. W. Mittag, Isoproterenol inhibits rod outer segment phagocytosis by both cAMP-dependent and independent pathways, *Invest. Ophthalmol. Vis. Sci.*, 36 (1995) 730–736.

[91] L. Low, J. P. Law, J. Hodson, R. McAlpine, U. O'Colmain and C. MacEwen, Impact of socioeconomic deprivation on the development of diabetic retinopathy: A population-based, cross-sectional and longitudinal study over 12 years, BMJ Open, 5 (2015) e007290; doi: 10.1136/bmjopen-2014-007290.

[92] H. Kimura, The physiological role of hydrogen sulfide and beyond, *Nitric Oxide*, 41 (2014) 410.

[93] R. C. Maurya and J. M. Mir, Nitric Oxide, carbon monoxide, and hydrogen sulfide as biologically important signaling molecules with the significance of their respective donors in ophthalmic diseases, in *Chemistry of biologically potent natural products and synthetic compounds*, S. Ul-islam and J. A. Banday Eds., USA, Scrivener Publishing LLC, 343–378, © (2021).

[94] R. Wang, Gasotransmitters: growing pains and joys, Trends Biochem. Sci., 39 (2014) 227–232.

[95] B. Olas, Hydrogen sulfide in signaling pathways, *Clinica Chimica Acta*, 439 (2015) 212–218.

[96] B. D. Paul and S. H. Snyder, Modes of physiologic H_2S signaling in the brain and peripheral tissues, *Antioxid. Redox Signal.*, 22 (2015) 411–423.

[97] R. Rodriguez-Kabana, J. W. Jordan and J. P. Hollis, Nematodes: biological control in rice fields: role of hydrogen sulfide, *Science*, 148 (1965) 524–526.

[98] C. R. Thompson and G. Kats, Effects of continuous hydrogen sulfide fumigation on crop and forest plants, *Environ. Sci. Technol.*, 12 (1978) 550–553.

[99] A. Aroca, C. Gotor and L. C. Romero, Hydrogen Sulfide Signaling in Plants: Emerging Roles of Protein Persulfidation, *Front. Plant Sci.*, 9 (2018) 1369–1377. and references therein.

[100] H. Zhang, L.-Y. Hu, K.-D. Hu, Y.-D. He, S.-H. Wang and J.-P. Luo, Hydrogen sulfide promotes wheat seed germination and alleviates oxidative damage against copper stress, *J. Integr. Plant Biol.*, 50 (2008) 1518–1529.

[101] H. Zhang, W. Dou, C.-X. Jiang, Z.-J. Wei, J. Liu and R. L. Jones, Hydrogen sulfide stimulates β-amylase activity during early stages of wheat grain germination, *Plant Signal. Behav.*, 5 (2010) 1031–1033.
[102] H. Zhang, J. Tang, X.-P. Liu, Y. Wang, W. Yu, W.-Y. Peng, F. Fang, D.-F. Ma, Z.-J. Wei and L.-Y. Hu, Hydrogen sulfide promotes root organogenesis in *Ipomoea batatas, Salix matsudana* and *Glycine max*, *J. Integr. Plant Biol.*, 51 (2009) 1086–1094.
[103] J. M. Carter, E. M. Brown, E. E. Irish and N. B. Bowden, Characterization of dialkyldithiophosphates as slow hydrogen sulfide releasing chemicals and their effect on the growth of maize, *J. Agric. Food Chem.*, (2019); doi: 10.1021/acs.jafc.9b04398.
[104] J. Chen, F.-H. Wu, W.-H. Wang, C.-J. Zheng, G.-H. Lin, X.-J. Dong, J.-X. He, Z.-M. Pei and H.-L. Zheng, Hydrogen sulphide enhances photosynthesis through promoting chloroplast biogenesis, photosynthetic enzyme expression, and thiol redox modification in *Spinacia oleracea* seedlings, *J. Exp. Botany*, 62 (2011) 4481–4493.
[105] B. Duan, Y. Ma, M. Jiang, F. Yang, L. Ni and W. Lu, Improvement of photosynthesis in rice (*Oryza sativa* L.) as a result of an increase in stomatal aperture and density by exogenous hydrogen sulfide treatment, *Plant Growth Regul.*, 75 (2014) 33–44.
[106] H. Zhang, Y.-K. Ye, S.-H. Wang, J.-P. Luo, J. Tang and D.-F. Ma, Hydrogen sulfide counteracts chlorophyll loss in sweetpotato seedling leaves and alleviates oxidative damage against osmotic stress, *Plant Growth Regul.*, 58 (2009) 243–250.
[107] H. Zhang, S.-L. Hu, Z.-J. Zhang, L.-Y. Hu, C.-X. Jiang, Z.-J. Wei, J. Liu, H.-L. Wang and S.-T. Jiang, Hydrogen sulfide acts as a regulator of flower senescence in plants, *Postharvest Biol. and Techno.*, 60 (2011) 251–257.
[108] F. D. Dooley, S. P. Nair and P. D. Ward, Increased growth and germination success in plants following hydrogen sulfide administration, PLoS ONE, 8 (2013) e62048; doi: 10.1371/journal.pone.0062048.
[109] F. J. Corpas and J. M. Palma, H_2S signaling in plants and applications in agriculture, *J. Adv. Res.*, 24 (2020) 131–137.
[110] M. A. Muñoz-Vargas, S. González-Gordo, A. Cañas, J. López-Jaramillo, J. M. Palma and F. J. Corpas, Endogenous hydrogen sulfide (H_2S) is up-regulated during sweet pepper (Capsicum annuum L.) fruit ripening. *In vitro* analysis shows that NADP-dependent isocitrate dehydrogenase (ICDH) activity is inhibited by H_2S and NO, *Nitric Oxide*, 81 (2018) 36–45.
[111] V. Ziogas, A. Molassiotis, V. Fotopoulo and G. Tanou, Hydrogen sulfide: A potent tool in postharvest fruit biology and possible mechanism of action, *Front. Plant Sci.*, 9 (2018) 1375–1383.
[112] M. S. Aghdam, R. Mahmoudi, F. Razavi, V. Rabiei and A. Soleimani, Hydrogen sulfide treatment confers chilling tolerance in hawthorn fruit during cold storage by triggering endogenous H_2S accumulation enhancing antioxidant enzymes activity and promoting phenols accumulation, *Sci. Hortic.*, 238 (2018) 264–271.
[113] D. Liu, S. Xu, H. Hu, J. Pan, P. Li and W. Shen, Endogenous hydrogen sulfide homeostasis is responsible for the alleviation of senescence of postharvest daylily flower via increasing antioxidant capacity and maintained energy status, *J. Agric. Food Chem.*, 65 (2017) 718–726.
[114] P. Ahmad, *Plant Metal Interaction: Emerging Remediation Techniques*, Elsevier, Netherlands, (2016).
[115] M. G. Mostofa, A. Rahman, M. M. U. Ansary, A. Watanabe, M. Fujita and L.-S. P. Tran, Hydrogen sulfide modulates cadmium-induced physiological and biochemical responses to alleviate cadmium toxicity in rice, *Sci. Rep.*, 5 (2015) 14078–14094; doi: 10.1038/srep14078.
[116] L. Zhang, Y. Pei, H. Wang, Z. Jin, Z. Liu, Z. Qiao, H. Fang and Y. Zhang, Hydrogen sulfide alleviates cadmium-induced cell death through restraining ROS accumulation in roots of *Brassica rapa* L. ssp. Pekinensis, *Oxid. Med. Cell Longev.*, (2015), 1–11.

[117] D. Lai, Y. Mao, H. Zhou, F. Li, M. Wu, J. Zhang, Z. He, W. Cui and Y. Xie, Endogenous hydrogen sulfide enhances salt tolerance by coupling the reestablishment of redox homeostasis and preventing salt-induced K^+ loss in seedlings of *Medicago sativa*, *Plant Sci.*, 225 (2014) 117–129.
[118] A. Christou, G. A. Manganaris, I. Papadopoulos and V. Fotopoulos, Hydrogen sulfide induces systemic tolerance to salinity and non-ionic osmotic stress in strawberry plants through modification of reactive species biosynthesis and transcriptional regulation of multiple defence pathways, *J. Exp. Bot.*, 64 (2013) 1953–1966.
[119] J. Shen, T. Xing, H. Yuan, Z. Liu, Z. Jin, L. Zhang and Y. Pei1, Hydrogen Sulfide Improves Drought Tolerance in Arabidopsis thaliana by MicroRNA Expressions, PLoS ONE, 8 (2013) e77047; doi: 10.1371/journal.pone.-0077047.
[120] H. Zhang, W. Dou, C. X. Jiang, Z. J. Wei, J. Liu and R. L. Jones, Hydrogen sulfide stimulatesβ-amylase activity during early stages of wheat grain germination, *Plant Signal. Behav.*, 5 (2010) 1031–1033; doi: 10.4161/psb.5.8.12297.
[121] P. N. Fu, W. J. Wang, L. X. Hou and X. Liu, Hydrogen sulfide is involved in the chilling stress response in *Vitisv inifera* L, *Acta Soc. Bot. Pol.*, 82 (2013) 295–302.
[122] H. Shi, T. Ye and Z. Chan, Exogenous application of hydrogen sulfide donor sodium hydrosulfide enhanced multiple abiotic stress tolerance in Bermuda grass {*Cynodon dactylon* (L). Pers.}, *Plant Physiol. Biochem.*, 71 (2013) 226–234.
[123] L. Ma, L. Yang, Z. Jingjie, J. Wei, X. Kong, C. Wang, X. Zhang, Y. Yang and X. Hu, Comparative proteomic analysis reveals the role of hydrogen sulfide in the adaptation of the alpine plant Lamiophlomis rotate to altitude gradient in the Northern Tibetan Plateau, *Planta*, 241 (2015) 887906.
[124] X. Chen, Q. Chen, X. Zhang, R. Li, Y. Jia, A. Ef, A. Jia, L. Hu and X. Hu, Hydrogen sulfide mediates nicotine biosynthesis in tobacco (*Nicotiana tabacum*) under high temperature conditions, *Plant Physiol. Biochem.*, 104 (2016) 174–179.
[125] A. Christou, P. Filippou, G. A. Manganaris and V. Fotopoulos, Sodium hydrogen sulfide induces systemic thermo tolerance to strawberry plants through transcriptional regulation of heat shock proteins and aquaporin, *BMC Plant Biol.*, 14 (2014) 42–52.
[126] Q. Li, Z. Wang, Y. Zhao, X. Zhang, S. Zhang, L. Bo1, Y. Wang, Y. Ding and L. An, Putrescine protects hulless barley from damage due to UV-B stress via H_2S- and H_2O_2.mediated signaling pathways, Plant Cell Rep., 35 (2016) 1155–1168.
[127] W. Cheng, L. Zhang, C. J. Jiao, M. Su, T. Yang and L. N. Zhou, 2013. Hydrogen sulfide alleviates hypoxia-induced root tip death in *Pisum sativum.*, *Plant Physiol.Biochem.*, 70 (2013) 278–286.
[128] E. Bloem, S. Haneklaus, J. Kesselmeier and E. Schnug, Sulfur fertilization and fungal infections affect the exchange of H_2S and COS from agricultural crops, *J. Agric.Food Chem.*, 60 (2012) 7588–7596.
[129] Z. Jin, J. Shen, Z. Qiao, G. Yang, R. Wang and Y. Pei, Hydrogen sulfide improves drought resistance in *Arabidopsis thaliana*, *Biochem. Biophy. Res. Commun.*, 414 (2011) 481–486.
[130] P. Ahmad, *Plant Metal Interaction: Emerging Remediation Techniques*, Elsevier (2016), Netherlands.
[131] C. H. Foyer and G. Noctor, Redox regulation in photosynthetic organisms: Signaling, acclimation, and practical implications, *Antioxid. Redox. Signal.*, 11 (2009) 861–905.
[132] C. H. Foyer and G. Noctor, Ascorbate and glutathione: The heart of the redox hub, *Plant Physiol.*, 155 (2011) 2–18.

[133] V. P. Singh, S. Singh, J. Kumar and S. M. Prasad, Hydrogen sulfide alleviates toxic effects of arsenate in pea seedlings through up-regulation of the ascorbate–glutathione cycle: possible involvement of nitric oxide, *J. Plant Physiol.*, 181 (2015) 20–29.

[134] H. Zhang, L. Y. Hu, P. Li, K. D. Hu, C. X. Jiang and J. P. Luo, Hydrogen sulfide alleviated chromium toxicity in wheat, *Biol. Plant*, 54 (2010) 743–747.

[135] H. Zhang, L. Y. Hu, K. D. Hu, Y. D. He, S. H. Wang and J. P. Luo, Hydrogen sulfide promotes wheat seed germination and alleviates oxidative damage against copper stress, *J. Integr. Plant Biol.*, 50 (2008) 1518–1529.

[136] J. Chen, W. H. Wang, F. H. Wu, C. Y. You, T. W. Liu, X. J. Dong, J. X. He and H. L. Zheng, Hydrogen sulfide alleviates aluminium toxicity in barley seedlings, *Plant Soil*, 362 (2013) 301–318.

[137] H. Zhang, Z. Q. Tan, L. Y. Hu, S. H. Wang, J. P. Luo and R. L. Jones, Hydrogen sulfide alleviates aluminium toxicity in germinating wheat seedlings, *J. Integr. Plant Biol.*, 52 (2010) 556–567.

[138] X. Liu, J. Chen, G.-H. Wang, W.-H. Wang, Z.-J. Shen, M.-R. Luo, G.-F. Gao, M. Simon, K. Ghoto and H.-L. Zheng, Hydrogen sulfide alleviates zinc toxicity by reducing zinc uptake and regulating genes expression of antioxidative enzymes and metallothioneins in roots of the cadmium/zinc hyperaccumulator *Solanum nigrum* L., *Plant Soil*, 400 (2016) 177–192.

[139] P. Ahmad, M. M. Azooz and M. N. V. Prasad, *Ecophysiology and Responses of Plants under Salt Stress*, Springer (2013), New York, NY.

[140] J. Bao, T. L. Ding, W. J. Jia, L. Y. Wang and B. S. Wang, Effect of exogenous hydrogen sulfide on wheat seed germination under salt stress, *Modern Agric. Sci. Technol.*, 20 (2011) 40–42.

[141] Y. Q. Wang, L. Li, W. T. Cui, S. Xu, W. B. Shen and R. Wang, Hydrogen sulfide enhances alfalfa (*Medicago sativa*) tolerance against salinity during seed germination by nitric oxide pathway, *Plant Soil*, 351 (2012) 107–119.

[142] N. Iqbal, R. Nazar and N. A. Khan, *Osmolytes and Plants Acclimation to Changing Environment: Emerging Omics Technologies*, Springer, London, (2016).

[143] H. Zhang, M. F. Wang, L. Y. Hua, S. H. Wang, K. D. Hua, L. J. Bao and J. P. Luo, Hydrogen sulfide promotes wheat seed germination under osmotic stress, *Russ. J. Plant Physiol.*, 57 (2010) 532–539.

[144] Z. Jin, J. Shen, Z. Qiao, G. Yang, R. Wang and Y. Pei, Hydrogen sulfide improves drought resistance in *Arabidopsis thaliana*, *Biochem. Biophy. Res. Commun*, 414 (2011) 481–486.

[145] H. Hemmati, D. Gupta and C. Basu, Molecular physiology of heat stress responses in plants, in *Elucidation of Abiotic Stress Signaling in Plants: Functional Genomics Gerspectives*, 2 Ed., G. K. Pandey, 109–142, New York, NY, Springer, (2015).

[146] A. Christou, G. A. Manganaris, I. Papadopoulos and V. Fotopou-los, "The Importance of hydrogen sulfide as a systemic priming agent in strawberry plants grown under key abiotic stress factors," in *Proceedings of the 4th International Conference: Plant Abiotic Stress: From Systems Biology to Sustainable Agriculture*, Limassol, 47 (2011).

[147] J. T. Van Dongen and F. Licausi, *Low-Oxygen Stress in Plants: Oxygen Sensing and Adaptive Responses to Hypoxia*, Springer (2014), Wien.

[148] W. Cheng, L. Zhang, C. Jiao, M. Su, T. Yang, L. Zhou, R. Peng, R. Wang and C. Wang, Hydrogen sulfide alleviates hypoxia-induced root tip death in *Pisum sativum*, *Plant Physiol. Biochem.*, 70 (2013) 278–286.

[149] Z.-G. Li, X. Min and Z.-H. Zhou, Hydrogen sulfide: A signal molecule in plant cross-adaptation, *Front. Plant Sci.*, 7 (2016) 1621–1632.

Exercises

Multiple choice questions

1. The concentration of H_2S capable of causing immediate death of humans is:
   (a) 5 ppm (b) 100 ppm (c) 5,000 ppm (d) >1,000 ppm

2. In mammalian species, the endogenous generation of H_2S by L-cysteine is catalysed by enzyme(s):
   (a) CBS
   (b) CSE
   (c) CAT/ 3-MST
   (d) All of these

3. H_2S is a endogenous gaseous signalling molecule called:
   (a) Neurotransmitter
   (b) Gasotransmitter
   (c) Both (a) and (b)
   (d) None of these

4. IOP is related to:
   (a) Heart (b) Lungs (c) Eye (d) None of these

5. The term "gasotransmitter" to a gaseous messenger molecule involved in any signalling process, was first time coined in 2002 by:
   (a) R. F. Furchgott
   (b) J. V. Zawadzki
   (c) R. Wang
   (d) None of them

6. Which one of the following is not a gasotransmitter?
   (a) NH_3 (b) NO (c) CO (d) H_2S

7. Which enzyme(s) pathway is dominating way to produce H_2S in mammalian retina because of being located in the retinal neurons?
   (a) CBS (b) CSE (c) 3MST/CAT (d) All of these

8. Which H_2S producing enzyme is most highly expressed in the mammalian cornea, conjunctiva and iris?
   (a) CSE (b) CBS (c) 3MST (d) CAT

9. Which one data is correct regarding the existence of H_2S under physiological conditions:
   (a) HS^- (82%), H_2S (18%), S^{2-} (<0.1%)
   (b) HS^- (72%), H_2S (27%), S^{2-} (1%)

(c) HS^- (27%), H_2S (72%), S^{2-} (1%)
(d) HS^- (18%), H_2S (82%), S^{2-} (<0.1%)

10. H_2S is:
(a) Strong acid (b) Weak acid (c) Weak base (d) Strong base

11. The abnormal level of H_2S is associated with human disease, such as:
(a) Neurodegenerative disease
(b) Myocardial injury
(c) Ophthalmic disease
(d) All of these

12. Sort out the third gasotransmitter from the following gaseous molecules:
(a) NO (b) CO (c) H_2S (d) N_2

13. The first report of H_2S effects in plants to influence the growth of vegetative plants and to affect disease resistance was made in year:
(a) 1975 (b) 1965 (c) 1960 (d) 1978

14. In plants hypotaurine serves as:
(a) NO donor
(b) H_2S donor
(c) H_2S scavenger
(d) None of these

15. Sort out the biotic stress from the following stresses:
(a) Drought (b) Salinity (c) Radiation (d) Nematodes

16. Which one of the following comes under abiotic stress?
(a) Extremes in temperature
(b) Fungi
(c) Virus
(d) Bacteria

Short answer type questions

1. What are gasotransmitters? What are criteria for establishing signalling molecules as gasotransmitter?
2. Differentiate between neurotransmitters and gasotransmitters.
3. Draw a diagram showing mechanistic action of neurotransmitters and gasotransmitters.
4. Make a diagram showing an overview of H_2S generation in higher plants.

5. Briefly describe the ocular drug delivery with schematic presentation of the ocular structure and the model showing drug movement into the eye from a topical formulation.
6. Present a brief account of ocular bioavailability of ophthalmic solution.
7. Describe the role of H_2S in reduction of IOP in case of increased IOP which is the major cause for optic neuropathy in glaucoma patients that damages the retinal neurons and optic nerve heads.
8. Describe the applications of H_2S on fruit ripening and post-harvest storage preventing the loss of fresh produce caused by fungi, bacteria and viruses.
9. Briefly describe the damaging effect of biotic stresses in crop plants and how can these be overcome.
10. Highlight the H_2S signalling triggered by exogenously applied H_2S donor or up-regulating the expression of genes involved in H_2S biosynthesis to exert its physiological role in plants.

Long answer type questions

1. Describe in detail the biosynthesis of H_2S in ocular tissues.
2. Present a detailed view of the catalytic aspects of metal carbonyls.
3. Present a detailed view of enzymatic and non-enzymatic pathways of H_2S generation in plants.
4. Describe the biochemistry of H_2S in detail.
5. Present the beneficial role of H_2S donors in in patients of glaucoma with special reference to: (i) Reduction of IOP (ii) Effect on ocular blood supply and (iii) Protection on neurons.
6. Describe the role of H_2S donors in DR in diabetic milieu with special reference to: (i) Reduction of the effects of advanced glycation end products (ii) Inhibition of oxidative stress and inflammation (iii) Protective effect on retinal neurons and (iv) Multiple effects on retinal blood vessels.
7. Present the beneficial role of H_2S donors in retinal degenerative diseases, which are, (i) RP (ii) AMD and (iii) Glaucomatous retinal neuropathy.
8. Describe the role of exogenous H_2S treatments in improvement of seed germination and plant growth.
9. What are abiotic stresses in plants? Describe the role of endogenous H_2S triggering under various abiotic stresses in plants to alleviate the damaging effect of such stresses taking suitable examples of plants.
10. What do you understand by cross-adaptation in plants? Describe the H_2S-induced cross-adaptation with special reference to: (i) H_2S-induced metal and metalloid tolerance (ii) H_2S-induced salt tolerance (iii) H_2S-induced drought tolerance (iv) H_2S-Induced cold tolerance (v) H_2S-induced heat tolerance.

Appendix I
The International System of Units, fundamental physical constants and conversion factors

(a) The International System of Units

In order to represent quantities at entire range of scales and within every area of science and technology, the International System of Units (SI) is the globally accepted foundation. This system has two groups of units. The first one is called as "base or basic units" and the second one is known as "derived units". In this system, there are seven base units and their basic quantities. These provide the process of mentioning used to express all the measurements of this system.

(i) Basic/base units and physical quantities

The seven fundamental basic units are given in Table I.1.

Table I.1: Some physical quantities and their respective base SI units with symbols.

Physical quantity	Designating symbol	Unit (SI)	Designating symbol
Time	t	Second	s
Length	l	Metre	m
Mass	m	Kilogram	kg
Amount of substance	lv	Mole	mol
Electric current	I	Ampere	A
Luminous intensity	n	Candela	cd
Thermodynamic temperature	T	Kelvin	K

https://doi.org/10.1515/9783110727302-005

(ii) Derived units

Some derived SI units are listed in Table I.2.

Table I.2: Some derived units along with names and respective symbols.

Physical quantity	Derived unit		Representation in terms of base or derived units
	Name	Designating symbol	
Force	Newton	N	1 N = 1 kg m s^{-2}
Frequency	Hertz	Hz	1 Hz = 1 s^{-1}
Electric charge	Coulomb	C	1 C = l A s
Pressure and stress	Pascal	Pa	1 P a = 1 Nm^{-2}
Energy, work, quantity of heat	Joule	J	1 J = l N m
Electric resistance	Ohm	Ω	1 Ω = 1 V A^{-1}
Electric potential	Volt	V	1 V = l J C^{-1}
Electric capacitance	Farad	F	1 F = l C V^{-1}
Electric conductance	Siemens	S	1 S = l Ω^{-1}
Magnetic flux	Weber	Wb	1 Wb = 1 V s
Luminous flux	Lumen	lm	1 lm = 1 cd sr
Magnetic flux density	Tesla	T	1 T = l Wb m^{-2}
Inductance	Henry	H	1 H = l Wb A^{-1}
Temperature	Degree Celsius	°C	t [°C] = T [K] – 273.15
Illuminance	Lux	lx	1 lx = 1 lm m^{-2}
Activity (of a radionuclide)	Becquerel	Bq	1 Bq = 1 s^{-1}

(b) Prefixes used for SI units

Multiples of both base and derived units are shown by one of the following prefixes given in Table I.3.

Table I.3: Prefixes having symbols.

Prefix (factor)	Designating symbol
Deci (10^{-1})	d
Centi (10^{-2})	c
Milli (10^{-3})	m
Micro (10^{-6})	μ
Nano (10^{-9})	n
Pico (10^{-12})	p
Femto (10^{-15})	f
Atto (10^{-18})	a
Zepto (10^{-21})	z
Deca (10)	da
Hector (10^{2})	h
Kilo (10^{3})	K

Table I.3 (continued)

Prefix (factor)	Designating symbol
Mega (10^6)	M
Giga (10^9)	G
Tera (10^{12})	T
Peta (10^{15})	P
Exa (10^{18})	E
Zetta (10^{21})	Z

(c) Physical constants: fundamental

The principal physical constants include light velocity, Planck's constant and electron mass. In fact, these offer a scheme of physical/natural units. Moreover, these provide the connection/relation among the theory, SI units, and scientific fields. The values of these constants are adopted from the latest recommendations provided by the CODATA Task Group on Physical Constants. Some important physical constants with symbols and values are given in Table I.4.

Table I.4: Some of the fundamental physical constants with symbols, recommended values and SI units.

Constant	Symbol	Value	SI unit
Velocity of light in vacuum	C	$2.99792458 \times l0^{8}$	$m\ s^{-1l}$
Electron charge	e	$1.602176487(40) \times 10^{-19}$	C
Planck's constant (in eVs)	h	$6.62606896\ (33) \times 10^{-34}$	J s
Avogadro constant	N_A	$6.02214179\ (30) \times 10^{23}$	mol^{-1}
Atomic mass constant (unified atomic mass unit)	m_u	$1.660538782(83) \times 10^{-27}$	kg
$m_u = 1/12\ m(^{12}C) = 1$ u			
Electron mass	m_e	$9.10938215(45) \times 10^{-31}$	kg
(in u)		$5.4857990943(23) \times 10^{-4}$	u
Proton mass	m_p	$1.672621637(83) \times 10^{-27}$	kg
(in u)		1.00727646677(10)	u
Neutron mass	m_n	$1.674927211(84) \times 10^{-27}$	kg
(in u)		1.00866491597(43)	u
Gas constant: molar	R	8.314472(15)	$J\ mol^{-1}\ K^{-1}$
Boltzmann's constant	k	$1.3806504(24) \times 10^{-23}$	$J\ K^{-1}$
Rydberg's constant	R_∞	1.097372×10^{7}	m^{-1}
Bohr's radius	a_0	0.529177×10^{-10}	m
Electron magnetic moment	μ_e	9.28483×10^{-24}	$J\ T^{-1}$
Proton magnetic moment	μ_p	1.410617×10^{-26}	$J\ T^{-1}$
Bohr's magneton	μ_B	9.27408×10^{-24}	$J\ T^{-1}$
Nuclear magneton	μ_N	5.05082×10^{-27}	$J\ T^{-1}$
Molar volume of ideal gas (stp)	V_m	0.0224138	$M^3\ mol^{-1}$

Table I.4 (continued)

Constant	Symbol	Value	SI unit
Acceleration of free fall (acceleration due to gravity)	g	9.80665	$m\ s^{-2}$
Gravitational constant	G	6.67259×10^{-11}	$m^2\ s^{-2}\ kg$
Debye (electric dipole moment of molecules)	D	1.0×10^{-18}	esu cm
		3.33×10^{-32}	C cm

(d) Conversion factors

Some of the useful conversion factors are given in Table I.5.

Table I.5: Some useful conversion factors.

1 eV	=	1.602×10^{-12} erg	=	1.602×10^{-19} J		
1 eV $molecule^{-1}$	≈	23.063 kcal mol^{-1}	=	96.496 kJ mol^{-1}	≈	8,065.46 cm^{-1}
1 kcal mol^{-1}	=	349.76 cm^{-1}				
1 kJ mol^{-1}	=	83.54 cm^{-1}				
10,000 cm^{-1}	=	1,000 nanometres (nm)	≈	1.24 eV $molecule^{-1}$		
20,000 cm^{-1}		500 nm				
1 (wavenumber) (cm^{-1})	=	2.8591×10^{-3} kcal mol^{-1}				
1 nanometre (nm)	=	10Å	=	10^{-9} m	=	1 millimicron (mμ)
1 picometre (pm)	=	10^{-2} Å				
1 cm	=	10^8 Å	=	10^7nm		
1 erg	=	2.390×10^{-11} kcal				
1 J		10^7 ergs				
1 L	=	1 dm^3	=	$10^3\ cm^3$		
1 dyne	=	1 g cm s^{-2}	=	10^{-5} N		
1 cal	=	4.1840 J				
1 esu	=	1 $dyn^{1/2}$ cm				
1 atmosphere		760 torr	=	1.101325×10^5 Pascal (Pa)		

(e) Chemical numerical prefixes

Some of the chemical numerical prefixes are given below:
Mono (1)
Di (bis) (2)
Tri (tris) (3)
Tetra (tetrakis) (4)
Penta (pentakis) (5)
Hexa (hexakis) (6)
Hepta (heptakis) (7)
Octa (octakis) (8)
Nona (nonakis) (9)
Deka (dekakis) (10)
Undeca (11)
Dodeca (12)
Pentadeca (15)
Hexadeca (16)
Icosa (20)
Triaconta (30)
Pentaconta (50)
Hecta (100)

Appendix II
Body mass index (BMI): an indicator of our body fat

Introduction

BMI (body mass index) is a scientific measurement which is used to determine a person's risk for weight-related health problem. Although the term may be new, the method has been used for many years by physicians and researchers who study obesity, one of the main causes of diabetes and other health-related problems.

We are familiar with the use of "weight-for-height table" on the walls of hospitals and doctors' clinics. There are many versions of these tables, but they are not especially effective because they do not take into account a person'ss frame size when considering an ideal weight. Although BMI is not a perfect system, it provides a useful general guideline for judging a healthy weight and an appropriate amount of body fat.

Calculation of BMI

The calculation of BMI makes use of a mathematical formula that accounts for a person's weight and body area (or "bulk"), both to indicate the extent of fat storage. The formula uses "mks" system of units:

$$\text{BMI} = \frac{\text{Weight in kilograms}}{(\text{Height in metres})^2}$$

The following conversion factors will be helpful in calculation of BMI:

$$1\,\text{inch} = 2.54\,\text{cm};\ 1\,\text{metre} = 100\,\text{cm};\ 1\,\text{lb(pound)} = 0.454\,\text{kg}$$

For example, the BMI of a person with a weight of 150 lb and a height of 5′ (5 feet) is calculated as follows:

Weight (kg) = 150 lb × 0.454 kg lb^{-1} = 68.1 kg

5′ = 5 × 12 = 60 inches (in)

Height (m) = 60 in × 2.54 cm in^{-1} × 1 m/100 cm = 1.52 m

$$\text{BMI} = \frac{68.1}{(1.52)^2} = \frac{68.1}{2.31} = 29.5\ \text{kg}\,\text{m}^{-2}$$

https://doi.org/10.1515/9783110727302-006

The following guidelines may be used by individuals who are 34 years of age and younger to determine the extent of body fat:

(i) If your BMI is in between 20 and 25, this shows a relatively lean body and a healthy amount of body fat.

(ii) If your BMI is in between 25 and 30, you may consider your body to be "hefty" or perhaps even overweight and you may want to lower your health risk through dieting and/or exercising after discussion with a physician.

(iii) If your BMI is over 30, it is an indication of moderate-to-severe obesity and a physician should be consulted.

Appendix III
Amino acids, the building blocks of proteins: names, symbols, structures, properties and some physical constants

Introduction

In the middle of the nineteenth century (1838), the Dutch chemist G. J. Mulder extracted a substance common to animal tissues and the juices of plants, which he believed to be "without doubt the most important of all substances of the organic kingdom, and without it life on the planate would probably not exist". At the suggestion of the famous Swedish chemist Berzelius, Mulder named this substance protein from the Greek word *proteios*, meaning "of first importance". He assigned to it a specific chemical formula, $C_{40}H_{62}N_{10}O_{12}$. Although he was wrong about the chemistry of proteins, he was right about their being indispensable to living organisms. The term "protein" survives.

Chemists of those times were not aware about proteins that are actually comprised of minor constituents of amino acids. Even though the amino acid was sequestered for the first time long back (1830), for several decades, it was thought that constituents of plants comprising proteins are merged totally within the issues of animals. The misconception was ended when the method of digestion could reveal. After that, it has become known that ingested proteins are split into smaller components comprising amino acids.

Studies carried out experimentally conclude that the amount of amino acids present in a protein decides the nutritional value of the protein in animal diets. For instance, some cereal proteins do not contain lysine (an amino acid). When animals are fed with these cereal proteins, they do not grow well. After determination of the structures of twentieth common amino acids in 1925, the relative nutritional significance of different amino acids in proteins has been systematically recognized.

It is now known that amino acids are vital for life. Moreover, these are the components of protein molecules. Among organic compounds, a number of amino acids are supposed to have appeared early in the history of the Earth. As ancient and omnipresent molecules, amino acids have been adopted in developmental processes for numerous functions in living organizations.

Structure of amino acids

Every protein molecule can be considered as a polymer of amino acids. The analytical data of several of proteins from different sources have revealed that all proteins are composed of 20 common amino acids. Each molecule of protein does not contain all

https://doi.org/10.1515/9783110727302-007

20 amino acids. But majority of proteins contain maximum number of amino acids if not all the 20 types.

Notably, the common amino acids are known as **α-amino acids**. This is owing to the fact that they contain a primary amino group (–**NH_2**) and a carboxylic group (–**COOH**) bonded as at the same central tetrahedral carbon atom called the **alpha carbon (C_α)** (Figure III.1). But proline is an exception. This contains an (–**NH**–) group along with carboxylic group. However, for keeping uniformity, it is also considered as **α-amino acid**. Moreover, the third bond is always hydrogen **(H)** while the fourth bond is of a variable side chain **(R)**. The bonded side chain (**R**) differentiates the 20 amino acids in their respective structures.

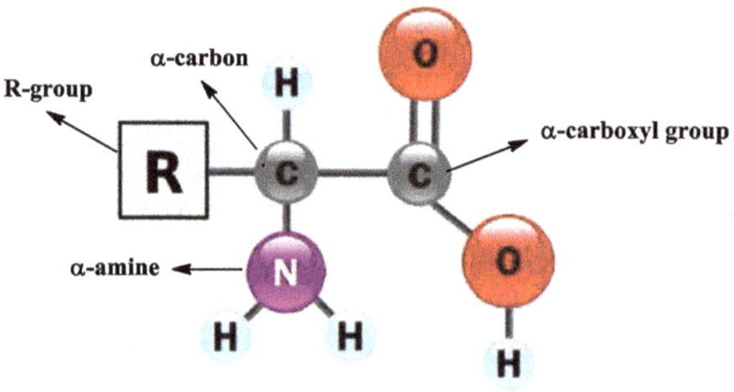

Figure III.1: Amino acid's structure.

Protein's amino acids are often abbreviated by three-letter symbols. When this proves to be too cumbersome, one-letter symbols are also used. Both designations are given in Table III.1.

Table III.1: Abbreviations of protein's amino acids.

Serial number	Name	One-letter abbreviation	Three-letter abbreviation
1.	Alanine	A	Ala
2.	Glycine	G	Gly
3.	Isoleucine	I	Ile
4.	Valine	V	Val
5.	Leucine	L	Leu
6.	Serine	S	Ser
7.	Methionine	M	Met
8.	Threonine	T	Thr
9.	Phenylalanine	F	Phe
10.	Proline	P	Pro
11.	Cysteine	C	Cys

Table III.1 (continued)

Serial number	Name	One-letter abbreviation	Three-letter abbreviation
12.	Asparagine	N	Asn
13.	Glutamine	Q	Gln
14.	Tyrosine	Y	Tyr
15.	Tryptophan	W	Trp
16.	Aspartate/aspartic acid	D	Asp
17.	Glutamate/glutamic acid	E	Glu
18.	Histidine	H	His
19.	Lysine	K	Lys
20.	Arginine	R	Arg

The complete chemical structures along with residue mass and average occurrence in proteins of the 20 amino acids are given in Table III.2.

Table III.2: Complete structures of α-amino acids discovered in proteins along with residue mass and average occurrence in proteins.

Name	Structure	Mass: residual (D)	Occurrence in proteins (%)
Category (i): Amino acids: nonpolar side chains (hydrophobic)			
1. Glycine	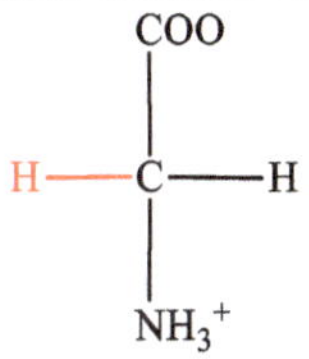	57.0	7.2
2. Alanine	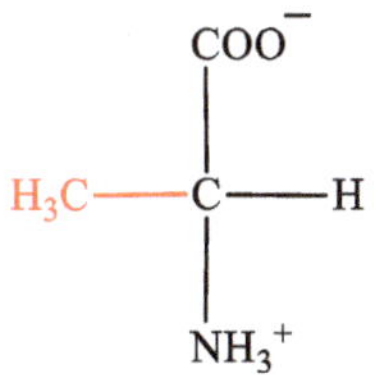	71.1	7.8
3. Valine	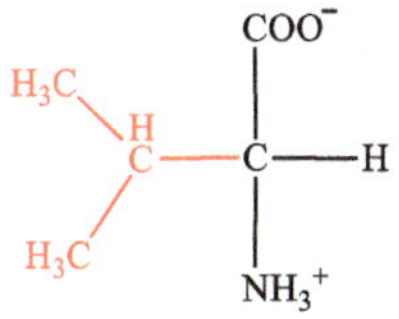	99.1	6.6

Table III.2 (continued)

Name	Structure	Mass: residual (D)	Occurrence in proteins (%)
4. Leucine	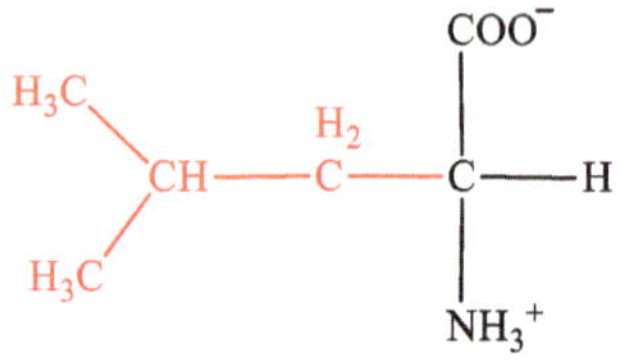	113.2	9.1
5. Isoleucine	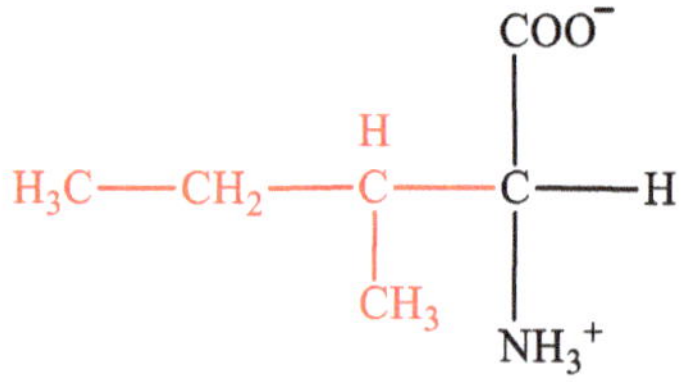	113.2	5.3
6. Methionine	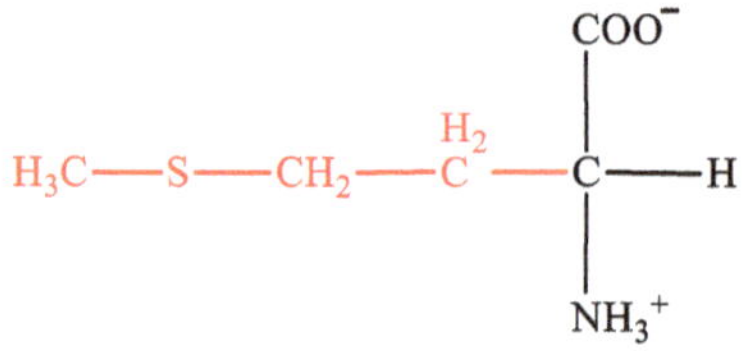	131.2	2.2
7. Proline	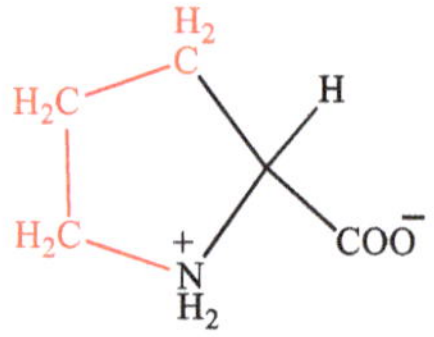	97.1	5.2
8. Phenylalanine	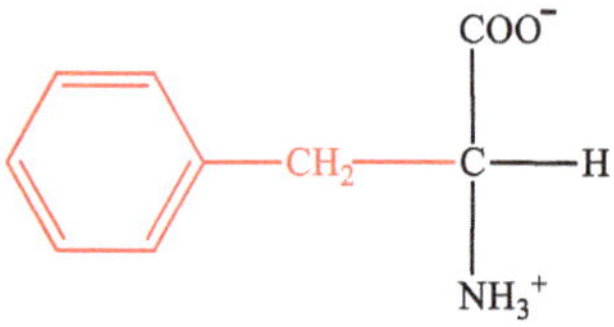	147.1	3.9
9. Tryptophan	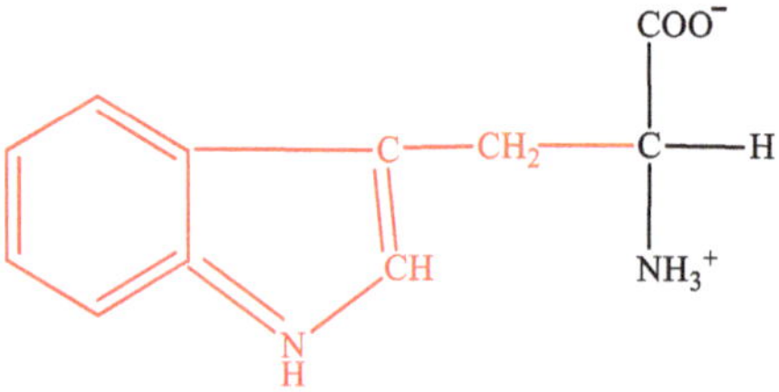	186.2	1.4

Table III.2 (continued)

Name	Structure	Mass: residual (D)	Occurrence in proteins (%)
Category (ii): Amino acids: uncharged polar side chains (hydrophilic)			
1. Serine	$HO-CH_2-\overset{\displaystyle COO^-}{\underset{\displaystyle NH_3^+}{C}}-H$	87.1	6.8
2. Threonine	$H_3C-\overset{}{\underset{\displaystyle OH}{CH}}-\overset{\displaystyle COO^-}{\underset{\displaystyle NH_3^+}{C}}-H$	101.1	5.9
3. Asparagine	$\overset{\displaystyle H_2N}{\underset{\displaystyle O}{C}}-CH_2-\overset{\displaystyle COO^-}{\underset{\displaystyle NH_3^+}{C}}-H$	114.1	4.3
4. Glutamine	$\overset{\displaystyle H_2N}{\underset{\displaystyle O}{C}}-CH_2-CH_2-\overset{\displaystyle COO^-}{\underset{\displaystyle NH_3^+}{C}}-H$	128.1	4.3
5. Tyrosine	$HO-\langle\text{benzene ring}\rangle-CH_2-\overset{\displaystyle COO^-}{\underset{\displaystyle NH_3^+}{C}}-H$	163.2	3.2
6. Cysteine	$HS-CH_2-\overset{\displaystyle COO}{\underset{\displaystyle NH_3^+}{C}}-H$	103.1	1.9

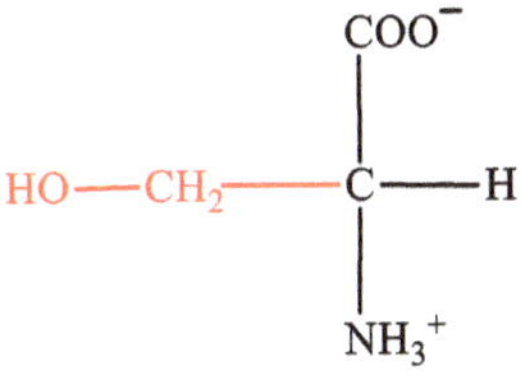

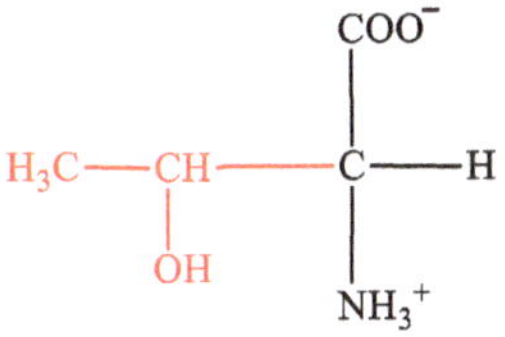

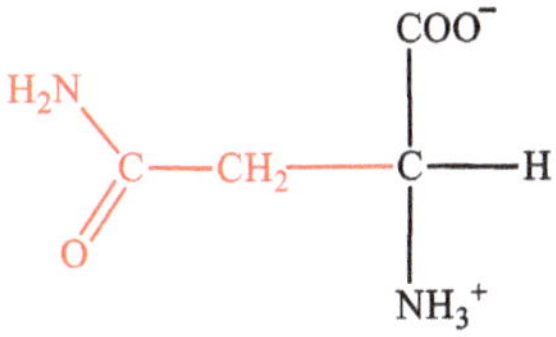

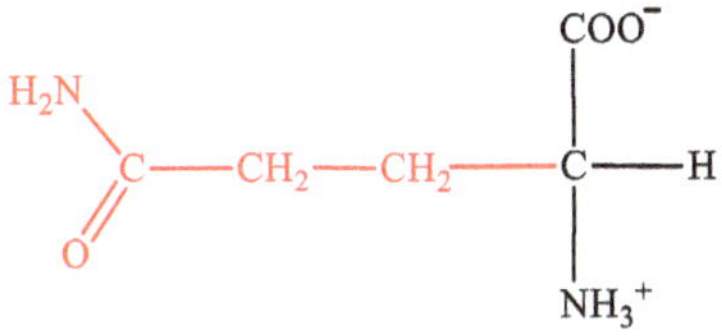

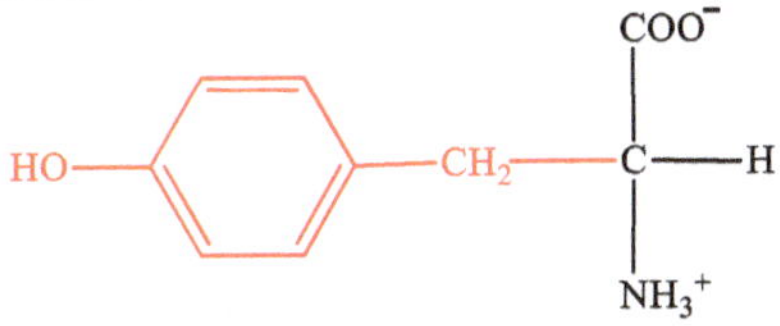

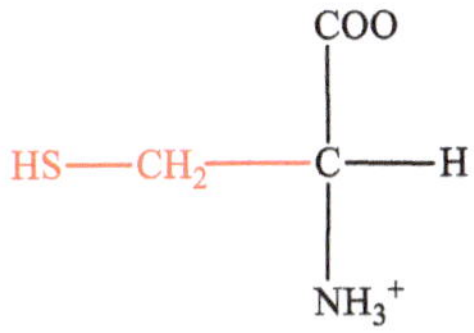

Table III.2 (continued)

Name	**Structure**	**Mass: residual (D)**	**Occurrence in proteins (%)**
Category (ii): Amino acids: charged polar side chains (hydrophilic)			
1. Lysine	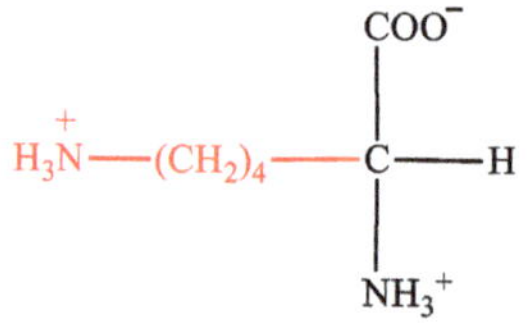	128.2	5.9
2. Arginine	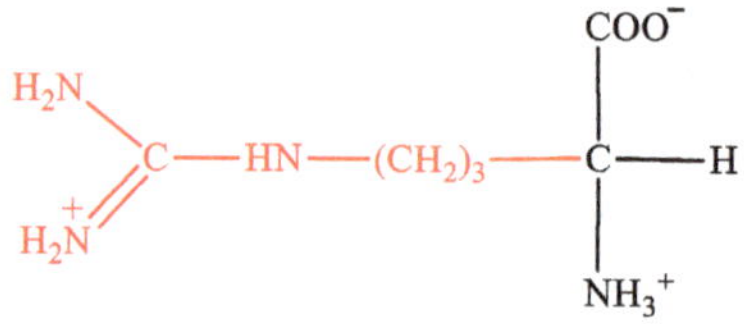	156.2	5.1
3. Histidine	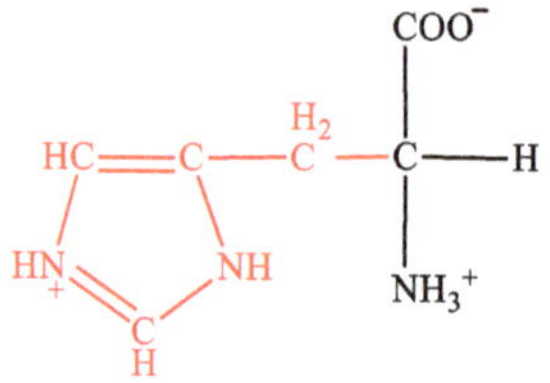	137.1	2.3
4. Aspartic acid	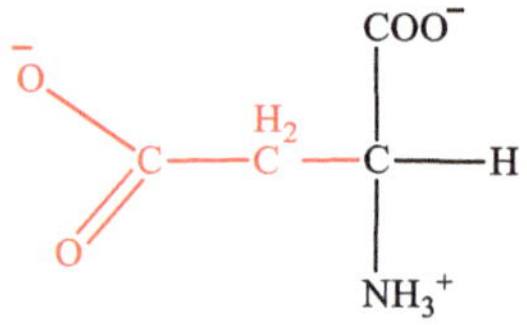	115.1	5.3
5. Glutamic acid	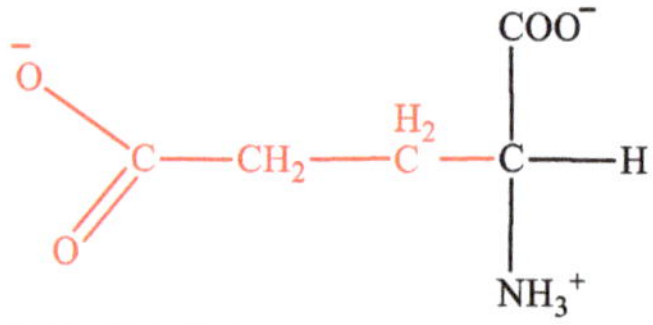	129.1	6.3

They are shown as they would exist at physiological pH and divided into three categories based on chemical reactivity and polarity of the side chain R.

Properties of amino acids

(i) Notably, α carbon in all the common amino acids contain four different atoms/groups [except glycine (R = H)] attached to it. Therefore, as usual, it is considered a chiral centre. An important consequence of this arrangement is the existence of two non-superimposable stereoisomers (enantiomers) called D- and L-amino acids (Figure III.2).

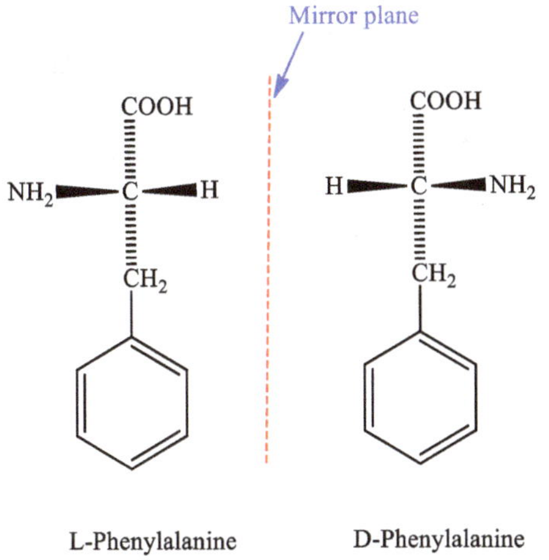

Figure III.2: D- and L-enantiomers for the amino acid phenylalanine.

(ii) In pure form, all of the 20 α-amino acids are solids. These are white in colour and crystalline in nature. Their melting points are found to be high. These are highly soluble in water, and the solutions conduct electric current. Instead, these are insoluble in organic solvents, namely, CH_3COCH_3, $CHCl_3$ and ethers. The above properties are those of ionic compounds or salts but such features are not present in the general structural formulas as shown in Figures III.1 and III.2. At physiological pH (~7.4), amino acids exist as dipolar ionic species (Figure III.3). This is because amino groups are present as NH_3^+ and the carboxylic acid groups are in the form of their conjugate base form, that is, as COO^-. Therefore, there is a positive and negative charge in a molecule of an amino acid. These forms of amino acids are sometimes known as **zwitterions**. The meaning of this German world is "**inner salts**".

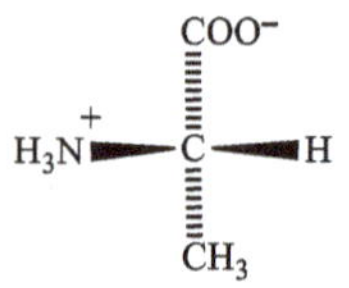

L-Alanine

Figure III.3: The zwitterionic form for amino acids. L-Alanine is shown in the figure. It is assumed here that α-carbon lies in the plane of the paper.

(iii) As outlined in Figure III.3, and the reaction shown below, the carboxylic group with a pk_1 value of 2.3 will dissociate a proton and the NH_3^+ group with a pk_2 of 9.7 will also dissociate a proton. Assuming that R side chain has no charge, the net charge on the entire zwitterion form (structure B) (at neutral pH) is zero. Understanding of acid and base properties of amino acids is important because it allows one to predict the major ionic form of an amino acid at any pH value. Predicting amino acid electrical charge becomes especially important when we discuss about the structure/function of protein. The pk values of the 20 amino acids found in proteins are given in Table III.3.

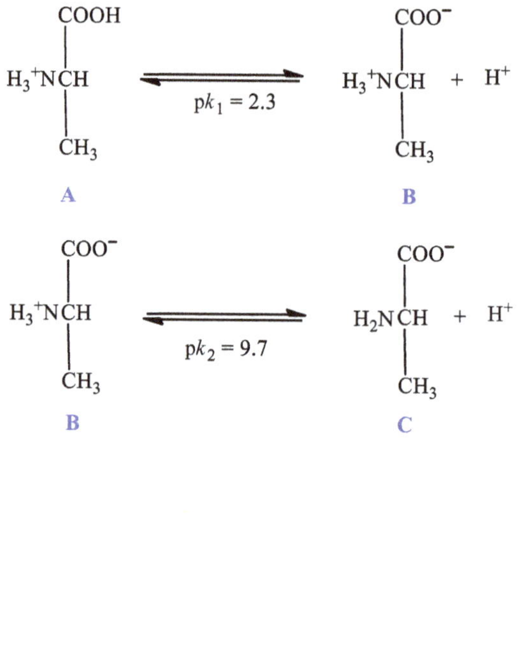

Table III.3: Amino acids with their p*k* values.

Name	pk_1 (α-COOH)	pk_2 (α-COOH)	pk_R (side chain)
Isoleucine	2.4	9.7	–
Alanine	2.3	9.9	–
Valine	2.3	9.6	–
Glycine	2.4	9.8	–
Leucine	2.4	9.6	–
Threonine	2.6	10.4	–
Methionine	2.3	9.2	–
Phenylalanine	1.8	9.1	–
Cysteine	1.8	10.8	8.3
Proline	2.0	10.6	–
Serine	2.1	9.2	–
Asparagine	2.0	8.8	–
Glutamine	2.2	9.1	–
Tryptophan	2.4	9.4	–
Aspartate/aspartic acid	2.0	10.0	3.9
Glutamate/glutamic acid	2.2	9.7	4.3
Histidine	1.8	9.2	6.0
Tyrosine	2.2	9.1	10.9
Lysine	2.2	9.2	10.8
Arginine	1.8	9.0	12.5

(iv) Amino acids undergo polymerization and form chains. The process can be shown as a condensation reaction. In this reaction, removal of a water molecule takes place as shown in Figure III.4. The CO–NH linkage (or amide linkage) so formed is known as a peptide bond. Polymers may contain 2, 3, more than 3 and up to 10, and many amino acid units. Such polymers are known as **dipeptide**, **tripeptide**, **oligopeptides** and **polypeptides, respectively**. These polymers are often denoted to simply as **"peptides"**. In peptide, the individual amino acids (the monomeric units) are known as **amino acid residues**.

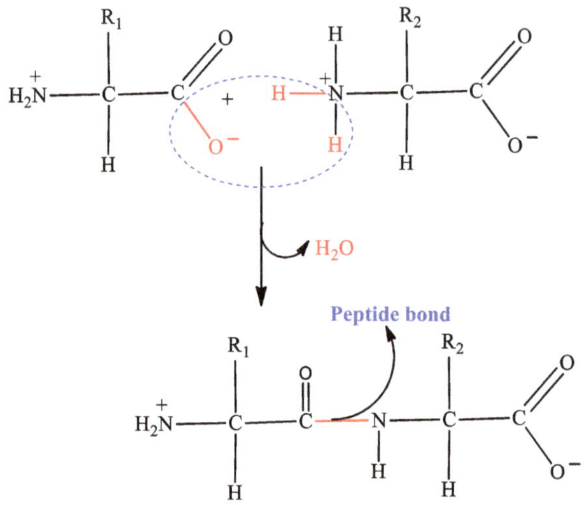

Figure III.4: Condensation of two amino acids forming peptide bond shown as red.

Polypeptides are linear polymers, that is, each amino acid residue participates in two peptide bonds and is linked to its neighbours in a head-to-tail fashion rather than forming branched chain. The residues at the two ends of the polypeptide each participate in just one peptide bond. The residue with the free amino group (by convention, the leftmost residue, as shown in Figure III.4) is called the **N-terminus.** On the other hand, the residue with a free carboxylate group (at the right) is called the **C-terminus.**

Proteins are molecules that contain one or more polypeptide chains. Variation in the length and the amino acid sequence of polypeptide contribute to the diversity in the shape and biological functions of proteins.

Amino acids: classification

These are classified taking into consideration of the polarities of their side chains. Accordingly, there are three major types of amino acids (Table III.2):
(i) With non-polar side chains
(ii) Having polar uncharged side chains
(iii) Including charged polar side chains

(i) Amino acids: non-polar side chains

Amino acids belonging to category (i) are glycine, alanine, valine, leucine, isoleucine, proline, phenylalanine, methionine and tryptophan (**9 in number**). All the side chains in this category of amino acids have aliphatic or aromatic groups and therefore have hydrophobic character. Because of the side chains are mainly hydrocarbons, there is little important chemical reactivity. When proteins dissolved in an aqueous solution, they fold into a three-dimensional shape in order to bury hydrophobic group (i) amino acid residues into the interior.

(ii) Amino acids: polar uncharged side chains at physiological pH

The amino acids of category (ii) are serine, cysteine, threonine, tyrosine, asparagine and glutamine (**6 in number**). When one observes the side chains of this category, they have a wide array of functional groups but all have at least one heteroatom (N, O or S) with electron pair. This electron pair is available for hydrogen bonding with water and other molecules. Perhaps the most interesting amino acid in this category is cysteine with –SH (sulphhydryl group) on the side chain. The –SH group of a cysteine may interact with an –SH group of another cysteine forming a disulphide bond in oxidizing environments (Figure III.5). Reducing agents transform the disulphide bond back to sulphhydryl group. This covalent bonding between cysteines becomes important in protein's three-dimensional structure.

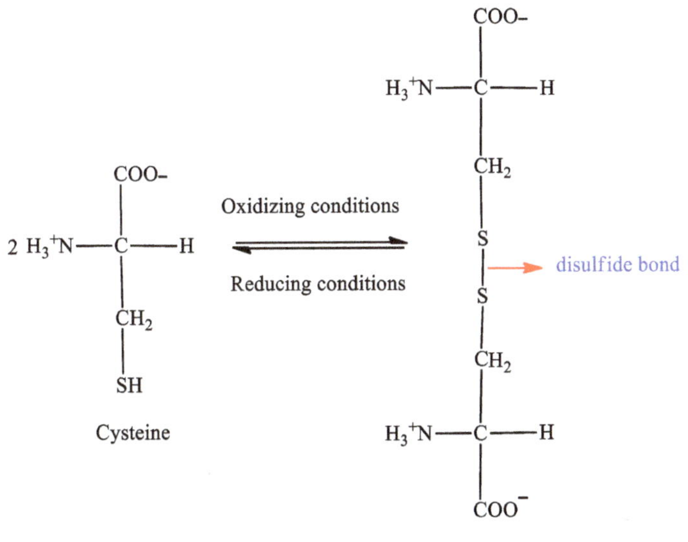

Figure III.5: Two cysteine residues under oxidizing conditions form a disulphide bond. Reducing agents result in cleavage of the disulphide bond and reverse the reaction giving cysteine.

(iii) Amino acids: polar, charged side chains at physiological pH

The amino acids of category (iii) are aspartate, glutamate, histidine, lysine and arginine (**5 in number**). Observation of the side chains reveals that functional groups may be either acidic or basic. The side chains of aspartate and glutamate dissociate protons to form carboxylate anions under physiological pH. The net charge on the molecules predominant at pH 7.4 is –1. The amino acid lysine having amino group on the side chain, and arginine with a guanidine group, exists primarily in the +1 ionic state under physiological pH. In histidine, the imidazole side chain has a pk_R value near physiological pH, so two major ionic forms are present at concentrations depending on the actual in vivo conditions (Figure III.6). At a physiological pH of 7.4, there will be a slight abundance of the zwitterionic form. The amino acids in categories (ii) and (iii) are hydrophilic and tend to cluster on the exterior of a protein in an aqueous environment.

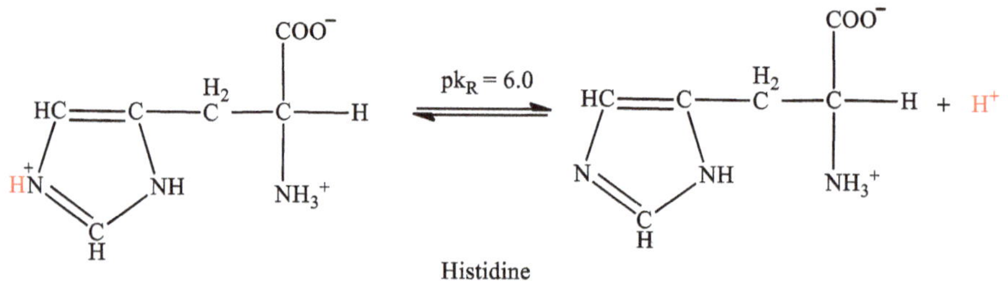

Figure III.6: Presence of two ionic forms of histidine in vivo.

Bibliography

The following books/reviews/research papers are recommended for further reading:

[1] C. R. Powell, K. M. Dillon and J. B. Matson, A review of hydrogen sulfide (H_2S) donors: Chemistry and potential therapeutic applications, *Biochem. Pharmacol.*, 149 (2018) 110–123 doi: 10.1016/j.bcp.2017.11.014

[2] *Nitric Oxide (NO) and Cancer; Prognosis, Prevention, and Therapy*, Edited by B. Bonavida, Springer, New York, Dordrecht, Heidelberg, London, (2010).

[3] J. A. McCleverty, Chemistry of nitric oxide relevant to biology, *Chem. Rev.*, 104 (2004) 403–418.

[4] M. A. Gonzales and P. K. Mascharak, Photoactive metal carbonyl complexes as potential agents for targeted CO delivery, *J. Inorg. Biochem.*, 133 (2014) 127–135.

[5] L. Wu and R. Wang, Carbon monoxide: endogenous production, physiological functions, and pharmacological applications, *Pharmacol. Rev.*, 57 (2005) 585–630.

[6] D. Mancardi, C. Penna, A. Merlino, P. D. Soldato, D. A. Wink and P. Pagliaro, Physiological and pharmacological features of the novel gasotransmitter: hydrogen sulfide, *Biochim. Biophys. Acta.*, 1787 (2009) 864–872.

[7] N. K. Lata, S. Sharma, Shalini Gupta and Radhe Shyam, Structural studies on mixed ligand nitrosyl complexes of transition metals chromium and iron, *Int. J. Curr. Res. Life Sci.*, 7 (2018) 2455–2468.

[8] M. Kaushika, A. Singh and M. Kumar, The chemistry of Group-VIb metal carbonyls, *Eur. J. Chem.*, 3 (2012) 367-394.

[9] P. Domingos, A. M. Prado, A. Wong, C. Gehring and J. A. Feijo, Nitric oxide: a multitasked signaling gas in plants, *Mol Plant*, 8 (2015) 506–520.

[10] G. L. Zubay, *Biochemistry*, 4th Edition Wm. C. Brown Publishers, Chicago, Mexico City, Sydney, Toronto, (1998).

[11] W. Durante, F. K. Johnson and R. A. Johnson, Role of carbon monoxide in cardiovascular function, *J. Cell. Mol. Med.*, 10 (2006) 672–686.

[12] A. N. Misra, M. Misra and R. Singh, Nitric oxide biochemistry, mode of action and signaling in plants, *J. Med. Plants Res.*, 4 (2010) 2729–2739.

[13] M. Wang and W. Liao, Carbon monoxide as a signaling molecule in plants, *Front. Plant Sci.*, 7 572 doi: 10.3389/fpls.2016.00572.

[14] R. Foresti, M. G. Bani-Hani and R. Motterlini, Use of carbon monoxide as a therapeutic agent: Promises and challenges, *Intensive Care Med.*, 34 (2008) 649–658.

[15] K. Ling, F. Men, W.-C. Wang, Y.-Q. Zhou, H.-W. Zhang and D.-W. Ye, Carbon monoxide and its controlled release: therapeutic application, detection, and development of carbon monoxide releasing molecules (CORMs), *J. Med. Chem.*, 61 (2018) 2611–2635.

[16] B. E. Mann and R. Motterlini, CO and NO in medicine, *Chem. Commun.*, (2007) 4197–4208.

[17] J. Du, H. Jin and L. Yang, Role of hydrogen sulfide in retinal diseases, *Front. Pharmacol.*, 8 2017 588–593 doi: 10.3389/fphar.2017.00588

[18] *Nitrosyl Complexes in Inorganic Chemistry, Biochemistry and Medicine II*, Edited by D. Michael and P. Mingos, Springer, Heidelberg, New York, Dordrecht, London, (2014).

[19] K. Bian and F. Murad, What is next in nitric oxide research? From cardiovascular system to cancer biology, *Nitric Oxide*, 43 (2014) 3–7.

[20] J. M. Mir, N. Jain, P. S. Jaget, W. Khan, P. K. Vishwakarma, D. K. Rajak, B. A. Malik and R. C. Maurya, Urinary tract anti-infectious potential of DFT-experimental composite analyzed ruthenium nitrosyl complex of N-dehydroacetic acid-thiosemicarbazide, *J. King Saud Univ. Sci.*, 31 (2019) 89–100.

https://doi.org/10.1515/9783110727302-008

[21] J. H. Enemark and R. D. Feltham, Principles of structure, bonding, and reactivity for metal nitrosyl complexes, *Coord. Chem. Rev.*, 13 (1974) 339–406.
[22] R. C. Maurya, J. M. Mir and I. Medicinal, Environmental relevance of metal nitrosyl complexes: A review, *Int. J. Sci. Eng. Res.*, 5 (2015) 307–321.
[23] R. Boyer, *Concepts in Biochemistry*, 2nd Edition Brooks/Cole, Thomson Learning, Australia, Canada, Mexico, Singapore, Spain, U.K., United States (2002).
[24] R. C. Mehrotra and A. Singh, *Organometallic Chemistry: A Unified Approach*, Wiley Eastern Limited, New Delhi (India).
[25] A. L. Lehninger, *Principles of Biochemistry*, CBS Publishers & Distributors Pvt. Ltd., Delhi (India), (1990).
[26] J. M. Mir and R. C. Maurya, Physiological and pathophysiological implications of hydrogen sulfide: a persuasion to change the fate of the dangerous molecule, *J. Chin. Adv. Mater. Soc.*, 6 2018 434–458 doi: doi.org/10.1080/-22243682.2018.1493951
[27] H. He and L. He, The role of carbon monoxide signaling in the responses of plants to abiotic stresses, *Nitric Oxide*, 42 (2014) 40–43.
[28] J. M. Mir and R. C. Maurya, NO news is good news for eyes: a mini review, *Ann. Ophthalmol. Vis. Sci.*, 1003 (2018) 1–4.
[29] K. K. Pandey, Transition metal nitrosyls in organic synthesis and in pollution control, *Coord. Chem. Rev.*, 51 (1983) 69–98.
[30] R. C. Maurya, A. Pandey, J. Chaurasia and H. Martin, Metal nitrosyl complexes of bioinorganic, catalytic, and environmental relevance: a novel single-step synthesis of dinitrosylmolybdenum(0) complexes of $\{Mo(NO)_2\}^6$ electron configuration involving Schiff bases derived from 4-acyl-3-methyl-1-phenyl-2-pyrazolin-5-one and 4-aminoantipyrine, directly from molybdate(VI) and their characterization, *J. Mol. Struct.*, 798 (2006) 89–101.
[31] X. Zhou, J. Zhang, G. Feng, J. Shen, D. Kong and Q. Zhao, Nitric oxide-releasing biomaterials for biomedical applications, *Curr. Med. Chem.*, 23 (2016) 1–23.
[32] *Applications of Density Functional Theory to Chemical Reactivity*, Edited by M. V. Putz, D. Michael and P. Mingos, Springer, Heidelberg, New York, Dordrecht, London (2012).
[33] L. Stryer, *Biochemistry*, 4th Edition W. H. Freeman and Company, New York (1995).
[34] F. Becquet, Y. Courtois and O. Goureau, Nitric oxide in the eye: multifaceted roles and diverse outcomes, *Surv. Ophthalmol.*, 142 (1997) 71–82.
[35] C. C. Romão, W. A. Blättler, J. D. Seixas and G. J. L. Bernardes, Developing drug molecules for therapy with carbon monoxide, *Chem. Soc. Rev.*, 41 (2012) 3571–3583.
[36] T. R. Johnson, B. E. Mann, J. E. Clark, R. Foresti, C. J. Green and R. Motterlini, Metal carbonyls: A new class of pharmaceuticals?, *Angew. Chem. Int. Ed.*, 42 (2003) 3722–3729.
[37] K. G. Caulton, Synthetic methods in transition metal nitrosyl chemistry, *Coord. Chem. Rev.*, 14 (1975) 317–355.
[38] R. C. Maurya, *Synthesis and Characterization of Some Novel Nitrosyl* Compounds, Pioneer Publications, Jabalpur (M.P.), India (2000).
[39] A. Ismailova, D. Kuter, D. S. Bohle and I. S. Butle, An overview of the potential therapeutic applications of CO-releasing molecules, Hindawi, *Bioinorg. Chem. Appl.* (2018) Article ID 8547364; doi: https://doi.org/10.1155/-2018/8547364.
[40] *New Trends in CO Activation*, Edited by B. Delrnon and J. T. Yates (Chapters 9 and 10) in "*Studies in Surface Science and Catalysis* Vol. 64 (1991) 1–490.
[41] C. Bucolo and F. Drago, Carbon monoxide and the eye: implications for glaucoma therapy, Pharmacol. Ther., 130 (2011) 191–201.
[42] A. W. Carpenter and M. H. Schoenfisch, Nitric oxide release part II. Therapeutic applications, *Chem. Soc. Rev.*, 41 (2012) 3742–3752.
[43] A. K. Das, *Bioinorganic Chemistry*, Books and Allied (P) Ltd., Kolkata (India), (2012).

[44] *Nitrosyl Complexes in Inorganic Chemistry, Biochemistry and Medicine I*, Edited by D. Michael and P. Mingos, Springer, Heidelberg, New York, Dordrecht, London (2014).

[45] L. K. Wareham, E. S. Buys and R. M. Sappington, The nitric oxide-guanylate cyclase pathway and glaucoma, Nitric Oxide, 77 (2018) 75–87.

[46] *Radicals for Life: The Various Forms of Nitric Oxide*, First Edition Edited by E. Van Faassen and A. F. Vanin, Elsevier B.V., New York, 2007.

[47] M. Faizan, N. Muhammad, K. U. Khan Niazi, Y. Hu, Y. Wang, Y. Wu, H. Sun, R. Liu, W. Dong, W. Zhang and Z. Gao, CO-releasing materials: an emphasis on therapeutic implications, as release and subsequent cytotoxicity are the part of therapy, *Materials*, 12 (2019) 1643–1684.

[48] B. D. Gupta and A. J. Elias, *Basic Organometallic Chemistry: Concepts, Synthesis and Applications*, 2nd Edition University Press (India) Pvt. Ltd, Hyderabad (India).

[49] M. R. Filipovic and V. M. Jovanović, More than just an intermediate: hydrogen sulfide signalling in plants, *J. Exp. Botany*, 68 2017 4733–4736 doi: 10.1093/jxb/erx352

[50] J. G. V. Donald Voet and C. W. Pratt, *Fundamentals of Biochemistry*, John Wiley & Sons, Inc, New York (1999).

[51] E. Stagni, M. G. Privitera, C. Bucolo, G. M. Leggio, R. Motterlini and F. Drago, A water-soluble carbon monoxide-releasing molecule (CORM-3) lowers intraocular pressure in rabbits, *Br. J. Ophthalmol.*, 93 2009 254–257 doi: 10.1136/bjo.2007.-137034

[52] G. C. Y. Chiou, Review: effects of nitric oxide on eye diseases and their treatment, *J. Ocul. Pharmacol. Therap.*, 17 (2001) 189–198.

[53] S. K. Choudhari, M. Chaudhary, S. Bagde, A. R. Gadbail and V. Joshi, Nitric oxide and cancer: a review, *World J. Surg. Oncol.*, 11 (2013) 118–127.

[54] M. Wrighton, The photochemistry of metal carbonyls, *Chem. Rev.*, 74 (1974) 401–430.

[55] J. C. Toledo Jr. and O. Augusto, Connecting the chemical and biological properties of nitric oxide, *Chem. Res. Toxicol.*, 25 (2012) 975–989.

[56] Z. Jin and Y. Pei, Physiological implications of hydrogen sulfide in plants: Pleasant exploration behind its unpleasant odour, *Oxid. Med. Cell. Longev.* (2015) Article ID 397502; doi: doi.org/10.1155/2015/397502.

[57] J. A. McCleverty, Reactions of nitric oxide coordinated to transition metals, *Chem. Rev.*, 79 (1979) 53–76.

[58] G. N. Mukherjee and A. Das, *Elements of Bioinorganic Chemistry*, First Edition U. N. Dhur & Sons Private Limited, Kolkata (India) (1993).

[59] S. W. Ryter, D. Morse and A. M. K. Choi, Carbon monoxide: to boldly go where NO has gone before, *Sci. Signal*, 230 (2004) 1–10. re6; doi: 10.1126/stke.2302004re6.

Index

https://doi.org/10.1515/9783110727302-009

www.ingramcontent.com/pod-product-compliance
Lightning Source LLC
Chambersburg PA
CBHW080655220326
41598CB00033B/5219
* 9 7 8 3 1 1 0 7 2 7 2 9 6 *